IMPORTANT NAMES, DATES AND ADDRESSES

This book belongs to

_____

_____

_____

phone _____

examination date: _____ hours: _____

examination location: _____

_____

_____

_____

phone number of your registration board: _____

address of your registration board: _____

_____

_____

names of contacts at your registration board: _____

_____

_____

date you sent your application: _____

registered/certified mail receipt number: _____

date confirmation was received: _____

names of examination proctors: _____

_____

_____

booklet number: _____ (A.M.) _____ (P.M.)

problems you disagreed with on the examination

problem no.   reason

_____

_____

_____

CIVIL ENGINEERING REVIEW MANUAL

Third Edition

A complete review course for the P.E. examination
for Civil Engineers

Michael R. Lindeburg, P. E.
Director, Professional Engineering Institute

Professional Publications
San Carlos, CA 94070

DISCLAIMER: This CIVIL ENGINEERING REVIEW MANUAL is intended as a review of the subjects included. Despite the completeness and depth of material covered, code-related subjects have been greatly summarized so this book should not be used for design purposes. The author, publisher, and distributors of this book do not assume or accept any responsibility or liability, including liability for negligence, for errors or oversights, for the use of this book in preparing engineering plans.

Much of the material that appears in this manual is based on original research done by others. Acknowledgement is given wherever possible to those original sources. Data, figures, and important formulas which originally appeared in sources listed in the chapter bibliographies are referenced to those sources. For example, (4:397) appearing adjacent to the figure means that the figure originally appeared on page 397 of the fourth entry in that chapter's bibliography.

In the *ENGINEERING REVIEW MANUAL SERIES*

Engineer-In-Training Review Manual
Quick Reference Cards for the E-I-T Exam
Mini-Exams for the E-I-T Exam
Civil Engineering Review Manual
Seismic Design for the Civil P.E. Exam
Timber Design for the Civil P.E. Exam
Structural Engineering Practice Problem Manual
Mechanical Engineering Review Manual
Electrical Engineering Review Manual
Chemical Engineering Review Manual
Chemical Engineering Practice Exam Set
Land Surveyor Reference Manual
Expanded Interest Tables
Engineering Law, Design Liability, and Professional Ethics

Distributed by: Professional Publications, Inc.
Post Office Box 199
Department 77
San Carlos, CA 94070
(415) 593-9119

CIVIL ENGINEERING REVIEW MANUAL

3rd Edition

Copyright © 1981 by the Professional Engineering Registration Program. All rights are reserved. No part of this publication may be reproduced, stored in a retrieval system, or transmitted, in any form or by any means, electronic, mechanical, photocopying, recording, or otherwise, without the prior written permission of the publisher.

Printed in the United States of America

Library of Congress catalog card number: 81-81682

ISBN: 0-932276-28-8

Professional Engineering Registration Program
Post Office Box 911, San Carlos, CA 94070
(415) 593-9731

Current printing of this edition (last number)   9   8   7   6

PREFACE

With this CIVIL ENGINEERING REVIEW MANUAL, you will be able to prepare effectively and efficiently for the NCEE Civil Engineering license examination. The contents, examples, and problems have been extensively tested in actual classroom review courses. They will help greatly during your preparation.

This manual serves as a reference for the majority of problem types that have appeared during the last 15 years. Thus, even low-probability subjects, such as statistical analysis and project scheduling, are included.

This third edition differs from the previous editions in several ways. The steel design chapter has been rewritten to reflect the 8th edition (1980) of the AISC manual. Similarly, the concrete design chapter now reflects the 1983 ACI 318-83 concrete code.

Also in this edition are complete solutions to all end-of-chapter problems, as well as an 8-hour practice exam with solutions. These new additions will make classroom and homestudy use of this book considerably easier.

Michael R. Lindeburg, P.E.
May 1984

OUTLINE OF SUBJECTS FOR SELF-STUDY

Week	Subject	Date to be started	Date to be completed	Check when complete
1.	Mathematics	_____	_____	[ ]
2.	Engineering Economy	_____	_____	[ ]
3.	Hydraulics	_____	_____	[ ]
4.	Hydraulic Machines	_____	_____	[ ]
5.	Open Channel Flow	_____	_____	[ ]
6.	Hydrology	_____	_____	[ ]
7.	Water Supply Engineering	_____	_____	[ ]
8.	Waste Water Engineering	_____	_____	[ ]
9.	Soils	_____	_____	[ ]
10.	Foundations and Retaining Walls	_____	_____	[ ]
11.	Statics: Determinate	_____	_____	[ ]
12.	Mechanics of Materials	_____	_____	[ ]
13.	Indeterminate Structures	_____	_____	[ ]
14.	Concrete Design	_____	_____	[ ]
15.	Steel Design	_____	_____	[ ]
16.	Traffic Analysis	_____	_____	[ ]
17.	Surveying	_____	_____	[ ]

TABLE OF CONTENTS

THE EXAMINATION PROCESS

Purpose of Registration

As an engineer, you may have to obtain your Professional Engineering license through procedures established by your state. These procedures are designed to protect the public by preventing unqualified individuals from legally practicing as engineers.

There are many excellent reasons for becoming a registered Professional Engineer. Among them are the following:

1.   You may wish to be an independent consultant, and by law, a consulting engineer must be registered.

2.   Your company may require a professional engineering license as a requirement for employment or advancement.

3.   Your state may require registration as a Professional Engineer before you are authorized to use the title 'engineer.'

The Registration Process

The registration procedure is similar in most states. You will probably take two 8-hour written examinations. The first examination is the Engineer-In-Training examination. This examination is known in some states as the Intern Engineer or Fundamentals of Engineering exam. The initials E-I-T, I.E., and F.E. are also commonly used. The second examination is the Professional Engineering (P.E.) exam, the format and subject matter of which differ from those of the E-I-T exam.

If you have significant experience in engineering, you may be allowed to skip the E-I-T examination. However, actual details of registration, experience requirements, minimum education levels, fees, and examination schedules vary from state to state. You should contact your state Board of Registration for Professional Engineers. The address and phone number of your board may be obtained from the Professional Engineering Institute by calling (415) 593-9731.

The National Council of Engineering Examiners

Few states write their own examination questions. Most use the exams prepared by the National Council of Engineering Examiners (NCEE). These exams are given simultaneously in all of the participating states. This use of identical examinations has led to the term 'Uniform Examination.' However, the Uniform Examination does not automatically ensure reciprocity among states since each state may choose its own minimum passing score.

NCEE has, thusfar, not made current examinations available for review purposes. The formats of its examinations are continually changing and are usually considerably different than examinations previously developed by individual state boards. Therefore, exam review books not based on the current NCEE exam format are out-of-date.

Examination Format

The NCEE Professional Engineering examination in Civil Engineering consists of two 4-hour sessions separated by a 1-hour lunch period. Both the morning and afternoon sessions contain 10 problems. Most states do not have required problems.

Each examinee is given an exam booklet which contains problems for civil, mechanical, electrical, and chemical engineers. Some states, such as California, will allow you to work problems only from the civil part of the booklet. Other states will allow you to work any problems you like from the entire booklet. Read the examination instructions carefully on this point.

Problems on the examination come from the following areas:

C.E. Examination Subjects	Probable No. of Problems
Mathematics and Scheduling	0
Economic Analysis	1
Statics	1
Mechanics of Materials	2
Timber Design	0
Concrete Design	2
Steel Design	2
Soils and Earthwork	1
Foundations and Retaining Walls	1
Conduit Fluid Flow	1
Fluid Machinery	1
Open Channel Flow	1
Hydrology	1
Sanitary Engineering	3
Surveying	1
Traffic Engineering and Roadway Construction	2

Since the examination structure is not rigid, it is not possible to give the exact number of problems that will appear in each subject area. Only economic analysis can be considered a permanent part of the examination. NCEE has announced that the number of economic analysis problems will be reduced to one starting with the November 1980 examination. There is no guarantee that any other single subject will appear.

California adds a required seismic design problem to the morning part of the examination, making a total of 11 morning problems. There is no minimum score that you must achieve on the seismic problem, but you must work it. This special problem is located in the back of the exam booklet, so make sure that you do not overlook it. (A special publication for California examinees is available from the Professional Engineering Registration Program: SEISMIC DESIGN FOR THE CALIFORNIA CIVIL PROFES-SIONAL ENGINEERING EXAMINATION.)

Grading the Examination

Full credit is achieved by correctly working four problems in the morning and four problems in the afternoon. You may not claim credit for more than eight worked problems or for more than four per session. All solutions are recorded in official solution booklets.

Each problem is worth ten points. Solutions are graded subjectively by hand. Scoring of each problem is on two scales, one recognizing method, the other accounting for mathematical errors. The methods scale runs from 0 to 10; the errors scale runs from 0 to minus 5. The sum of the two scales is your score for the problem.

The total number of possible points for both sessions is 80. A score of 56 (70%) is generally considered passing.

The examination is open book. Usually, all forms of solutions aids are allowed in the examination, including nomographs, specialty slide-rules, and pre-programmed and programmable calculators. Since their use says little about the depth of your knowledge, such aids should be used only to check your work. For example, it would not be appropriate to solve a surveying problem entirely with a pre-programmed calculator.

Grading and reporting of your score will take approximately four months. You probably will not be told your score if you pass the exam. If you fail, you will be told your score and it will be possible to review the grading. If you feel that you are entitled to more points than were awarded, you will be able to request that your solutions be re-evaluated.

Preparing for the Exam

You should develop an examination strategy early in the preparation process. This strategy will depend on your background. One of the following two general strategies is recommended.

    (1)   A broad approach has been successful for examinees who have recently completed academic studies. Their strategy has been to review the fundamentals of a broad range of undergraduate civil engineering subjects. The examination includes enough fundamental problems to give merit to this strategy.

    (2)   Working engineers who have been away from class work for a long time have found it better to concentrate on the subjects in which they have had extensive professional experience. By studying the list of subjects, they have been able to choose those which will give them a good probability of finding enough problems that they can solve.

Although this manual will provide an excellent review of examination subjects, you might want to enroll in a classroom review course to help pace your studies. Such courses will also enable you to obtain answers to your questions about difficult subjects.

Whether or not you enroll in a classroom review course, you should work sample problems (such as those contained in this book) in order to become familiar with the problem types and solution methods. You will usually not have sufficient time during the examination to derive solution methods; you must know them instinctively.

Steps to Take Before the Exam

The engineers who have taken the P.E. exam in previous years have developed the suggestions listed below. These suggestions will make your examination experience as enjoyable and successful as possible.

1. Keep a copy of your examination application. Send the original copy by certified mail and request a receipt of delivery.

2. Keep the application copy, receipt of delivery, and your cancelled check with the materials you will taking to the examination.

3. Visit the exam site the day before the examination. Find the examination room, the parking area, and the rest rooms.

4. If you live a long distance from the examination site, consider obtaining a hotel room in which to spend the night before.

5. Get a good night's sleep the night before. Don't cram the last night.

6. Make arrangements for babysitters and transportation. Allow for a delayed completion.

7. Organize the materials you are taking to the examinaton. You may want to start with the following checklist.

    [ ] The documents listed in suggestion #2 above
    [ ] The letter admitting you to the examination
    [ ] Photographic identification
    [ ] This book
    [ ] Other reference books, including: [ ] Uniform Building Code
                                          [ ] AISC Manual
                                          [ ] ACI Standard 318

       [ ] Course notes
       [ ] English dictionary
       [ ] Dictionary of scientific terms
       [ ] A cardboard box cut to fit your books
       [ ] Pad of scratch paper with holes in a 3-ring binder
       [ ] Ruler, compass, protractor, french curves
       [ ] A collection of all required types of graph paper
       [ ] Calculators (primary and back-up)
       [ ] Extra batteries or battery pack
       [ ] Battery charger and extension cord
       [ ] Chair cushions
       [ ] Earplugs
       [ ] Twist-to-advance pencils
       [ ] Extra leads
       [ ] Large eraser

[　] Thermos filled with hot chocolate
[　] A light lunch
[　] Interim snacks (i.e., raisins, nuts, trail mix)
[　] Scotch and masking tape
[　] Scissors, stapler, and staple remover
[　] Extra prescription glasses
[　] Sunglasses and/or visored cap
[　] Aspirin
[　] $2 in change
[　] A light sweater
[　] Comfortable shoes or slippers
[　] Raincoat, boots, gloves, hat, umbrella
[　] Street maps of the examination site area
[　] Note to the parking patrol for your windshield
[　] Battery-power desk lamp
[　] Watch
[　] Travel pack of kleenex

What to Do During the Examination

Previous examinees have reported that the following strategies and techniques helped them considerably.

1. Read through all of the problems before starting your first solution. In order to save you from rereading and re-evaluating each problem later in the day, you should classify each problem at the beginning of the four hour session. The following categories are suggested:

    a. problems you can easily do
    b. problems you can do with effort
    c. problems you can get partial credit on
    d. problems you cannot do

2. Do all of the problems in order of increasing difficulty. All problems on the examination are worth 10 points. There is nothing to be gained by attempting the difficult or long problems if easier or shorter problems are available. Do not spend an inordinate amount of time on difficult mandatory questions. Mandatory questions must be attempted--they do not have to be correctly solved.

3. Follow these guidelines when solving a problem:

    a. Do not rewrite the problem statement.
    b. Do not unnecessarily redraw any figures.
    c. Use pencil only.
    d. Be neat. (Print all text. Use a straight-edge or template where possible.)
    e. Draw a box around each answer.
    f. Label each answer with a symbol.
    g. Give the units.
    h. List your sources whenever you use obscure solution methods or data.
    i. Write on one side of the page only.
    j. Use one page per problem, no matter how short the solution is.
    k. Go through all calculations a second time and check for mathematical errors.

Things to Do Between Now and the Examination

1. Make extra copies of the worksheets contained on pages 13-25, 16-21, 16-22, and 16-23 in this book.

2. Obtain a package of tracing paper (approximately 20# weight).

3. Obtain ten sheets of each of the following types of graph paper:
    - 10 squares to the inch grid
    - semi-log: 3 cycles x 10 squares to the inch grid
    - full-log: 3 cycles x 3 cycles

4. Obtain a flexible clear plastic ruler marked in tenths of an inch.

5. Send for the publications listed below. (A minimal charge is involved unless noted otherwise. The current price should be obtained from the publisher.)

    - DIMENSIONS AND PROPERTIES: NEW W, HP, AND WT SHAPES, American Institute of Steel Construction, 400 North Michigan Avenue, Chicago, IL 60611 (312) 670-2400.

    - RECOMMENDED STANDARDS FOR SEWAGE WORKS, Health Education Service, Inc., P.O. Box 7126, Albany, NY 12224 (518)392-3951. This is commonly known as the "10-States' Standards."

    - HIGHWAY CAPACITY MANUAL: SPECIAL REPORT 87, National Academy of Sciences, National Research Council, 2101 Constitution Avenue, Washington, DC 20418 (202) 334-2000.

    - SIMPLIFIED LABORATORY PROCEDURES FOR WASTEWATER EXAMINATION, Water Pollution Control Federation, 2626 Pennsylvania Avenue, Washington, DC 20037 (202) 337-2500.

    - SEISMIC DESIGN FOR THE CIVIL P.E. EXAM, Professional Publications, Inc., P.O. Box 199, Department 77, San Carlos, CA 94070 (415) 593-9119.

    - TIMBER DESIGN FOR THE CIVIL P.E. EXAM, Professional Publications, Inc., P.O. Box 199, Department 77, San Carlos, CA 94070 (415) 593-9119.

    - ENGINEERING LAW, DESIGN LIABILITY, AND PROFESSIONAL ETHICS, Professional Publications, Inc., P.O. Box 199, Department 77, San Carlos, CA 94070 (415) 593-9119.

    - EXPANDED INTEREST TABLES FOR THE E-I-T AND P.E. EXAMS, Professional Publications, Inc., P.O. Box 199, Department 77, San Carlos, CA 94070 (415) 593-9119.

MATHEMATICS AND RELATED SUBJECTS

1. Introduction

The probability of a purely mathematical problem appearing on the civil engineering exam is extremely small. You will not be asked for the roots of a cubic equation or the determinant of a matrix. However, you may have to use these techniques as part of a solution to a much more complex problem. This chapter is designed as a reference for formulas and techniques required to solve most of the civil engineering problems.

2. Symbols Used in This Book

Many symbols, letters, and Greek characters are used to represent variables in the formulas used throughout this book. These symbols and characters are defined in the nomenclature sections of each chapter. However, some of the symbols which are used in this book as operators are listed below.

Symbol	Name	Use	Example
Σ	sigma	series addition	$\sum_{i=1}^{5} x_i = x_1 + x_2 + x_3 + x_4 + x_5$
$\prod$	pi	series multiplication	$\prod^{3} x_i = x_1 x_2 x_3$
Δ	delta	change in quantity	$\Delta h = h_2 - h_1$
$-$	over bar	average value	$\bar{x}$
$\cdot$	over dot	per unit time	$\dot{Q}$ = quantity flowing per second
$\simeq$	approximately equal to		$x \simeq 1.5$
$\propto$	proportional to		$x \propto y$
∞	infinity		$x \to \infty$
log	base 10 logarithm		$\log(5.74)$
ln	natural logarithm		$\ln(5.74)$
EE	scientific notation		EE-4

3. Mensuration

Nomenclature

A	total surface area
d	a distance
h	height
p	perimeter
r	radius
s	side (edge) length, arc length
V	volume
θ	vertex angle, in radians
ϕ	central angle, in radians

Circle

$p = 2\pi r$

$A = \pi r^2 = \dfrac{p^2}{4\pi}$

Circular Sector

$A = \dfrac{1}{2}\phi r^2 = \dfrac{1}{2}sr$

$\phi = s/r$

Circular Segment

$A \simeq \dfrac{1}{2}r^2(\phi - \sin\phi)$

$\phi = s/r = 2(\arccos\dfrac{r-d}{r})$

Ellipse

$A = \pi ab$

$p = 2\pi\sqrt{\dfrac{1}{2}(a^2 + b^2)}$

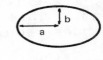

Triangle

$A = \dfrac{1}{2}bh$

Trapezoid

$p = a + b + c + d$

$A = \dfrac{1}{2}h(a + b)$

The trapezoid is isosceles if c = d.

Parabola

$A = \dfrac{2bh}{3}$

$A = \dfrac{1}{3}bh$

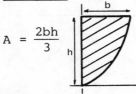
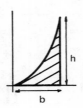

Regular Polygon
(n equal sides)

$\phi = 2\pi/n$

$\theta = \dfrac{\pi(n-2)}{n}$

$p = ns$

$s = 2r(\tan(\dfrac{1}{2}\phi))$

$A = \dfrac{1}{2}nsr$

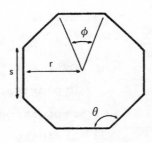

Parallelogram

$p = 2(a + b)$

$d_1 = \sqrt{(a^2 + b^2 - 2ab(\cos\phi))}$

$d_2 = \sqrt{(a^2 + b^2 + 2ab(\cos\phi))}$

$d_1^2 + d_2^2 = 2(a^2 + b^2)$

$A = ah = ab(\sin\phi)$

If a = b, the parallelogram is a rhombus.

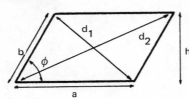

Right Circular Cone

$V = \dfrac{\pi r^2 h}{3}$

$A = \pi r\sqrt{r^2 + h^2}$ (does not include base area)

Sphere

$V = \dfrac{4\pi r^3}{3}$

$A = 4\pi r^2$

Right Circular Cylinder

$V = \pi h r^2$

$A = 2\pi rh$ (does not include end area)

Example 1.1

What is the hydraulic radius of a 6" pipe filled to a depth of 2"?

The hydraulic radius is defined as

$$r_h = \frac{\text{area in flow}}{\text{length of wetted perimeter}} = \frac{A}{s}$$

Points o, a, and b may be used to find the central angle of the circular segment.

$$\tfrac{1}{2}(\text{angle aob}) = \arccos(\tfrac{1}{3}) = 70.53°$$

$$\emptyset = 141.06° = 2.46 \text{ radians}$$
$$\sin(2.46 \text{ rad}) = .63$$

Then,
$$A \simeq \tfrac{1}{2}(3)^2(2.46 - .63) = 8.235$$

$$s = (3)(2.46) = 7.38$$

$$r_h = \frac{8.235}{7.38} = 1.11$$

Note that table 3.9 can also be used to solve this example.

4. Significant Digits

The significant digits in a number include the left-most, non-zero digits to the right-most digit written. Final answers from computations should be rounded off to the number of decimal places justified by the data. The answer can be no more accurate than the least-accurate number in the data. Of course, rounding should be done on final calculation results only. It should not be done on interim results.

number as written	number of significant digits	implied range
341	3	340.5 to 341.5
34.1	3	34.05 to 34.15
.00341	3	.003405 to .003415
3410	4	3409.5 to 3410.5
341 EE7	3	340.5 EE7 to 341.5 EE7
3.41 EE-2	3	3.405 EE-2 to 3.415 EE-2

5. Algebra

A. Polynomial Equations

1. Standard Forms

$$(a + b)(a - b) = a^2 - b^2$$

$$(a \pm b)^2 = a^2 \pm 2ab + b^2$$

$$(a \pm b)^3 = a^3 \pm 3a^2b + 3ab^2 \pm b^3$$

$$(a^3 \pm b^3) = (a \pm b)(a^2 \mp ab + b^2)$$

$$(a^n + b^n) = (a + b)(a^{n-1} - a^{n-2}b + \cdots + b^{n-1}) \quad \text{(for n odd)}$$

$$(a^n - b^n) = (a - b)(a^{n-1} + a^{n-2}b + \cdots + b^{n-1})$$

2. Quadratic Equations

Given a quadratic equation with form $ax^2 + bx + c = 0$, the roots x_1^* and x_2^* may be found from

$$x_1^*, x_2^* = \frac{-b \pm \sqrt{b^2 - 4ac}}{2a}$$

$$x_1^* + x_2^* = -\frac{b}{a}$$

$$x_1^* x_2^* = \frac{c}{a}$$

3. Cubic Equations

Cubic and higher order equations occur infrequently in most engineering problems. However, they usually are difficult to factor when they do occur. Trial and error solutions are usually unsatisfactory except for finding the general region in which a root occurs. Graphical means can only be used to obtain a fair approximation to the root.

Numerical analysis techniques must be used if extreme accuracy is needed. The more efficient numerical analysis techniques are too complicated to present here. However, the bisection method illustrated in example 1.2 can usually provide the required accuracy with only a few simple iterations.

The bisection method starts out with two values of the independent variable, L_O and R_O, which straddle a root. Since the function has a value of zero at a root, $f(L_O)$ and $f(R_O)$ will have opposite signs. The following algorithm describes the remainder of the bisection method.

Let n be the iteration number. Then, for $n = 0, 1, 2,\ldots$ perform the following steps until sufficient accuracy is attained.

Set $m = \frac{1}{2}(L_n + R_n)$

Calculate $f(m)$

If $f(L_n)f(m) \leq 0$, set $L_{n+1} = L_n$ and $R_{n+1} = m$

Otherwise, $L_{n+1} = m$ and $R_{n+1} = R_n$

$f(x)$ has at least one root in the interval (L_{n+1}, R_{n+1})

The estimated value of that root, x^*, is

$$x^* \simeq \frac{1}{2}(L_{n+1} + R_{n+1})$$

The maximum error is $\frac{1}{2}(R_{n+1} - L_{n+1})$. The iterations continue until the maximum error is reasonable for the accuracy of the problem.

Example 1.2

Use the bisection method to find the roots of $f(x) = x^3 - 2x - 7$.

The first step is to find L_0 and R_0, which are the values of x which straddle a root and have opposite signs. A table can be made and values of $f(x)$ calculated for random values of x:

x	-2	-1	0	+1	+2	+3
f(x)	-11	-6	-7	-8	-3	+14

Since $f(x)$ changes sign between $x = 2$ and $x = 3$,

$$L_0 = 2 \quad \text{and} \quad R_0 = 3$$

Iteration 0: $m = \frac{1}{2}(2 + 3) = 2.5$

$f(2.5) = (2.5)^3 - 2(2.5) - 7 = 3.625$

Since $f(2.5)$ is positive, a root must exist in the interval $(2, 2.5)$. Therefore,

$$L_1 = 2 \quad \text{and} \quad R_1 = 2.5$$

At this point, the best estimate of the root is

$$x^* \simeq \frac{1}{2}(2 + 2.5) = 2.25$$

The maximum error is $\frac{1}{2}(2.5 - 2) = .25$

Iteration 1: $m = \frac{1}{2}(2 + 2.5) = 2.25$

$f(2.25) = -.1094$

Since f(m) is negative, a root must exist in the interval (2.25, 2.5).   Therefore,

$L_2 = 2.25$ and $R_2 = 2.5$

The best estimate of the root is

$x^* \simeq \frac{1}{2}(2.25 + 2.5) = 2.375$

The maximum error is $\frac{1}{2}(2.5 - 2.25) = .125$

This procedure continues until the maximum error is acceptable.   Of course, this method does not automatically find any other roots that may exist on the real number line.

◆

B. Simultaneous Linear Equations

Given n independent equations and n unknowns, the n values which simultaneously solve all n equations can be found by the methods illustrated below.

1. By Substitution (shown by example)

Example 1.3

Solve $2x + 3y = 12$ (a)

$3x + 4y = 8$ (b)

step 1:   From equation (a), solve for $x = 6 - 1.5y$

step 2:   Substitute $(6 - 1.5y)$ into equation (b) wherever x appears.
$3(6 - 1.5y) + 4y = 8$ or $y^* = 20$

step 3:   Solve for x^* from either equation: $x^* = 6 - 1.5(20) = -24$

step 4:   Check that (-24, 20) solves both original equations.

◆

2. By Reduction (same example)

step 1:   Multiply each equation by a number chosen to make the coefficient of one of the variables the same in each equation.

3 times equation (a): $6x + 9y = 36$ (c)

2 times equation (b): $6x + 8y = 16$ (d)

step 2:   Subtract one equation from the other.   Solve for one of the variables.

(c) - (d): $y^* = 20$

step 3: Solve for the remaining variable.

step 4: Check that the calculated values of (x^*, y^*) solve both original equations.

3. By Cramer's Rule

This method is best for 3 or more simultaneous equations. (The calculation of determinants is covered later in this chapter.)

To find x^* and y^* which satisfy

$$a_1 x + b_1 y = c_1$$

$$a_2 x + b_2 y = c_2$$

Calculate the determinants

$$D_1 = \begin{vmatrix} a_1 & b_1 \\ a_2 & b_2 \end{vmatrix} \qquad D_2 = \begin{vmatrix} c_1 & b_1 \\ c_2 & b_2 \end{vmatrix} \qquad D_3 = \begin{vmatrix} a_1 & c_1 \\ a_2 & c_2 \end{vmatrix}$$

Then, if $D_1 \neq 0$, the unique numbers satisfying the two simultaneous equations are:

$$x^* = D_2/D_1 \qquad y^* = D_3/D_1$$

This method may be extended to equations in 3 or more unknowns as illustrated in example 1.4

Example 1.4

Solve the following system of simultaneous equations:

$$2x + 3y - 4z = 1$$
$$3x - y - 2z = 4$$
$$4x - 7y - 6z = -7$$

Calculate the determinants:

$$D_1 = \begin{vmatrix} 2 & 3 & -4 \\ 3 & -1 & -2 \\ 4 & -7 & -6 \end{vmatrix} = 82 \qquad D_2 = \begin{vmatrix} 1 & 3 & -4 \\ 4 & -1 & -2 \\ -7 & -7 & -6 \end{vmatrix} = 246$$

$$D_3 = \begin{vmatrix} 2 & 1 & -4 \\ 3 & 4 & -2 \\ 4 & -7 & -6 \end{vmatrix} = 82 \qquad D_4 = \begin{vmatrix} 2 & 3 & 1 \\ 3 & -1 & 4 \\ 4 & -7 & -7 \end{vmatrix} = 164$$

Then, $x^* = D_2/D_1 = 3$ $\qquad$ $y^* = D_3/D_1 = 1$ $\qquad$ $z^* = D_4/D_1 = 2$

C. Exponentiation (x is any variable or constant)

$$x^m x^n = x^{(n+m)} \quad x^m/x^n = x^{(m-n)} \quad (x^n)^m = x^{(mn)} \quad (a/b)^n = a^n/b^n$$

$$x^0 = 1 \qquad x^{-n} = 1/(x^n) \qquad \sqrt[n]{x} = (x)^{1/n} \qquad a^{m/n} = \sqrt[n]{a^m}$$

D. Logarithms and Log Identities

Logarithms are exponents. Therefore, the exponent x in the expression $b^x = n$ is the base b logarithm of n. Thus, $(\log_b n = x)$ is equivalent to $(b^x = n)$.

$$x^a = \text{antilog}[a \log(x)] \qquad \ln(x) = (\log_{10} x)/(\log_{10} e)$$

$$\simeq 2.3[\log_{10}(x)]$$

$$\log(x^a) = a[\log(x)] \qquad \log_b(b) = 1$$

$$\log(xy) = \log(x) + \log(y) \qquad \log(1) = 0$$

$$\log(x/y) = \log(x) - \log(y) \qquad \log_b(b^n) = n$$

Example 1.5

The surviving fraction, x, of a radioactive isotope is given by

$$x = e^{-.005t}$$

For what value of t will the surviving fraction be 7%?

$$.07 = e^{-.005t}$$

Taking the natural log of both sides,

$$\ln(.07) = \ln(e^{-.005t})$$

$$-2.66 = -.005t$$

$$t = 532$$

◆

E. Partial Fractions

Given some rational fraction $H(x) = P(x)/Q(x)$ where $P(x)$ and $Q(x)$ are polynomials, the polynomials and constants A_i and $Y_i(x)$ are needed such that

$$H(x) = \sum_i \frac{A_i}{Y_i(x)}$$

case 1: $Q(x)$ factors into n different linear terms. That is,

$$Q(x) = (x-a_1)(x-a_2)\cdots(x-a_n)$$

Set
$$H(x) = \sum_{i=1}^{n} \frac{A_i}{(x-a_i)}$$

case 3: $Q(x)$ factors into n different quadratic terms $(x^2 + p_i x + q_i)$

Set $\qquad H(x) = \sum_{i=1}^{n} \dfrac{A_i x + B_i}{x^2 + p_i x + q_i}$

case 4: $Q(x)$ factors into n identical quadratic terms, $(x^2 + px + q)$

Set $\qquad H(x) = \sum_{i=1}^{n} \dfrac{A_i x + B_i}{(x^2 + px + q)^i}$

case 5: $Q(x)$ factors into any combination of the above. The solution is illustrated by example 1.6.

Example 1.6

Resolve $H(x) = \dfrac{x^2 + 2x + 3}{x^4 + x^3 + 2x^2}$ into partial fractions.

Here, $Q(x) = x^4 + x^3 + 2x^2$ which factors into $x^2(x^2 + x + 2)$. This is a combination of cases 2 and 3. We set

$$H(x) = \frac{A_1}{x} + \frac{A_2}{x^2} + \frac{A_3 + A_4 x}{x^2 + x + 2}$$

Cross multiplying to obtain a common denominator yields

$$H(x) = \frac{(A_1 + A_4)x^3 + (A_1 + A_2 + A_3)x^2 + (2A_1 + A_2)x + 2A_2}{x^4 + x^3 + 2x^2}$$

Since the original numerator is known, the following simultaneous equations result:

$$A_1 + A_4 = 0 \qquad\qquad 2A_1 + A_2 = 2$$

$$A_1 + A_2 + A_3 = 1 \qquad\qquad 2A_2 = 3$$

The solutions are: $A_1^* = .25 \quad A_2^* = 1.5 \quad A_3^* = -.75 \quad A_4^* = -.25$

So that $H(x) = \dfrac{1}{4x} + \dfrac{3}{2x^2} - \dfrac{x + 3}{4(x^2 + x + 2)}$

◆

F. Linear and Matrix Algebra

A matrix is a rectangular collection of variables or scalars contained within a set of square or round brackets. In the discussions below, matrix A will be assumed to have m rows and n columns.

Matrixes are used to simplify the presentation and solution of sets of linear equations (hence the name 'linear algebra'). For example, the system of equations in example 1.3 can be written in matrix form as:

$$\begin{pmatrix} 2 & 3 \\ 3 & 4 \end{pmatrix} \begin{pmatrix} x \\ y \end{pmatrix} = \begin{pmatrix} 12 \\ 8 \end{pmatrix}$$

The above expression implies that there is a set of algebraic operations that can be performed with matrixes. The important algebraic operations are listed here, along with their extensions to linear algebra.

(a)   Equality of Matrixes: For two matrixes to be equal, they must have the same number of rows and columns. Corresponding entries must all be the same.

(b)   Inequality of Matrixes: There are no 'less-than' or 'greater-than' relationships in linear algebra.

(c)   Addition and Subtraction of Matrixes: Addition (or subtraction) of two matrixes may be accomplished by adding (or subtracting) the corresponding entries of two matrixes which have the same shape.

(d)   Multiplication of Matrixes: Multiplication can only be done if the left-hand matrix has the same number of columns as the right-hand matrix has rows. Multiplication is accomplished by multiplying the elements in each left-hand matrix row by the elements in each right-hand matrix column, adding the products, and then placing the sum at the intersection point of the involved row and column. This is illustrated by example 1.7.

(e)   Division of Matrixes: Division can only be accomplished by multiplying by the inverse of the denominator matrix.

Example 1.7

$$\begin{pmatrix} 1 & 4 & 3 \\ 5 & 2 & 6 \end{pmatrix} \begin{pmatrix} 7 & 12 \\ 11 & 8 \\ 9 & 10 \end{pmatrix} = [C]$$

$$[(1)(7) + (4)(11) + (3)(9)] = 78$$

$$[(1)(12) + (4)(8) + (3)(10)] = 74$$

$$[(5)(7) + (2)(11) + (6)(9)] = 111$$

$$[(5)(12) + (2)(8) + (6)(10)] = 136$$

$$[C] = \begin{pmatrix} 78 & 74 \\ 111 & 136 \end{pmatrix}$$

Other operations which can be performed on a matrix are described and illustrated below.

1. The TRANSPOSE is an (n x m) matrix formed from the original (m x n) matrix by taking the ith row and making it the ith column. The diagonal is unchanged in this operation. The transpose of a matrix A is indicated as A^t.

Example 1.8

What is the transpose of $A = \begin{pmatrix} 1 & 6 & 9 \\ 2 & 3 & 4 \\ 7 & 1 & 5 \end{pmatrix}$?

$$A^t = \begin{pmatrix} 1 & 2 & 7 \\ 6 & 3 & 1 \\ 9 & 4 & 5 \end{pmatrix}$$

◆

2. The DETERMINANT, D, is a scalar calculated from a square matrix. The determinant of a matrix is indicated by enclosing the matrix in vertical lines.

For a (2 x 2) matrix,

$$A = \begin{pmatrix} a & b \\ c & d \end{pmatrix} \qquad D = \begin{vmatrix} a & b \\ c & d \end{vmatrix} = ad - bc$$

For a (3 x 3) matrix,

$$A = \begin{pmatrix} a & b & c \\ d & e & f \\ g & h & i \end{pmatrix}$$

$$D = a\begin{vmatrix} e & f \\ h & i \end{vmatrix} - d\begin{vmatrix} b & c \\ h & i \end{vmatrix} + g\begin{vmatrix} b & c \\ e & f \end{vmatrix}$$

There are several rules governing the calculation of determinants:

- If A has a row or column of zeros, the determinant is zero.

- If A has two identical rows or columns, the determinant is zero.

- If A is triangular, the determinant is equal to the product of the diagonal entries.

- If B is obtained from A by multiplying a row or column by a scalar k, then $D_B = k(D_A)$.

- If B is obtained from A by switching two rows or columns, then $D_B = -D_A$.

- If B is obtained from A by adding a multiple of a row or column to another, then $D_B = D_A$.

Example 1.9

What is the determinant of $\begin{pmatrix} 2 & 3 & -4 \\ 3 & -1 & -2 \\ 4 & -7 & -6 \end{pmatrix}$?

$$D = 2\begin{vmatrix} -1 & -2 \\ -7 & -6 \end{vmatrix} - 3\begin{vmatrix} 3 & -4 \\ -7 & -6 \end{vmatrix} + 4\begin{vmatrix} 3 & -4 \\ -1 & -2 \end{vmatrix} = 2(6-14) - 3(-18-28) + 4(-6-4)$$

$$= 82$$

3. The COFACTOR of an entry in a matrix is the determinant of the matrix formed by omitting the entry's row and column in the original matrix. The sign of the cofactor is determined from the following positional matrixes:

For a (2 x 2) matrix, $\begin{pmatrix} + & - \\ - & + \end{pmatrix}$

For a (3 x 3) matrix, $\begin{pmatrix} + & - & + \\ - & + & - \\ + & - & + \end{pmatrix}$

Example 1.10

What is the cofactor of the (-3) in the following matrix? $\begin{pmatrix} 2 & 9 & 1 \\ -3 & 4 & 0 \\ 7 & 5 & 9 \end{pmatrix}$

The resulting matrix is $\begin{pmatrix} 9 & 1 \\ 5 & 9 \end{pmatrix}$ with determinant 76. The cofactor is -76.

4. The CLASSICAL ADJOINT is a matrix formed from the transposed cofactor matrix with the conventional sign arrangement. The resulting matrix is represented as A_{adj}.

Example 1.11

What is the classical adjoint of $\begin{pmatrix} 2 & 3 & -4 \\ 0 & -4 & 2 \\ 1 & -1 & 5 \end{pmatrix}$?

The matrix of cofactors (considering the sign convention) is

$$\begin{pmatrix} -18 & 2 & 4 \\ -11 & 14 & 5 \\ -10 & -4 & -8 \end{pmatrix}$$

The transposed cofactor matrix is

$$A_{adj} = \begin{pmatrix} -18 & -11 & -10 \\ 2 & 14 & -4 \\ 4 & 5 & -8 \end{pmatrix}$$

5. The INVERSE, A^{-1}, of A is a matrix such that $(A)(A^{-1}) = I$. (I is a square matrix with ones along the left-to-right diagonal and zeros elsewhere.)

For a (2 x 2) matrix, $\begin{pmatrix} a & b \\ c & d \end{pmatrix}$, the inverse is $\frac{1}{D} \begin{pmatrix} d & -b \\ -c & a \end{pmatrix}$

For larger matrixes, the inverse is best calculated by dividing every entry in the classical adjoint by the determinant of the original matrix.

Example 1.12

What is the inverse of $\begin{pmatrix} 4 & 5 \\ 2 & 3 \end{pmatrix}$?

The determinant is 2. The inverse is $\frac{1}{2} \begin{pmatrix} 3 & -5 \\ -2 & 4 \end{pmatrix} = \begin{pmatrix} 3/2 & -5/2 \\ -1 & 2 \end{pmatrix}$

5. Trigonometry

   A. Degrees and Radians

         360 degrees = one complete circle = 2π radians

         90 degrees = right angle = $\frac{1}{2}\pi$ radians

         one radian = 57.3 degrees
         one degree = .0175 radians

         multiply degrees by $(\pi/180)$ to obtain radians
         multiply radians by $(180/\pi)$ to obtain degrees

   B. Right Triangles

      1) Pythagorean Theorem
         $$x^2 + y^2 = r^2$$

      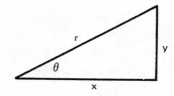

      2) Trigonometric Functions

         $\sin\theta = y/r$ = opposite side/hypotenuse

         $\cos\theta = x/r$ = adjacent side/hypotenuse

         $\tan\theta = y/x$ = opposite side/adjacent side

         $\csc\theta = r/y$

         $\sec\theta = r/x$

         $\cot\theta = x/y$

3) Relationship of the Trigonometric Functions to the Unit Circle

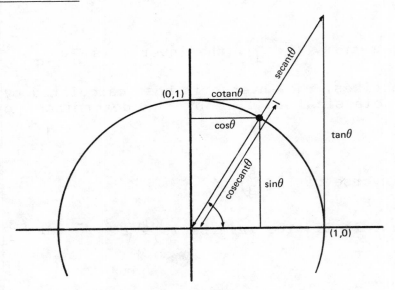

4) Signs of the Trigonometric Functions

quadrants		quadrant	I	II	III	IV
II	I	sin	+	+	–	–
		cos	+	–	–	+
III	IV	tan	+	–	+	–

5) Functions of the Related Angles

	$-\theta$	$90-\theta$	$90+\theta$	$180-\theta$	$180+\theta$
sin	$-\sin\theta$	$\cos\theta$	$\cos\theta$	$\sin\theta$	$-\sin\theta$
cos	$\cos\theta$	$\sin\theta$	$-\sin\theta$	$-\cos\theta$	$-\cos\theta$
tan	$-\tan\theta$	$\cot\theta$	$-\cot\theta$	$-\tan\theta$	$\tan\theta$

6) Trigonometric Identities

$$\sin^2\theta + \cos^2\theta = 1$$

$$1 + \tan^2\theta = \sec^2\theta$$

$$1 + \cot^2\theta = \csc^2\theta$$

$$\sin 2\theta = 2(\sin\theta)(\cos\theta)$$

$$\cos 2\theta = \cos^2\theta - \sin^2\theta$$

$$= 1 - 2\sin^2\theta$$

$$\sin\theta = 2\left[\sin\left(\tfrac{1}{2}\theta\right)\cos\left(\tfrac{1}{2}\theta\right)\right]$$

$$\sin\left(\tfrac{1}{2}\theta\right) = \pm\sqrt{\tfrac{1}{2}(1-\cos\theta)}$$

7) Two-Angle Formulas

$$\sin(\theta + \phi) = [\sin\theta][\cos\phi] + [\cos\theta][\sin\phi]$$
$$\sin(\theta - \phi) = [\sin\theta][\cos\phi] - [\cos\theta][\sin\phi]$$
$$\cos(\theta + \phi) = [\cos\theta][\cos\phi] - [\sin\theta][\sin\phi]$$
$$\cos(\theta - \phi) = [\cos\theta][\cos\phi] + [\sin\theta][\sin\phi]$$

C. General Triangles

Law of Sines: $\dfrac{\sin A}{a} = \dfrac{\sin B}{b} = \dfrac{\sin C}{c}$

Law of Cosines: $a^2 = b^2 + c^2 - 2bc(\cos A)$

Area $= \dfrac{1}{2}ab(\sin C)$

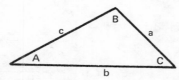

6. Straight Line Analytic Geometry

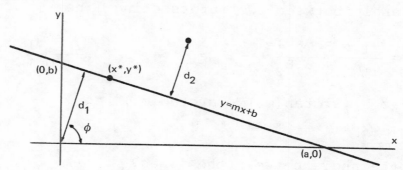

A. Equations of a Straight Line

General Form: $Ax + By + C = 0$

Slope Form: $y = mx + b$

Point-Slope Form:
(x^*, y^*) is any point on the line
$(y - y^*) = m(x - x^*)$

Intercept Form: $\dfrac{x}{a} + \dfrac{y}{b} = 1$

Two-Point Form: $\dfrac{y - y_1^*}{x - x_1^*} = \dfrac{y_2^* - y_1^*}{x_2^* - x_1^*}$

Normal Form: $x\cos\phi + y\sin\phi - d_1 = 0$

B. Points, Lines, and Distances

Given a linear equation in general form, the following apply:

The distance d_2 between a point and a line is:

$$d_2 = \frac{|Ax_1 + By_1 + C|}{\sqrt{A^2 + B^2}}$$

The distance between two points is:

$$d = \sqrt{(x_2-x_1)^2 + (y_2-y_1)^2}$$

Parallel lines: $A_1/A_2 = B_1/B_2$ or $m_1 = m_2$

Perpendicular lines: $A_1A_2 = -B_1B_2$ or $m_1 = \frac{-1}{m_2}$

Point of intersection of two lines:

$$x_1 = \frac{B_2C_1 - B_1C_2}{A_2B_1 - A_1B_2} \qquad\qquad y_1 = \frac{A_1C_2 - A_2C_1}{A_2B_1 - A_1B_2}$$

Smaller angle between two intersecting lines:

$$\tan\phi = \frac{A_1B_2 - A_2B_1}{A_1A_2 + B_1B_2} \qquad = \qquad \frac{m_2 - m_1}{1 + m_1m_2}$$

Also, $\phi = |\arctan(m_1) - \arctan(m_2)|$

Example 1.13

What is the angle between the lines?

$$y_1 = -.577x + 2$$

$$y_2 = +.577x - 5$$

method 1: $\arctan\left(\dfrac{m_2 - m_1}{1+m_1m_2}\right) = \arctan\left(\dfrac{.577-(-.577)}{1 + (.577)(-.577)}\right) = 60°$

method 2:   Write both equations in general form:

$$-.577x - y_1 + 2 = 0$$

$$.577x - y_2 - 5 = 0$$

$$\arctan\left(\frac{A_1B_2 - A_2B_1}{A_1A_2 + B_1B_2}\right) = \arctan\left(\frac{(-.577)(-1)-(.577)(-1)}{(-.577)(.577)+(-1)(-1)}\right) = 60°$$

C. Linear Regression

If it is necessary to draw a straight line through n data points (x_1, y_1), $(x_2, y_2), \ldots, (x_n, y_n)$, the following method based on the theory of least squares may be used:

<u>step 1</u>:   Calculate the following quantities:

$$\Sigma x_i \qquad \Sigma x_i^2 \qquad (\Sigma x_i)^2 \qquad \overline{x} = (\Sigma x_i / n) \qquad \Sigma x_i y_i$$

$$\Sigma y_i \qquad \Sigma y_i^2 \qquad (\Sigma y_i)^2 \qquad \overline{y} = (\Sigma y_i / n)$$

<u>step 2</u>:   Calculate the slope of the line, $y = mx + b$.

$$m = \frac{n\Sigma(x_i y_i) - (\Sigma x_i)(\Sigma y_i)}{n\Sigma x_i^2 - (\Sigma x_i)^2}$$

<u>step 3</u>:   Calculate the y intercept.

$$b = \overline{y} - m\overline{x}$$

<u>step 4</u>:   To determine the goodness of fit, calculate the correlation coefficient.

$$r = \frac{n\Sigma(x_i y_i) - (\Sigma x_i)(\Sigma y_i)}{\sqrt{[n\Sigma x_i^2 - (\Sigma x_i)^2][n\Sigma y_i^2 - (\Sigma y_i)^2]}}$$

If m is positive, r will be positive. If m is negative, r will be negative. As a general rule, if the absolute value of r exceeds .85, the fit is good. Otherwise, the fit is poor. r equals one if the fit is a perfect straight line.

Example 1.14

An experiment is performed in which the dependent variable (y) is measured against the independent variable (x). The results are as follows:

x	y
1.2	.602
4.7	5.107
8.3	6.984
20.9	10.031

What is the least squares straight line equation which represents this data? Is the fit good or poor?

<u>step 1</u>: $\Sigma x = 35.1$ $\Sigma x_i^2 = 529.23$ $(\Sigma x_i)^2 = 1232.01$ $\overline{x} = 8.775$

$\Sigma y = 22.72$ $\Sigma y_i^2 = 175.84$ $(\Sigma y_i)^2 = 516.19$ $\overline{y} = 5.681$

$\Sigma xy = 292.34$ $n = 4$

step 2: $m = \dfrac{(4)(292.34) - (35.1)(22.72)}{(4)(529.23) - (35.1)^2} = .42$

step 3: $b = 5.681 - (.42)(8.775) = 2.0$

step 4: From the above equation, r = .91. This is not a particularly good correlation coefficient for the small number of data points. ◆

A low value of r does not eliminate the possibility of a non-linear relationship existing between x and y. It is possible that the data describe a parabolic, logarithmic, or other non-linear relationship. (Usually this will be apparent if the data are graphed.) It may be necessary to convert one or both variables to new variables by taking squares, square roots, cubes, or logs to name a few of the possibilities.

The apparent shape of the line through the data will give a clue to the type of variable transformation that is required. The following curves may be used as guides to some of the more simple variable transformations.

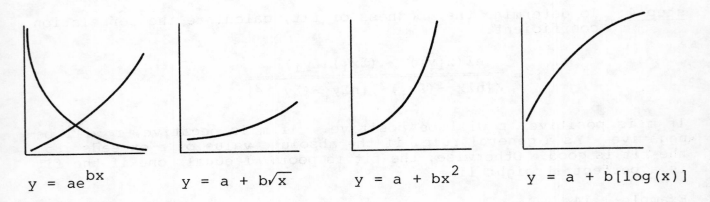

$y = ae^{bx}$ $y = a + b\sqrt{x}$ $y = a + bx^2$ $y = a + b[\log(x)]$

Example 1.15

Repeat example 1.14 assuming that the relationship between the variables is non-linear.

The first step is to graph the data. Since the graph has the appearance of the fourth case above, it can be assumed that the relationship between the variables has the form of $y = a + b[\log(x)]$. Therefore, the variable change $z = \log(x)$ is made, resulting in the following set of data:

z	y
.0792	.602
.672	5.107
.919	6.984
1.32	10.031

If the regression analysis is performed on this set of data, the resulting equation and correlation coefficient are:

$$y = -.036 + 7.65z$$

$$r = .999$$

This is a very good fit. We can conclude that the relationship between the variable x and y is

$$y = -.036 + 7.65[\log(x)]$$

D. Vector Operations

A vector is a directed straight line of a given magnitude. Two directed straight lines with the same magnitudes and directions are said to be equivalent. Thus, the actual end-points of a vector are often irrelevant as long as the direction and magnitude are known.

A vector defined by its end-points and direction is designated as

$$V = \overrightarrow{P_1P_2}$$

Usually, P_1 will be the origin, in which case V will be designated by its end-point, $P_2 = (x,y)$. Such a zero-based vector is equivalent to all other vectors of the same magnitude and direction. Any vector P_1P_2 can be transformed into a zero-based vector by subtracting (x_1,y_1) from all points along the vector line.

A vector may also be specified in terms of the unit vectors $(\vec{i},\vec{j},\vec{k})$. Thus,

$$V = (x,y) = (x\vec{i} + y\vec{j})$$

Or,

$$V = (x,y,z) = (x\vec{i} + y\vec{j} + z\vec{k})$$

Important operations on vectors based at the origin are:

$$cV = (cx, cy) \quad \text{(vector multiplication by a scalar)}$$

$$V_1 + V_2 = (x_1 + x_2, y_1 + y_2)$$

$$|V| = \sqrt{x^2 + y^2} \quad \text{(vector magnitude)}$$

$$\alpha = \text{angle between vector V and x axis}$$

$$= \arccos(x/|V|) = \arcsin(y/|V|)$$

$$m = \text{slope of vector} = y/x$$

$$\theta = \begin{Bmatrix} \text{angle between} \\ \text{two vectors} \end{Bmatrix} = \arccos\left(\frac{x_1x_2 + y_1y_2}{|V_1||V_2|}\right)$$

$$V_1 \cdot V_2 = \text{dot product} = |V_1||V_2|\cos\theta = x_1 x_2 + y_1 y_2$$

$$V_1 \times V_2 = \text{cross product}$$

$$= \begin{vmatrix} i & x_1 & x_2 \\ j & y_1 & y_2 \\ k & z_1 & z_2 \end{vmatrix}$$

$$V_1 \times V_2 = -V_2 \times V_1$$

$$|V_1 \times V_2| = |V_1||V_2|\sin\theta$$

Example 1.16

What is the angle between the vectors $V_1 = (-\sqrt{3}, 1)$ and $V_2 = (2\sqrt{3}, 2)$?

$$\cos\theta = \frac{V_1 \cdot V_2}{|V_1||V_2|} = \frac{(-\sqrt{3})(2\sqrt{3}) + (1)(2)}{\sqrt{3+1}\quad\sqrt{12+4}} = \frac{-1}{2}$$

$$\theta = 120°$$

(Graph and compare this result to example 1.13 in which the lines were not directed.) ◆

Example 1.17

Find a unit vector orthogonal to $V_1 = i - j + 2k$ and $V_2 = 3j - k$.

The cross product is orthogonal to V_1 and V_2, although its length may not be equal to one.

$$V_1 \times V_2 = \begin{vmatrix} i & 1 & 0 \\ j & -1 & 3 \\ k & 2 & -1 \end{vmatrix} = -5i + j + 3k$$

Since the length of $|V_1 \times V_2|$ is $\sqrt{35}$, it is necessary to divide $|V_1 \times V_2|$ by this amount to obtain a unit vector. Thus,

$$V_3 = \frac{-5i + j + 3k}{\sqrt{35}}$$

The orthogonality can be proved from

$$V_1 \cdot V_3 = 0 \quad \text{and} \quad V_2 \cdot V_3 = 0$$

That V_3 is a unit vector can be proved from

$$V_3 \cdot V_3 = +1$$

◆

E. Direction Numbers, Direction Angles, and Direction Cosines

case 1: Given a directed line from (x_1, y_1, z_1) to (x_2, y_2, z_2),

The direction numbers are: $L = x_2 - x_1$; $M = y_2 - y_1$; $N = z_2 - z_1$

The distance between the two points is: $d = \sqrt{L^2 + M^2 + N^2}$

The direction cosines are: $\cos\alpha = L/d$; $\cos\beta = M/d$; $\cos\gamma = N/d$

Note that $\cos^2\alpha + \cos^2\beta + \cos^2\gamma = 1$

The direction angles are the angles between the axes and the line. They are found from the inverse functions of the direction cosines.

case 2: Given a directed line from the origin to a point.

Set $(x_1, y_1, z_1) = (0, 0, 0)$ and use case 1 above.

case 3: Given two directed lines L_1 and L_2.

The angle between L_1 and L_2 is defined as the angle between the two arrow heads.

$$\cos\phi = [\cos\alpha_1][\cos\alpha_2] + [\cos\beta_1][\cos\beta_2] + [\cos\gamma_1][\cos\gamma_2]$$

$$= \frac{L_1 L_2 + M_1 M_2 + N_1 N_2}{d_1 d_2}$$

If L_1 and L_2 are parallel and in the same direction, then

$$\alpha_1 = \alpha_2; \quad \beta_1 = \beta_2; \quad \gamma_1 = \gamma_2$$

If L_1 and L_2 are parallel but in opposite directions, then

$$\alpha_1 = -\alpha_2 \quad \text{(etc.)} \quad \text{and} \quad \cos\alpha_1 = -\cos\alpha_2 \quad \text{(etc.)}$$

case 4: Given two undirected lines, L_1 and L_2.

The angle between L_1 and L_2 is the smaller of the two angles.

F. Curvilinear Interpolation

A situation which occurs frequently is one in which a function value must be interpolated from other data along the curve. Straight-line interpolation is typically used because of its simplicity and speed. However, straight-line interpolation ignores all but two of the points on the curve, and is, therefore, unable to include any effects of curvature.

A more powerful technique is the Lagrangian Interpolating Polynominal, as illustrated on the following page. It is assumed that $(n + 1)$ values of $f(x)$ are known (for x_0, x_1, $x_2, \cdots, x_n$) and that $f(x)$ is a continuous, real-valued function on the interval (x_0, x_n). The value of $f(x)$ at x^* can be estimated from the following equations:

$$f(x^*) = \sum_{k=0}^{n} f(x_k) L_k(x^*)$$

where the Lagrangian Interpolating Polynomial is

$$L_k(x^*) = \prod_{\substack{i=0 \\ i \neq k}}^{n} \frac{x^* - x_i}{x_k - x_i}$$

Example 1.18

A real-valued function has the following values:

$$f(1) = 1.5709 \qquad f(4) = 1.5727 \qquad f(6) = 1.5751$$

What is $f(3.5)$?

$$
\begin{array}{llll}
 & i = 0 & i = 1 & i = 2 \\
\underline{k = 0:} & L_0(3.5) = (\cancel{\frac{3.5-1}{1-1}}) & (\frac{3.5-4}{1-4}) & (\frac{3.5-6}{1-6}) = .08333 \\
\underline{k = 1:} & L_1(3.5) = (\frac{3.5-1}{4-1}) & (\cancel{\frac{3.5-4}{4-4}}) & (\frac{3.5-6}{4-6}) = 1.04167 \\
\underline{k = 2:} & L_2(3.5) = (\frac{3.5-1}{6-1}) & (\frac{3.5-4}{6-4}) & (\cancel{\frac{3.5-6}{6-6}}) = -.12500
\end{array}
$$

$$
\begin{aligned}
f(x^*) &= (1.5709)(.08333) + (1.5727)(1.04167) \\
&\quad + (1.5751)(-.12500) \\
&= 1.57225
\end{aligned}
$$

7. Probability and Statistics

A. Probability Rules:

The following rules are applied to sample spaces A and B:

$A = (A_1, A_2, A_3, \cdots, A_n)$ and $B = (B_1, B_2, B_3, \cdots, B_n)$ where the A_i and B_i are independent.

1) $p(\emptyset)$ = probability of an impossible event = 0

Example 1.19

An urn contains 5 white balls, 2 red balls, and 3 green balls. What is the probability of drawing a blue ball from the urn?

$$p(\text{blue ball}) = p(\emptyset) = 0$$

2) $p(A_1 \text{ or } A_2 \text{ or } \cdots \text{ or } A_n) = p(A_1) + p(A_2) + \cdots + p(A_n)$

Example 1.20

Returning to the urn described in example 1.19, what is the probability of getting either a white ball or a red ball in one draw from the urn?

$$p(\text{red or white}) = p(\text{red}) + p(\text{white}) = .5 + .2 = .7$$

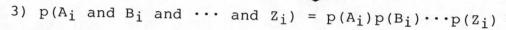

3) $p(A_i \text{ and } B_i \text{ and } \cdots \text{ and } Z_i) = p(A_i)p(B_i)\cdots p(Z_i)$

Example 1.21

Given two identical urns (as described in example 1.19), what is the probability of getting a red ball from the first urn and a green ball from the second urn, given one draw from each urn?

$$p(\text{red and green}) = [p(\text{red})][p(\text{green})] = (.2)(.3) = .06$$

4) $p(\text{not } A) = $ probability of event A not occurring $= 1 - p(A)$

Example 1.22

Given the urn of example 1.19, what is the probability of not getting a red ball from the urn in one draw?

$$p(\text{not red}) = 1 - p(\text{red}) = 1 - .2 = .8$$

5) $p(A_i \text{ or } B_i) = p(A_i) + p(B_i) - p(A_i)p(B_i)$

Example 1.23

Given one urn as described in example 1.19 and a second urn containing 8 red balls and 2 black balls, what is the probability of drawing either a white ball from the first urn or a red ball from the second urn, given one draw from each urn?

$$p(\text{white or red}) = p(\text{white}) + p(\text{red}) - [p(\text{white})][p(\text{red})]$$

$$= .5 + .8 - (.5)(.8) = .9$$

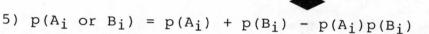

B. Probability Density Functions

Probability density functions are mathematical functions giving the probabilities of numerical events. A numerical event is any occurrence that can be described by an integer or real number. For example, the probability of obtaining a heads in a coin-toss is not a numerical event. However, a concrete sample having a compressive strength less than 5000 psi is a numerical event.

Discrete density functions give the probability that the event x will occur. That is,

$$f(x) = \text{probability of a process having a value of } x$$

Important discrete functions are the binomial and poisson distributions.

1) Binomial

n   is the number of trials

x   is the number of successes

p   is the probability of a success in a single trial

q   is the probability of failure, 1 - p

$\binom{n}{x}$ is the binomial coefficient $= \dfrac{n!}{(n-x)!\,x!}$

$x! = x(x-1)(x-2)\cdots(2)(1)$

Then, the probability of obtaining x successes in n trials is

$$f(x) = \binom{n}{x} p^x q^{(n-x)}$$

Example 1.24

In a large quantity of items, 5% are defective. If 7 items are sampled, what is the probability that exactly three will be defective?

$$f(3) = \binom{7}{3}(.05)^3(.95)^4 = .0036$$

2) Poisson

Suppose an event occurs, on the average, λ times per period. The probability that the event will occur x times per period is

$$f(x) = \frac{e^{-\lambda}\lambda^x}{x!}$$

Example 1.25

The number of customers arriving in some period is a poisson distribution with a mean of 8. What is the probability that 6 customers will arrive in any given period?

$$f(6) = \frac{e^{-8}8^6}{6!} = .122$$

Continuous probability density functions are used to find the cumulative distribution functions, F(x). Cumulative distribution functions give the probability of event x or less occurring.

x = any value, not necessarily an integer

$$f(x) = \frac{dF(x)}{dx}$$

F(x) = probability of x or less occurring

3) Exponential

$$f(x) = u(e^{-ux}) \text{ where } (1/u) \text{ is the mean}$$

$$F(x) = 1 - e^{-ux}$$

Example 1.26

The reliabiltiy of a unit is exponentially distributed with mean time to failure (MTBF) of 1000 hours. What is the probability that the unit will be operational at t = 1200 hours?

The reliability of an item is (1 - probability of failing before time t). Therefore,

$$R(t) = 1 - F(t) = 1 - (1 - e^{-ux}) = e^{-ux}$$

$$u = 1/MTBF = 1/1000 = .001$$

$$R(1200) = e^{-(.001)(1200)} = .3$$

◆

4) Normal

Although f(x) may be expressed mathematically for the normal distribution, tables are used to evaluate F(x) since f(x) cannot be easily integrated. Since the x axis of the normal distribution will seldom correspond to actual sample variables, the sample values are converted into standard values. Given the mean, u, and the standard deviation, σ, the standard normal variable is

$$z = \frac{(\text{sample value} - u)}{\sigma}$$

Then, the probability of a sample exceeding the given sample value is equal to the area in the tail past point z.

AREAS UNDER THE STANDARD NORMAL CURVE
(0 to z)

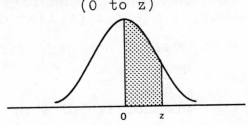

z	0	1	2	3	4	5	6	7	8	9
0.0	.0000	.0040	.0080	.0120	.0160	.0199	.0239	.0279	.0319	.0359
0.1	.0398	.0438	.0478	.0517	.0557	.0596	.0636	.0675	.0714	.0754
0.2	.0793	.0832	.0871	.0910	.0948	.0987	.1026	.1064	.1103	.1141
0.3	.1179	.1217	.1255	.1293	.1331	.1368	.1406	.1443	.1480	.1517
0.4	.1554	.1591	.1628	.1664	.1700	.1736	.1772	.1808	.1844	.1879
0.5	.1915	.1950	.1985	.2019	.2054	.2088	.2123	.2157	.2190	.2224
0.6	.2258	.2291	.2324	.2357	.2389	.2422	.2454	.2486	.2518	.2549
0.7	.2580	.2612	.2642	.2673	.2704	.2734	.2764	.2794	.2823	.2852
0.8	.2881	.2910	.2939	.2967	.2996	.3023	.3051	.3078	.3106	.3133
0.9	.3159	.3186	.3212	.3238	.3264	.3289	.3315	.3340	.3365	.3389
1.0	.3413	.3438	.3461	.3485	.3508	.3531	.3554	.3577	.3599	.3621
1.1	.3643	.3665	.3686	.3708	.3729	.3749	.3770	.3790	.3810	.3830
1.2	.3849	.3869	.3888	.3907	.3925	.3944	.3962	.3980	.3997	.4015
1.3	.4032	.4049	.4066	.4082	.4099	.4115	.4131	.4147	.4162	.4177
1.4	.4192	.4207	.4222	.4236	.4251	.4265	.4279	.4292	.4306	.4319
1.5	.4332	.4345	.4357	.4370	.4382	.4394	.4406	.4418	.4429	.4441
1.6	.4452	.4463	.4474	.4484	.4495	.4505	.4515	.4525	.4535	.4545
1.7	.4554	.4564	.4573	.4582	.4591	.4599	.4608	.4616	.4625	.4633
1.8	.4641	.4649	.4656	.4664	.4671	.4678	.4686	.4693	.4699	.4706
1.9	.4713	.4719	.4726	.4732	.4738	.4744	.4750	.4756	.4761	.4767
2.0	.4772	.4778	.4783	.4788	.4793	.4798	.4803	.4808	.4812	.4817
2.1	.4821	.4826	.4830	.4834	.4838	.4842	.4846	.4850	.4854	.4857
2.2	.4861	.4864	.4868	.4871	.4875	.4878	.4881	.4884	.4887	.4890
2.3	.4893	.4896	.4898	.4901	.4904	.4906	.4909	.4911	.4913	.4916
2.4	.4918	.4920	.4922	.4925	.4927	.4929	.4931	.4932	.4934	.4936
2.5	.4938	.4940	.4941	.4943	.4945	.4946	.4948	.4949	.4951	.4952
2.6	.4953	.4955	.4956	.4957	.4959	.4960	.4961	.4962	.4963	.4964
2.7	.4965	.4966	.4967	.4968	.4969	.4970	.4971	.4972	.4973	.4974
2.8	.4974	.4975	.4976	.4977	.4977	.4978	.4979	.4979	.4980	.4981
2.9	.4981	.4982	.4982	.4983	.4984	.4984	.4985	.4985	.4986	.4986
3.0	.4987	.4987	.4987	.4988	.4988	.4989	.4989	.4989	.4990	.4990
3.1	.4990	.4991	.4991	.4991	.4992	.4992	.4994	.4995	.4995	.4995
3.2	.4993	.4993	.4994	.4994	.4994	.4994	.4994	.4995	.4995	.4995
3.3	.4995	.4995	.4995	.4996	.4996	.4996	.4996	.4996	.4996	.4997
3.4	.4997	.4997	.4997	.4997	.4997	.4997	.4997	.4997	.4997	.4998
3.5	.4998	.4998	.4998	.4998	.4998	.4998	.4998	.4998	.4998	.4998
3.6	.4998	.4998	.4999	.4999	.4999	.4999	.4999	.4999	.4999	.4999
3.7	.4999	.4999	.4999	.4999	.4999	.4999	.4999	.4999	.4999	.4999
3.8	.4999	.4999	.4999	.4999	.4999	.4999	.4999	.4999	.4999	.4999
3.9	.5000	.5000	.5000	.5000	.5000	.5000	.5000	.5000	.5000	.5000

Example 1.27

Given a population that is normally distributed with mean of 66 and standard deviation of 5, what per cent of the population exceeds 72?

$$z = \frac{(72 - 66)}{5} = 1.2$$

Then, from the normal table on page 1-26,

$$p(\text{exceeding } 72) = .5 - .3849 = .1151 \quad \text{or } 11.5\%$$

C. Statistical Analysis of Experimental Data

Experiments can take on many forms. An experiment might consist of measuring the weight of one cubic foot of concrete. Or, an experiment might consist of measuring the speed of a car on a roadway. Generally, such experiments are performed more than once to increase the precision and accuracy of the results.

Of course, the intrinsic variablity of the process being measured will cause the observations to vary, and we would not expect the experiment to yield the same result each time it was performed. Eventually, a collection of experimental outcomes (observations) will be available for analysis.

One fundamental technique for organizing random observations is the frequency distribution. The frequency distribution is a systematic method for ordering the observations from small to large, according to some convenient numerical characteristic.

Example 1.28

The number of cars that travel through an intersection between 12 noon and 1 p.m. is measured for 30 consecutive working days. The results of the 30 observations are:

79, 66, 72, 70, 68, 66, 68, 76, 73, 71, 74, 70, 71, 69, 67, 74, 70, 68, 69, 64, 75, 70, 68, 69, 64, 69, 62, 63, 63, 61

What is the frequency distribution using an interval of 2 cars per hour?

cars per hour	frequency of occurrence
60-61	1
62-63	3
64-65	2
66-67	3
68-69	8
70-71	6
72-73	2
74-75	3
76-77	1
78-79	1

In example 1.28, 2 cars per hour is known as the step interval. The step interval should be chosen so that the data is presented in a meaningful manner. If there are too many intervals, many of them will have zero frequencies. If there are too few intervals, the frequency distribution will have little value. Generally, 10 to 15 intervals are used.

Once the frequency distribution is complete, it may be represented graphically as a histogram. The procedure in drawing a histogram is to mark off the interval limits on a number line and then draw bars with lengths that are proportional to the frequencies in the intervals. If it is necessary to show the continuous nature of the data, a frequency polygon can be drawn.

Example 1.29

Draw the frequency histogram and frequency polygon for the data given in example 1.28.

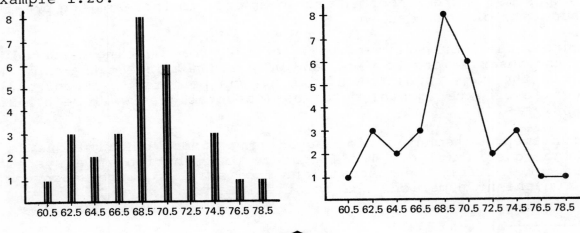

If it is necessary to know the number or percentage of observations that occur up to and including some value, the cumulative frequency table can be formed. This procedure is illustrated in the following example.

Example 1.30

Form the cumulative frequency distribution and graph for the data given in example 1.28.

cars per hour	frequency	cumulative frequency	cumulative per cent
60-61	1	1	3
62-63	3	4	13
64-65	2	6	20
66-67	3	9	30
68-69	8	17	57
70-71	6	23	77
72-73	2	25	83
74-75	3	28	93
76-77	1	29	97
78-79	1	30	100

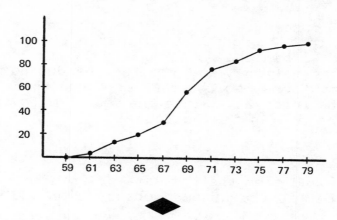

It is often unnecessary to present the experimental data in its entirety, either in tabular or graphical form. In such cases, the data and distribution can be represented by various parameters. One type of parameter is a measure of central tendency. The mode, median, and mean are measures of central tendency. The other type of parameter is a measure of dispersion. Standard deviation and variance are measures of dispersion.

The mode is the observed value which occurs most frequently. The mode may vary greatly between series of observations. Therefore, its main use is as a quick measure of the central value since no computation is required to find it. Beyond this, the usefulness of the mode is limited.

The median is the point in the distribution which divides the total observations into two parts containing equal numbers of observations. It is not influenced by the extremity of scores on either side of the distribution. The median is found by counting up (from either end of the frequency distribution) until half of the observations have been accounted for. The procedure is more difficult if the median falls within an interval, as is illustrated in example 1.31.

The arithmetic mean is the arithmetic average of the observations. The mean may be found without ordering the data (which was necessary to find the mode and median). The mean can be found from the following formula:

$$\overline{x} = \left(\frac{1}{n}\right)(x_1 + x_2 + \cdots + x_n) = \frac{\Sigma x_i}{n}$$

The geometric mean is occasionally used when it is necessary to average ratios. The primary application of the geometric mean in this book involves coliform counting. The geometric mean is calculated as

$$\text{geometric mean} = \sqrt[n]{x_1 x_2 x_3 \cdots x_n}$$

Example 1.31

Find the mode, median, and arithmetic mean of the distribution represented by the data given in example 1.28.

The mode is the interval 68-69, since this interval has the highest frequency. If 68.5 is taken as the interval center, then 68.5 would be the mode.

Since there are 30 observations, the median is the value which separates the observations into 2 groups of 15. From example 1.30, the median occurs some place within the 68-69 interval. Up through interval 66-67, there are 9 observations, so 6 more are needed to make 15. Interval 68-69 has 8 observations, so the mean is found to be (6/8) or (3/4) of the way through the interval. Since the real limits of the interval are 67.5 and 69.5, the median is located at

$$67.5 + \frac{3}{4}(69.5 - 67.5) = 69$$

The mean can be found from the raw data or from the grouped data using the interval center as the assumed observation value. Using the raw data,

$$\overline{x} = \frac{\Sigma x}{n} = \frac{2069}{30} = 68.97$$

◆

Similar in concept to the median are percentile ranks, quartiles, and deciles. The median could also have been called the 50th percentile observation. Similarly, the 80th percentile would be the number of cars per hour for which the cumulative frequency was 80%. The quartile and decile points on the distribution divide the observations or distribution into segments of 25% and 10% respectively.

The most simple statistical parameter which describes the variation in observed data is the range. The range is found by subtracting the smallest value from the largest. Since the range is influenced by extreme (low probability) observations, its use is limited as a measure of variability.

The standard deviation is a better estimate of variability because it considers every observation. The standard deviation can be found from:

$$\sigma = \sqrt{\frac{\Sigma(x_i - \overline{x})^2}{n}} = \sqrt{\frac{\Sigma x_i^2}{n} - (\overline{x})^2}$$

The above formula assumes that n is a large number, such as above 50. Theoretically, n is the size of the entire population. If a small sample (less than 50) is used to calculate the standard deviation of the distribution, the formulas are changed. The sample standard deviation is

$$s = \sqrt{\frac{\Sigma(x_i - \overline{x})^2}{n - 1}} = \sqrt{\frac{\Sigma x_i^2 - (\Sigma x_i)^2/n}{n - 1}}$$

The difference is small when n is large, but care must be taken in reading the problem. If the 'standard deviation of the sample' is requested, calculate σ. If an estimate of the 'population standard deviation' or 'sample standard deviation' is requested, calculate s. (Note that the standard deviation of the sample is not the same as the sample standard deviation.)

Example 1.32

Calculate the range, standard deviation of the sample, and population variance from the data given in example 1.28.

$$\Sigma x = 2069 \qquad (\Sigma x)^2 = 4280761 \qquad \Sigma x^2 = 143225 \qquad n = 30 \qquad \bar{x} = 68.97$$

$$\sigma = \sqrt{\frac{143225}{30} - (68.97)^2} = 4.16$$

$$s = \sqrt{\frac{143225 - (4280761)/30}{29}} = 4.29$$

$$s^2 = 18.4$$

$$R = 79 - 61 = 18$$

◆

Referring again to example 1.28, suppose that the hourly through-put for 15 similar intersections is measured over a 30 day period. At the end of the 30 day period, there will be 15 ranges, 15 medians, 15 means, 15 standard deviations, and so on. These parameters themselves constitute distributions.

The mean of the sample means is an excellent estimator of the average hourly through-put of an intersection:

$$\bar{\bar{x}} = (\frac{1}{15})\Sigma\bar{x}$$

The standard deviation of the sample means is known as the standard error of the mean to distinguish it from the standard deviation of the raw data. The standard error is written as $\sigma_{\bar{x}}$.

The standard error is not a good estimator of the population standard deviation.

In general, if k sets of n observations each are used to estimate the population mean (u) and the population standard deviation (σ'), then

$$u \approx (\frac{1}{k})\Sigma\bar{x}$$

$$\sigma' \approx \sqrt{k}\ \sigma_{\bar{x}}$$

8. Critical Path Techniques

Critical path techniques are used to graphically represent the multiple relationships between stages in complicated projects. The graphical networks show the dependencies or precedence relationships between the various activities and can be used to control and monitor the progress, cost, and resources of projects. Critical path techniques will also identify the most critical activities in projects.

Definitions:

Activity: Any subdivision of a project whose execution requires
                     time and other resources.

Critical path: A path connecting all activities which have minimum or
                     zero slack times. The critical path is the longest path
                     through the network.

Duration: The time required to perform an activity. All durations
                     are normal durations unless otherwise referred to as
                     crash durations.

Event: The beginning or completion of an activity.

Event time: Actual time at which an event occurs.

Float: Same as slack time.

Slack time: The maximum time that an activity may be delayed without
                     causing the project to fall behind schedule. Slack time
                     is always minimum or zero along the critical path.

Critical path techniques use directed graphs to represent a project.
These graphs are made up of arcs (arrows) and nodes (junctions). The
placement of the arcs and nodes completely specifies the precedences of
the project. Durations and precedences are usually given in a precedence table or matrix.

One specific technique is known as the Critical Path Method, CPM. This
deterministic method is applicable when all activity durations are
known in advance. CPM is usually represented as an Activity-on-Node
model. Arcs are used to specify precedence and the nodes actually
represent the activities. Events are not present on the graph, other
than as the heads and tails of the arcs. Two dummy nodes taking zero
time may be used to specify the start and finish of the project.

Example 1.33

Given the project listed in the precedence table below, construct the
precedence matrix and draw an activity-on-node-network.

Activity	Time (days)	Predecessors
A, Start	0	-
B	7	A
C	6	A
D	3	B
E	9	B,C
F	1	D,E
G	4	C
H, Finish	0	F,G

The precedence matrix is given below.

successor

	A	B	C	D	E	F	G	H
A		X	X					
B				X	X			
C					X		X	
D						X		
E						X		
F								X
G								X
H								

predecessor

The activity-on-node network is given below.

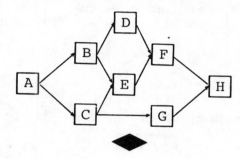

Solving a CPM Problem

The solution of a critical path problem results in a knowledge of the earliest and latest times that an activity may be started and finished. It also identifies the critical path and generates the slack time for each activity.

To facilitate the solution method, each node may be replaced by a square which has been quartered. The compartments have the meanings indicated by the following key.

ES	EF
LS	LF

Key

ES: Earliest Start
EF: Earliest Finish
LS: Latest Start
LF: Latest Finish

Then, the following procedure may be used to find the earliest and latest starts and finishes of each node.

1. Place the project start time or date in the ES and EF positions of the start activity. The start time is zero for relative calculations.

2. Consider any unmarked activity whose predecessors have all been marked in the EF and ES positions. (Go to step 4 if there is none.) Mark in its ES position the largest number marked in the EF position of those predecessors.

3.   Add the activity time to the ES time and write this in the EF box. Return to step 2.

4.   Place the value of the latest finish date in the LS and LF boxes of the finish node.

5.   Consider any unmarked activity whose successors have all been marked in the LS and LF positions. The LF is the smallest LS of the successors. Go to step 7 if there are no unmarked activities.

6.   The LS for the new node is LF minus its activity time. Return to step 5.

7.   The slack for each node is (LS-ES) or (LF-EF).

8.   The critical path encompasses nodes for which the slack equals (LS-ES) from the start node. There may be more than one critical path.

Example 1.34

Complete the network for the previous example and find the critical path. Assume the desired completion date is in 19 days.

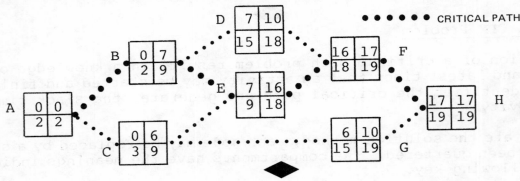

Probabilistic Critical Path Models

Probabilistic networks differ from deterministic networks only in the way in which the activity durations are found. Whereas durations are known explicit for a deterministic network, the time for a probabilistic activity is distributed as a random variable.

This variable nature complicates the problem greatly since the actual distribution of times is often unknown. For this reason, such a problem is usually solved as a deterministic model using the mean of the duration distribution as the activity duration.

The most common probabilistic critical path model is PERT, which stands for Program Evaluation and Review Technique. In PERT, all duration variables are assumed to come from a beta distribution, with mean and standard deviation given as:

$$t_{mean} = (1/6)(t_{minimum} + 4t_{most\ likely} + t_{maximum})$$

$$\sigma = (1/6)(t_{maximum} - t_{minimum})$$

The project completion time is assumed for large projects to be normally distributed with mean (μ) equal to the critical path length and overall variance (σ^2) equal to the sum of the variances along the critical path.

If necessary, the probability that a project duration will exceed some length (D) can be found from the normal table and the following relationship:

$$P\{duration > D\} = p\{X > z\}$$

where z is the standard normal variable equal to

$$z = \frac{D - \mu}{\sigma}$$

Practice Problems: MATHEMATICS

<u>Required</u>

1. California law requires a statistical analysis of the average speed driven by motorists on a road prior to the use of radar speed control. The following speeds were observed in a random sample of 40 cars:

44, 48, 26, 25, 20, 43, 40, 42, 29, 39, 23, 26, 24, 47, 45, 28, 29, 41, 38, 36, 27, 44, 42, 43, 29, 37, 34, 31, 33, 30, 42, 43, 28, 41, 29, 36, 35, 30, 32, 31 (all in mph)

a) Tabulate the frequency distribution of the above data
b) Draw the frequency histogram
c) Draw the frequency polygon
d) Tabulate the cumulative frequency distribution
e) Draw the cumulative frequency graph
f) What is the upper quartile speed?
g) What are the mode, median, and mean speeds?
h) What is the standard deviation of the sample data?
i) What is the sample standard deviation?
j) What is the sample variance?

2. Activities constituting a project are listed below. The project starts at time zero.

Activity	Predecessors	Successors	Duration
start	–	A	0
A	start	B,C,D	7
B	A	G	6
C	A	E,F	5
D	A	G	2
E	C	H	13
F	C	H,I	4
G	D,B	I	18
H	E,F	finish	7
I	F,G	finish	5
finish	H,I	–	0

a) Draw the CPM network
b) Indicate the critical path
c) What is the earliest finish?
d) What is the latest finish?
e) What is the slack along the critical path?
f) What is the float along the critical path?

3. Activities constituting a short project are listed below.

Activity	Predecessors	Successors	t_{min}	t_{likely}	t_{max}
start	–	A	0	0	0
A	start	B,D	1	2	5
B	A	C	7	9	20
C	B	D	5	12	18
D	A,C	finish	2	4	7
finish	D	–	0	0	0

If the project starts at t=15, what is the probability that the project will be completed by t=42 or sooner?

Optional

4. A pipe with an inside diameter of 18.812" contains fluid to a depth of 15.7". What is the hydraulic radius?

5. What is the determinant of the following flexibility matrix?

$$\begin{bmatrix} 8 & 2 & 0 & 0 \\ 2 & 8 & 2 & 0 \\ 0 & 2 & 8 & 2 \\ 0 & 0 & 2 & 4 \end{bmatrix}$$

6. The number of cars entering a toll plaza on a bridge during the hour following midnight is distributed as poisson with a mean of 20. What is the probability that 17 cars will pass through the toll plaza during that hour on any given night? What is the probability that 3 or fewer cars will pass through the toll plaza at that hour on any given night?

7. The time taken by a toll taker to collect the toll from vehicles crossing a bridge is an exponential distribution with mean of 23 seconds when a line of vehicles exists waiting to enter the toll booth. What is the probability that a random vehicle will be processed in 25 seconds or more (i.e., will take longer than 25 seconds?)

8. The number of vehicles lining up behind a flashing railroad crossing has been observed for five trains of different lengths, as given below. What is the mathematical formula which relates the two variables?

# cars in train	# vehicles
2	14.8
5	18.0
8	20.4
12	23.0
27	29.9

9. Holes drilled in structural steel parts for bolts are normally distributed with a mean of .502"" and standard deviation of .005". Holes are defective if their diameters are less than .497" or more than .507". (a) What is the probability that a hole chosen at random will be defective? (b) What is the probability that 2 holes out of a sample of 15 will be defective?

10. The oscillation exhibited by a certain 1-story building in free motion is given by the following differential equation:

$$x'' + 2x' + 2x = 0 \qquad x(0) = 0; \; x'(0) = 1$$

(a) What is x as a function of time? (b) What is the building's natural frequency of vibration? (c) What is the amplitude of oscillation? (d) What is x as a function of time if a lateral wind load is applied with form of sin(t)?

NOTES

ENGINEERING ECONOMY

Nomenclature

A	annual amount or annuity	$
B	present worth of all benefits	$
BV_j	book value at the end of the jth year	$
d	declining balance depreciation rate	decimal
D_j	depreciation in year j	$
D.R.	present worth of after-tax depreciation recovery	$
e	natural logarithm base (2.718)	–
EUAC	equivalent uniform annual cost	$
f	federal income tax rate	decimal
F	future amount or future worth	$
G	uniform gradient amount	$
i	annual effective rate per period	decimal
k	number of compounding periods per year	–
n	number of compounding periods, or life of asset	–
P	present worth or present value	$
P_t	present worth after taxes	$
ROR	rate of return	decimal
ROI	return on investment	decimal
r	nominal rate per year (rate per annum)	decimal
s	state income tax rate	decimal
S_n	expected salvage value in year n	$
t	composite tax rate	decimal
z	a factor equal to $\frac{(1+i)}{(1-d)}$	decimal
∅	nominal rate per period	decimal

1. Equivalence

Unlike most individuals involved with personal financial affairs, industrial decision makers using engineering economics are not as much concerned with the timing of a project's cash flows as with the total profitability of that project. In this situation, a method is required to compare projects involving receipts and disbursements occurring at different times.

By way of illustration, consider $100 placed in a bank account which pays 5% effective annual interest at the end of each year. After the first year, the account will have grown to $105. After the second year, the account will have grown to $110.25.

Assume that you will have no need for money during the next two years and that any money received would immediately go into your 5% bank account. Then, which of the following options would be more desirable?

      option a: $100 now

      option b: $105 to be delivered in one year

      option c: $110.25 to be delivered in two years

In light of the previous illustration, none of the options is superior under the assumptions given. If the first option is chosen, you will immediately place $100 into a 5% account, and in two years the account will have grown to $110.25. In fact, the account will contain $110.25 at the end of two years regardless of the option chosen. Therefore, these alternatives are said to be equivalent.

This is the way that many industrial economic decisions are made. Economically stable companies with many income sources evaluate alternative investments, not on the basis of when income is received, but rather on the basis of which alternative has the largest eventual profitability.

2. Cash Flow Diagrams

Although they are not always necessary in simple problems (and they are often unwieldly in very complex problems), cash flow diagrams may be drawn to help visualize and simplify problems having diverse receipts and disbursements.

The conventions below are used to standardize cash flow diagrams.

a.   The horizontal (time) axis is marked off in equal increments, one per period, up to the duration or horizon of the project.

b.   All disbursements and receipts (cash flows) are assumed to take place at the end of the year in which they occur. This is known as the year-end convention. The exception to the year-end convention is any initial cost (purchase cost) which occurs at t = 0.

c.   Disbursements are represented by downward arrows. Receipts are represented by upward arrows. Arrow lengths are approximately proportional to the magnitude of the cash flow.

d.   Two or more transfers in the same year are placed end-to-end, and these may be combined.

e.   Expenses incurred before t = 0 are called sunk costs. Sunk costs are not relevant to the problem.

Example 2.1

A mechanical device will cost $20,000 when purchased. Maintenance will cost $1,000 each year. The device will generate revenues of $5,000 each year for 5 years after which the salvage value is expected to be $7,000. Draw and simplify the cash flow diagram.

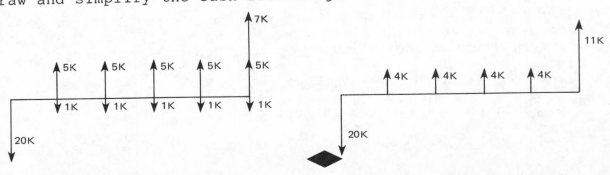

3. Typical Problem Format

With the exception of some investment and rate of return problems, the typical problem involving engineering economics will have the following characteristics:

a.   An interest rate will be given.

b.   Two or more alternatives will be competing for funding.

c.   Each alternative will have its own cash flows.

d.   It is necessary to select the best alternative.

Example 2.2

Investment A costs $10,000 today and pays back $11,500 two years from now. Investment B costs $8,000 today and pays back $4,500 each year for two years. If an interest rate of 5% is used, which alternative is superior?

The solution to this example is not difficult, but it will be postponed until methods of calculating equivalence have been covered.

4. Calculating Equivalence

It was previously illustrated that $100 now is equivalent at 5% to $105 in one year. The equivalence of any present amount, P, at t = 0 to any future amount, F, at t = n is called the future worth and can be calculated from equation 2.1.

$$F = P(1 + i)^n$$

2.1

The factor $(1 + i)^n$ is known as the compound amount factor and has been tabulated at the end of this chapter for various combinations of i and n. Rather than actually write the formula for the compound amount factor, the convention is to use the standard functional notation (F/P,i%,n). Thus,

$$F = P(F/P,i\%,n)$$

2.2

Similarly, the equivalence of any future amount to any present amount is called the present worth and can be calculated from

$$P = F(1 + i)^{-n} = F(P/F,i\%,n)$$

2.3

The factor $(1 + i)^{-n}$ is known as the present worth factor, with functional notation (P/F,i%,n). Tabulated values are also given for this factor at the end of this chapter.

Example 2.3

How much should you put into a 10% savings account in order to have $10,000 in 5 years?

This problem could also be stated: What is the equivalent present worth of $10,000 5 years from now if money is worth 10%?

$$P = F(1 + i)^{-n} = 10,000(1+.10)^{-5} = 6,209$$

The factor .6209 would usually be obtained from the tables.

A cash flow which repeats regularly each year is known as an annual amount. Although the equivalent value for each of the n annual amounts could be calculated and then summed, it is much easier to use one of the uniform series factors, as illustrated in example 2.4 below.

Example 2.4

Maintenance costs for a machine are $250 each year. What is the present worth of these maintenance costs over a 12 year period if the interest rate is 8%?

Notice that $(P/A,8\%,12) = (P/F,8\%,1) + (P/F,8\%,2) + \cdots + (P/F,8\%,12)$

$$P = A(P/A,i\%,n) = -250(7.536) = -1,884$$

A common complication involves a uniformly increasing cash flow. Such an increasing cash flow should be handled with the uniform gradient factor, $(P/G,i\%,n)$. The uniform gradient factor finds the present worth of a uniformly increasing cash flow which starts in year 2 (not year 1) as shown in example 2.5.

Example 2.5

Maintenance on an old machine is $100 this year, but is expected to increase by $25 each year thereafter. What is the present worth of 5 years of maintenance? Use an interest rate of 10%.

In this problem, the cash flow must be broken down into parts. Notice that the 5-year gradient factor is used even though there are only 4 non-zero gradient cash flows.

$$P = A(P/A,10\%,5) + G(P/G,10\%,5)$$

$$= -100(3.791) - 25(6.8618) = -551$$

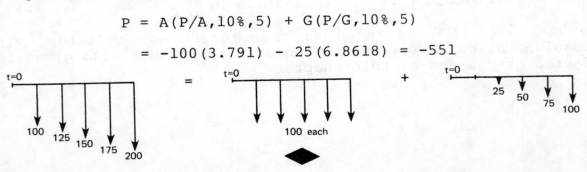

Various combinations of the compounding and discounting factors are possible. For instance, the annual cash flow that would be equivalent to a uniform gradient may be found from

$$A = G(P/G,i\%,n)(A/P,i\%,n)$$ 2.4

Formulas for all of the compounding and discounting factors are contained in table 2.1. Normally, it will not be necessary to calculate factors from the formulas. The tables at the end of this chapter are adequate for solving most problems.

TABLE 2.1

Discount Factors for Discrete Compounding

Factor Name	Converts	Symbol	Formula
Single Payment Compound Amount	P to F	$(F/P,i\%,n)$	$(1+i)^n$
Present Worth	F to P	$(P/F,i\%,n)$	$(1+i)^{-n}$
Uniform Series Sinking Fund	F to A	$(A/F,i\%,n)$	$\dfrac{i}{((1+i)^n-1)}$
Capital Recovery	P to A	$(A/P,i\%,n)$	$\dfrac{i(1+i)^n}{((1+i)^n-1)}$
Compound Amount	A to F	$(F/A,i\%,n)$	$\dfrac{(1+i)^n-1}{i}$
Equal Series Present Worth	A to P	$(P/A,i\%,n)$	$\dfrac{(1+i)^n-1}{i(1+i)^n}$
Uniform Gradient	G to P	$(P/G,i\%,n)$	$\dfrac{(1+i)^n-1}{i^2(1+i)^n} - \dfrac{n}{i(1+i)^n}$

5. The Meaning of 'Present Worth' and 'i'

It is clear that $100 invested in a 5% bank account will allow you to remove $105 one year from now. If this investment is made, you will clearly receive a return on your investment (ROI) of $5. The cash flow diagram and the present worth of the two transactions are:

$$P = -100 + 105(P/F,5\%,1)$$
$$= -100 + 105(.95238) = 0$$

Notice that the present worth is zero even though you did receive a 5% return on your investment.

However, if you were offered $120 for use of $100 over a one-year period, the cash flow diagram and present worth (at 5%) would be

$$P = -100 + 120(P/F,5\%,1)$$

$$= -100 + 120(.95238) = 14.29$$

Therefore, it appears that the present worth of an alternative is equal to the equivalent value at t = 0 of the increase in return above that which you would be able to earn in an investment offering i% per period. In the above case, $14.29 is the present worth of ($20-$5), the difference in the two ROI's.

Alternatively, the actual earned interest rate, called rate of return (ROR), can be defined as the rate which makes the present worth of the alternative zero.

The present worth is also the amount that you would have to be given to dissuade you from making an investment, since placing the initial investment amount along with the present worth into a bank account earning i% will yield the same eventual ROI. Relating this to the previous paragraphs, you could be dissuaded against investing $100 in an alternative which would return $120 in one year by a t = 0 payment of $14.29. Clearly, ($100 + $14.29) invested at t = 0 will also yield $120 in one year at 5%.

The selection of the interest rate is difficult in engineering economics problems. Usually it is taken as the average rate of return that an individual or business organization has realized in past investments. Fortunately, choosing an interest rate is seldom required as it is usually given.

It should be obvious that alternatives with negative present worths are undesirable, and alternatives with positive present worths are desirable since they increase the average earning power of invested capital.

6.   Choice Between Alternatives

A variety of methods exists for selecting a superior alternative from among a group of proposals. Each method has its own merits and applications.

The Present Worth Method has already been implied. When two or more alternatives are capable of performing the same functions, the superior alternative will have the largest present worth. This method is suitable for ranking the desirability of alternatives. The present worth method is restricted to evaluating alternatives that are mutually exclusive and which have the same lives.

Returning to example 2.2, the present worth of each alternative should be found in order to determine which alternative is superior.

Example 2.2, continued

$$P(A) = -10,000 + 11,500(P/F,5\%,2) = 431$$

$$P(B) = -8,000 + 4,500(P/A,5\%,2) = 367$$

Alternative A is superior and should be chosen.

The present worth of a project with an infinite life is known as the capitalized cost. Capitalized cost is the amount of money at t = 0 needed to perpetually support the project on the earned interest only. Capitalized cost is a positive number when expenses exceed income.

$$\text{Capitalized Cost} = \text{Initial Cost} + \frac{\text{Annual Costs}}{i} \qquad 2.5$$

If disbursements occur irregularly instead of annually, the capitalized cost is

$$\text{Capitalized Cost} = \text{Initial Cost} + \frac{EUAC}{i} \qquad 2.6$$

In comparing two alternatives, each of which is infinitely lived, the superior alternative will have the lowest capitalized cost.

Alternatives which accomplish the same purpose but which have unequal lives must be compared by the Annual Cost Method. The annual cost method assumes that each alternative will be replaced by an identical twin at the end of its useful life (infinite renewal). This method, which may also be used to rank alternatives according to their desirability, is also called the Annual Return Method and Capital Recovery Method. Restrictions are that the alternatives must be mutually exclusive and infinitely renewed up to the duration of the longest-lived alternative. The calculated annual cost is known as the equivalent uniform annual cost, EUAC. Cost is a positive number when expenses exceed income.

Example 2.6

Which of the following alternatives is superior over a 30 year period if the interest rate is 7%?

	A	B
type	brick	wood
life	30 years	10 years
cost	$1800	$450
maintenance	$5/year	$20/year

$$EUAC(A) = 1800(A/P,7\%,30) + 5 = 150$$
$$EUAC(B) = 450(A/P,7\%,10) + 20 = 84$$

Alternative B is superior since its annual cost of operation is the lowest. It is assumed that three wood facilities, each with a life of 10 years and a cost of $450, will be built to span the 30 year period.

The Benefit-Cost Ratio Method is often used in municipal project evaluations where benefits and costs accrue to different segments of the community. With this method, the present worth of all benefits (regardless of the beneficiary) is divided by the present worth of all costs. The project is considered acceptable if the ratio exceeds one.

When the benefit-cost ratio method is used, disbursements by the initiators or sponsors are costs. Disbursements by the users of the project are known as disbenefits. It is often difficult to determine whether a cash flow is a cost or a disbenefit (whether to place it in the numerator or denominator of the benefit-cost ratio calculation).

However, regardless of where the cash flow is placed, an acceptable project will always have a benefit-cost ratio greater than one, although the actual numerical result will depend on the placement. For this reason, the benefit-cost ratio method should not be used to rank competing projects.

The benefit-cost ratio method may be used to rank alternative proposals only if an incremental analysis is used. First, determine that the ratio is greater than one for each alternative. Then, calculate the ratio $(B_2-B_1)/(C_2-C_1)$ for each possible pair of alternatives. If the ratio exceeds one, alternative 2 is superior to alternative 1. Otherwise, alternative 1 is superior.

Perhaps no method of analysis is less understood than the Rate of Return (ROR) Method. As was stated previously, the ROR is the interest rate that would yield identical profits if all money was invested at that rate. The present worth of any such investment is zero.

The ROR is defined as the interest rate that will discount all cash flows to a total present worth equal to the initial required investment. This definition is used to determine the ROR of an alternative. The ROR should not be used to rank or compare alternatives unless an incremental analysis is used. The advantage of the ROR method is that no knowledge of an interest rate is required.

To find the ROR of an alternative, proceed as follows:

1. Set up the problem as if to calculate the present worth.

2. Arbitrarily select a reasonable value for i. Calculate the present worth.

3. Choose another value of i (not too close to the original value) and again solve for the present worth.

4. Interpolate or extrapolate the value of i which gives a zero present worth.

5. For increased accuracy, repeat steps (2) and (3) with two more values that straddle the value found in step (4).

A common, although incorrect, method of calculating the ROR involves dividing the annual receipts or returns by the initial investment. However, this technique ignores such items as salvage, depreciation, taxes, and the time value of money. This technique also fails when the annual returns vary.

Example 2.7

What is the return on invested capital if $1000 is invested now with $500 being returned in year 4 and $1000 being returned in year 8?

First, set up the problem as a present worth calculation.

$$P = -1000 + 500(P/F,i\%,4) + 1000(P/F,i\%,8)$$

Arbitrarily select i = 5%. The present worth is then found to be $88.50. Next take a higher value of i to reduce the present worth. If i = 10%, the present worth is -$192. The ROR is found from simple interpolation to be approximately 6.6%.

7. Treatment of Salvage Value in Replacement Studies

An investigation into the retirement of an existing process or piece of equipment is known as a replacement study. Replacement studies are similar in most respects to other alternative comparison problems: an interest rate is given, two alternatives exist, and one of the previously mentioned methods of comparing alternatives is used to choose the superior alternative.

In replacement studies, the existing process or piece of equipment is known as the defender. The new process or piece of equipment being considered for purchase is known as the challenger.

Because most defenders still have some market value when they are retired, the problem of what to do with the salvage value arises. It seems logical to use the salvage value of the defender to reduce the initial purchase cost of the challenger. This is consistent with what would actually happen if the defender were to be retired.

By convention, however, the salvage value is subtracted from the defender's present value. This does not seem logical but is done to keep all costs and benefits related to the defender with the defender. In this case, the salvage value is treated as an opportunity cost which would be incurred if the defender is not retired.

If the defender and the challenger have the same lives and a present worth study is used to choose the superior alternative, the placement of the salvage value will have no effect on the net difference between present worths for the challenger and defender. Although the values of the two present worths will be different depending on the placement, the difference in present worths will be the same.

If the defender and challenger have different lives, an annual cost comparison must be made. Since the salvage value would be 'spread over' a different number of years depending on its placement, it is important to abide by the conventions listed in this section. If the problem is to decide on replacing a defender now versus waiting an additional year, the best way is to think of the EUAC of the defender as the cost of keeping the defender from now until next year. In addition to the usual operating and maintenance costs, that cost would include an opportunity interest cost incurred by not selling the defender and also a drop in the salvage value if the defender is kept for one additional year. Specifically,

$$\begin{aligned}
\text{EUAC(defender)} = \ &\text{maintenance costs} \\
&+ i(\text{current salvage value}) \\
&+ (\text{current salvage} - \text{next year's salvage})
\end{aligned} \qquad 2.7$$

It is important in retirement studies not to double count the salvage value. That is, it would be incorrect to add the salvage value to the defender and at the same time subtract it from the challenger.

8. Basic Income Tax Considerations

Assume that an organization pays f% of its profits to the federal government as income taxes. If the organization also pays a state income tax of s%, and if state taxes paid are recognized by the federal government as expenses, then the composite tax rate is

$$t = s + f - sf \qquad 2.8$$

The basic principles used to incorporate taxation into economic analyses are listed below.

  a. Initial purchase cost is unaffected by income taxes.

  b. Salvage value is unaffected by income taxes.

  c. Deductible expenses, such as operating costs, maintenance costs, and interest payments, are reduced by t% (e.g., multiplied by the quantity (1-t)).

  d. Revenues are reduced by t% (e.g., multiplied by the quantity (1-t)).

  e. Depreciation is multiplied by the quantity (t) and added to the appropriate year's cash flow, increasing that year's present worth.

Income taxes and depreciation have no bearing on municipal or governmental projects since municipalities, states, and the U.S. Government pay no taxes.

Example 2.8

A corporation which pays 53% of its revenues in income taxes invests $10,000 in a project which will result in $3000 annual revenues for 8 years. If the annual expenses are $700, salvage after 8 years is $500, and 9% interest is used, what is the after-tax present worth? Disregard depreciation.

$$P_t = -10,000 + 3000(P/A,9\%,8)(1-.53) - 700(P/A,9\%,8)(1-.53)$$
$$+ 500(P/F,9\%,8)$$
$$= -3766$$

◆

It is interesting that the alternative evaluated in example 2.8 is undesirable if income taxes are considered but is desirable if income taxes are omitted.

9. Depreciation (Also see p. 2-29.)

Although depreciation calculations may be considered independently in examination questions, it is important to recognize that depreciation has no effect on engineering economic calculations unless income taxes are also considered.

Generally, tax regulations do not allow the cost of equipment to be treated as a deductible expense in the year of purchase. Rather, portions of the cost may be allocated to each of the years of the item's economic life (which may be different from the actual useful life). Each year, the book value (which is initially equal to the purchase price) is reduced by the depreciation in that year. Theoretically, the book value of an item will equal the market value at any time within the economic life of that item.

Since tax regulations allow the depreciation in any year to be handled as if it were an actual operating expense, and since operating expenses are deductible from the income base prior to taxation, the after-tax profits will be increased. If D is the depreciation, the net result to the after-tax cash flow will be the addition of $(t)(D)$.

The present worth of all depreciation over the economic life of the item is called the depreciation recovery. Although originally established to do so, depreciation recovery can never fully replace an item at the end of its life.

Depreciation is often confused with amortization and depletion. While depreciation spreads the cost of a fixed asset over a number of years, amortization spreads the cost of an intangible asset (e.g., a patent) over some basis such as time or expected units of production.

Depletion is another artificial deductible operating expense designed to compensate mining òrganizations for decreasing mineral reserves. Since original and remaining quantities of minerals are seldom known accurately, the depletion allowance is calculated as a fixed percentage of the organization's gross income. These percentages are usually in the 10% - 20% range and apply to such mineral deposits as oil, natural gas, coal, uranium, and most metal ores.

There are four common methods of calculating depreciation. The book value of an asset depreciated with the Straight Line (SL) Method (also known as the Fixed Percentage Method) decreases linearly from the inital purchase at $t = 0$ to the estimated salvage at $t = n$. The depreciated amount is the same each year. The quantity $(C-S_n)$ in equation 2.9 is known as the depreciation base.

$$D_j = \frac{(C-S_n)}{n} \qquad 2.9$$

Double Declining Balance (DDB) depreciation (also called Declining Balance) is independent of salvage value. Furthermore, the book value never stops decreasing, although the depreciation decreases in magnitude. Usually, any remaining book value is written off in the last year of the asset's estimated life. Unlike any of the other depreciation methods, DDB depends on accumulated depreciation.

$$D_j = \frac{2(C - \sum_{i=1}^{j-1} D_i)}{n} \qquad 2.10$$

In Sum-of-the-Year's-Digits (SOYD) depreciation, the digits from 1 to n inclusive are summed. The total, T, can also be calculated from

$$T = \frac{1}{2}n(n + 1) \qquad 2.11$$

The depreciation can be found from

$$D_j = \frac{(C-S_n)(n-j+1)}{T} \qquad 2.12$$

The Sinking Fund Method is seldom used in industry because the initial depreciation is low. The formula for sinking fund depreciation is

$$D_j = (C-S_n)(A/F,i\%,n) \qquad 2.13$$

The depreciation allowed by the Internal Revenue Service in the first year is usually less than the value D_1 as calculated from formulas 2.9, 2.10, 2.12, and 2.13. This reduction occurs whenever an asset is purchased during the fiscal year (as compared to a purchase made at the beginning of the fiscal year). In such an instance, D_1 must be prorated in proportion to the amount of time remaining until the end of the fiscal year. For example, only $(11/12)D_1$ would be allowed if an asset was purchased on February 1.

Three other depreciation methods should be mentioned, not because they are currently accepted or in widespread use, but because they have seen recent use on the licensing examinations.

The 'sinking-fund plus interest on first cost' depreciation method, like the other two methods, is an attempt to include the opportunity interest cost on the purchase price with the depreciation. That is, the purchasing company not only incurs an annual loss due to the drop in book value, but it also loses the interest on the purchase price. The formula for this method is

$$D_j = (C-S_n)(A/F,i\%,n) + (C)(i) \qquad 2.14$$

The 'straight-line plus interest on first cost' method is similar. Its formula is

$$D_j = (\frac{1}{n})(C-S_n) + (C)(i) \qquad\qquad 2.15$$

The 'straight-line plus average interest' assumes that the opportunity interest cost should be based on the book value only, not on the full purchase price. Since the book value changes each year, an average value is used. The depreciation formula is

$$D_j = (\frac{1}{n})(C-S_n) + \frac{1}{2}(i)(C-S_n)(\frac{n+1}{n}) \qquad\qquad 2.16$$

These three depreciation methods are not to be used in the usual manner, e.g., in conjunction with the income tax rate. These methods are attempts to calculate a more accurate annual cost of an alternative. Sometimes they work, and sometimes they give misleading answers. Their use cannot be recommended. They are included in this chapter only for the sake of completeness.

Example 2.9

An asset is purchased for $9000. Its estimated economic life is 10 years, after which it will be sold for $1000. Find the depreciation in the first three years using SL, DDB, SOYD, and Sinking Fund at 6%.

SL:	$D = (9000-1000)/10$	= 800 each year
DDB:	$D_1 = 2(9000)/10$	= 1800 in year 1
	$D_2 = 2(9000-1800)/10$	= 1440 in year 2
	$D_3 = 2(9000-3240)/10$	= 1152 in year 3
SOYD:	$T = \frac{1}{2}(10)(11) = 55$	
	$D_1 = (10/55)(9000-1000)$	= 1454 in year 1
	$D_2 = (9/55)(8000)$	= 1309 in year 2
	$D_3 = (8/55)(8000)$	= 1164 in year 3
Sinking Fund	$D = (9000-1000)(A/F,6\%,10)$	= 607 each year

◆

Example 2.10

For the asset described in example 2.9, calculate the book value during the first three years if SOYD depreciation is used.

The book value at the beginning of year 1 is $9000. Then,

$$BV_1 = 9000-1454 = 7546$$

$$BV_2 = 7546-1309 = 6327$$

$$BV_3 = 6327-1164 = 5073$$

◆

Example 2.11

For the asset described in example 2.9, calculate the after-tax depreciation recovery with SL and SOYD depreciation methods. Use 6% interest with 48% income taxes.

SL: D.R. = .48(800)(P/A,6%,10) = 2826

SOYD: The depreciation series can be thought of as a constant 1454 term with a negative 145 gradient.

D.R. = .48(1454)(P/A,6%,10) - .48(145)(P/G,6%,10)

= 3076

Finding book values, depreciation, and depreciation recovery is particularly difficult with DDB depreciation, since all previous years' quantities seem to be required. It appears that the depreciation in the 6th year cannot be calculated unless the values of depreciation for the first five years are calculated first. Questions asking for depreciation or book value in the middle or at the end of an asset's economic life may be solved from the following equations:

$$d = 2/n \tag{2.17}$$

$$z = \frac{(1+i)}{(1-d)} \tag{2.18}$$

$$(P/EG) = \frac{z^n - 1}{z^n(z-1)} \tag{2.19}$$

Then, assuming that the remaining book value at t = n is written off in one lump sum, the present worth of the depreciation recovery is

$$D.R. = t\left[\frac{(d)(C)}{(1-d)}(P/EG) + (1-d)^n(C)(P/F,i\%,n)\right] \tag{2.20}$$

$$D_j = (d)(C)(1-d)^{j-1} \tag{2.21}$$

$$BV_j = C(1-d)^j \tag{2.22}$$

Example 2.12

What is the after-tax present worth of the asset described in example 2.8 if SL, SOYD, and DDB depreciation methods are used?

The after-tax present worth, neglecting depreciation, was previously found to be -3766.

Using SL, the depreciation recovery is

$$D.R. = (.53)\frac{(10,000-500)}{8}(P/A,9\%,8)$$

$$= 3483$$

Using SOYD, the depreciation recovery is calculated as follows:

$$T = \frac{1}{2}(8)(9) = 36$$

$$\text{Depreciation base} = (10,000-500) = 9500$$

$$D_1 = \frac{8}{36}(9500) = 2111$$

$$G = \text{gradient} = \frac{1}{36}(9500) = 264$$

$$D.R. = (.53)[2111(P/A,9\%,8) - 264(P/G,9\%,8)]$$
$$= 3829$$

Using DDB, the depreciation recovery is calculated as follows:

$$d = \frac{2}{8} = .25$$

$$z = \frac{(1.09)}{.75} = 1.453$$

$$(P/EG) = \frac{(1.453)^8-1}{(1.453)^8(.453)} = 2.096$$

$$D.R. = .53[\frac{(.25)(10,000)}{.75}(2.096)$$

$$+ (.75)^8(10,000)(P/F,9\%,8)]$$

$$= 3969$$

The after-tax present worths including depreciation recovery are:

SL: $P_t = -3766 + 3483 = -283$

SOYD: $P_t = -3766 + 3829 = 63$

DDB: $P_t = -3766 + 3969 = 203$

◆

10. Advanced Income Tax Considerations

There are a number of specialized techniques that are infrequently needed. These techniques are related more to the accounting profession than the engineering profession. Nevertheless, it is occasionally necessary to use these techniques.

A. Additional First-Year Depreciation

The Internal Revenue Service permits the purchaser of business-related equipment to take an additional depreciation allowance in the first year. This additional allowance is known as 'additional first-year depreciation.' Additional first-year depreciation is completely separate from the normal depreciation calculated in the first year (as determined by formulas 2.9, 2.10, 2.12, and 2.13).

The additional first-year depreciation is calculated as 20% of the actual cost of qualifying property. The following special rules apply:

1. Only tangible property qualifies for additional first-year depreciation. Land and buildings do not qualify.

2. The property must have a life of 6 years or greater.

3. Salvage value is not subtracted from the cost of acquisition to find the depreciation base for additional first-year depreciation.

4. The actual cost used as the depreciation base is the actual amount paid or signed for. Therefore, the list price must be reduced by the allowed value of any trade-in equipment.

5. The maximum additional first-year depreciation allowed is $2000.

6. Property acquired during the year (any time before year-end) qualifies for the entire 20% allowance.

7. The additional first-year depreciation must be subtracted from the actual purchase price when calculating the basis to be used with normal depreciation.

Example 2.13

A corporation whose fiscal year ends on December 31 purchases a pile-driver for $14,500 on July 1. The estimated useful life is 10 years. The estimated salvage after 10 years is $500. What is the total depreciation in the first and second years? Use the straight-line method to calculate the normal depreciation.

20% of the actual cost is $(.2)(14,500) = 2900$. But, additional first-year depreciation is limited to 2000.

The normal depreciation is $(\frac{1}{10})(14,500-2000-500)(1/2 \text{ year}) = 600$.

The total first-year depreciation is $(2000 + 600) = 2600$.

The second year depreciation is $(\frac{1}{10})(14,500 - 2000 - 500) = 1200$.

◆

B. Salvage Value Reduction

If an asset is to be kept longer than 3 years, the IRS will allow a reduction in salvage value. Specifically, the salvage value may be reduced by any amount up to 10% of the purchase price.

Example 2.14

A new in-place sewer pipe forming unit is purchased for $85,000. The unit has an expected life of 10 years and a salvage of $12,500. What will be the normal straight-line depreciation?

10% of the purchase price is 8,500. Therefore, the salvage can be taken as (12,500 - 8,500) = 4,000. The normal straight-line depreciation is (1/10)(85,000 - 4,000) = 8,100.

C. Investment Tax Credit

Equipment which is purchased and which will be kept for at least 3 years is eligible for a special investment tax credit. This credit is used to directly reduce the income tax paid. It is not a deductible expense as is depreciation.

The credit allowed is

1)   10% of the purchase price for items kept 7 or more years

2)   6.67% of the purchase price for items kept 5 to 7 years

3)   3.33% of the purchase price for items kept 3 to 5 years

4)   0% of the purchase price for items kept less than 3 years

D. Gain or Loss on the Sale of a Depreciated Asset

Each year an asset is depreciated, its book value declines. Theoretically, this book value corresponds to the actual market value of the asset. If the asset is sold at any time after purchase (even after the asset has been fully depreciated) for an amount different from the book value, an income tax adjustment is required.

If the amount received from the sale exceeds the book value, the difference must be taken as regular taxable income in that year. If the sale price was less than the book value, the difference may be taken as the depreciation in that year.

11. Rate and Period Changes

All of the foregoing calculations were based on compounding once a year at an effective interest rate, i. However, some problems specify compounding more frequently than annually. In such cases, a nominal interest rate, r, will be given. The nominal rate does not include the effect of compounding and is not the same as the effective rate, i. A nominal rate may be used to calculate the effective rate by using equation 2.23 or 2.24.

$$i = (1 + \frac{r}{k})^k - 1 \qquad\qquad 2.23$$

$$= (1 + \emptyset)^k - 1 \qquad\qquad 2.24$$

A problem may also specify an effective rate per period, $\emptyset$, (e.g., per month). However, that will be a simple problem since compounding for n periods at an effective rate per period is not affected by the definition or length of the period.

The following rules may be used to determine which interest rate is given in a problem:

- Unless specifically qualified in the problem, the interest rate given is an annual rate.

- If the compounding is annually, the rate given is the effective rate. If compounding is other than annually, the rate given is the nominal rate.

- If the type of compounding is not specified, assume annual compounding.

In the case of continuous compounding, the appropriate discount factors may be calculated from the formulas in the following table:

Table 2.2

Discount Factors for Continuous Compounding

(F/P)	e^{rn}
(P/F)	e^{-rn}
(A/F)	$(e^r-1)/(e^{rn}-1)$
(F/A)	$(e^{rn}-1)/(e^r-1)$
(A/P)	$(e^r-1)/(1-e^{-rn})$
(P/A)	$(1-e^{-rn})/(e^r-1)$

Example 2.15

A savings and loan offers $5\frac{1}{4}\%$ compounded daily. What is the annual effective rate?

method 1: $r = .0525$, $k = 365$

$$i = (1 + \frac{.0525}{365})^{365} - 1 = .0539$$

method 2: Assume daily compounding is the same as continuous compounding.

$$i = (F/P) - 1$$

$$= e^{.0525} - 1 = .0539$$

12. Probabilistic Problems

Thus far, all of the cash flows included in the examples have been known exactly. If the cash flows are not known exactly but are given by some implicit or explicit probability distribution, the problem is probabilistic.

Probabilistic problems typically possess the following characteristics:

a. There is a chance of extreme loss that must be minimized.

b. There are multiple alternatives that must be chosen from. Each alternative gives a different degree of protection against the loss or failure.

c. The outcome is independent of the alternative chosen. Thus, as illustrated in example 2.17, the size of the dam that is chosen for construction will not alter the rainfall in successive years. However, it will alter the effects on the downstream watershed areas.

Probabilistic problems are typically found using annual costs and expected values. An expected value is similar to an 'average value' since it is calculated as the mean of the given probability distribution. If cost 1 has a probability of occurrence of p_1, cost 2 has a probability of occurrence of p_2, and so on, the expected value is

$$E(cost) = p_1(cost\ 1) + p_2(cost\ 2) + \cdots \qquad 2.25$$

Example 2.16

Flood damage in any year is given according to the table below. What is the present worth of flood damage for a 10-year period? Use 6%.

Damage	Probability
0	.75
$10,000	.20
$20,000	.04
$30,000	.01

The expected value of flood damage is

$$E(damage) = (0)(.75)+(10,000)(.20)+(20,000)(.04)+(30,000)(.01)$$
$$= \$3,100$$

present worth = $3,100(P/A,6\%,10) = 22,816$

◆

Probabilities in probabilistic problems may be given to you in the problem (as in the example above) or you may have to obtain them from some named probability distribution. In either case, the probabilities are known explicitly and such problems are known as explicit probability problems.

Example 2.17

A dam is being considered on a river which periodically overflows and causes $600,000 damage. The damage is essentially the same each time the river causes flooding. The project horizon is 40 years. A 10% interest rate is being used.

Three different designs are available, each with different costs and storage capacities.

design alternative	cost	maximum capacity
A	500,000	1 unit
B	625,000	1.5 units
C	900,000	2.0 units

The U.S. Weather Service has provided a statistical analysis of annual rainfall in the area draining into the river.

units annual rainfall	probability
0	.10
.1 - .5	.60
.6 - 1.0	.15
1.1 - 1.5	.10
1.6 - 2.0	.04
2.0 or more	.01

Which design alternative would you choose assuming the dam is essentially empty at the start of each rainfall season?

The sum of the construction cost and the expected damage needs to be minimized. If alternative A is chosen, it will have a capacity of 1 unit. Its capacity will be exceeded (causing $600,000 damage) when the annual rainfall exceeds 1 unit. Therefore, the annual cost of A is

EUAC(A) = 500,000(A/P,10%,40) + 600,000(.10 + .04 + .01) = 141,150

Similarly,

EUAC(B) = 625,000(A/P,10%,40) + 600,000(.04 + .01) = 93,940

EUAC(C) = 900,000(A/P,10%,40) + 600,000(.01) = 98,070

Alternative B should be chosen.

In other problems, a probability distribution will not be given even though some parameter (such as the life of an alternative) is not known with certainty. Such problems are known as implicit probability problems since they require a reasonable assumption about the probability distribution.

Implicit probability problems typically involve items whose average lives are known. (Another term for average life is 'mean time before failure,' MTBF.) The key to such problems is in recognizing that an average life is not the same as a fixed life. An item may belong to a class which has an average life of 10 years, but it may fail after only 1 year of service. Conversely, it may not fail until after 25 or more years.

Obviously, it is not possible to 'read the mind' of a grader or question writer. However, if reasonable assumptions are made in writing at the start of the solution, you should receive full credit. One such reasonable assumption is that of a rectangular distribution. A rectangular distribution is one which is assumed to give an equal probability of failure in each year. Such an assumption is illustrated in example 2.18.

Example 2.18

A bridge is needed for 20 years. Failure of the bridge at any time will require a 50% reinvestment. Evaluate the two design alternatives below using 6% interest.

design alternative	initial cost	MTBF	annual costs	salvage at t=20
A	15,000	9 years	1,200	0
B	22,000	43 years	1,000	0

It will be assumed that each of the alternatives has an annual probability of failure that is inversely proportional to its average life. Then, for alternative A, the probability of failure in any year is (1/9). Similarly, the annual failure probability for alternative B is (1/43).

$$EUAC(A) = 15,000(A/P,6\%,20) + 15,000(.5)(1/9) + 1,200 = 3,341$$

$$EUAC(B) = 22,000(A/P,6\%,20) + 22,000(.5)(1/43) + 1,000 = 3,174$$

Alternative B should be chosen.

13. Estimating Economic Life

As assets grow older, their operating and maintenance costs typically increase each year. Eventually, the cost to keep an asset in operation becomes prohibitive, and the asset is retired or replaced. However, it is not always obvious when an asset should be retired or replaced.

As the asset's maintenance is increasing each year, the amortized cost of its initial purchase is decreasing. It is the sum of these two costs that should be evaluated to determine the point at which the asset should be retired or replaced. Since an asset's initial purchase price is likely to be high, the amortized cost will be the controlling factor in those years when the maintenance costs are low. Therefore, the EUAC of the asset will decrease in the initial part of its life.

However, as the asset grows older, the change in its amortized cost decreases while maintenance increases. Eventually the sum of the two costs reaches a minimum and then starts to increase. The age of the asset at the minimum cost point is known as the economic life of the asset.

The determination of an asset's economic life is illustrated by example 2.19.

Example 2.19

A bus in a municipal transit system has the characteristics listed below. When should the city replace its buses if money can be borrowed at 8%?

Initial cost: $120,000

year	maintenance cost	salvage value
1	35,000	60,000
2	38,000	55,000
3	43,000	45,000
4	50,000	25,000
5	65,000	15,000

If the bus is kept for 1 year and then sold, the annual cost will be

$$EUAC(1) = 120,000(A/P,8\%,1) + 35,000(A/F,8\%,1) - 60,000(A/F,8\%,1)$$

$$= 104,600$$

If the bus is kept for 2 years and then sold, the annual cost will be

$$EUAC(2) = [120,000 + 35,000(P/F,8\%,1)](A/P,8\%,2)$$
$$+ (38,000 - 55,000)(A/F,8\%,2)$$

$$= 77,300$$

If the bus is kept for 3 years and then sold, the annual cost will be

$$EUAC(3) = [120,000 + 35,000(P/F,8\%,1) + 38,000(P/F,8\%,2)](A/P,8\%,3)$$
$$+ (43,000 - 45,000)(A/F,8\%,3)$$

$$= 71,200$$

This process is continued until EUAC begins to increase. In this example, EUAC(4) is 71,700. Therefore, the buses should be retired after 3 years.

◆

14. Basic Cost Accounting

Cost accounting is the system which determines the cost of manufactured products. Cost accounting is called job cost accounting if costs are accumulated by part number or contract. It is called process cost accounting if costs are accumulated by departments or manufacturing processes.

Three types of costs make up the total manufacturing cost of a product. Those are: direct material, direct labor, and all indirect costs.

Direct material costs are the costs of all materials that go into the product, priced at the original purchase cost.

Direct labor costs are the costs of all labor required to assemble or shape the product.

Indirect material and labor costs are generally limited to costs incurred in the factory, excluding costs incurred in the office area. Examples of indirect materials are cleaning fluids, assembly lubricants, and temporary routing tags. Examples of indirect labor are stock-picking, inspection, expediting, and supervision labor.

Here are some important points concerning basic cost accounting:

a. The sum of direct material and direct labor costs is known as the prime cost.

b. Indirect costs may be called indirect manufacturing expenses (IME).

c. Indirect costs may also include the overhead sector of the company (i.e., secretaries, engineers, and corporate administration). In this case, the indirect cost is usually called burden or overhead. Burden may also include the EUAC of non-regular costs which must be spread evenly over several years.

d. The cost of a product is usually known in advance, either from previous manufacturing runs or by estimation. Any deviation from this known cost is called a variance. Variance may be broken down into labor variance and material variance.

e. Indirect cost per item is not easily measured. The method of allocating indirect costs to a product is as follows:

step 1: Estimate the total expected indirect (and overhead) costs for the upcoming year.

step 2: Decide on some convenient vehicle for allocating the overhead to production. Usually, this vehicle is either the number of units expected to be produced or the number of direct hours expected to be worked in the upcoming year.

step 3: Estimate the quantity or size of the overhead vehicle.

step 4: Divide the expected overhead costs by the expected overhead vehicle to obtain the unit overhead.

step 5: Regardless of the true size of the overhead vehicle during the upcoming year, one unit of overhead cost is allocated per product.

f. Although estimates of production for the next year are always somewhat inaccurate, the cost of the product is assumed to be independent of forecasting errors. Any difference between true cost and calculated cost goes into a variance account.

g. Burden (overhead) variance will be caused by errors in forecasting both the actual overhead for the upcoming year and the vehicle size. In the former case, the variance is called burden budget variance; in the latter, it is called burden capacity variance.

Example 2.20

A small company expects to produce 8,000 items in the upcoming year. The current material cost is $4.54 each. 16 minutes of direct labor are required per unit. Workers are paid $7.50 per hour. 2,133 direct labor hours are forecasted for the product. Miscellaneous overhead costs are estimated at $45,000.

Find the expected direct material cost, the direct labor cost, the prime cost, the burden as a function of production and direct labor, and the total cost.

- The direct material cost was given as $4.54.

- The direct labor cost is (16/60) ($7.50) = $2.00

- The prime cost is $4.54 + $2.00 = $6.54.

- If the burden vehicle is production, the burden rate is ($45,000/8,000) = $5.63 per item, making the total cost $4.54 + $2.00 + $5.63 = $12.17.

- If the burden vehicle is direct labor hours, the burden rate is (45,000/2,133) = $21.10 per hour, making the total cost $4.54 + $2.00 + 16/60($21.10) = $12.17.

◆

Example 2.21

The actual performance of the company in example 2.20 is given by the following figures:

actual production: 7,560
actual overhead costs: $47,000

What are the burden budget variance and the burden capacity variance?

The burden capacity variance is

$$\$45,000 - 7,560(\$5.63) = \$2,437$$

The burden budget variance is

$$\$47,000 - \$45,000 = \$2,000$$

The overall burden variance is

$$\$47,000 - 7,560(\$5.63) = \$4,437$$

15. Break-Even Analysis

Break-even analysis is a method of determining when costs exactly equal revenue. If the manufactured quantity is less than the break-even quantity, a loss is incurred. If the manufactured quantity is greater than the break-even quantity, a profit is incurred.

Consider the following special variables:

f   a fixed cost which does not vary with production

a   an incremental cost, which is the cost to produce one additional item. It may also be called the marginal cost or differential cost.

Q   the quantity sold

p   the incremental revenue

R   the total revenue

C   the total cost

Assuming no change in the inventory, the break-even point can be found from C = R, where

$$C = f + aQ \qquad \text{2.26}$$

$$R = pQ \qquad \text{2.27}$$

An alternate form of the break-even problem is to find the number of units per period for which two alternatives have the same total costs. Fixed costs are to be spread over a period longer than one year. One of the alternatives will have a lower cost if production is less than the break-even point. The other will have a lower cost for production greater than the break-even point.

Example 2.22

Two plans are available for a company to obtain automobiles for its salesmen. How many miles must the cars be driven each year for the two plans to have the same costs? Use an interest rate of 10%.

> Plan A: Lease the cars and pay $.15 per mile
>
> Plan B: Purchase the cars for $5,000. Each car has an economic life of 3 years, after which it can be sold for $1,200. Gas and oil cost $.04 per mile. Insurance is $500 per year.

Let x be the number of miles driven per year. Then, the EUAC for both alternatives are:

EUAC(A) = .15x

EUAC(B) = .04x + 500 + 5,000(A/P,10%,3) - 1,200(A/F,10%,3)

= .04x + 2,148

Setting EUAC(A) and EUAC(B) equal and solving for x yields 19,527 miles per year as the break-even point.

16. Handling Inflation

It is important to perform economic studies in terms of 'constant value dollars.' One method of converting all cash flows to constant value dollars is to divide the flows by some annual economic indicator or price index. Such indicators would normally be given to you as part of a problem.

If indicators are not available, this method can still be used by assuming that inflation is relatively constant at a decimal rate (e) per year. Then, all cash flows can be converted to 't = 0' dollars by dividing by $(1 + e)^n$ where n is the year of the cash flow.

Example 2.23

What is the uninflated present worth of $2,000 in 2 years if the average inflation rate is 6% and i is 10%?

$$P = \frac{\$2,000}{(1.10)^2(1.06)^2} = \$1,471.07$$

An alternative is to replace (i) with a value corrected for inflation. This corrected value, i', is

$$i' = i + e + ie$$

2.28

This method has the advantage of simplifying the calculations. However, pre-calculated factors may not be available for the non-integer values of i'. Therefore, table 2.1 will have to be used to calculate the factors.

Example 2.24

Repeat example 2.23 using i'.

$$i' = .10 + .06 + (.10)(.06) = .166$$

$$P = \frac{\$2,000}{(1.166)^2} = \$1,471.07$$

◆

17. Learning Curves

The more products that are made, the more efficient the operation becomes due to experience gained. Therefore, direct labor costs decrease. Usually, a learning curve is specified by the decrease in cost each time the quantity produced doubles. If there is a 20% decrease per doubling, the curve is said to be an 80% learning curve.

Consider the following special variables:

T_1 the time or cost for the first item

T_n the time or cost for the nth item

n the total number of items produced

b the learning curve constant

learning curve	b
75%	.415
80	.322
85	.234
90	.152
95	.074

Then, the time to produce the nth item is given by

$$T_n = T_1 (n)^{-b} \qquad\qquad 2.29$$

The total time to produce units from quantity n_1 to n_2 inclusive is

$$\int_{n_1}^{n_2} T_n \, dn \approx \frac{T_1}{(1-b)} \left[(n_2 + \tfrac{1}{2})^{1-b} - (n_1 - \tfrac{1}{2})^{1-b} \right] \qquad\qquad 2.30$$

The average time per unit over the production from n_1 to n_2 is the above total time from equation 2.30 divided by the quantity produced, $(n_2 - n_1 + 1)$.

It is important to remember that learning curve reductions apply only to direct labor costs. They are not applied to indirect labor or direct material costs.

Example 2.25

A 70% learning curve is used with an item whose first production time was 1.47 hours. How long will it take to produce the 11th item? How long will it take to produce the 11th through 27th items?

First, find b.

$$\frac{T_2}{T_1} = .7 = (2)^{-b} \qquad \text{or } b = .515$$

Then, $\qquad T_{11} = 1.47(11)^{-.515} = .428 \text{ hours}$

The time to produce the 11th item through 27th items is approximately

$$T = \frac{1.47}{1-.515}[(27.5)^{1-.515} - (10.5)^{1-.515}]$$

$$= 5.643 \text{ hours}$$

Bibliography

1. DeGarmo, E.P., and Canada, J.R., _Engineering Economy_, 5th edition, The MacMillan Company, New York, NY, 1973

2. Grant, Eugene L., and Bell, Lawrence F., _Basic Accounting and Cost Accounting_, 2nd edition, McGraw-Hill Book Company, New York, NY, 1964

3. Grant, Eugene L., Ireson, W. Grant, and Leavenworth, Richard, _Principles of Engineering Economy_, 6th edition, Ronald Press, New York, NY, 1976

4. Newnan, Donald G., _Engineering Economic Analysis_, revised edition, Engineering Press, San Jose, CA, 1977

5. Oakford, Robert V., _Capital Budgeting_, Ronald Press, New York, NY, 1970

6. Thuesen, H.G., and Fabrycky, W.G., _Engineering Economy_, 3rd edition, Prentice-Hall, Englewood Cliffs, NJ, 1964

Highlights of the ECONOMIC RECOVERY ACT of 1981

Depreciation

Property placed into service in 1981 or after must use the *Accelerated Cost Recovery System (ACRS).* Other methods (straight-line, declining balance, etc.) cannot be used except in special cases. The term *depreciation* is replaced with the term *cost recovery amount.* The term *useful life* is replaced with the term *cost recovery period.*

Property placed into service in 1980 or before must continue to be depreciated according to the method originally chosen (e.g., straight-line, declining balance, etc.) ACRS cannot be used.

Under ACRS, the cost recovery amount in the *j*th year of an asset's cost recovery period is calculated by multiplying the initial cost by a factor found from a table.

$$D_j = \text{(initial cost)(factor)}$$

The initial cost used is not reduced by the asset's salvage value for either the regular or alternate ACRS calculations. The factor found from the table depends on the asset's cost recovery period.

Although the cost recovery factors are based on declining balance calculations, the tables incorporate changeovers to straight-line and sum-of-year's-digits calculations. These changeovers have been timed to maximize the cost recovery amounts realized by the asset owner. Therefore, the factors cannot be easily calculated.

The ACRS groups tangible property into four categories depending on the cost recovery period. Tangible property is divided into *real property* and *personal property.* Real property is real estate (i.e., land and its structures). Land is never eligible for cost recovery. Personal property consists of all assets (other than real assets) used for the production of business income. Personal property is not for personal use.

3-Year Personal Property

The cost of cars, light trucks, special tools, all R&D equipment, and other assets with short lives should be recovered over 3 years. The full amount applies regardless of when the asset is placed into service during the first year.

If desired, the cost of the asset can be recovered over a longer period. Straight-line depreciation can be used with either a 5- or 12-year life. This is known as the *alternate ACRS* calculation. No other lives or depreciation methods are allowed. If this alternate method is used, only one-half of the cost recovery amount can be claimed in the first year. The remaining one-half can be claimed in the year following the end of the recovery period.

5-Year Personal Property

Machinery, single-purpose agricultural structures, and equipment not recovered over 3 years should be recovered over 5 years. The full percentage applies regardless of when the asset is placed into service during the first year.

The alternate method uses 12- and 25-year periods with straight-line calculations. No other lives or depreciation methods can be used. As with 3-year property, the half-year rule applies to amounts calculated with this alternate method.

10-Year Personal and Real Property

Selected heavy equipment and machinery (e.g., rail cars), theme park structures, and some public utility property should be recovered over 10 years. The full percentage applies regardless of when the asset is placed into service during the first year.

If desired, the cost of the asset can be recovered over 35 or 45 years using straight-line calculations for the alternate method. No other lives or depreciation methods can be used. The half-year rule applies to amounts calculated for the first year.

15-Year Real Property

Real property, including residential real estate, should be recovered over 15 years. Unlike the previous asset categories, however, the factors used depend on the month in which the asset was placed into service.

If desired, the asset can be recovered over 35 or 45 years using straight-line calculations. The half-year rule limiting first-year deductions does not apply to 15-year assets.

TABLE OF RECOVERY FACTORS

year	recovery period 3 yrs	5 yrs	10 yrs
1	.25	.15	.08
2	.38	.22	.14
3	.37	.21	.12
4		.21	.10
5		.21	.10
6			.10
7			.09
8			.09
9			.09
10			.09

RECOVERY FACTORS FOR 15-YEAR PROPERTY

year	month placed in service 1	2	3	4	5	6	7	8	9	10	11	12
1	.12	.11	.10	.09	.08	.07	.06	.05	.04	.03	.02	.01
2	.10	.10	.11	.11	.11	.11	.11	.11	.11	.11	.11	.12
3	.09	.09	.09	.09	.10	.10	.10	.10	.10	.10	.10	.10
4	.08	.08	.08	.08	.08	.08	.09	.09	.09	.09	.09	.09

Refer to IRS publication 534 for a complete table of recovery factors for 15-year assets.

STANDARD CASH FLOW FACTORS

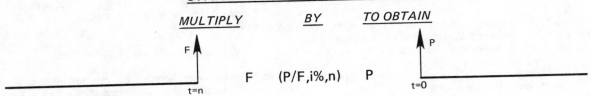

MULTIPLY	*BY*	*TO OBTAIN*

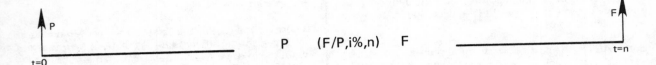

F   (P/F, i%, n)   P

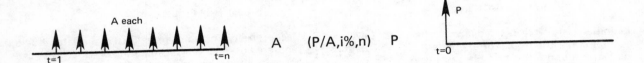

P   (F/P, i%, n)   F

A   (P/A, i%, n)   P

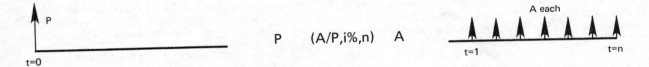

P   (A/P, i%, n)   A

A   (F/A, i%, n)   F

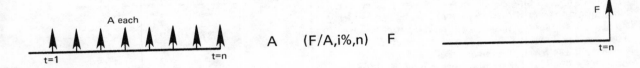

F   (A/F, i%, n)   A

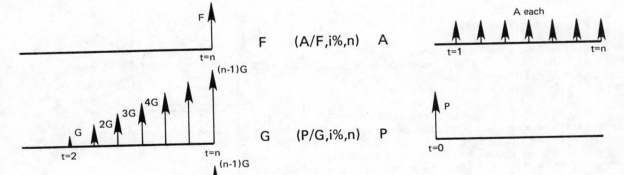

G   (P/G, i%, n)   P

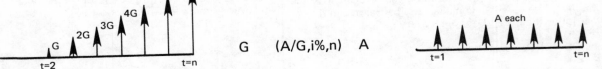

G   (A/G, i%, n)   A

I = 0.50 %

N	(P/F)	(P/A)	(P/G)	(F/P)	(F/A)	(A/P)	(A/F)	(A/G)	N
1	.9950	0.9950	-0.0000	1.0050	1.0000	1.0050	1.0000	-0.0000	1
2	.9901	1.9851	0.9901	1.0100	2.0050	0.5038	0.4988	0.4988	2
3	.9851	2.9702	2.9604	1.0151	3.0150	0.3367	0.3317	0.9967	3
4	.9802	3.9505	5.9011	1.0202	4.0301	0.2531	0.2481	1.4938	4
5	.9754	4.9259	9.8026	1.0253	5.0503	0.2030	0.1980	1.9900	5
6	.9705	5.8964	14.6552	1.0304	6.0755	0.1696	0.1646	2.4855	6
7	.9657	6.8621	20.4493	1.0355	7.1059	0.1457	0.1407	2.9801	7
8	.9609	7.8230	27.1755	1.0407	8.1414	0.1278	0.1228	3.4738	8
9	.9561	8.7791	34.8244	1.0459	9.1821	0.1139	0.1089	3.9668	9
10	.9513	9.7304	43.3865	1.0511	10.2280	0.1028	0.0978	4.4589	10
11	.9466	10.6770	52.8526	1.0564	11.2792	0.0937	0.0887	4.9501	11
12	.9419	11.6189	63.2136	1.0617	12.3356	0.0861	0.0811	5.4406	12
13	.9372	12.5562	74.4602	1.0670	13.3972	0.0796	0.0746	5.9302	13
14	.9326	13.4887	86.5835	1.0723	14.4642	0.0741	0.0691	6.4190	14
15	.9279	14.4166	99.5743	1.0777	15.5365	0.0694	0.0644	6.9069	15
16	.9233	15.3399	113.4238	1.0831	16.6142	0.0652	0.0602	7.3940	16
17	.9187	16.2586	128.1231	1.0885	17.6973	0.0615	0.0565	7.8803	17
18	.9141	17.1728	143.6634	1.0939	18.7858	0.0582	0.0532	8.3658	18
19	.9096	18.0824	160.0360	1.0994	19.8797	0.0553	0.0503	8.8504	19
20	.9051	18.9874	177.2322	1.1049	20.9791	0.0527	0.0477	9.3342	20
21	.9006	19.8880	195.2434	1.1104	22.0840	0.0503	0.0453	9.8172	21
22	.8961	20.7841	214.0611	1.1160	23.1944	0.0481	0.0431	10.2993	22
23	.8916	21.6757	233.6768	1.1216	24.3104	0.0461	0.0411	10.7806	23
24	.8872	22.5629	254.0820	1.1272	25.4320	0.0443	0.0393	11.2611	24
25	.8828	23.4456	275.2686	1.1328	26.5591	0.0427	0.0377	11.7407	25
26	.8784	24.3240	297.2281	1.1385	27.6919	0.0411	0.0361	12.2195	26
27	.8740	25.1980	319.9523	1.1442	28.8304	0.0397	0.0347	12.6975	27
28	.8697	26.0677	343.4332	1.1499	29.9745	0.0384	0.0334	13.1747	28
29	.8653	26.9330	367.6625	1.1556	31.1244	0.0371	0.0321	13.6510	29
30	.8610	27.7941	392.6324	1.1614	32.2800	0.0360	0.0310	14.1265	30
31	.8567	28.6508	418.3348	1.1672	33.4414	0.0349	0.0299	14.6012	31
32	.8525	29.5033	444.7618	1.1730	34.6086	0.0339	0.0289	15.0750	32
33	.8482	30.3515	471.9055	1.1789	35.7817	0.0329	0.0279	15.5480	33
34	.8440	31.1955	499.7583	1.1848	36.9606	0.0321	0.0271	16.0202	34
35	.8398	32.0354	528.3123	1.1907	38.1454	0.0312	0.0262	16.4915	35
40	.8191	36.1722	681.3347	1.2208	44.1588	0.0276	0.0226	18.8359	40
50	.7793	44.1428	1035.6966	1.2832	56.6452	0.0227	0.0177	23.4624	50
60	.7414	51.7256	1448.6458	1.3489	69.7700	0.0193	0.0143	28.0064	60
70	.7053	58.9394	1913.6427	1.4178	83.5661	0.0170	0.0120	32.4680	70
80	.6710	65.8023	2424.6455	1.4903	98.0677	0.0152	0.0102	36.8474	80
90	.6383	72.3313	2976.0769	1.5666	113.3109	0.0138	0.0088	41.1451	90
100	.6073	78.5426	3562.7934	1.6467	129.3337	0.0127	0.0077	45.3613	100

I = 0.75 %

N	(P/F)	(P/A)	(P/G)	(F/P)	(F/A)	(A/P)	(A/F)	(A/G)	N
1	.9926	0.9926	-0.0000	1.0075	1.0000	1.0075	1.0000	-0.0000	1
2	.9852	1.9777	0.9852	1.0151	2.0075	0.5056	0.4981	0.4981	2
3	.9778	2.9556	2.9408	1.0227	3.0226	0.3383	0.3308	0.9950	3
4	.9706	3.9261	5.8525	1.0303	4.0452	0.2547	0.2472	1.4907	4
5	.9633	4.8894	9.7058	1.0381	5.0756	0.2045	0.1970	1.9851	5
6	.9562	5.8456	14.4866	1.0459	6.1136	0.1711	0.1636	2.4782	6
7	.9490	6.7946	20.1808	1.0537	7.1595	0.1472	0.1397	2.9701	7
8	.9420	7.7366	26.7747	1.0616	8.2132	0.1293	0.1218	3.4608	8
9	.9350	8.6716	34.2544	1.0696	9.2748	0.1153	0.1078	3.9502	9
10	.9280	9.5996	42.6064	1.0776	10.3443	0.1042	0.0967	4.4384	10
11	.9211	10.5207	51.8174	1.0857	11.4219	0.0951	0.0876	4.9253	11
12	.9142	11.4349	61.8740	1.0938	12.5076	0.0875	0.0800	5.4110	12
13	.9074	12.3423	72.7632	1.1020	13.6014	0.0810	0.0735	5.8954	13
14	.9007	13.2430	84.4720	1.1103	14.7034	0.0755	0.0680	6.3786	14
15	.8940	14.1370	96.9876	1.1186	15.8137	0.0707	0.0632	6.8606	15
16	.8873	15.0243	110.2973	1.1270	16.9323	0.0666	0.0591	7.3413	16
17	.8807	15.9050	124.3887	1.1354	18.0593	0.0629	0.0554	7.8207	17
18	.8742	16.7792	139.2494	1.1440	19.1947	0.0596	0.0521	8.2989	18
19	.8676	17.6468	154.8671	1.1525	20.3387	0.0567	0.0492	8.7759	19
20	.8612	18.5080	171.2297	1.1612	21.4912	0.0540	0.0465	9.2516	20
21	.8548	19.3628	188.3253	1.1699	22.6524	0.0516	0.0441	9.7261	21
22	.8484	20.2112	206.1420	1.1787	23.8223	0.0495	0.0420	10.1994	22
23	.8421	21.0533	224.6682	1.1875	25.0010	0.0475	0.0400	10.6714	23
24	.8358	21.8891	243.8923	1.1964	26.1885	0.0457	0.0382	11.1422	24
25	.8296	22.7188	263.8029	1.2054	27.3849	0.0440	0.0365	11.6117	25
26	.8234	23.5422	284.3888	1.2144	28.5903	0.0425	0.0350	12.0800	26
27	.8173	24.3595	305.6387	1.2235	29.8047	0.0411	0.0336	12.5470	27
28	.8112	25.1707	327.5416	1.2327	31.0282	0.0397	0.0322	13.0128	28
29	.8052	25.9759	350.0867	1.2420	32.2609	0.0385	0.0310	13.4774	29
30	.7992	26.7751	373.2631	1.2513	33.5029	0.0373	0.0298	13.9407	30
31	.7932	27.5683	397.0602	1.2607	34.7542	0.0363	0.0288	14.4028	31
32	.7873	28.3557	421.4675	1.2701	36.0148	0.0353	0.0278	14.8636	32
33	.7815	29.1371	446.4746	1.2796	37.2849	0.0343	0.0268	15.3232	33
34	.7757	29.9128	472.0712	1.2892	38.5646	0.0334	0.0259	15.7816	34
35	.7699	30.6827	498.2471	1.2989	39.8538	0.0326	0.0251	16.2387	35
40	.7416	34.4469	637.4693	1.3483	46.4465	0.0290	0.0215	18.5058	40
50	.6883	41.5664	953.8486	1.4530	60.3943	0.0241	0.0166	22.9476	50
60	.6387	48.1734	1313.5189	1.5657	75.4241	0.0208	0.0133	27.2665	60
70	.5927	54.3046	1708.6065	1.6872	91.6201	0.0184	0.0109	31.4634	70
80	.5500	59.9944	2132.1472	1.8180	109.0725	0.0167	0.0092	35.5391	80
90	.5104	65.2746	2577.9961	1.9591	127.8790	0.0153	0.0078	39.4946	90
100	.4737	70.1746	3040.7453	2.1111	148.1445	0.0143	0.0068	43.3311	100

I = 1.00 %

N	(P/F)	(P/A)	(P/G)	(F/P)	(F/A)	(A/P)	(A/F)	(A/G)	N
1	.9901	0.9901	-0.0000	1.0100	1.0000	1.0100	1.0000	-0.0000	1
2	.9803	1.9704	0.9803	1.0201	2.0100	0.5075	0.4975	0.4975	2
3	.9706	2.9410	2.9215	1.0303	3.0301	0.3400	0.3300	0.9934	3
4	.9610	3.9020	5.8044	1.0406	4.0604	0.2563	0.2463	1.4876	4
5	.9515	4.8534	9.6103	1.0510	5.1010	0.2060	0.1960	1.9801	5
6	.9420	5.7955	14.3205	1.0615	6.1520	0.1725	0.1625	2.4710	6
7	.9327	6.7282	19.9168	1.0721	7.2135	0.1486	0.1386	2.9602	7
8	.9235	7.6517	26.3812	1.0829	8.2857	0.1307	0.1207	3.4478	8
9	.9143	8.5660	33.6959	1.0937	9.3685	0.1167	0.1067	3.9337	9
10	.9053	9.4713	41.8435	1.1046	10.4622	0.1056	0.0956	4.4179	10
11	.8963	10.3676	50.8067	1.1157	11.5668	0.0965	0.0865	4.9005	11
12	.8874	11.2551	60.5687	1.1268	12.6825	0.0888	0.0788	5.3815	12
13	.8787	12.1337	71.1126	1.1381	13.8093	0.0824	0.0724	5.8607	13
14	.8700	13.0037	82.4221	1.1495	14.9474	0.0769	0.0669	6.3384	14
15	.8613	13.8651	94.4810	1.1610	16.0969	0.0721	0.0621	6.8143	15
16	.8528	14.7179	107.2734	1.1726	17.2579	0.0679	0.0579	7.2886	16
17	.8444	15.5623	120.7834	1.1843	18.4304	0.0643	0.0543	7.7613	17
18	.8360	16.3983	134.9957	1.1961	19.6147	0.0610	0.0510	8.2323	18
19	.8277	17.2260	149.8950	1.2081	20.8109	0.0581	0.0481	8.7017	19
20	.8195	18.0456	165.4664	1.2202	22.0190	0.0554	0.0454	9.1694	20
21	.8114	18.8570	181.6950	1.2324	23.2392	0.0530	0.0430	9.6354	21
22	.8034	19.6604	198.5663	1.2447	24.4716	0.0509	0.0409	10.0998	22
23	.7954	20.4558	216.0660	1.2572	25.7163	0.0489	0.0389	10.5626	23
24	.7876	21.2434	234.1800	1.2697	26.9735	0.0471	0.0371	11.0237	24
25	.7798	22.0232	252.8945	1.2824	28.2432	0.0454	0.0354	11.4831	25
26	.7720	22.7952	272.1957	1.2953	29.5256	0.0439	0.0339	11.9409	26
27	.7644	23.5596	292.0702	1.3082	30.8209	0.0424	0.0324	12.3971	27
28	.7568	24.3164	312.5047	1.3213	32.1291	0.0411	0.0311	12.8516	28
29	.7493	25.0658	333.4863	1.3345	33.4504	0.0399	0.0299	13.3044	29
30	.7419	25.8077	355.0021	1.3478	34.7849	0.0387	0.0287	13.7557	30
31	.7346	26.5423	377.0394	1.3613	36.1327	0.0377	0.0277	14.2052	31
32	.7273	27.2696	399.5858	1.3749	37.4941	0.0367	0.0267	14.6532	32
33	.7201	27.9897	422.6291	1.3887	38.8690	0.0357	0.0257	15.0995	33
34	.7130	28.7027	446.1572	1.4026	40.2577	0.0348	0.0248	15.5441	34
35	.7059	29.4086	470.1583	1.4166	41.6603	0.0340	0.0240	15.9871	35
40	.6717	32.8347	596.8561	1.4889	48.8864	0.0305	0.0205	18.1776	40
50	.6080	39.1961	879.4176	1.6446	64.4632	0.0255	0.0155	22.4363	50
60	.5504	44.9550	1192.8061	1.8167	81.6697	0.0222	0.0122	26.5333	60
70	.4983	50.1685	1528.6474	2.0068	100.6763	0.0199	0.0099	30.4703	70
80	.4511	54.8882	1879.8771	2.2167	121.6715	0.0182	0.0082	34.2492	80
90	.4084	59.1609	2240.5675	2.4486	144.8633	0.0169	0.0069	37.8724	90
100	.3697	63.0289	2605.7758	2.7048	170.4814	0.0159	0.0059	41.3426	100

I = 1.50 %

N	(P/F)	(P/A)	(P/G)	(F/P)	(F/A)	(A/P)	(A/F)	(A/G)	N
1	.9852	0.9852	-0.0000	1.0150	1.0000	1.0150	1.0000	-0.0000	1
2	.9707	1.9559	0.9707	1.0302	2.0150	0.5113	0.4963	0.4963	2
3	.9563	2.9122	2.8833	1.0457	3.0452	0.3434	0.3284	0.9901	3
4	.9422	3.8544	5.7098	1.0614	4.0909	0.2594	0.2444	1.4814	4
5	.9283	4.7826	9.4229	1.0773	5.1523	0.2091	0.1941	1.9702	5
6	.9145	5.6972	13.9956	1.0934	6.2296	0.1755	0.1605	2.4566	6
7	.9010	6.5982	19.4018	1.1098	7.3230	0.1516	0.1366	2.9405	7
8	.8877	7.4859	25.6157	1.1265	8.4328	0.1336	0.1186	3.4219	8
9	.8746	8.3605	32.6125	1.1434	9.5593	0.1196	0.1046	3.9008	9
10	.8617	9.2222	40.3675	1.1605	10.7027	0.1084	0.0934	4.3772	10
11	.8489	10.0711	48.8568	1.1779	11.8633	0.0993	0.0843	4.8512	11
12	.8364	10.9075	58.0571	1.1956	13.0412	0.0917	0.0767	5.3227	12
13	.8240	11.7315	67.9454	1.2136	14.2368	0.0852	0.0702	5.7917	13
14	.8118	12.5434	78.4994	1.2318	15.4504	0.0797	0.0647	6.2582	14
15	.7999	13.3432	89.6974	1.2502	16.6821	0.0749	0.0599	6.7223	15
16	.7880	14.1313	101.5178	1.2690	17.9324	0.0708	0.0558	7.1839	16
17	.7764	14.9076	113.9400	1.2880	19.2014	0.0671	0.0521	7.6431	17
18	.7649	15.6726	126.9435	1.3073	20.4894	0.0638	0.0488	8.0997	18
19	.7536	16.4262	140.5084	1.3270	21.7967	0.0609	0.0459	8.5539	19
20	.7425	17.1686	154.6154	1.3469	23.1237	0.0582	0.0432	9.0057	20
21	.7315	17.9001	169.2453	1.3671	24.4705	0.0559	0.0409	9.4550	21
22	.7207	18.6208	184.3798	1.3876	25.8376	0.0537	0.0387	9.9018	22
23	.7100	19.3309	200.0006	1.4084	27.2251	0.0517	0.0367	10.3462	23
24	.6995	20.0304	216.0901	1.4295	28.6335	0.0499	0.0349	10.7881	24
25	.6892	20.7196	232.6310	1.4509	30.0630	0.0483	0.0333	11.2276	25
26	.6790	21.3986	249.6065	1.4727	31.5140	0.0467	0.0317	11.6646	26
27	.6690	22.0676	267.0002	1.4948	32.9867	0.0453	0.0303	12.0992	27
28	.6591	22.7267	284.7958	1.5172	34.4815	0.0440	0.0290	12.5313	28
29	.6494	23.3761	302.9779	1.5400	35.9987	0.0428	0.0278	12.9610	29
30	.6398	24.0158	321.5310	1.5631	37.5387	0.0416	0.0266	13.3883	30
31	.6303	24.6461	340.4402	1.5865	39.1018	0.0406	0.0256	13.8131	31
32	.6210	25.2671	359.6910	1.6103	40.6883	0.0396	0.0246	14.2355	32
33	.6118	25.8790	379.2691	1.6345	42.2986	0.0386	0.0236	14.6555	33
34	.6028	26.4817	399.1607	1.6590	43.9331	0.0378	0.0228	15.0731	34
35	.5939	27.0756	419.3521	1.6839	45.5921	0.0369	0.0219	15.4882	35
40	.5513	29.9158	524.3568	1.8140	54.2679	0.0334	0.0184	17.5277	40
50	.4750	34.9997	749.9636	2.1052	73.6828	0.0286	0.0136	21.4277	50
60	.4093	39.3803	988.1674	2.4432	96.2147	0.0254	0.0104	25.0930	60
70	.3527	43.1549	1231.1658	2.8355	122.3638	0.0232	0.0082	28.5290	70
80	.3039	46.4073	1473.0741	3.2907	152.7109	0.0215	0.0065	31.7423	80
90	.2619	49.2099	1709.5439	3.8189	187.9299	0.0203	0.0053	34.7399	90
100	.2256	51.6247	1937.4506	4.4320	228.8030	0.0194	0.0044	37.5295	100

I = 2.00 %

N	(P/F)	(P/A)	(P/G)	(F/P)	(F/A)	(A/P)	(A/F)	(A/G)	N
1	.9804	0.9804	-0.0000	1.0200	1.0000	1.0200	1.0000	-0.0000	1
2	.9612	1.9416	0.9612	1.0404	2.0200	0.5150	0.4950	0.4950	2
3	.9423	2.8839	2.8458	1.0612	3.0604	0.3468	0.3268	0.9868	3
4	.9238	3.8077	5.6173	1.0824	4.1216	0.2626	0.2426	1.4752	4
5	.9057	4.7135	9.2403	1.1041	5.2040	0.2122	0.1922	1.9604	5
6	.8880	5.6014	13.6801	1.1262	6.3081	0.1785	0.1585	2.4423	6
7	.8706	6.4720	18.9035	1.1487	7.4343	0.1545	0.1345	2.9208	7
8	.8535	7.3255	24.8779	1.1717	8.5830	0.1365	0.1165	3.3961	8
9	.8368	8.1622	31.5720	1.1951	9.7546	0.1225	0.1025	3.8681	9
10	.8203	8.9826	38.9551	1.2190	10.9497	0.1113	0.0913	4.3367	10
11	.8043	9.7868	46.9977	1.2434	12.1687	0.1022	0.0822	4.8021	11
12	.7885	10.5753	55.6712	1.2682	13.4121	0.0946	0.0746	5.2642	12
13	.7730	11.3484	64.9475	1.2936	14.6803	0.0881	0.0681	5.7231	13
14	.7579	12.1062	74.7999	1.3195	15.9739	0.0826	0.0626	6.1786	14
15	.7430	12.8493	85.2021	1.3459	17.2934	0.0778	0.0578	6.6309	15
16	.7284	13.5777	96.1288	1.3728	18.6393	0.0737	0.0537	7.0799	16
17	.7142	14.2919	107.5554	1.4002	20.0121	0.0700	0.0500	7.5256	17
18	.7002	14.9920	119.4581	1.4282	21.4123	0.0667	0.0467	7.9681	18
19	.6864	15.6785	131.8139	1.4568	22.8406	0.0638	0.0438	8.4073	19
20	.6730	16.3514	144.6003	1.4859	24.2974	0.0612	0.0412	8.8433	20
21	.6598	17.0112	157.7959	1.5157	25.7833	0.0588	0.0388	9.2760	21
22	.6468	17.6580	171.3795	1.5460	27.2990	0.0566	0.0366	9.7055	22
23	.6342	18.2922	185.3309	1.5769	28.8450	0.0547	0.0347	10.1317	23
24	.6217	18.9139	199.6305	1.6084	30.4219	0.0529	0.0329	10.5547	24
25	.6095	19.5235	214.2592	1.6406	32.0303	0.0512	0.0312	10.9745	25
26	.5976	20.1210	229.1987	1.6734	33.6709	0.0497	0.0297	11.3910	26
27	.5859	20.7069	244.4311	1.7069	35.3443	0.0483	0.0283	11.8043	27
28	.5744	21.2813	259.9392	1.7410	37.0512	0.0470	0.0270	12.2145	28
29	.5631	21.8444	275.7064	1.7758	38.7922	0.0458	0.0258	12.6214	29
30	.5521	22.3965	291.7164	1.8114	40.5681	0.0446	0.0246	13.0251	30
31	.5412	22.9377	307.9538	1.8476	42.3794	0.0436	0.0236	13.4257	31
32	.5306	23.4683	324.4035	1.8845	44.2270	0.0426	0.0226	13.8230	32
33	.5202	23.9886	341.0508	1.9222	46.1116	0.0417	0.0217	14.2172	33
34	.5100	24.4986	357.8817	1.9607	48.0338	0.0408	0.0208	14.6083	34
35	.5000	24.9986	374.8826	1.9999	49.9945	0.0400	0.0200	14.9961	35
40	.4529	27.3555	461.9931	2.2080	60.4020	0.0366	0.0166	16.8885	40
50	.3715	31.4236	642.3606	2.6916	84.5794	0.0318	0.0118	20.4420	50
60	.3048	34.7609	823.6975	3.2810	114.0515	0.0288	0.0088	23.6961	60
70	.2500	37.4986	999.8343	3.9996	149.9779	0.0267	0.0067	26.6632	70
80	.2051	39.7445	1166.7868	4.8754	193.7720	0.0252	0.0052	29.3572	80
90	.1683	41.5869	1322.1701	5.9431	247.1567	0.0240	0.0040	31.7929	90
100	.1380	43.0984	1464.7527	7.2446	312.2323	0.0232	0.0032	33.9863	100

I = 3.00 %

N	(P/F)	(P/A)	(P/G)	(F/P)	(F/A)	(A/P)	(A/F)	(A/G)	N
1	.9709	0.9709	-0.0000	1.0300	1.0000	1.0300	1.0000	-0.0000	1
2	.9426	1.9135	0.9426	1.0609	2.0300	0.5226	0.4926	0.4926	2
3	.9151	2.8286	2.7729	1.0927	3.0909	0.3535	0.3235	0.9803	3
4	.8885	3.7171	5.4383	1.1255	4.1836	0.2690	0.2390	1.4631	4
5	.8626	4.5797	8.8888	1.1593	5.3091	0.2184	0.1884	1.9409	5
6	.8375	5.4172	13.0762	1.1941	6.4684	0.1846	0.1546	2.4138	6
7	.8131	6.2303	17.9547	1.2299	7.6625	0.1605	0.1305	2.8819	7
8	.7894	7.0197	23.4806	1.2668	8.8923	0.1425	0.1125	3.3450	8
9	.7664	7.7861	29.6119	1.3048	10.1591	0.1284	0.0984	3.8032	9
10	.7441	8.5302	36.3088	1.3439	11.4639	0.1172	0.0872	4.2565	10
11	.7224	9.2526	43.5330	1.3842	12.8078	0.1081	0.0781	4.7049	11
12	.7014	9.9540	51.2482	1.4258	14.1920	0.1005	0.0705	5.1485	12
13	.6810	10.6350	59.4196	1.4685	15.6178	0.0940	0.0640	5.5872	13
14	.6611	11.2961	68.0141	1.5126	17.0863	0.0885	0.0585	6.0210	14
15	.6419	11.9379	77.0002	1.5580	18.5989	0.0838	0.0538	6.4500	15
16	.6232	12.5611	86.3477	1.6047	20.1569	0.0796	0.0496	6.8742	16
17	.6050	13.1661	96.0280	1.6528	21.7616	0.0760	0.0460	7.2936	17
18	.5874	13.7535	106.0137	1.7024	23.4144	0.0727	0.0427	7.7081	18
19	.5703	14.3238	116.2788	1.7535	25.1169	0.0698	0.0398	8.1179	19
20	.5537	14.8775	126.7987	1.8061	26.8704	0.0672	0.0372	8.5229	20
21	.5375	15.4150	137.5496	1.8603	28.6765	0.0649	0.0349	8.9231	21
22	.5219	15.9369	148.5094	1.9161	30.5368	0.0627	0.0327	9.3186	22
23	.5067	16.4436	159.6566	1.9736	32.4529	0.0608	0.0308	9.7093	23
24	.4919	16.9355	170.9711	2.0328	34.4265	0.0590	0.0290	10.0954	24
25	.4776	17.4131	182.4336	2.0938	36.4593	0.0574	0.0274	10.4768	25
26	.4637	17.8768	194.0260	2.1566	38.5530	0.0559	0.0259	10.8535	26
27	.4502	18.3270	205.7309	2.2213	40.7096	0.0546	0.0246	11.2255	27
28	.4371	18.7641	217.5320	2.2879	42.9309	0.0533	0.0233	11.5930	28
29	.4243	19.1885	229.4137	2.3566	45.2189	0.0521	0.0221	11.9558	29
30	.4120	19.6004	241.3613	2.4273	47.5754	0.0510	0.0210	12.3141	30
31	.4000	20.0004	253.3609	2.5001	50.0027	0.0500	0.0200	12.6678	31
32	.3883	20.3888	265.3993	2.5751	52.5028	0.0490	0.0190	13.0169	32
33	.3770	20.7658	277.4642	2.6523	55.0778	0.0482	0.0182	13.3616	33
34	.3660	21.1318	289.5437	2.7319	57.7302	0.0473	0.0173	13.7018	34
35	.3554	21.4872	301.6267	2.8139	60.4621	0.0465	0.0165	14.0375	35
40	.3066	23.1148	361.7499	3.2620	75.4013	0.0433	0.0133	15.6502	40
50	.2281	25.7298	477.4803	4.3839	112.7969	0.0389	0.0089	18.5575	50
60	.1697	27.6756	583.0526	5.8916	163.0534	0.0361	0.0061	21.0674	60
70	.1263	29.1234	676.0869	7.9178	230.5941	0.0343	0.0043	23.2145	70
80	.0940	30.2008	756.0865	10.6409	321.3630	0.0331	0.0031	25.0353	80
90	.0699	31.0024	823.6302	14.3005	443.3489	0.0323	0.0023	26.5667	90
100	.0520	31.5989	879.8540	19.2186	607.2877	0.0316	0.0016	27.8444	100

I = 4.00 %

N	(P/F)	(P/A)	(P/G)	(F/P)	(F/A)	(A/P)	(A/F)	(A/G)	N
1	.9615	0.9615	-0.0000	1.0400	1.0000	1.0400	1.0000	-0.0000	1
2	.9246	1.8861	0.9246	1.0816	2.0400	0.5302	0.4902	0.4902	2
3	.8890	2.7751	2.7025	1.1249	3.1216	0.3603	0.3203	0.9739	3
4	.8548	3.6299	5.2670	1.1699	4.2465	0.2755	0.2355	1.4510	4
5	.8219	4.4518	8.5547	1.2167	5.4163	0.2246	0.1846	1.9216	5
6	.7903	5.2421	12.5062	1.2653	6.6330	0.1908	0.1508	2.3857	6
7	.7599	6.0021	17.0657	1.3159	7.8983	0.1666	0.1266	2.8433	7
8	.7307	6.7327	22.1806	1.3686	9.2142	0.1485	0.1085	3.2944	8
9	.7026	7.4353	27.8013	1.4233	10.5828	0.1345	0.0945	3.7391	9
10	.6756	8.1109	33.8814	1.4802	12.0061	0.1233	0.0833	4.1773	10
11	.6496	8.7605	40.3772	1.5395	13.4864	0.1141	0.0741	4.6090	11
12	.6246	9.3851	47.2477	1.6010	15.0258	0.1066	0.0666	5.0343	12
13	.6006	9.9856	54.4546	1.6651	16.6268	0.1001	0.0601	5.4533	13
14	.5775	10.5631	61.9618	1.7317	18.2919	0.0947	0.0547	5.8659	14
15	.5553	11.1184	69.7355	1.8009	20.0236	0.0899	0.0499	6.2721	15
16	.5339	11.6523	77.7441	1.8730	21.8245	0.0858	0.0458	6.6720	16
17	.5134	12.1657	85.9581	1.9479	23.6975	0.0822	0.0422	7.0656	17
18	.4936	12.6593	94.3498	2.0258	25.6454	0.0790	0.0390	7.4530	18
19	.4746	13.1339	102.8933	2.1068	27.6712	0.0761	0.0361	7.8342	19
20	.4564	13.5903	111.5647	2.1911	29.7781	0.0736	0.0336	8.2091	20
21	.4388	14.0292	120.3414	2.2788	31.9692	0.0713	0.0313	8.5779	21
22	.4220	14.4511	129.2024	2.3699	34.2480	0.0692	0.0292	8.9407	22
23	.4057	14.8568	138.1284	2.4647	36.6179	0.0673	0.0273	9.2973	23
24	.3901	15.2470	147.1012	2.5633	39.0826	0.0656	0.0256	9.6479	24
25	.3751	15.6221	156.1040	2.6658	41.6459	0.0640	0.0240	9.9925	25
26	.3607	15.9828	165.1212	2.7725	44.3117	0.0626	0.0226	10.3312	26
27	.3468	16.3296	174.1385	2.8834	47.0842	0.0612	0.0212	10.6640	27
28	.3335	16.6631	183.1424	2.9987	49.9676	0.0600	0.0200	10.9909	28
29	.3207	16.9837	192.1206	3.1187	52.9663	0.0589	0.0189	11.3120	29
30	.3083	17.2920	201.0618	3.2434	56.0849	0.0578	0.0178	11.6274	30
31	.2965	17.5885	209.9556	3.3731	59.3283	0.0569	0.0169	11.9371	31
32	.2851	17.8736	218.7924	3.5081	62.7015	0.0559	0.0159	12.2411	32
33	.2741	18.1476	227.5634	3.6484	66.2095	0.0551	0.0151	12.5396	33
34	.2636	18.4112	236.2607	3.7943	69.8579	0.0543	0.0143	12.8324	34
35	.2534	18.6646	244.8768	3.9461	73.6522	0.0536	0.0136	13.1198	35
40	.2083	19.7928	286.5303	4.8010	95.0255	0.0505	0.0105	14.4765	40
50	.1407	21.4822	361.1638	7.1067	152.6671	0.0466	0.0066	16.8122	50
60	.0951	22.6235	422.9966	10.5196	237.9907	0.0442	0.0042	18.6972	60
70	.0642	23.3945	472.4789	15.5716	364.2905	0.0427	0.0027	20.1961	70
80	.0434	23.9154	511.1161	23.0498	551.2450	0.0418	0.0018	21.3718	80
90	.0293	24.2673	540.7369	34.1193	827.9833	0.0412	0.0012	22.2826	90
100	.0198	24.5050	563.1249	50.5049	1237.6237	0.0408	0.0008	22.9800	100

I = 5.00 %

N	(P/F)	(P/A)	(P/G)	(F/P)	(F/A)	(A/P)	(A/F)	(A/G)	N
1	.9524	0.9524	-0.0000	1.0500	1.0000	1.0500	1.0000	-0.0000	1
2	.9070	1.8594	0.9070	1.1025	2.0500	0.5378	0.4878	0.4878	2
3	.8638	2.7232	2.6347	1.1576	3.1525	0.3672	0.3172	0.9675	3
4	.8227	3.5460	5.1028	1.2155	4.3101	0.2820	0.2320	1.4391	4
5	.7835	4.3295	8.2369	1.2763	5.5256	0.2310	0.1810	1.9025	5
6	.7462	5.0757	11.9680	1.3401	6.8019	0.1970	0.1470	2.3579	6
7	.7107	5.7864	16.2321	1.4071	8.1420	0.1728	0.1228	2.8052	7
8	.6768	6.4632	20.9700	1.4775	9.5491	0.1547	0.1047	3.2445	8
9	.6446	7.1078	26.1268	1.5513	11.0266	0.1407	0.0907	3.6758	9
10	.6139	7.7217	31.6520	1.6289	12.5779	0.1295	0.0795	4.0991	10
11	.5847	8.3064	37.4988	1.7103	14.2068	0.1204	0.0704	4.5144	11
12	.5568	8.8633	43.6241	1.7959	15.9171	0.1128	0.0628	4.9219	12
13	.5303	9.3936	49.9879	1.8856	17.7130	0.1065	0.0565	5.3215	13
14	.5051	9.8986	56.5538	1.9799	19.5986	0.1010	0.0510	5.7133	14
15	.4810	10.3797	63.2880	2.0789	21.5786	0.0963	0.0463	6.0973	15
16	.4581	10.8378	70.1597	2.1829	23.6575	0.0923	0.0423	6.4736	16
17	.4363	11.2741	77.1405	2.2920	25.8404	0.0887	0.0387	6.8423	17
18	.4155	11.6896	84.2043	2.4066	28.1324	0.0855	0.0355	7.2034	18
19	.3957	12.0853	91.3275	2.5270	30.5390	0.0827	0.0327	7.5569	19
20	.3769	12.4622	98.4884	2.6533	33.0660	0.0802	0.0302	7.9030	20
21	.3589	12.8212	105.6673	2.7860	35.7193	0.0780	0.0280	8.2416	21
22	.3418	13.1630	112.8461	2.9253	38.5052	0.0760	0.0260	8.5730	22
23	.3256	13.4886	120.0087	3.0715	41.4305	0.0741	0.0241	8.8971	23
24	.3101	13.7986	127.1402	3.2251	44.5020	0.0725	0.0225	9.2140	24
25	.2953	14.0939	134.2275	3.3864	47.7271	0.0710	0.0210	9.5238	25
26	.2812	14.3752	141.2585	3.5557	51.1135	0.0696	0.0196	9.8266	26
27	.2678	14.6430	148.2226	3.7335	54.6691	0.0683	0.0183	10.1224	27
28	.2551	14.8981	155.1101	3.9201	58.4026	0.0671	0.0171	10.4114	28
29	.2429	15.1411	161.9126	4.1161	62.3227	0.0660	0.0160	10.6936	29
30	.2314	15.3725	168.6226	4.3219	66.4388	0.0651	0.0151	10.9691	30
31	.2204	15.5928	175.2333	4.5380	70.7608	0.0641	0.0141	11.2381	31
32	.2099	15.8027	181.7392	4.7649	75.2988	0.0633	0.0133	11.5005	32
33	.1999	16.0025	188.1351	5.0032	80.0638	0.0625	0.0125	11.7566	33
34	.1904	16.1929	194.4168	5.2533	85.0670	0.0618	0.0118	12.0063	34
35	.1813	16.3742	200.5807	5.5160	90.3203	0.0611	0.0111	12.2498	35
40	.1420	17.1591	229.5452	7.0400	120.7998	0.0583	0.0083	13.3775	40
50	.0872	18.2559	277.9148	11.4674	209.3480	0.0548	0.0048	15.2233	50
60	.0535	18.9293	314.3432	18.6792	353.5837	0.0528	0.0028	16.6062	60
70	.0329	19.3427	340.8409	30.4264	588.5285	0.0517	0.0017	17.6212	70
80	.0202	19.5965	359.6460	49.5614	971.2288	0.0510	0.0010	18.3526	80
90	.0124	19.7523	372.7488	80.7304	1594.6073	0.0506	0.0006	18.8712	90
100	.0076	19.8479	381.7492	131.5013	2610.0252	0.0504	0.0004	19.2337	100

I = 6.00 %

N	(P/F)	(P/A)	(P/G)	(F/P)	(F/A)	(A/P)	(A/F)	(A/G)	N
1	.9434	0.9434	-0.0000	1.0600	1.0000	1.0600	1.0000	-0.0000	1
2	.8900	1.8334	0.8900	1.1236	2.0600	0.5454	0.4854	0.4854	2
3	.8396	2.6730	2.5692	1.1910	3.1836	0.3741	0.3141	0.9612	3
4	.7921	3.4651	4.9455	1.2625	4.3746	0.2886	0.2286	1.4272	4
5	.7473	4.2124	7.9345	1.3382	5.6371	0.2374	0.1774	1.8836	5
6	.7050	4.9173	11.4594	1.4185	6.9753	0.2034	0.1434	2.3304	6
7	.6651	5.5824	15.4497	1.5036	8.3938	0.1791	0.1191	2.7676	7
8	.6274	6.2098	19.8416	1.5938	9.8975	0.1610	0.1010	3.1952	8
9	.5919	6.8017	24.5768	1.6895	11.4913	0.1470	0.0870	3.6133	9
10	.5584	7.3601	29.6023	1.7908	13.1808	0.1359	0.0759	4.0220	10
11	.5268	7.8869	34.8702	1.8983	14.9716	0.1268	0.0668	4.4213	11
12	.4970	8.3838	40.3369	2.0122	16.8699	0.1193	0.0593	4.8113	12
13	.4688	8.8527	45.9629	2.1329	18.8821	0.1130	0.0530	5.1920	13
14	.4423	9.2950	51.7128	2.2609	21.0151	0.1076	0.0476	5.5635	14
15	.4173	9.7122	57.5546	2.3966	23.2760	0.1030	0.0430	5.9260	15
16	.3936	10.1059	63.4592	2.5404	25.6725	0.0990	0.0390	6.2794	16
17	.3714	10.4773	69.4011	2.6928	28.2129	0.0954	0.0354	6.6240	17
18	.3503	10.8276	75.3569	2.8543	30.9057	0.0924	0.0324	6.9597	18
19	.3305	11.1581	81.3062	3.0256	33.7600	0.0896	0.0296	7.2867	19
20	.3118	11.4699	87.2304	3.2071	36.7856	0.0872	0.0272	7.6051	20
21	.2942	11.7641	93.1136	3.3996	39.9927	0.0850	0.0250	7.9151	21
22	.2775	12.0416	98.9412	3.6035	43.3923	0.0830	0.0230	8.2166	22
23	.2618	12.3034	104.7007	3.8197	46.9958	0.0813	0.0213	8.5099	23
24	.2470	12.5504	110.3812	4.0489	50.8156	0.0797	0.0197	8.7951	24
25	.2330	12.7834	115.9732	4.2919	54.8645	0.0782	0.0182	9.0722	25
26	.2198	13.0032	121.4684	4.5494	59.1564	0.0769	0.0169	9.3414	26
27	.2074	13.2105	126.8600	4.8223	63.7058	0.0757	0.0157	9.6029	27
28	.1956	13.4062	132.1420	5.1117	68.5281	0.0746	0.0146	9.8568	28
29	.1846	13.5907	137.3096	5.4184	73.6398	0.0736	0.0136	10.1032	29
30	.1741	13.7648	142.3588	5.7435	79.0582	0.0726	0.0126	10.3422	30
31	.1643	13.9291	147.2864	6.0881	84.8017	0.0718	0.0118	10.5740	31
32	.1550	14.0840	152.0901	6.4534	90.8898	0.0710	0.0110	10.7988	32
33	.1462	14.2302	156.7681	6.8406	97.3432	0.0703	0.0103	11.0166	33
34	.1379	14.3681	161.3192	7.2510	104.1838	0.0696	0.0096	11.2276	34
35	.1301	14.4982	165.7427	7.6861	111.4348	0.0690	0.0090	11.4319	35
40	.0972	15.0463	185.9568	10.2857	154.7620	0.0665	0.0065	12.3590	40
50	.0543	15.7619	217.4574	18.4202	290.3359	0.0634	0.0034	13.7964	50
60	.0303	16.1614	239.0428	32.9877	533.1282	0.0619	0.0019	14.7909	60
70	.0169	16.3845	253.3271	59.0759	967.9322	0.0610	0.0010	15.4613	70
80	.0095	16.5091	262.5493	105.7960	1746.5999	0.0606	0.0006	15.9033	80
90	.0053	16.5787	268.3946	189.4645	3141.0752	0.0603	0.0003	16.1891	90
100	.0029	16.6175	272.0471	339.3021	5638.3681	0.0602	0.0002	16.3711	100

$$I = 7.00 \%$$

N	(P/F)	(P/A)	(P/G)	(F/P)	(F/A)	(A/P)	(A/F)	(A/G)	N
1	.9346	0.9346	-0.0000	1.0700	1.0000	1.0700	1.0000	-0.0000	1
2	.8734	1.8080	0.8734	1.1449	2.0700	0.5531	0.4831	0.4831	2
3	.8163	2.6243	2.5060	1.2250	3.2149	0.3811	0.3111	0.9549	3
4	.7629	3.3872	4.7947	1.3108	4.4399	0.2952	0.2252	1.4155	4
5	.7130	4.1002	7.6467	1.4026	5.7507	0.2439	0.1739	1.8650	5
6	.6663	4.7665	10.9784	1.5007	7.1533	0.2098	0.1398	2.3032	6
7	.6227	5.3893	14.7149	1.6058	8.6540	0.1856	0.1156	2.7304	7
8	.5820	5.9713	18.7889	1.7182	10.2598	0.1675	0.0975	3.1465	8
9	.5439	6.5152	23.1404	1.8385	11.9780	0.1535	0.0835	3.5517	9
10	.5083	7.0236	27.7156	1.9672	13.8164	0.1424	0.0724	3.9461	10
11	.4751	7.4987	32.4665	2.1049	15.7836	0.1334	0.0634	4.3296	11
12	.4440	7.9427	37.3506	2.2522	17.8885	0.1259	0.0559	4.7025	12
13	.4150	8.3577	42.3302	2.4098	20.1406	0.1197	0.0497	5.0648	13
14	.3878	8.7455	47.3718	2.5785	22.5505	0.1143	0.0443	5.4167	14
15	.3624	9.1079	52.4461	2.7590	25.1290	0.1098	0.0398	5.7583	15
16	.3387	9.4466	57.5271	2.9522	27.8881	0.1059	0.0359	6.0897	16
17	.3166	9.7632	62.5923	3.1588	30.8402	0.1024	0.0324	6.4110	17
18	.2959	10.0591	67.6219	3.3799	33.9990	0.0994	0.0294	6.7225	18
19	.2765	10.3356	72.5991	3.6165	37.3790	0.0968	0.0268	7.0242	19
20	.2584	10.5940	77.5091	3.8697	40.9955	0.0944	0.0244	7.3163	20
21	.2415	10.8355	82.3393	4.1406	44.8652	0.0923	0.0223	7.5990	21
22	.2257	11.0612	87.0793	4.4304	49.0057	0.0904	0.0204	7.8725	22
23	.2109	11.2722	91.7201	4.7405	53.4361	0.0887	0.0187	8.1369	23
24	.1971	11.4693	96.2545	5.0724	58.1767	0.0872	0.0172	8.3923	24
25	.1842	11.6536	100.6765	5.4274	63.2490	0.0858	0.0158	8.6391	25
26	.1722	11.8258	104.9814	5.8074	68.6765	0.0846	0.0146	8.8773	26
27	.1609	11.9867	109.1656	6.2139	74.4838	0.0834	0.0134	9.1072	27
28	.1504	12.1371	113.2264	6.6488	80.6977	0.0824	0.0124	9.3289	28
29	.1406	12.2777	117.1622	7.1143	87.3465	0.0814	0.0114	9.5427	29
30	.1314	12.4090	120.9718	7.6123	94.4608	0.0806	0.0106	9.7487	30
31	.1228	12.5318	124.6550	8.1451	102.0730	0.0798	0.0098	9.9471	31
32	.1147	12.6466	128.2120	8.7153	110.2182	0.0791	0.0091	10.1381	32
33	.1072	12.7538	131.6435	9.3253	118.9334	0.0784	0.0084	10.3219	33
34	.1002	12.8540	134.9507	9.9781	128.2588	0.0773	0.0078	10.4987	34
35	.0937	12.9477	138.1353	10.6766	138.2369	0.0772	0.0072	10.6687	35
40	.0668	13.3317	152.2928	14.9745	199.6351	0.0750	0.0050	11.4233	40
50	.0339	13.8007	172.9051	29.4570	406.5289	0.0725	0.0025	12.5287	50
60	.0173	14.0392	185.7677	57.9464	813.5204	0.0712	0.0012	13.2321	60
70	.0088	14.1604	193.5185	113.9894	1614.1342	0.0706	0.0006	13.6662	70
80	.0045	14.2220	198.0748	224.2344	3189.0627	0.0703	0.0003	13.9273	80
90	.0023	14.2533	200.7042	441.1030	6287.1854	0.0702	0.0002	14.0812	90
100	.0012	14.2693	202.2001	867.7163	12381.6618	0.0701	0.0001	14.1703	100

I = 8.00 %

N	(P/F)	(P/A)	(P/G)	(F/P)	(F/A)	(A/P)	(A/F)	(A/G)	N
1	.9259	0.9259	-0.0000	1.0800	1.0000	1.0800	1.0000	-0.0000	1
2	.8573	1.7833	0.8573	1.1664	2.0800	0.5608	0.4808	0.4808	2
3	.7938	2.5771	2.4450	1.2597	3.2464	0.3880	0.3080	0.9487	3
4	.7350	3.3121	4.6501	1.3605	4.5061	0.3019	0.2219	1.4040	4
5	.6806	3.9927	7.3724	1.4693	5.8666	0.2505	0.1705	1.8465	5
6	.6302	4.6229	10.5233	1.5869	7.3359	0.2163	0.1363	2.2763	6
7	.5835	5.2064	14.0242	1.7138	8.9228	0.1921	0.1121	2.6937	7
8	.5403	5.7466	17.8061	1.8509	10.6366	0.1740	0.0940	3.0985	8
9	.5002	6.2469	21.8081	1.9990	12.4876	0.1601	0.0801	3.4910	9
10	.4632	6.7101	25.9768	2.1589	14.4866	0.1490	0.0690	3.8713	10
11	.4289	7.1390	30.2657	2.3316	16.6455	0.1401	0.0601	4.2395	11
12	.3971	7.5361	34.6339	2.5182	18.9771	0.1327	0.0527	4.5957	12
13	.3677	7.9038	39.0463	2.7196	21.4953	0.1265	0.0465	4.9402	13
14	.3405	8.2442	43.4723	2.9372	24.2149	0.1213	0.0413	5.2731	14
15	.3152	8.5595	47.8857	3.1722	27.1521	0.1168	0.0368	5.5945	15
16	.2919	8.8514	52.2640	3.4259	30.3243	0.1130	0.0330	5.9046	16
17	.2703	9.1216	56.5883	3.7000	33.7502	0.1096	0.0296	6.2037	17
18	.2502	9.3719	60.8426	3.9960	37.4502	0.1067	0.0267	6.4920	18
19	.2317	9.6036	65.0134	4.3157	41.4463	0.1041	0.0241	6.7697	19
20	.2145	9.8181	69.0898	4.6610	45.7620	0.1019	0.0219	7.0369	20
21	.1987	10.0168	73.0629	5.0338	50.4229	0.0998	0.0198	7.2940	21
22	.1839	10.2007	76.9257	5.4365	55.4568	0.0980	0.0180	7.5412	22
23	.1703	10.3711	80.6726	5.8715	60.8933	0.0964	0.0164	7.7786	23
24	.1577	10.5288	84.2997	6.3412	66.7648	0.0950	0.0150	8.0066	24
25	.1460	10.6748	87.8041	6.8485	73.1059	0.0937	0.0137	8.2254	25
26	.1352	10.8100	91.1842	7.3964	79.9544	0.0925	0.0125	8.4352	26
27	.1252	10.9352	94.4390	7.9881	87.3508	0.0914	0.0114	8.6363	27
28	.1159	11.0511	97.5687	8.6271	95.3388	0.0905	0.0105	8.8289	28
29	.1073	11.1584	100.5738	9.3173	103.9659	0.0896	0.0096	9.0133	29
30	.0994	11.2578	103.4558	10.0627	113.2832	0.0888	0.0088	9.1897	30
31	.0920	11.3498	106.2163	10.8677	123.3459	0.0881	0.0081	9.3584	31
32	.0852	11.4350	108.8575	11.7371	134.2135	0.0875	0.0075	9.5197	32
33	.0789	11.5139	111.3819	12.6760	145.9506	0.0869	0.0069	9.6737	33
34	.0730	11.5869	113.7924	13.6901	158.6267	0.0863	0.0063	9.8208	34
35	.0676	11.6546	116.0920	14.7853	172.3168	0.0858	0.0058	9.9611	35
40	.0460	11.9246	126.0422	21.7245	259.0565	0.0839	0.0039	10.5699	40
50	.0213	12.2335	139.5928	46.9016	573.7702	0.0817	0.0017	11.4107	50
60	.0099	12.3766	147.3000	101.2571	1253.2133	0.0808	0.0008	11.9015	60
70	.0046	12.4428	151.5326	218.6064	2720.0801	0.0804	0.0004	12.1783	70
80	.0021	12.4735	153.8001	471.9548	5886.9354	0.0802	0.0002	12.3301	80
90	.0010	12.4877	154.9925	1018.9151	12723.9386	0.0801	0.0001	12.4116	90
100	.0005	12.4943	155.6107	2199.7613	27484.5157	0.0800	0.0000	12.4545	100

I = 9.00 %

N	(P/F)	(P/A)	(P/G)	(F/P)	(F/A)	(A/P)	(A/F)	(A/G)	N
1	.9174	0.9174	-0.0000	1.0900	1.0000	1.0900	1.0000	-0.0000	1
2	.8417	1.7591	0.8417	1.1881	2.0900	0.5685	0.4785	0.4785	2
3	.7722	2.5313	2.3860	1.2950	3.2781	0.3951	0.3051	0.9426	3
4	.7084	3.2397	4.5113	1.4116	4.5731	0.3087	0.2187	1.3925	4
5	.6499	3.8897	7.1110	1.5386	5.9847	0.2571	0.1671	1.8282	5
6	.5963	4.4859	10.0924	1.6771	7.5233	0.2229	0.1329	2.2498	6
7	.5470	5.0330	13.3746	1.8280	9.2004	0.1987	0.1087	2.6574	7
8	.5019	5.5348	16.8877	1.9926	11.0285	0.1807	0.0907	3.0512	8
9	.4604	5.9952	20.5711	2.1719	13.0210	0.1668	0.0768	3.4312	9
10	.4224	6.4177	24.3728	2.3674	15.1929	0.1558	0.0658	3.7978	10
11	.3875	6.8052	28.2481	2.5804	17.5603	0.1469	0.0569	4.1510	11
12	.3555	7.1607	32.1590	2.8127	20.1407	0.1397	0.0497	4.4910	12
13	.3262	7.4869	36.0731	3.0658	22.9534	0.1336	0.0436	4.8182	13
14	.2992	7.7862	39.9633	3.3417	26.0192	0.1284	0.0384	5.1326	14
15	.2745	8.0607	43.8069	3.6425	29.3609	0.1241	0.0341	5.4346	15
16	.2519	8.3126	47.5849	3.9703	33.0034	0.1203	0.0303	5.7245	16
17	.2311	8.5436	51.2821	4.3276	36.9737	0.1170	0.0270	6.0024	17
18	.2120	8.7556	54.8860	4.7171	41.3013	0.1142	0.0242	6.2687	18
19	.1945	8.9501	58.3868	5.1417	46.0185	0.1117	0.0217	6.5236	19
20	.1784	9.1285	61.7770	5.6044	51.1601	0.1095	0.0195	6.7674	20
21	.1637	9.2922	65.0509	6.1088	56.7645	0.1076	0.0176	7.0006	21
22	.1502	9.4424	68.2048	6.6586	62.8733	0.1059	0.0159	7.2232	22
23	.1378	9.5802	71.2359	7.2579	69.5319	0.1044	0.0144	7.4357	23
24	.1264	9.7066	74.1433	7.9111	76.7898	0.1030	0.0130	7.6384	24
25	.1160	9.8226	76.9265	8.6231	84.7009	0.1018	0.0118	7.8316	25
26	.1064	9.9290	79.5863	9.3992	93.3240	0.1007	0.0107	8.0156	26
27	.0976	10.0266	82.1241	10.2451	102.7231	0.0997	0.0097	8.1906	27
28	.0895	10.1161	84.5419	11.1671	112.9682	0.0989	0.0089	8.3571	28
29	.0822	10.1983	86.8422	12.1722	124.1354	0.0981	0.0081	8.5154	29
30	.0754	10.2737	89.0280	13.2677	136.3075	0.0973	0.0073	8.6657	30
31	.0691	10.3428	91.1024	14.4618	149.5752	0.0967	0.0067	8.8083	31
32	.0634	10.4062	93.0690	15.7633	164.0370	0.0961	0.0061	8.9436	32
33	.0582	10.4644	94.9314	17.1820	179.8003	0.0956	0.0056	9.0718	33
34	.0534	10.5178	96.6935	18.7284	196.9823	0.0951	0.0051	9.1933	34
35	.0490	10.5668	98.3590	20.4140	215.7108	0.0946	0.0046	9.3083	35
40	.0318	10.7574	105.3762	31.4094	337.8824	0.0930	0.0030	9.7957	40
50	.0134	10.9617	114.3251	74.3575	815.0836	0.0912	0.0012	10.4295	50
60	.0057	11.0480	118.9683	176.0313	1944.7921	0.0905	0.0005	10.7683	60
70	.0024	11.0844	121.2942	416.7301	4619.2232	0.0902	0.0002	10.9427	70
80	.0010	11.0998	122.4306	986.5517	10950.5741	0.0901	0.0001	11.0299	80
90	.0004	11.1064	122.9758	2335.5266	25939.1842	0.0900	0.0000	11.0726	90
100	.0002	11.1091	123.2335	5529.0408	61422.6755	0.0900	0.0000	11.0930	100

I = 10.00 %

N	(P/F)	(P/A)	(P/G)	(F/P)	(F/A)	(A/P)	(A/F)	(A/G)	N
1	.9091	0.9091	-0.0000	1.1000	1.0000	1.1000	1.0000	-0.0000	1
2	.8264	1.7355	0.8264	1.2100	2.1000	0.5762	0.4762	0.4762	2
3	.7513	2.4869	2.3291	1.3310	3.3100	0.4021	0.3021	0.9366	3
4	.6830	3.1699	4.3781	1.4641	4.6410	0.3155	0.2155	1.3812	4
5	.6209	3.7908	6.8618	1.6105	6.1051	0.2638	0.1638	1.8101	5
6	.5645	4.3553	9.6842	1.7716	7.7156	0.2296	0.1296	2.2236	6
7	.5132	4.8684	12.7631	1.9487	9.4872	0.2054	0.1054	2.6216	7
8	.4665	5.3349	16.0287	2.1436	11.4359	0.1874	0.0874	3.0045	8
9	.4241	5.7590	19.4215	2.3579	13.5795	0.1736	0.0736	3.3724	9
10	.3855	6.1446	22.8913	2.5937	15.9374	0.1627	0.0627	3.7255	10
11	.3505	6.4951	26.3963	2.8531	18.5312	0.1540	0.0540	4.0641	11
12	.3186	6.8137	29.9012	3.1384	21.3843	0.1468	0.0468	4.3884	12
13	.2897	7.1034	33.3772	3.4523	24.5227	0.1408	0.0408	4.6988	13
14	.2633	7.3667	36.8005	3.7975	27.9750	0.1357	0.0357	4.9955	14
15	.2394	7.6061	40.1520	4.1772	31.7725	0.1315	0.0315	5.2789	15
16	.2176	7.8237	43.4164	4.5950	35.9497	0.1278	0.0278	5.5493	16
17	.1978	8.0216	46.5819	5.0545	40.5447	0.1247	0.0247	5.8071	17
18	.1799	8.2014	49.6395	5.5599	45.5992	0.1219	0.0219	6.0526	18
19	.1635	8.3649	52.5827	6.1159	51.1591	0.1195	0.0195	6.2861	19
20	.1486	8.5136	55.4069	6.7275	57.2750	0.1175	0.0175	6.5081	20
21	.1351	8.6487	58.1095	7.4002	64.0025	0.1156	0.0156	6.7189	21
22	.1228	8.7715	60.6893	8.1403	71.4027	0.1140	0.0140	6.9189	22
23	.1117	8.8832	63.1462	8.9543	79.5430	0.1126	0.0126	7.1085	23
24	.1015	8.9847	65.4813	9.8497	88.4973	0.1113	0.0113	7.2881	24
25	.0923	9.0770	67.6964	10.8347	98.3471	0.1102	0.0102	7.4580	25
26	.0839	9.1609	69.7940	11.9182	109.1818	0.1092	0.0092	7.6186	26
27	.0763	9.2372	71.7773	13.1100	121.0999	0.1083	0.0083	7.7704	27
28	.0693	9.3066	73.6495	14.4210	134.2099	0.1075	0.0075	7.9137	28
29	.0630	9.3696	75.4146	15.8631	148.6309	0.1067	0.0067	8.0489	29
30	.0573	9.4269	77.0766	17.4494	164.4940	0.1061	0.0061	8.1762	30
31	.0521	9.4790	78.6395	19.1943	181.9434	0.1055	0.0055	8.2962	31
32	.0474	9.5264	80.1078	21.1138	201.1378	0.1050	0.0050	8.4091	32
33	.0431	9.5694	81.4856	23.2252	222.2515	0.1045	0.0045	8.5152	33
34	.0391	9.6086	82.7773	25.5477	245.4767	0.1041	0.0041	8.6149	34
35	.0356	9.6442	83.9872	28.1024	271.0244	0.1037	0.0037	8.7086	35
40	.0221	9.7791	88.9525	45.2593	442.5926	0.1023	0.0023	9.0962	40
50	.0085	9.9148	94.8889	117.3909	1163.9085	0.1009	0.0009	9.5704	50
60	.0033	9.9672	97.7010	304.4816	3034.8164	0.1003	0.0003	9.8023	60
70	.0013	9.9873	98.9870	789.7470	7887.4696	0.1001	0.0001	9.9113	70
80	.0005	9.9951	99.5606	2048.4002	20474.0021	0.1000	0.0000	9.9609	80
90	.0002	9.9981	99.8118	5313.0226	53120.2261	0.1000	0.0000	9.9831	90
100	.0001	9.9993	99.9202	13780.6123	137796.1234	0.1000	0.0000	9.9927	100

I = 12.00 %

N	(P/F)	(P/A)	(P/G)	(F/P)	(F/A)	(A/P)	(A/F)	(A/G)	N
1	.8929	0.8929	-0.0000	1.1200	1.0000	1.1200	1.0000	-0.0000	1
2	.7972	1.6901	0.7972	1.2544	2.1200	0.5917	0.4717	0.4717	2
3	.7118	2.4018	2.2208	1.4049	3.3744	0.4163	0.2963	0.9246	3
4	.6355	3.0373	4.1273	1.5735	4.7793	0.3292	0.2092	1.3589	4
5	.5674	3.6048	6.3970	1.7623	6.3528	0.2774	0.1574	1.7746	5
6	.5066	4.1114	8.9302	1.9738	8.1152	0.2432	0.1232	2.1720	6
7	.4523	4.5638	11.6443	2.2107	10.0890	0.2191	0.0991	2.5515	7
8	.4039	4.9676	14.4714	2.4760	12.2997	0.2013	0.0813	2.9131	8
9	.3606	5.3282	17.3563	2.7731	14.7757	0.1877	0.0677	3.2574	9
10	.3220	5.6502	20.2541	3.1058	17.5487	0.1770	0.0570	3.5847	10
11	.2875	5.9377	23.1288	3.4785	20.6546	0.1684	0.0484	3.8953	11
12	.2567	6.1944	25.9523	3.8960	24.1331	0.1614	0.0414	4.1897	12
13	.2292	6.4235	28.7024	4.3635	28.0291	0.1557	0.0357	4.4683	13
14	.2046	6.6282	31.3624	4.8871	32.3926	0.1509	0.0309	4.7317	14
15	.1827	6.8109	33.9202	5.4736	37.2797	0.1468	0.0268	4.9803	15
16	.1631	6.9740	36.3670	6.1304	42.7533	0.1434	0.0234	5.2147	16
17	.1456	7.1196	38.6973	6.8660	48.8837	0.1405	0.0205	5.4353	17
18	.1300	7.2497	40.9080	7.6900	55.7497	0.1379	0.0179	5.6427	18
19	.1161	7.3658	42.9979	8.6128	63.4397	0.1358	0.0158	5.8375	19
20	.1037	7.4694	44.9676	9.6463	72.0524	0.1339	0.0139	6.0202	20
21	.0926	7.5620	46.8188	10.8038	81.6987	0.1322	0.0122	6.1913	21
22	.0826	7.6446	48.5543	12.1003	92.5026	0.1308	0.0108	6.3514	22
23	.0738	7.7184	50.1776	13.5523	104.6029	0.1296	0.0096	6.5010	23
24	.0659	7.7843	51.6929	15.1786	118.1552	0.1285	0.0085	6.6406	24
25	.0588	7.8431	53.1046	17.0001	133.3339	0.1275	0.0075	6.7708	25
26	.0525	7.8957	54.4177	19.0401	150.3339	0.1267	0.0067	6.8921	26
27	.0469	7.9426	55.6369	21.3249	169.3740	0.1259	0.0059	7.0049	27
28	.0419	7.9844	56.7674	23.8839	190.6989	0.1252	0.0052	7.1098	28
29	.0374	8.0218	57.8141	26.7499	214.5828	0.1247	0.0047	7.2071	29
30	.0334	8.0552	58.7821	29.9599	241.3327	0.1241	0.0041	7.2974	30
31	.0298	8.0850	59.6761	33.5551	271.2926	0.1237	0.0037	7.3811	31
32	.0266	8.1116	60.5010	37.5817	304.8477	0.1233	0.0033	7.4586	32
33	.0238	8.1354	61.2612	42.0915	342.4294	0.1229	0.0029	7.5302	33
34	.0212	8.1566	61.9612	47.1425	384.5210	0.1226	0.0026	7.5965	34
35	.0189	8.1755	62.6052	52.7996	431.6635	0.1223	0.0023	7.6577	35
40	.0107	8.2438	65.1159	93.0510	767.0914	0.1213	0.0013	7.8988	40
50	.0035	8.3045	67.7624	289.0022	2400.0182	0.1204	0.0004	8.1597	50
60	.0011	8.3240	68.8100	897.5969	7471.6411	0.1201	0.0001	8.2664	60
70	.0004	8.3303	69.2103	2787.7998	23223.3319	0.1200	0.0000	8.3082	70
80	.0001	8.3324	69.3594	8658.4831	72145.6925	0.1200	0.0000	8.3241	80
90	.0000	8.3330	69.4140	26891.9342	224091.1185	0.1200	0.0000	8.3300	90
100	.0000	8.3332	69.4336	83522.2657	696010.5477	0.1200	0.0000	8.3321	100

I = 15.00 %

N	(P/F)	(P/A)	(P/G)	(F/P)	(F/A)	(A/P)	(A/F)	(A/G)	N
1	.8696	0.8696	-0.0000	1.1500	1.0000	1.1500	1.0000	-0.0000	1
2	.7561	1.6257	0.7561	1.3225	2.1500	0.6151	0.4651	0.4651	2
3	.6575	2.2832	2.0712	1.5209	3.4725	0.4380	0.2880	0.9071	3
4	.5718	2.8550	3.7864	1.7490	4.9934	0.3503	0.2003	1.3263	4
5	.4972	3.3522	5.7751	2.0114	6.7424	0.2983	0.1483	1.7228	5
6	.4323	3.7845	7.9368	2.3131	8.7537	0.2642	0.1142	2.0972	6
7	.3759	4.1604	10.1924	2.6600	11.0668	0.2404	0.0904	2.4498	7
8	.3269	4.4873	12.4807	3.0590	13.7268	0.2229	0.0729	2.7813	8
9	.2843	4.7716	14.7548	3.5179	16.7858	0.2096	0.0596	3.0922	9
10	.2472	5.0188	16.9795	4.0456	20.3037	0.1993	0.0493	3.3832	10
11	.2149	5.2337	19.1289	4.6524	24.3493	0.1911	0.0411	3.6549	11
12	.1869	5.4206	21.1849	5.3503	29.0017	0.1845	0.0345	3.9082	12
13	.1625	5.5831	23.1352	6.1528	34.3519	0.1791	0.0291	4.1438	13
14	.1413	5.7245	24.9725	7.0757	40.5047	0.1747	0.0247	4.3624	14
15	.1229	5.8474	26.6930	8.1371	47.5804	0.1710	0.0210	4.5650	15
16	.1069	5.9542	28.2960	9.3576	55.7175	0.1679	0.0179	4.7522	16
17	.0929	6.0472	29.7828	10.7613	65.0751	0.1654	0.0154	4.9251	17
18	.0808	6.1280	31.1565	12.3755	75.8364	0.1632	0.0132	5.0843	18
19	.0703	6.1982	32.4213	14.2318	88.2118	0.1613	0.0113	5.2307	19
20	.0611	6.2593	33.5822	16.3665	102.4436	0.1598	0.0098	5.3651	20
21	.0531	6.3125	34.6448	18.8215	118.8101	0.1584	0.0084	5.4883	21
22	.0462	6.3587	35.6150	21.6447	137.6316	0.1573	0.0073	5.6010	22
23	.0402	6.3988	36.4988	24.8915	159.2764	0.1563	0.0063	5.7040	23
24	.0349	6.4338	37.3023	28.6252	184.1678	0.1554	0.0054	5.7979	24
25	.0304	6.4641	38.0314	32.9190	212.7930	0.1547	0.0047	5.8834	25
26	.0264	6.4906	38.6918	37.8568	245.7120	0.1541	0.0041	5.9612	26
27	.0230	6.5135	39.2890	43.5353	283.5688	0.1535	0.0035	6.0319	27
28	.0200	6.5335	39.8283	50.0656	327.1041	0.1531	0.0031	6.0960	28
29	.0174	6.5509	40.3146	57.5755	377.1697	0.1527	0.0027	6.1541	29
30	.0151	6.5660	40.7526	66.2118	434.7451	0.1523	0.0023	6.2066	30
31	.0131	6.5791	41.1466	76.1435	500.9569	0.1520	0.0020	6.2541	31
32	.0114	6.5905	41.5006	87.5651	577.1005	0.1517	0.0017	6.2970	32
33	.0099	6.6005	41.8184	100.6998	664.6655	0.1515	0.0015	6.3357	33
34	.0086	6.6091	42.1033	115.8048	765.3654	0.1513	0.0013	6.3705	34
35	.0075	6.6166	42.3586	133.1755	881.1702	0.1511	0.0011	6.4019	35
40	.0037	6.6418	43.2830	267.8635	1779.0903	0.1506	0.0006	6.5168	40
50	.0009	6.6605	44.0958	1083.6574	7217.7163	0.1501	0.0001	6.6205	50
60	.0002	6.6651	44.3431	4383.9987	29219.9916	0.1500	0.0000	6.6530	60
70	.0001	6.6663	44.4156	17735.7200	118231.4669	0.1500	0.0000	6.6627	70
80	.0000	6.6666	44.4364	71750.8794	478332.5293	0.1500	0.0000	6.6656	80
90	.0000	6.6666	44.4422	290272.3252	1935142.1680	0.1500	0.0000	6.6664	90
100	.0000	6.6667	44.4438	1174313.4507	7828749.6713	0.1500	0.0000	6.6666	100

I = 20.00 %

N	(P/F)	(P/A)	(P/G)	(F/P)	(F/A)	(A/P)	(A/F)	(A/G)	N
1	.8333	0.8333	-0.0000	1.2000	1.0000	1.2000	1.0000	-0.0000	1
2	.6944	1.5278	0.6944	1.4400	2.2000	0.6545	0.4545	0.4545	2
3	.5787	2.1065	1.8519	1.7280	3.6400	0.4747	0.2747	0.8791	3
4	.4823	2.5887	3.2986	2.0736	5.3680	0.3863	0.1863	1.2742	4
5	.4019	2.9906	4.9061	2.4883	7.4416	0.3344	0.1344	1.6405	5
6	.3349	3.3255	6.5806	2.9860	9.9299	0.3007	0.1007	1.9788	6
7	.2791	3.6046	8.2551	3.5832	12.9159	0.2774	0.0774	2.2902	7
8	.2326	3.8372	9.8831	4.2998	16.4991	0.2606	0.0606	2.5756	8
9	.1938	4.0310	11.4335	5.1598	20.7989	0.2481	0.0481	2.8364	9
10	.1615	4.1925	12.8871	6.1917	25.9587	0.2385	0.0385	3.0739	10
11	.1346	4.3271	14.2330	7.4301	32.1504	0.2311	0.0311	3.2893	11
12	.1122	4.4392	15.4667	8.9161	39.5805	0.2253	0.0253	3.4841	12
13	.0935	4.5327	16.5883	10.6993	48.4966	0.2206	0.0206	3.6597	13
14	.0779	4.6106	17.6008	12.8392	59.1959	0.2169	0.0169	3.8175	14
15	.0649	4.6755	18.5095	15.4070	72.0351	0.2139	0.0139	3.9588	15
16	.0541	4.7296	19.3208	18.4884	87.4421	0.2114	0.0114	4.0851	16
17	.0451	4.7746	20.0419	22.1861	105.9306	0.2094	0.0094	4.1976	17
18	.0376	4.8122	20.6805	26.6233	128.1167	0.2078	0.0078	4.2975	18
19	.0313	4.8435	21.2439	31.9480	154.7400	0.2065	0.0065	4.3861	19
20	.0261	4.8696	21.7395	38.3376	186.6880	0.2054	0.0054	4.4643	20
21	.0217	4.8913	22.1742	46.0051	225.0256	0.2044	0.0044	4.5334	21
22	.0181	4.9094	22.5561	55.2061	271.0307	0.2037	0.0037	4.5941	22
23	.0151	4.9245	22.8867	66.2474	326.2369	0.2031	0.0031	4.6475	23
24	.0126	4.9371	23.1760	79.4968	392.4842	0.2025	0.0025	4.6943	24
25	.0105	4.9476	23.4276	95.3962	471.9811	0.2021	0.0021	4.7352	25
26	.0087	4.9563	23.6460	114.4755	567.3773	0.2018	0.0018	4.7709	26
27	.0073	4.9636	23.8353	137.3706	681.8528	0.2015	0.0015	4.8020	27
28	.0061	4.9697	23.9991	164.8447	819.2233	0.2012	0.0012	4.8291	28
29	.0051	4.9747	24.1406	197.8136	984.0680	0.2010	0.0010	4.8527	29
30	.0042	4.9789	24.2628	237.3763	1181.8816	0.2008	0.0008	4.8731	30
31	.0035	4.9824	24.3681	284.8516	1419.2579	0.2007	0.0007	4.8908	31
32	.0029	4.9854	24.4588	341.8219	1704.1095	0.2006	0.0006	4.9061	32
33	.0024	4.9878	24.5368	410.1863	2045.9314	0.2005	0.0005	4.9194	33
34	.0020	4.9898	24.6038	492.2235	2456.1176	0.2004	0.0004	4.9308	34
35	.0017	4.9915	24.6614	590.6682	2948.3411	0.2003	0.0003	4.9406	35
40	.0007	4.9966	24.8469	1469.7716	7343.8578	0.2001	0.0001	4.9728	40
50	.0001	4.9995	24.9698	9100.4382	45497.1908	0.2000	0.0000	4.9945	50
60	.0000	4.9999	24.9942	56347.5144	281732.5718	0.2000	0.0000	4.9989	60
70	.0000	5.0000	24.9989	348888.9569	1744439.7847	0.2000	0.0000	4.9998	70

I = 25.00 %

N	(P/F)	(P/A)	(P/G)	(F/P)	(F/A)	(A/P)	(A/F)	(A/G)	N
1	.8000	0.8000	0.0	1.2500	1.0000	1.2500	1.0000	0.0	1
2	.6400	1.4400	0.6400	1.5625	2.2500	0.6944	0.4444	0.4444	2
3	.5120	1.9520	1.6640	1.9531	3.8125	0.5123	0.2623	0.8525	3
4	.4096	2.3616	2.8928	2.4414	5.7656	0.4234	0.1734	1.2249	4
5	.3277	2.6893	4.2035	3.0518	8.2070	0.3718	0.1218	1.5631	5
6	.2621	2.9514	5.5142	3.8147	11.2588	0.3388	0.0888	1.8683	6
7	.2097	3.1611	6.7725	4.7684	15.0735	0.3163	0.0663	2.1424	7
8	.1678	3.3289	7.9469	5.9605	19.8419	0.3004	0.0504	2.3872	8
9	.1342	3.4631	9.0207	7.4506	25.8023	0.2888	0.0388	2.6048	9
10	.1074	3.5705	9.9870	9.3132	33.2529	0.2801	0.0301	2.7971	10
11	.0859	3.6564	10.8460	11.6415	42.5661	0.2735	0.0235	2.9663	11
12	.0687	3.7251	11.6020	14.5519	54.2077	0.2684	0.0184	3.1145	12
13	.0550	3.7801	12.2617	18.1899	68.7596	0.2645	0.0145	3.2437	13
14	.0440	3.8241	12.8334	22.7374	86.9495	0.2615	0.0115	3.3559	14
15	.0352	3.8593	13.3260	28.4217	109.6868	0.2591	0.0091	3.4530	15
16	.0281	3.8874	13.7482	35.5271	138.1085	0.2572	0.0072	3.5366	16
17	.0225	3.9099	14.1085	44.4089	173.6357	0.2558	0.0058	3.6084	17
18	.0180	3.9279	14.4147	55.5112	218.0446	0.2546	0.0046	3.6698	18
19	.0144	3.9424	14.6741	69.3889	273.5558	0.2537	0.0037	3.7222	19
20	.0115	3.9539	14.8932	86.7362	342.9447	0.2529	0.0029	3.7667	20
21	.0092	3.9631	15.0777	108.4202	429.6809	0.2523	0.0023	3.8045	21
22	.0074	3.9705	15.2326	135.5253	538.1011	0.2519	0.0019	3.8365	22
23	.0059	3.9764	15.3625	169.4066	673.6264	0.2515	0.0015	3.8634	23
24	.0047	3.9811	15.4711	211.7582	843.0329	0.2512	0.0012	3.8861	24
25	.0038	3.9849	15.5618	264.6978	1054.7912	0.2509	0.0009	3.9052	25
26	.0030	3.9879	15.6373	330.8722	1319.4890	0.2508	0.0008	3.9212	26
27	.0024	3.9903	15.7002	413.5903	1650.3612	0.2506	0.0006	3.9346	27
28	.0019	3.9923	15.7524	516.9879	2063.9515	0.2505	0.0005	3.9457	28
29	.0015	3.9938	15.7957	646.2349	2580.9394	0.2504	0.0004	3.9551	29
30	.0012	3.9950	15.8316	807.7936	3227.1743	0.2503	0.0003	3.9628	30
31	.0010	3.9960	15.8614	1009.7420	4034.9678	0.2502	0.0002	3.9693	31
32	.0008	3.9968	15.8859	1262.1774	5044.7098	0.2502	0.0002	3.9746	32
33	.0006	3.9975	15.9062	1577.7218	6306.8872	0.2502	0.0002	3.9791	33
34	.0005	3.9980	15.9229	1972.1523	7884.6091	0.2501	0.0001	3.9828	34
35	.0004	3.9984	15.9367	2465.1903	9856.7613	0.2501	0.0001	3.9858	35
40	.0001	3.9995	15.9766	7523.1638	30088.6554	0.2500	0.0000	3.9947	40
50	.0000	3.9999	15.9969	70064.9232	280255.6929	0.2500	0.0000	3.9993	50
60	.0000	4.0000	15.9996	652530.4468	2610117.7872	0.2500	0.0000	3.9999	60

I = 30.00 %

N	(P/F)	(P/A)	(P/G)	(F/P)	(F/A)	(A/P)	(A/F)	(A/G)	N
1	.7692	0.7692	-0.0000	1.3000	1.0000	1.3000	1.0000	-0.0000	1
2	.5917	1.3609	0.5917	1.6900	2.3000	0.7348	0.4348	0.4348	2
3	.4552	1.8161	1.5020	2.1970	3.9900	0.5506	0.2506	0.8271	3
4	.3501	2.1662	2.5524	2.8561	6.1870	0.4616	0.1616	1.1783	4
5	.2693	2.4356	3.6297	3.7129	9.0431	0.4106	0.1106	1.4903	5
6	.2072	2.6427	4.6656	4.8268	12.7560	0.3784	0.0784	1.7654	6
7	.1594	2.8021	5.6218	6.2749	17.5828	0.3569	0.0569	2.0063	7
8	.1226	2.9247	6.4800	8.1573	23.8577	0.3419	0.0419	2.2156	8
9	.0943	3.0190	7.2343	10.6045	32.0150	0.3312	0.0312	2.3963	9
10	.0725	3.0915	7.8872	13.7858	42.6195	0.3235	0.0235	2.5512	10
11	.0558	3.1473	8.4452	17.9216	56.4053	0.3177	0.0177	2.6833	11
12	.0429	3.1903	8.9173	23.2981	74.3270	0.3135	0.0135	2.7952	12
13	.0330	3.2233	9.3135	30.2875	97.6250	0.3102	0.0102	2.8895	13
14	.0254	3.2487	9.6437	39.3738	127.9125	0.3078	0.0078	2.9685	14
15	.0195	3.2682	9.9172	51.1859	167.2863	0.3060	0.0060	3.0344	15
16	.0150	3.2832	10.1426	66.5417	218.4722	0.3046	0.0046	3.0892	16
17	.0116	3.2948	10.3276	86.5042	285.0139	0.3035	0.0035	3.1345	17
18	.0089	3.3037	10.4788	112.4554	371.5180	0.3027	0.0027	3.1718	18
19	.0068	3.3105	10.6019	146.1920	483.9734	0.3021	0.0021	3.2025	19
20	.0053	3.3158	10.7019	190.0496	630.1655	0.3016	0.0016	3.2275	20
21	.0040	3.3198	10.7828	247.0645	820.2151	0.3012	0.0012	3.2480	21
22	.0031	3.3230	10.8482	321.1839	1067.2796	0.3009	0.0009	3.2646	22
23	.0024	3.3254	10.9009	417.5391	1388.4635	0.3007	0.0007	3.2781	23
24	.0018	3.3272	10.9433	542.8008	1806.0026	0.3006	0.0006	3.2890	24
25	.0014	3.3286	10.9773	705.6410	2348.8033	0.3004	0.0004	3.2979	25
26	.0011	3.3297	11.0045	917.3333	3054.4443	0.3003	0.0003	3.3050	26
27	.0008	3.3305	11.0263	1192.5333	3971.7776	0.3003	0.0003	3.3107	27
28	.0006	3.3312	11.0437	1550.2933	5164.3109	0.3002	0.0002	3.3153	28
29	.0005	3.3317	11.0576	2015.3813	6714.6042	0.3001	0.0001	3.3189	29
30	.0004	3.3321	11.0687	2619.9956	8729.9855	0.3001	0.0001	3.3219	30
31	.0003	3.3324	11.0775	3405.9943	11349.9811	0.3001	0.0001	3.3242	31
32	.0002	3.3326	11.0845	4427.7926	14755.9755	0.3001	0.0001	3.3261	32
33	.0002	3.3328	11.0901	5756.1304	19183.7681	0.3001	0.0001	3.3276	33
34	.0001	3.3329	11.0945	7482.9696	24939.8985	0.3000	0.0000	3.3288	34
35	.0001	3.3330	11.0980	9727.8604	32422.8681	0.3000	0.0000	3.3297	35
40	.0000	3.3332	11.1071	36118.8648	120392.8827	0.3000	0.0000	3.3322	40
50	.0000	3.3333	11.1108	497929.2230	1659760.7433	0.3000	0.0000	3.3332	50

I = 35.00 %

N	(P/F)	(P/A)	(P/G)	(F/P)	(F/A)	(A/P)	(A/F)	(A/G)	N
1	.7407	0.7407	-0.0000	1.3500	1.0000	1.3500	1.0000	-0.0000	1
2	.5487	1.2894	0.5487	1.8225	2.3500	0.7755	0.4255	0.4255	2
3	.4064	1.6959	1.3616	2.4604	4.1725	0.5897	0.2397	0.8029	3
4	.3011	1.9969	2.2648	3.3215	6.6329	0.5008	0.1508	1.1341	4
5	.2230	2.2200	3.1568	4.4840	9.9544	0.4505	0.1005	1.4220	5
6	.1652	2.3852	3.9828	6.0534	14.4384	0.4193	0.0693	1.6698	6
7	.1224	2.5075	4.7170	8.1722	20.4919	0.3988	0.0488	1.8811	7
8	.0906	2.5982	5.3515	11.0324	28.6640	0.3849	0.0349	2.0597	8
9	.0671	2.6653	5.8886	14.8937	39.6964	0.3752	0.0252	2.2094	9
10	.0497	2.7150	6.3363	20.1066	54.5902	0.3683	0.0183	2.3338	10
11	.0368	2.7519	6.7047	27.1439	74.6967	0.3634	0.0134	2.4364	11
12	.0273	2.7792	7.0049	36.6442	101.8406	0.3598	0.0098	2.5205	12
13	.0202	2.7994	7.2474	49.4697	138.4848	0.3572	0.0072	2.5889	13
14	.0150	2.8144	7.4421	66.7841	187.9544	0.3553	0.0053	2.6443	14
15	.0111	2.8255	7.5974	90.1585	254.7385	0.3539	0.0039	2.6889	15
16	.0082	2.8337	7.7206	121.7139	344.8970	0.3529	0.0029	2.7246	16
17	.0061	2.8398	7.8180	164.3138	466.6109	0.3521	0.0021	2.7530	17
18	.0045	2.8443	7.8946	221.8236	630.9247	0.3516	0.0016	2.7756	18
19	.0033	2.8476	7.9547	299.4619	852.7483	0.3512	0.0012	2.7935	19
20	.0025	2.8501	8.0017	404.2736	1152.2103	0.3509	0.0009	2.8075	20
21	.0018	2.8519	8.0384	545.7693	1556.4838	0.3506	0.0006	2.8186	21
22	.0014	2.8533	8.0669	736.7886	2102.2532	0.3505	0.0005	2.8272	22
23	.0010	2.8543	8.0890	994.6646	2839.0418	0.3504	0.0004	2.8340	23
24	.0007	2.8550	8.1061	1342.7973	3833.7064	0.3503	0.0003	2.8393	24
25	.0006	2.8556	8.1194	1812.7763	5176.5037	0.3502	0.0002	2.8433	25
26	.0004	2.8560	8.1296	2447.2480	6989.2800	0.3501	0.0001	2.8465	26
27	.0003	2.8563	8.1374	3303.7848	9436.5280	0.3501	0.0001	2.8490	27
28	.0002	2.8565	8.1435	4460.1095	12740.3128	0.3501	0.0001	2.8509	28
29	.0002	2.8567	8.1481	6021.1478	17200.4222	0.3501	0.0001	2.8523	29
30	.0001	2.8568	8.1517	8128.5495	23221.5700	0.3500	0.0000	2.8535	30
31	.0001	2.8569	8.1545	10973.5418	31350.1195	0.3500	0.0000	2.8543	31
32	.0001	2.8569	8.1565	14814.2815	42323.6613	0.3500	0.0000	2.8550	32
33	.0001	2.8570	8.1581	19999.2800	57137.9428	0.3500	0.0000	2.8555	33
34	.0000	2.8570	8.1594	26999.0280	77137.2228	0.3500	0.0000	2.8559	34
35	.0000	2.8571	8.1603	36448.6878	104136.2508	0.3500	0.0000	2.8562	35
40	.0000	2.8571	8.1625	163437.1347	466960.3848	0.3500	0.0000	2.8569	40
50	.0000	2.8571	8.1632	3286157.8795	9389019.6556	0.3500	0.0000	2.8571	50

I = 40.00 %

N	(P/F)	(P/A)	(P/G)	(F/P)	(F/A)	(A/P)	(A/F)	(A/G)	N
1	.7143	0.7143	-0.0000	1.4000	1.0000	1.4000	1.0000	-0.0000	1
2	.5102	1.2245	0.5102	1.9600	2.4000	0.8167	0.4167	0.4167	2
3	.3644	1.5889	1.2391	2.7440	4.3600	0.6294	0.2294	0.7798	3
4	.2603	1.8492	2.0200	3.8416	7.1040	0.5408	0.1408	1.0923	4
5	.1859	2.0352	2.7637	5.3782	10.9456	0.4914	0.0914	1.3580	5
6	.1328	2.1680	3.4278	7.5295	16.3238	0.4613	0.0613	1.5811	6
7	.0949	2.2628	3.9970	10.5414	23.8534	0.4419	0.0419	1.7664	7
8	.0678	2.3306	4.4713	14.7579	34.3947	0.4291	0.0291	1.9185	8
9	.0484	2.3790	4.8585	20.6610	49.1526	0.4203	0.0203	2.0422	9
10	.0346	2.4136	5.1696	28.9255	69.8137	0.4143	0.0143	2.1419	10
11	.0247	2.4383	5.4166	40.4957	98.7391	0.4101	0.0101	2.2215	11
12	.0176	2.4559	5.6106	56.6939	139.2348	0.4072	0.0072	2.2845	12
13	.0126	2.4685	5.7618	79.3715	195.9287	0.4051	0.0051	2.3341	13
14	.0090	2.4775	5.8788	111.1201	275.3002	0.4036	0.0036	2.3729	14
15	.0064	2.4839	5.9688	155.5681	386.4202	0.4026	0.0026	2.4030	15
16	.0046	2.4885	6.0376	217.7953	541.9883	0.4018	0.0018	2.4262	16
17	.0033	2.4918	6.0901	304.9135	759.7837	0.4013	0.0013	2.4441	17
18	.0023	2.4941	6.1299	426.8789	1064.6971	0.4009	0.0009	2.4577	18
19	.0017	2.4958	6.1601	597.6304	1491.5760	0.4007	0.0007	2.4682	19
20	.0012	2.4970	6.1828	836.6826	2089.2064	0.4005	0.0005	2.4761	20
21	.0009	2.4979	6.1998	1171.3556	2925.8889	0.4003	0.0003	2.4821	21
22	.0006	2.4985	6.2127	1639.8978	4097.2445	0.4002	0.0002	2.4866	22
23	.0004	2.4989	6.2222	2295.8569	5737.1423	0.4002	0.0002	2.4900	23
24	.0003	2.4992	6.2294	3214.1997	8032.9993	0.4001	0.0001	2.4925	24
25	.0002	2.4994	6.2347	4499.8796	11247.1990	0.4001	0.0001	2.4944	25
26	.0002	2.4996	6.2387	6299.8314	15747.0785	0.4001	0.0001	2.4959	26
27	.0001	2.4997	6.2416	8819.7640	22046.9099	0.4000	0.0000	2.4969	27
28	.0001	2.4998	6.2438	12347.6696	30866.6739	0.4000	0.0000	2.4977	28
29	.0001	2.4999	6.2454	17286.7374	43214.3435	0.4000	0.0000	2.4983	29
30	.0000	2.4999	6.2466	24201.4324	60501.0809	0.4000	0.0000	2.4988	30
31	.0000	2.4999	6.2475	33882.0053	84702.5132	0.4000	0.0000	2.4991	31
32	.0000	2.4999	6.2482	47434.8074	118584.5185	0.4000	0.0000	2.4993	32
33	.0000	2.5000	6.2487	66408.7304	166019.3260	0.4000	0.0000	2.4995	33
34	.0000	2.5000	6.2490	92972.2225	232428.0563	0.4000	0.0000	2.4996	34
35	.0000	2.5000	6.2493	130161.1116	325400.2789	0.4000	0.0000	2.4997	35
40	.0000	2.5000	6.2498	700037.6966	1750091.7415	0.4000	0.0000	2.4999	40

I = 45.00 %

N	(P/F)	(P/A)	(P/G)	(F/P)	(F/A)	(A/P)	(A/F)	(A/G)	N
1	.6897	0.6897	-0.0000	1.4500	1.0000	1.4500	1.0000	-0.0000	1
2	.4756	1.1653	0.4756	2.1025	2.4500	0.8582	0.4082	0.4082	2
3	.3280	1.4933	1.1317	3.0486	4.5525	0.6697	0.2197	0.7578	3
4	.2262	1.7195	1.8103	4.4205	7.6011	0.5816	0.1316	1.0528	4
5	.1560	1.8755	2.4344	6.4097	12.0216	0.5332	0.0832	1.2980	5
6	.1076	1.9831	2.9723	9.2941	18.4314	0.5043	0.0543	1.4988	6
7	.0742	2.0573	3.4176	13.4765	27.7255	0.4861	0.0361	1.6612	7
8	.0512	2.1085	3.7758	19.5409	41.2019	0.4743	0.0243	1.7907	8
9	.0353	2.1438	4.0581	28.3343	60.7428	0.4665	0.0165	1.8930	9
10	.0243	2.1681	4.2772	41.0847	89.0771	0.4612	0.0112	1.9728	10
11	.0168	2.1849	4.4450	59.5728	130.1618	0.4577	0.0077	2.0344	11
12	.0116	2.1965	4.5724	86.3806	189.7346	0.4553	0.0053	2.0817	12
13	.0080	2.2045	4.6682	125.2518	276.1151	0.4536	0.0036	2.1176	13
14	.0055	2.2100	4.7398	181.6151	401.3670	0.4525	0.0025	2.1447	14
15	.0038	2.2138	4.7929	263.3419	582.9821	0.4517	0.0017	2.1650	15
16	.0026	2.2164	4.8322	381.8458	846.3240	0.4512	0.0012	2.1802	16
17	.0018	2.2182	4.8611	553.6764	1228.1699	0.4508	0.0008	2.1915	17
18	.0012	2.2195	4.8823	802.8308	1781.8463	0.4506	0.0006	2.1998	18
19	.0009	2.2203	4.8978	1164.1047	2584.6771	0.4504	0.0004	2.2059	19
20	.0006	2.2209	4.9090	1687.9518	3748.7818	0.4503	0.0003	2.2104	20
21	.0004	2.2213	4.9172	2447.5301	5436.7336	0.4502	0.0002	2.2136	21
22	.0003	2.2216	4.9231	3548.9187	7884.2638	0.4501	0.0001	2.2160	22
23	.0002	2.2218	4.9274	5145.9321	11433.1824	0.4501	0.0001	2.2178	23
24	.0001	2.2219	4.9305	7461.6015	16579.1145	0.4501	0.0001	2.2190	24
25	.0001	2.2220	4.9327	10819.3222	24040.7161	0.4500	0.0000	2.2199	25
26	.0001	2.2221	4.9343	15688.0172	34860.0383	0.4500	0.0000	2.2206	26
27	.0000	2.2221	4.9354	22747.6250	50548.0556	0.4500	0.0000	2.2210	27
28	.0000	2.2222	4.9362	32984.0563	73295.6806	0.4500	0.0000	2.2214	28
29	.0000	2.2222	4.9368	47826.8816	106279.7368	0.4500	0.0000	2.2216	29
30	.0000	2.2222	4.9372	69348.9783	154106.6184	0.4500	0.0000	2.2218	30
31	.0000	2.2222	4.9375	100556.0185	223455.5967	0.4500	0.0000	2.2219	31
32	.0000	2.2222	4.9378	145806.2269	324011.6152	0.4500	0.0000	2.2220	32
33	.0000	2.2222	4.9379	211419.0289	469817.8421	0.4500	0.0000	2.2221	33
34	.0000	2.2222	4.9380	306557.5920	681236.8710	0.4500	0.0000	2.2221	34
35	.0000	2.2222	4.9381	444508.5083	987794.4630	0.4500	0.0000	2.2221	35

I = 50.00 %

N	(P/F)	(P/A)	(P/G)	(F/P)	(F/A)	(A/P)	(A/F)	(A/G)	N
1	.6667	0.6667	0.0	1.5000	1.0000	1.5000	1.0000	0.0	1
2	.4444	1.1111	0.4444	2.2500	2.5000	0.9000	0.4000	0.4000	2
3	.2963	1.4074	1.0370	3.3750	4.7500	0.7105	0.2105	0.7368	3
4	.1975	1.6049	1.6296	5.0625	8.1250	0.6231	0.1231	1.0154	4
5	.1317	1.7366	2.1564	7.5938	13.1875	0.5758	0.0758	1.2417	5
6	.0878	1.8244	2.5953	11.3906	20.7813	0.5481	0.0481	1.4226	6
7	.0585	1.8829	2.9465	17.0859	32.1719	0.5311	0.0311	1.5648	7
8	.0390	1.9220	3.2196	25.6289	49.2578	0.5203	0.0203	1.6752	8
9	.0260	1.9480	3.4277	38.4434	74.8867	0.5134	0.0134	1.7596	9
10	.0173	1.9653	3.5838	57.6650	113.3301	0.5088	0.0088	1.8235	10
11	.0116	1.9769	3.6994	86.4976	170.9951	0.5058	0.0058	1.8713	11
12	.0077	1.9846	3.7842	129.7463	257.4927	0.5039	0.0039	1.9068	12
13	.0051	1.9897	3.8459	194.6195	387.2390	0.5026	0.0026	1.9329	13
14	.0034	1.9931	3.8904	291.9293	581.8585	0.5017	0.0017	1.9519	14
15	.0023	1.9954	3.9224	437.8939	873.7878	0.5011	0.0011	1.9657	15
16	.0015	1.9970	3.9452	656.8408	1311.6817	0.5008	0.0008	1.9756	16
17	.0010	1.9980	3.9614	985.2613	1968.5225	0.5005	0.0005	1.9827	17
18	.0007	1.9986	3.9729	1477.8919	2953.7838	0.5003	0.0003	1.9878	18
19	.0005	1.9991	3.9811	2216.8378	4431.6756	0.5002	0.0002	1.9914	19
20	.0003	1.9994	3.9868	3325.2567	6648.5135	0.5002	0.0002	1.9940	20
21	.0002	1.9996	3.9908	4987.8851	9973.7702	0.5001	0.0001	1.9958	21
22	.0001	1.9997	3.9936	7481.8276	14961.6553	0.5001	0.0001	1.9971	22
23	.0001	1.9998	3.9955	11222.7415	22443.4829	0.5000	0.0000	1.9980	23
24	.0001	1.9999	3.9969	16834.1122	33666.2244	0.5000	0.0000	1.9986	24
25	.0000	1.9999	3.9979	25251.1683	50500.3366	0.5000	0.0000	1.9990	25
26	.0000	1.9999	3.9985	37876.7524	75751.5049	0.5000	0.0000	1.9993	26
27	.0000	2.0000	3.9990	56815.1287	113628.2573	0.5000	0.0000	1.9995	27
28	.0000	2.0000	3.9993	85222.6930	170443.3860	0.5000	0.0000	1.9997	28
29	.0000	2.0000	3.9995	127834.0395	255666.0790	0.5000	0.0000	1.9998	29
30	.0000	2.0000	3.9997	191751.0592	383500.1185	0.5000	0.0000	1.9998	30
31	.0000	2.0000	3.9998	287626.5888	575251.1777	0.5000	0.0000	1.9999	31
32	.0000	2.0000	3.9998	431439.8833	862877.7665	0.5000	0.0000	1.9999	32
33	.0000	2.0000	3.9999	647159.8249	1294317.6498	0.5000	0.0000	1.9999	33
34	.0000	2.0000	3.9999	970739.7374	1941477.4747	0.5000	0.0000	2.0000	34
35	.0000	2.0000	3.9999	1456109.6060	2912217.2121	0.5000	0.0000	2.0000	35

Practice Problems: ENGINEERING ECONOMY

Required

1. A structure costing $10,000 has the operating costs and salvage values given below. (a) What is the economic life of the structure? (b) Assuming that the structure has been owned and operated for 4 years, what is the cost of owning the structure for exactly one more year? Use 20%.

Year 1: maintenance $2000, salvage $8000
Year 2: maintenance $3000, salvage $7000
Year 3: maintenance $4000, salvage $6000
Year 4: maintenance $5000, salvage $5000
Year 5: maintenance $6000, salvage $4000

2. A man purchases a car for $5000 for personal use of 15,000 miles a year. It costs him $200 per year for insurance and $150 per year for maintenance. He gets 15 mpg and gasoline costs $.60 per gallon. The resale value after 5 years is $1000. Because of unexpected business driving (5000 miles per year extra), his insurance is increased to $300 per year and maintenance to $200. Salvage is reduced to $500. Use 10% to answer the following questions. (a) The man's company offers $.10 per mile reimbursement. Is that adequate? (b) How many miles must be driven per year at $.10 per mile to justify the company buying a car for its use? The cost would be $5000, but insurance, maintenance, and salvage would be $250, $200, and $800 respectively.

3. A shredder installed at the entrance to a sewage treatment plant can remove 7 lb/hr of debris from the incoming flow. The economic life of this shredder is 20 years. Any debris left in the flow will cause $25,000 damage to the wet-well pumps. Several investments are available to increase the capacity of the shredder. At 10%, what should be done?

debris rate (lb/hr)	probability of exceeding debris rate	required investment to meet debris rate
7	.15 per year	no cost
8	.10 per year	$15,000
9	.07 per year	$20,000
10	.03 per year	$30,000

Optional

4. A new machine will cost $17,000 and will have a value of $14,000 in 5 years. Special tooling will cost $5,000 and it will have a resale value of $2500 after 5 years. Maintenance will be $200 per year. What will be the average cost of ownership during the next 5 years if interest is at 6%?

5. An old highway bridge may be strengthened at a cost of $9000 or it may be replaced for $40,000. The present salvage value of the old bridge is $13,000. It is estimated that the reinforced bridge will last for 20 years with an annual cost of $500 and will have a salvage value of $10,000 at the end of 20 years. The estimated salvage of the new bridge after 25 years is $15,000. The maintenance for the new bridge will be $100 annually. Which is the best alternative at 8%?

6. A firm expects to receive $32,000 each year for 15 years from the sale of a product. It will require an initial investment of $150,000. Expenses will run $7530 per year. Salvage is zero and straight-line depreciation is used. The tax rate is 48%. What is the after-tax rate of return?

7. A public works project has initial costs of $1,000,000, benefits of $1,5000,000, and disbenefits of $300,000. (a) What is the benefit/cost ratio? (b) What is the excess of benefits over costs?

8. An apartment complex is purchased for $500,000. What is the depreciation in each of the first 3 years if the salvage value is $100,000 in 25 years? Use (a) straight line, (b) sum-of-the-year's digits, and (c) double declining balance depreciations.

9. Equipment is purchased for $12,000 which is expected to be sold after 10 years for $2000. The estimated maintenance is $1000 the first year, but is expected to increase $200 each year thereafter. Using 10%, find the present worth and the annual cost.

10. One of 5 grades of pipe with average lives (in years) and costs (in dollars) of (9,1500), (14,1600), (30,1750), (52, 1900), and (86,2100) is to be chosen for a 20-year project. A failure of the pipe at any time during the project will result in a cost equal to 35% of the original cost. Annual costs are 4% of the initial cost, and the pipes are not recoverable. At 6%, which pipe is superior? HINT: The lives are average lives, not absolute replacement times. Assume a rectangular failure distribution.

NOTES

NOTES

Hydraulics

Nomenclature

a	speed of sound in an elastic pipe	ft/sec
A	area	ft^2
c	speed of sound in a pure medium	ft/sec
C	compressibility, or Hazen-Williams constant	ft^2/lb, -
d	diameter, or depth	inches, ft
D	diameter	ft
e	pipe wall thickness	inches
E	modulus of elasticity or bulk modulus	lb/ft^2
f	Darcy friction factor	-
F	factor, or force	-, lb
g	acceleration due to gravity (32.2)	ft/sec^2
h	height, or head	inches, ft
I	moment of inertia	ft^4
K	fitting loss coefficient	-
L	length	ft
$\dot{m}$	mass flow rate	slug/sec
n	Manning roughness constant, or friction exponent	-,-
N_{Re}	Reynolds number	-
p	pressure, or perimeter	psi or psf, ft
P	power	ft-lb/sec
Q	flow quantity	ft^3/sec
r_H	hydraulic radius	ft
R	resultant force	lb
S	slope	-
SG	specific gravity	-
t	time	seconds
v	velocity	ft/sec
V	volume	ft^3
$\dot{w}$	weight flow	lb/sec
z	height above datum	ft

Symbols

υ	specific volume	ft^3/lb
ρ	density	lb/ft^3
μ	absolute viscosity	lb-sec/ft^2
τ	shear stress	psf
ν	kinematic viscosity	ft^2/sec
ϵ	specific roughness	ft
δ	Hardy Cross correction	ft^3/sec or gpm
β	ratio of small diameter to large diameter	-

Subscripts

a	atmospheric
A	added by a pump
c	centroidal or contraction
d	discharge
e	equivalent
E	extracted by a turbine, or English
f	friction, or flow
i	inside
m	manometer fluid
M	metric
n	normal
o	outside, static, or orifice
p	pressure
R	resultant
s	stagnation
t	tank, or true
v	velocity
va	velocity of approach
w	wetted

CONVERSIONS

(see table 3.2 for viscosity conversions)

Multiply	By	To Obtain
cubic feet	7.4805	gallons
cfs	448.83	gpm
cfs	.64632	MGD
gallons	.1337	cubic feet
gpm	.002228	cfs
inches of mercury	.491	psi
inches of mercury	70.7	psf
inches of mercury	13.60	inches of water
inches of water	5.199	psf
inches of water	.0361	psi
inches of water	.0735	inches of mercury
psi	144	psf
psi	2.308	feet of water
psi	27.7	inches of water
psi	2.037	inches of mercury
psf	.006944	psi
feet of water	62.4	psf

1. Types of Fluids

Fluids are generally divided into two categories: ideal and real. Ideal fluids are those which have zero viscosity and shearing forces, are incompressible, and have uniform velocity distributions when flowing.

Real fluids are divided into Newtonian and non-Newtonian fluids. Newtonian fluids are typified by gases, thin liquids, and most fluids having a simple chemical formula. Non-Newtonian fluids are typified by gels, emulsions, and suspensions. Both Newtonian and non-Newtonian fluids exhibit finite viscosities and non-uniform distributions. However, Newtonian fluids exhibit viscosities which are independent of the rate of change of shear stress while non-Newtonian fluids exhibit viscosities dependent on the rate of change of shear stress.

Most fluid problems assume Newtonian fluid characteristics.

2. Fluid Properties

A. Density

Mass density and weight density are often confused. In the English system, mass density is measured in slugs/ft^3 and is mentioned for reference only. In this chapter, the density term ρ refers to weight density as measured in lb/ft^3. In the problems in this chapter, fluids are assumed to be incompressible.

B. Specific Volume

Specific volume is the volume occupied by a pound of fluid. It is the reciprocal of the weight density.

$$\upsilon = 1/\rho$$

3.1

C. Specific Gravity

Specific gravity is the ratio of a liquid's density to the density of pure water at 4°C and one atmosphere.

$$S.G. = \rho/62.4$$

3.2

D. Viscosity

Viscosity of a fluid is a measure of its resistance to flow. In liquids, this resistance to flow decreases as the liquid temperature increases. Viscosity may be measured in either absolute (dynamic) or kinematic systems of units.

Absolute viscosity is a measure of the ratio of the shear stresses between layers in laminar flow to the velocity gradient within the fluid.

$$\mu = \frac{\tau}{dv/dy}$$

3.3

Kinematic viscosity is the ratio of absolute viscosity to the fluid mass density.

$$\nu = \mu g/\rho$$

3.4

Common units of absolute and kinematic viscosities in the English and metric systems are given in Table 3.1

TABLE 3.1
Types of Viscosities

	Absolute (μ)	Kinematic (ν)
English	lb-sec/ft^2 (slug/ft-sec)	ft^2/sec
Metric	dyne-sec/cm^2 (poise)	cm^2/sec (stoke)

Conversions between the two types of viscosities and between the English and metric units may be done using Table 3.2.

TABLE 3.2
Viscosity Conversions

To Obtain	Multiply	By	and Divide By
ft^2/sec	lb-sec/ft^2	32.2	weight density
ft^2/sec	stokes	1.076 EE-3	1
lb-sec/ft^2	ft^2/sec	density	32.2
lb-sec/ft	poise	1	478.8
pascal-sec	lb-sec/ft^2	47.88	1
pascal-sec	poise	.1	1
poise	lb-sec/ft^2	478.8	1
poise	stokes	specific gravity	1
stokes	ft^2/sec	929	1
stokes	poise	1	specific gravity

Example 3.1

Water at 60°F has a specific gravity of .999 and a kinematic viscosity of 1.12 centistokes. What is the absolute viscosity in lb-sec/ft^2?

$$\nu_M = 1.12/100 = .0112 \text{ stokes}$$

$$\mu_M = (.0112)(.999) = .01119 \text{ poise}$$

$$\mu_E = .01119/478.8 = 2.34 \text{ EE-5 lb-sec/ft}^2$$

◆

E. Vapor Pressure

Vapor pressure is the pressure of a saturated vapor. If the ambient pressure is lower than the vapor pressure, the fluid will boil and vaporize. At pressures greater than the vapor pressure, the vapor will condense. Vapor pressure is a function solely of temperature.

F. Compressibility

Usually fluids are considered to be incompressible. However, fluids are actually somewhat compressible. Compressibility is the percentage change in a unit volume under a unit change in pressure.

$$C = \frac{\Delta V / V}{\Delta p}$$

3.5

G. Bulk Modulus

The reciprocal of compressibility is the bulk modulus. The bulk modulus is required to determine the sonic velocity in a fluid.

$$E = 1/C$$

3.6

$$c = \sqrt{Eg/\rho}$$

3.7

Example 3.2

Determine the velocity of sound in 150°F water.

From Table 3.6, $E = 328 \text{ EE3 psi}$ and $\rho = 61.2 \text{ lb/ft}^3$

From equation 3.7, $c = \sqrt{\dfrac{(328 \text{ EE3})(144)(32.2)}{61.2}} = 4985 \text{ ft/sec}$

3. Fluid Flow Parameters

A. Pressure and Head

Fluid pressure is usually measured as force per unit area. Another common practice is to measure pressure as a height of fluid column ('head'). Pressure is converted to a height of a fluid column by using equation 3.8.

$$h_p = p/\rho$$

3.8

Velocity pressure may also be expressed in terms of head:

$$h_v = \frac{v^2}{2g}$$

3.9

Gravitational or potential head, z, is the energy of a fluid due to elevation above a given datum.

The sum of the preceding three head terms is the total unit energy of a fluid stream. The total head, (h), is defined by equation 3.10.

$$h = \frac{p}{\rho} + \frac{v^2}{2g} + z$$

3.10

B. Reynolds Number

The Reynolds number is the ratio of the inertial flow forces to the viscous forces within the fluid. Two expressions for Reynolds number are used, one requiring absolute viscosity, the other kinematic viscosity:

$$N_{Re} = \frac{D_e v \rho}{\mu g} \qquad \qquad 3.11$$

$$= \frac{D_e v}{\nu} \qquad \qquad 3.12$$

C. Equivalent Diameter

In equations 3.11 and 3.12, the equivalent diameter (D_e) is used. For a round pipe, equivalent diameter is the inside diameter. The equivalent diameters for conduits with other shapes are given in Table 3.3.

TABLE 3.3

Conduit Cross Section	D_e
Flowing Full	
annulus	$D_o - D_i$
square	L
rectangle	$2L_1 L_2 / (L_1 + L_2)$
Flowing Partially Full	
half-filled circle	D
rectangle (h deep, L wide)	$4hL/(L+2h)$
wide, shallow stream (h deep)	$4h$
triangle (h deep, L broad, s side)	hL/s
trapezoid (h deep, a wide at top, b wide at bottom, s side)	$2h(a+b)/(b+2s)$

Example 3.3

Determine the equivalent diameter of the open trapezoidal channel shown below:

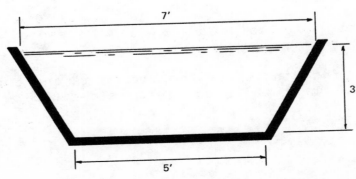

$$s = \sqrt{3^2 + 1^2} = 3.16 \text{ feet}$$

$$D_e = \frac{2(3)(7 + 5)}{5 + 2(3.16)} = 6.36 \text{ feet}$$

◆

D. Hydraulic Radius

An alternative expression related to equivalent diameter is the hydraulic radius. (Equation 3.14 is valid for all flow cross sections.)

$$r_H = \frac{\text{flow area}}{\text{wetted perimeter}} \qquad\qquad 3.13$$

$$= D_e/4 \qquad\qquad 3.14$$

Example 3.4

What is the hydraulic radius of the trapezoidal channel described in example 3.3?

From equation 3.13,

$$r_H = \frac{(5)(3) + (3)(1)}{3.16 + 5 + 3.16} = 1.59 \text{ feet}$$

Using the results of the previous example and equation 3.14,

$$r_H = 6.36/4 = 1.59 \text{ feet}$$

◆

4. Measuring Pressures

Pressure values are dependent on the reference pressure chosen. Two common references are zero absolute pressure and standard atmospheric pressure.

If standard atmospheric pressure (approximately 14.7 psia) is chosen as the reference, pressures are known as gage pressures. Positive gage pressures are always pressures above atmospheric pressure. Vacuum (negative gage pressure) is pressure below atmospheric pressure. According to this convention, maximum vacuum is -14.7 psig. The term 'gage' is somewhat misleading, since a mechanical gage may be used to indicate both gage and absolute pressures.

If zero absolute pressure is chosen as the reference, the pressures are known as absolute pressures. The barometer is one device for measuring the absolute pressure of the atmosphere. It is constructed by filling a long, hollow-bore tube open at one end with mercury and inverting it such that the open end is below the level of a mercury-filled container. If the vapor pressure is neglected, the mercury will be supported only by the atmospheric pressure transmitted through the container fluid at the lower, open end. The equation balancing the weight of the fluid against the atmospheric force is:

$$p_a = (.491)(h) \quad \text{(in psi)} \qquad 3.15$$

where h is the height of the mercury column in inches and .491 is the density of mercury in pounds per cubic inch.

Any fluid may be used to measure atmospheric pressure, although a correction for vapor pressure may be needed. For any fluid used in a barometer,

$$p_a = (.0361)(S.G.)(h) + p_v \quad \text{(in psi)} \qquad 3.16$$

where .0361 is the density of water in pounds per cubic inch.

Example 3.5

A vacuum pump is used to remove 70°F water from a flooded mine shaft. The pump is incapable of lifting the water beyond 400 inches. What is the atmospheric pressure?

From Table 3.6, the vapor pressure of 70°F water is

$$(.84)(12)(.0361) = .364 \text{ psi}$$

Then, the atmospheric pressure is

$$p_a = (.0361)(1)(400) + .364 = 14.8 \text{ psia}$$

◆

Manometers are frequently used to measure pressure differentials. Figure 3.1 shows a simple U-tube manometer whose ends are connected to two pressure vessels. Often, one end will be open to the atmosphere, which then defines that end's pressure.

Figure 3.1

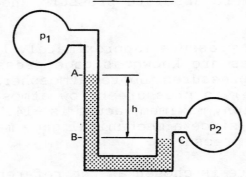

Since the pressure at point B is the same as at point C, the pressure differential produces the fluid column of height h.

$$\Delta p = p_2 - p_1 = \rho_m h \qquad 3.17$$

Equation 3.17 assumes that the manometer is small and that only low-density gases fill the tubes above the measuring fluid. If a high-density fluid (such as water) is present above the measuring fluid, or if the gas columns h_1 or h_2 are very long, corrections must be included, as is done in equation 3.18.

$$\Delta p = \rho_m h + \rho_1 h_1 - \rho_2 h_2 \qquad\qquad 3.18$$

Figure 3.2

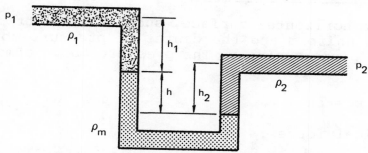

Equations 3.15 through 3.18 assume a standard gravity of 32.2 ft/sec^2. If the gravity is non-standard, it will be necessary to multiply the standard fluid density by the quantity (gravity/32.2).

Example 3.6

What is the pressure at the bottom of the tank?

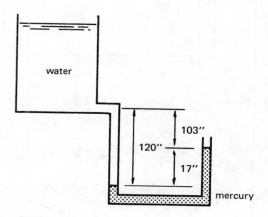

Using equation 3.18 and taking 4.34 EE-5 pounds per cubic inch as the density of air, the pressure differential is

$$\Delta p = p_{\text{tank bottom}} - p_a$$

$$= (.491)(17) - (.0361)(120) + (4.34 \text{ EE-5})(103)$$

$$= 8.347 - 4.332 + .004$$

$$= 4.019 \text{ psig}$$

Example 3.6 shows that corrections for short gas columns may be omitted.

5. Hydrostatic Pressure Due to Incompressible Fluids

Hydrostatic pressure is that pressure which a fluid exerts on an object or container walls. It always acts through the center of pressure, C.P., and is normal to the exposed surface regardless of the object's orientation or shape. It varies linearly with depth and is a function of depth and density only. There are four basic hydrostatic configurations.

A. Horizontal Plane Surface

In the case of a horizontal surface, such as the bottom of a container, the pressure is uniform and the center of pressure corresponds to the centroid of the plane surface. The pressure above atmospheric per unit area is

$$p = \rho h \qquad\qquad 3.19$$

The total vertical force is

$$R = pA \qquad\qquad 3.20$$

Figure 3.3

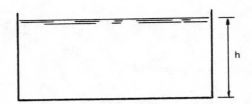

B. Vertical/Inclined Rectangular Plane Surface

Figure 3.4

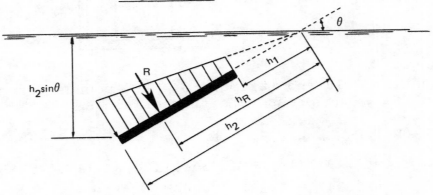

The linearity of the variation in pressure with depth is maintained. The pressures at the top and bottom of the plate are

$$P_1 = \rho h_1 \sin\theta \qquad\qquad 3.21$$

$$P_2 = \rho h_2 \sin\theta \qquad\qquad 3.22$$

The average pressure occurs at the average depth $(1/2)(h_1 + h_2)\sin\theta$. The average pressure over the entire vertical or inclined surface is

$$\bar{p} = \frac{1}{2}\rho(h_1 + h_2)\sin\theta \qquad\qquad 3.23$$

The total resultant force is

$$R = \bar{p}A \qquad\qquad 3.24$$

The center of pressure is not located at the average depth but is located at the centroid of the triangular or trapezoidal pressure distribution. That depth is

$$\frac{2}{3}\left[h_1 + h_2 - \frac{h_1 h_2}{(h_1 + h_2)}\right] \qquad\qquad 3.25$$

If the object is inclined, both h_C and h_R must be measured parallel to the object's surface (e.g., an inclined length).

Example 3.7

The tank shown below is filled with water. What is the force on a one-foot length of the inclined portion of the wall? Where is the resultant located on the inclined section?

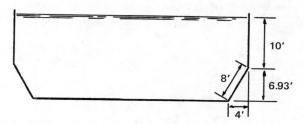

The average pressure on the inclined section is

$$\bar{p} = \frac{1}{2}(62.4)(10 + 16.93) = 840.2 \text{ psf}$$

The total force is $R = (840.2)(8)(1) = 6721.6 \text{ lbf}$

The resultant acts at

$$h_R = \frac{2}{3}\left[10 + 16.93 - \frac{(10)(16.93)}{(10+16.93)}\right] = 13.76 \text{ ft (vertical)}$$

$$h_R = \frac{13.76}{\sin 60°} = 15.89 \text{ ft (inclined)}$$

C. General Plane Surface

For any non-rectangular plane surface, the average pressure depends on the location of the surface's centroid.

$$\bar{p} = \rho h_c \tag{3.26}$$

$$R = \bar{p}A \tag{3.27}$$

<u>Figure 3.5</u>

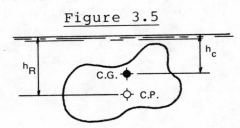

The resultant is normal to the surface, acting at depth h_R.

$$h_R = h_c + (I_c/Ah_c) \tag{3.28}$$

I_c is the moment of inertia about an axis parallel to the surface through the area's centroid. As with the previous case, h_c and h_R must be measured parallel to the area's surface.

Example 3.8

What is the force on a one-foot diameter circular sight-hole whose top edge is located 4' below the water surface? Where does the resultant act?

$$h_c = 4.5 \text{ ft}$$

$$A = \frac{\pi}{4}(1)^2 = .7854 \text{ ft}^2$$

$$I_c = \frac{1}{4}\pi r^4 = .049 \text{ ft}^4$$

$$\bar{p} = (62.4)(4.5) = 280.8 \text{ psf}$$

$$R = (280.8)(.7854) = 220.5 \text{ lb}$$

$$h_R = 4.5 + \frac{.049}{(.7854(4.5)} = 4.514 \text{ ft}$$

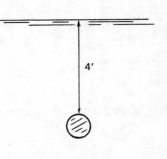

D. Curved Surfaces

The horizontal component of the resultant force acting on a curved surface can be found by the same method used for a plane surface.

The vertical component of force on an area will usually equal the weight of the liquid above it. In figure 3.6, the vertical force on length AB is the weight of area ABCD, with a line of action passing through the centroid of the area ABCD.

The resultant magnitude and direction may be found from conventional component composition.

Figure 3.6

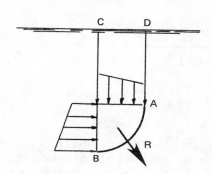

Example 3.9

What is the total force on a one-foot section of the wall described in example 3.7?

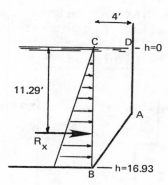

The average depth is $\frac{1}{2}(0 + 16.93) = 8.465$

The average pressure and horizontal component are

$$\bar{p} = 62.4(8.465) = 528.2 \text{ psf}$$

$$R_x = (16.93)(1)(528.2) = 8942.4 \text{ lb}$$

The horizontal component acts $(\frac{2}{3})(16.93) = 11.29$ ft from the top.

The volume of a one foot section of area ABCD is

$$(1)[(4)(10) + \frac{1}{2}(4)(6.93)] = 53.86 \ ft^3$$

Therefore, the vertical component is

$$R_y = (62.4)(53.86) = 3360.9 \ lb$$

The centroid of section ABCD (with point B serving as the reference) is

$$\bar{x} = \frac{\Sigma A_i \bar{x}_i}{\Sigma A_i} = \frac{(4)(10)(2) + (1/2)(4)(6.93)(1/3)(4)}{40 + 13.86} = 1.83$$

The resultant of R_x and R_y is

$$R = \sqrt{(8942.4)^2 + (3360.9)^2} = 9553.1 \ lb$$

$$\phi = \arctan(\frac{3360.9}{8942.4}) = 20.6°$$

◆

In general, it is not correct to calculate the vertical component of force on a submerged surface as being the weight of the fluid above it, as was done in example 3.9. This procedure is valid only when there is no change in the cross section of the tank area.

The 'hydrostatic paradox' is illustrated by figure 3.7. The pressure anywhere on the bottom of either container is the same. This pressure is dependent only on the maximum height of the fluid, not the volume.

Figure 3.7

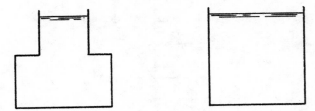

6. Buoyancy

The buoyancy theorem, also known as Archimedes' principle, states that the upward force on an immersed object is equal to the weight of the displaced fluid. A buoyant force due to displaced air is also relevant in the case of partially-submerged objects.

In the case of stationary floating or submerged objects, the buoyant force and weight are equal. If the forces are not in equilibrium, the object will rise or fall until an equilibrium is reached. The object will sink until it is supported by the bottom or until the density of the supporting fluid increases sufficiently. It will rise until the weight of the displaced fluid is reduced, either by a decrease in the fluid density or by breaking the surface.

7. Flow of Fluids in Pipes

A. Laws of Conservation

Many fluid flow problems can be solved using the principles of conservation of mass and energy.

When applied to fluid flow, the principle of mass conservation is known as the continuity equation:

$$\rho_1 A_1 v_1 = \rho_2 A_2 v_2 \qquad 3.29$$

$$\dot{w}_1 = \dot{w}_2 \qquad 3.30$$

If the fluid is incompressible,

$$A_1 v_1 = A_2 v_2 \qquad 3.31$$

$$Q_1 = Q_2 \qquad 3.32$$

When applied to fluid flow, the principle of energy conservation is known as the Bernoulli equation, equation 3.33.

$$\left(\frac{p_1}{\rho} + \frac{v_1^2}{2g} + z_1\right) + h_A = \left(\frac{p_2}{\rho} + \frac{v_2^2}{2g} + z_2\right) + h_E + h_f \qquad 3.33$$

Example 3.10

Water is pumped at a rate of 3 cfs through the piping system illustrated. If the pump has a discharge pressure of 150 psig, to what elevation can the tank be raised if the depth of the water in the tank is 10 feet? Assume the head loss due to friction is 10 feet.

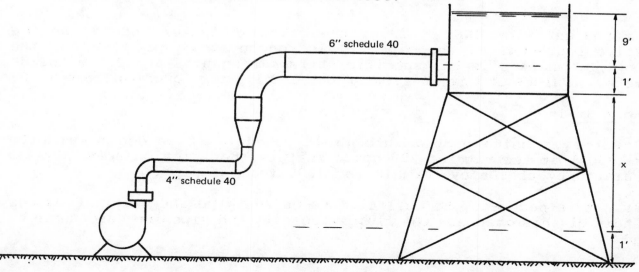

$$h_{p,1} = \frac{(150)(144)}{62.4} = 346.15 \text{ ft}$$

$$v_1 = 3/.0884 = 33.94 \text{ fps}$$

$$h_{v,1} = \frac{(33.94)^2}{(2)(32.2)} = 17.89 \text{ ft}$$

$$z_1 = 0$$

$$h_{p,2} = 0 \text{ (at free surface)}$$

$$v_2 = 0 \text{ (at free surface)}$$

From equation 3.33,

$$346.15 + 17.89 = z_2 + 10$$

$$z_2 = 354.0 \text{ ft}$$

The tank bottom may be raised to (354.0 - 10 + 1) = 345 feet above the ground.

◆

B. Fluid Friction in Pipeline Flow

The most common expression for calculating head loss due to friction (h_f) is the Darcy formula:

$$h_f = \frac{fL_e v^2}{2Dg} \qquad\qquad 3.34$$

The Moody friction factor chart (figure 3.28) is probably the most convenient method of determining the friction factor, f.

The basic parameter required to use the Moody friction factor chart is the Reynolds number. If the Reynolds number is less than 2000, the flow is laminar and the friction factor is given by equation 3.35.

$$f = 64/N_{Re} \qquad\qquad 3.35$$

For turbulent flow ($N_{Re} > 2000$) the friction factor depends on the relative roughness of the pipe. This roughness is expressed by the ratio ε/D, where ε is the specific surface roughness and D the inside diameter. Values of ε for various types of pipe are found in Table 3.10.

Example 3.11

50°F water is being pumped through 4" schedule 40 welded steel pipe ($\varepsilon = .0002$) at the rate of 300 gpm. What is the friction head loss as calculated by the Darcy formula for 1000 feet of pipe?

First, it is necessary to collect data on the pipe and water. From the appendix of this chapter, the fluid viscosity and pipe dimensions can be found.

kinematic viscosity = 1.41 EE-5 ft^2/sec

inside diameter = .3355 ft

flow area = .0884 ft^2

The flow quantity is

$$(300)(.002228) = .6684 \text{ cfs}$$

The velocity is
$$v = \frac{Q}{A} = \frac{.6684}{.0884} = 7.56 \text{ fps}$$

The Reynolds number is

$$N_{Re} = \frac{(.3355)(7.56)}{1.41 \text{ EE-5}} = 1.8 \text{ EE5}$$

$$\varepsilon/D = .0002/.3355 = .0006$$

From the Moody friction factor chart, f = .0195

From equation 3.34,

$$h_f = \frac{(.0195)(1000)(7.56)^2}{(2)(.3355)(32.2)} = 51.6 \text{ ft}$$

◆

Another method for finding the friction head loss is the Hazen-Williams formula. The Hazen-Williams formula gives good results for liquids which have kinematic viscosities around 1.2 EE-5 ft^2/sec (which corresponds to 60°F water). At extremely high and low temperatures, the Hazen-Williams formula can be as much as 20% in error for water. Of course, the Hazen-Williams formula should only be used for water in turbulent flow.

The Hazen-William formula for flow velocity as a function of slope is

$$v = (1.318)C(r_H)^{.63}(S)^{.54} \qquad \qquad 3.36a$$

$$= \frac{(.551)(C)(D)^{.63}}{(L)^{.54}}(h_f)^{.54} \qquad \qquad 3.36b$$

The friction head loss is

$$h_f = \frac{(3.012)(v)^{1.85}L}{(C)^{1.85}(D_i)^{1.165}} \qquad \qquad 3.37$$

Or, in terms of other units,

$$h_f = (10.44)(L)\frac{(gpm)^{1.85}}{(C)^{1.85}(d_{inches})^{4.8655}} \qquad \qquad 3.38$$

Use of these formulas requires a knowledge of the Hazen-Williams coefficient, C, which is assumed independent of the Reynolds number. Table 3.10 gives values of C for various types of pipe.

Example 3.12

Repeat example 3.11 using the Hazen-Williams formula. Assume C = 100.

Using equation 3.37,

$$h_f = \frac{(3.012)(7.56)^{1.85}(1000)}{(100)^{1.85}(.3355)^{1.165}} = 90.5 \text{ ft}$$

Using equation 3.38,

$$h_f = (10.44)(1000)\frac{(300)^{1.85}}{(100)^{1.85}(4.026)^{4.8655}} = 90.9 \text{ ft}$$

◆

Another set of equations that has seen widespread use, particularly with concrete irrigation pipe, was developed by Fred Scobey. Although equations for the friction loss in wood stave and riveted steel pipes were also developed, only the equation for concrete pipe is listed here.

$$h_f = \frac{Lv^2}{22,330(C_s)^2(D)^{1.25}}$$

3.39

C_s is the Scobey concrete pipe coefficient which has the following values:

.267 for cement pipes with mortar squeeze inside at pipe joints

.310 for modern machine-made concrete pipe made on rough or wooden forms

.345 for modern machine-made concrete pipe with smooth surface, such as made on steel forms

.370 for glazed interior pipes or other pipes with very smooth surfaces

C. Minor Losses

In addition to the head loss caused by friction between the fluid and the pipe wall, losses are also caused by obstructions in the line, changes in direction, and changes in flow area. Two methods are used to determine these losses: the method of equivalent length and the method of loss coefficients.

By use of Table 3.7, each fitting and valve can be converted into an equivalent length of straight pipe. The sum of these equivalent lengths is added to the actual pipeline length and substituted into the Darcy equation as L_e.

$$h_f = \frac{fL_e v^2}{2Dg}$$

3.40

Example 3.13

Using Table 3.7, determine the equivalent length of the piping network shown below.

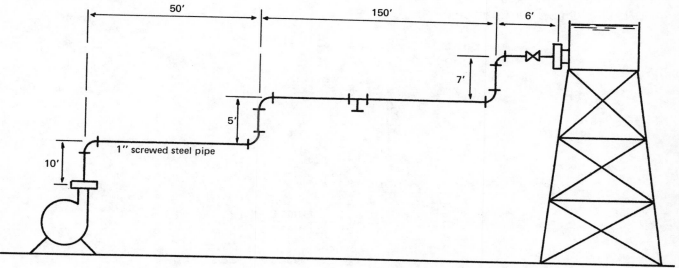

1" screwed steel pipe

Line consists of:

1 gate valve		.84
5 90° standard elbows		5.2 x 5
1 tee run		3.2
straight pipe		228
	◆	258 feet

In order to calculate a head loss due to turbulence, a loss coefficient can be determined which, when multiplied by the velocity head, will give the pressure drop in feet of head. (This method must be used to find exit losses.)

$$h_f = K\frac{v^2}{2g} \qquad\qquad 3.41$$

Values of K are widely tabulated, but they can also be calculated from the following formulas:

> Valves and Fittings: Refer to the manufacturer's data, or calculate from the equivalent length.
>
> $$K = fL_e/D \qquad\qquad 3.42$$

Sudden Enlargements:

$$K = [1 - (D_1/D_2)^2]^2 \qquad (D_1 < D_2) \qquad\qquad 3.43$$

Sudden Contractions:

$$K = \frac{1}{2}[1 - (D_1/D_2)^2] \qquad (D_1 > D_2) \qquad\qquad 3.44$$

Pipe Exit: (projecting exit, sharp edged, and rounded)

K = 1.0

Pipe Entrance:

Reentrant - .78
Sharp edged - .5
Rounded
(r/D) = .02, K = .28
          .04   .24
          .06   .15
          .10   .09
          .15   .04

Tapered Diameter Changes:

Given that $\beta = \dfrac{\text{small diameter}}{\text{large diameter}}$

ϕ = wall-to-horizontal angle

	Gradual, $\phi < 22°$	Sudden, $\phi > 22°$	
Enlargement	$2.6(\sin\phi)(1-\beta^2)^2$	$(1-\beta^2)^2$	3.45
Contraction	$.8(\sin\phi)(1-\beta^2)$	$\frac{1}{2}(1-\beta^2)\sqrt{\sin\phi}$	3.46

Example 3.14

A 6" schedule 40 steel pipe is 500 feet long. It enlarges suddenly to 12" schedule 40 steel pipe and continues for 200 feet more. What is the change in static pressure between the two ends of the pipe if the entire line is installed on the level? 300 cfm of 70°F water are flowing.

The pipe areas are: $A_{6"} = .2006$ ft^2

$$A_{12"} = .7773 \text{ ft}^2$$

The velocities are: $v_{6"} = \dfrac{300}{(.2006)(60)} = 24.93$ fps

$$v_{12"} = \dfrac{300}{(.7773)(60)} = 6.43 \text{ fps}$$

The kinematic viscosity of 70° water is 1.059 EE-5 ft^2/sec. So, the Reynolds numbers are

$$N_{Re,6"} = \frac{(24.93)(.5054)}{1.059 \text{ EE-5}} = 1.2 \text{ EE6}$$

$$N_{Re,12"} = \frac{(6.43)(.9948)}{1.059 \text{ EE-5}} = 6 \text{ EE5}$$

For the 6" pipe,

$$\varepsilon = .0002$$

$$\varepsilon/D = .0002/.5054 = .0004$$

$$f = .0164$$

$$h_f = \frac{(.0164)(500)(24.93)^2}{(2)(.5054)(32.2)} = 156.6$$

Similarly, for the 12" pipe,

$$\varepsilon = .0002$$

$$\varepsilon/D = .0002$$

$$f = .015$$

$$h_f = \frac{(0.15)(200)(6.43)^2}{(2)(\frac{11.938}{12})(32.2)} = 1.9 \text{ ft}$$

The enlargement loss from equation 3.43 is

$$h_e = K_e h_v = [1-(\frac{.5054}{.9948})^2]^2 \frac{(24.93)^2}{(2)(32.2)} = 5.3$$

Therefore, the friction loss (not counting any entrance or exit losses) is

$$156.6 + 1.9 + 5.3 = 163.8 \text{ ft}$$

However, the problem asked for the change in static pressure, not the friction loss. Since the velocity decreased at the sudden enlargement, Bernoulli's conservation of energy equation requires that the static head increase an amount equal to the decrease in velocity head. This is known as a static regain.

$$h_{regain} = \frac{(v_2^2 - v_1^2)}{2g} = \frac{(6.43)^2 - (24.92)^2}{(2)(32.2)} = -9.0 \text{ ft}$$

The total head change is 163.8 - 9 = 154.8 feet.

The static pressure change is

$$\Delta p = h\rho = \frac{(154.8)(62.4)}{144} = 67.1 \text{ psig}$$

◆

D. Power in Fluid Streams

Bernoulli's equation can also be used to calculate the power available in a fluid stream by multiplying the total energy by the mass flow rate. This is typically called the 'water horsepower.' (Also see Chapter 4.)

$$P = \dot{w}(\frac{p}{\rho} + \frac{v^2}{2g} + z) \tag{3.47}$$

$$\dot{w} = \rho A v \tag{3.48}$$

$$whp = P/550 \tag{3.49}$$

E. Special Considerations for Sewage and Sludge

Sewage and sludge are mechanical mixtures of water and solids. As such, particles in the mixture tend to settle out while the mixture is in motion. The critical velocity of a water-solid mixture is the velocity below which particles start to settle out. As they settle out, the particles form a sliding bed on the pipe bottom which eventually clogs the pipe.

The critical velocity is difficult to predict. For small particles under 50 microns in size, 3 to 7 ft/sec is the minimum range. A rough guide for larger particles (above 150 microns) is the Durand formula:

$$v_{critical} = 14\sqrt{D} \qquad\qquad 3.50$$

The actual velocity in digested sludge mains is typically in the 3 to 5 ft/sec range. Sewer mains are generally self-cleaning if a minimum velocity of 2 ft/sec is maintained. 1.5 ft/sec may even be used in sewer mains if the pipe is occasionally flushed out by peak flow.

8. Discharge from Tanks

A. General Discussion

Flow from a tank discharging liquid to the atmosphere through an opening in the tank wall (figure 3.8) is affected by both the area and shape of the opening. Correction factors for both velocity and flow rate are given in Table 3.4.

At the orifice, the total head of the fluid is converted into kinetic energy according to equation 3.51.

$$v_o = C_v\sqrt{2gh} \qquad\qquad 3.51$$

C_v is the coefficient of velocity which can be calculated from the coefficients of discharge and contraction.

$$C_v = C_d/C_c \qquad\qquad 3.52$$

The discharge is given by Toricelli's equation.

$$Q_o = (C_c A_o)v_o = C_c A_o C_v\sqrt{2gh} = C_d A_o\sqrt{2gh} \qquad\qquad 3.53$$

The head loss due to turbulence at the orifice is

$$h_f = \left(\frac{1}{C_v^2} - 1\right)\frac{v_o^2}{2g} \qquad\qquad 3.54$$

Figure 3.8

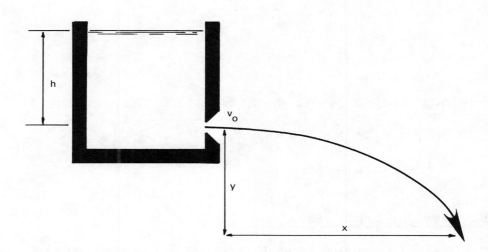

Table 3.4

Orifice Coefficients for Water

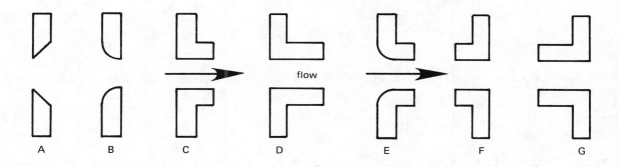

illustration	description	C_d	C_c	C_v
A	sharp-edged	.62	.63	.98
B	round-edged	.98	1.00	.98
C	short tube (fluid separates from walls)	.61	1.00	.61
D	short tube (no separation)	.82	1.00	.82
E	short tube with rounded entrance	.97	.99	.98
F	reentrant tube, length less than one-half of pipe diameter	.54	.55	.99
G	reentrant tube, length 2 to 3 pipe diameters	.72	1.00	.72
not shown	smooth, well-tapered nozzle	.98	.99	.99

The stream coordinates are

$$x = v_o t = v_o \sqrt{2y/g} = 2C_v \sqrt{hy}$$

3.55

$$y = \frac{gt^2}{2}$$

3.56

Fluid velocity at a point downstream of the orifice is

$$v_x = v_o$$

3.57

$$v_y = gt$$

3.58

B. Time to Empty Tank

If the liquid in a tank is not constantly being replenished, the static head forcing discharge through the orifice will decrease. For a tank with a constant cross-sectional area, the time required to lower the fluid level from level h_1 to h_2 is calculated from equation 3.59.

$$t = \frac{2A_t (\sqrt{h_1} - \sqrt{h_2})}{C_d A_o \sqrt{2g}}$$

3.59

If the tank has a varying cross-section, the following basic relationship holds:

$$Q dt = -A_t dh$$

3.60

An expression for the tank area, A_t, as a function of h must be determined. Then, the time to empty the tank from height h_1 to h_2 is

$$t = \int_{h_1}^{h_2} \frac{A_t \, dh}{C_d A_o \sqrt{2gh}}$$

3.61

For a tank being fed at a rate Q_{in}, which is less than the discharge through the orifice, the time to empty expression is

$$t = \int_{h_1}^{h_2} \frac{A_t \, dh}{(C_d A_o \sqrt{2gh}) - Q_{in}}$$

When a tank is being fed at a rate greater than the discharge, the above expression will become positive indicating a rising head. t will then be the time to raise the fluid level from h_1 to h_2.

Example 3.15

A tank constructed as a square-base truncated pyramid is discharging through a 2" re-entrant orifice as illustrated. What is the initial discharge rate? How long will it take to completely drain the tank?

$$A_o = \frac{\pi}{4}(\frac{2}{12})^2 = .0218 \text{ ft}^2$$

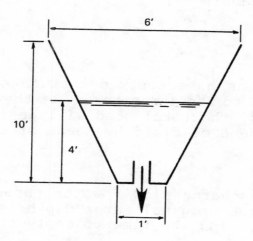

Using table 3.4 and equation 3.53,

$$Q = .72(.0218)\sqrt{2(32.2)(4)} = .252 \text{ cfs}$$

The expression for A_t in terms of h is

$$A_t = 4(\frac{2 + h}{4})^2$$

$$t = \int_0^4 \frac{\frac{1}{4}(h^2 + 4h + 4)}{(.72)(.0218)\sqrt{2(32.2)} \sqrt{h}}$$

$$= 1.985[\frac{2}{5}h^{5/2} + \frac{8}{3}h^{3/2} + 8h^{.5}]_0^4$$

$$= 99.6 \text{ seconds}$$

◆

C. Time to Establish Flow

Consider a tank filled to depth h whose discharge line is controlled by a valve. When this valve is opened, the fluid velocity will increase gradually until .it reaches a maximum given by the Bernoulli equation.

The time required for the flow to reach any intermediate velocity (v_i) up to its maximum (v) is

$$t = \frac{L_e v}{2gh} \ln\left(\frac{v + v_i}{v - v_i}\right)$$
3.63

The velocity at any time after the valve has opened is

$$v_i = v\left[\tanh\left(\frac{ght}{vL_e}\right)\right]$$
3.64

D. Pressurized Tanks

The preceding discussions have assumed that tanks discharging liquid have been open or vented to the atmosphere. If the fluid is discharging from a pressurized tank, the total head will be increased by the gage pressure converted to head of fluid by means of equation 3.19.

Example 3.16

A 15' diameter tank discharges 150°F water through a sharp edged 1" diameter orifice. If the original water depth is 12' and the tank is pressurized to 50 psig, find the time to empty the tank.

At 150°F, $\rho = 61.20$ lb/ft^3

For the orifice,

$$A_o = .00545 \text{ ft}^2, \quad C_d = .62$$

$$h_1 = 12 + \frac{(50)(144)}{61.2} = 129.65 \text{ ft}$$

$$h_2 = \frac{(50)(144)}{61.20} = 117.65 \text{ ft}$$

From equation 3.59,

$$t = \frac{2[\pi(7.5)^2](\sqrt{129.65} - \sqrt{117.65})}{(.62)(.00545)\sqrt{(2)(32.2)}} = 7035 \text{ seconds}$$

◆

9. Special Flow Problems

A. Culvert Design and Analysis

A culvert is a pathway that is used to drain a feature which collects runoff. Because of the numerous variables involved, no single formula can be given. Thus, culvert design is an empirical trial and error procedure.

A culvert can operate with any combination of the following exit and entrance conditions:

entrance	exit
submerged	submerged
partially exposed	free outfall

case 1: Submerged Entrance and Submerged Exit

The culvert will always flow full, and the discharge will be independent of the culvert barrel slope. For culverts less than 50 feet long, the friction loss can be ignored. Discharge can be calculated using the Bernoulli equation evaluated at the entrance and exit. The resulting equation is the familiar Torricelli equation:

$$Q = C_d A \sqrt{2gh} \qquad\qquad 3.65$$

In the above equation, h is the difference in surface levels of the headwater and the tailwater.

If the culvert length is greater than 50 feet or if the entrance is not smooth, the available energy will be divided between friction and velocity head. The effective head to be used in equation 3.65 is:

$$h' = h - h_{entrance} - h_f \qquad\qquad 3.66$$

The entrance head loss is calculated using loss coefficients:

$$h_{entrance} = K_e \left(\frac{v^2}{2g}\right) \qquad\qquad 3.67$$

Typical values of K_e are:

    .08 for a smooth and tapered entrance
    .10 for a flush concrete groove or bell design
    .15 for a projecting concrete groove or bell design
    .50 for a flush square-edged entrance
    .90 for a projecting square-edged entrance

The friction loss, h_f, can be found in the usual manner, either from the Darcy equation and Moody friction factor chart or from the Hazen-Williams equation. The Manning equation may also be used. The Manning equation is particularly useful since it eliminates the need for trial and error solutions.

If the Manning equation is used, the available energy is divided between velocity head, entrance loss, and friction according to the following equation:

$$h = \frac{v^2}{2g} + K_e \left(\frac{v^2}{2g}\right) + \frac{v^2 n^2 L}{(2.21)(r_H)^{4/3}}$$

$$3.68$$

This may be solved for v:

$$v = \sqrt{\cfrac{h}{\cfrac{(1 + K_e)}{2g} + \cfrac{n^2 L}{(2.21)(r_H)^{4/3}}}} \qquad 3.69$$

case 2: Entrance Controlled; Submerged Entrance, Free Outfall

The normal depth in a culvert is the depth of a uniform flow occurring in it. This is generally found by solving the Chezy-Manning equation for depth. The slope is said to be supercritical if the normal depth is less than the critical depth. (The critical depth maybe calculated from formulas given in Chapter 5. However, if the normal depth is greater than the culvert diameter, the flow is not supercritical.)

If the flow is supercritical, the culvert will not flow full regardless of the slope. Under these conditions, the culvert is said to be entrance-controlled because the entrance will not admit water fast enough to fill the barrel. The culvert entrance acts like an orifice, and Torricelli's equation is again used to determine the discharge.

$$Q = C_d A \sqrt{2gh} \qquad 3.70$$

h is measured from the headwater surface level to the centerline of the culvert entrance. Torricelli's equation is most accurate when h is considerably larger than the culvert diameter. When h is large, it is not important whether h is measured from the top of the barrel, centerline, or culvert invert. This has led to much confusion in choosing h for this case.

The head required for a given outflow is

$$h = \frac{Q^2}{(C_d A)^2 2g} \qquad 3.71$$

Since there are many possible entrance conditions, it is not possible to list here more than a few values of the discharge coefficient, C_d. Generally, Table 3.4 should be used. Although the coefficient is .62 for a sharp-edged entrance, it approaches 1.0 as the entrance becomes rounded and as contraction is suppressed. Contraction can be partially suppressed by placing the culvert invert on the stream bed level or by using flanged wing walls.

If the entrance loss coefficient and coefficient of contraction are known, the discharge coefficient may be found from them.

$$C_d = \frac{C_c}{\sqrt{1 + K_e}}$$

case 3: Entrance Controlled; Partially Exposed Entrance; Free Outfall

The culvert discharge is given by weir equations. h is measured from the headwater surface to the invert.

case 4: Entrance Controlled; Partially Exposed Entrance; Submerged Exit

Although there will be a hydraulic jump within the culvert, the discharge is determined by weir equations.

case 5: Exit Controlled, Long Culverts, Slightly Submerged Entrance; Free Outfall

If the culvert is horizontal or with a slope less than required to achieve a critical depth in the barrel, the discharge may be under outlet control. In such cases, the normal depth is greater than the critical depth. The equation relating the flow velocity and the actual available head is

$$h = K_e \left(\frac{v^2}{2g}\right) + \frac{v^2}{2g} + d_n \qquad 3.73$$

h is measured from the surface of the headwater to the culvert invert.

The Manning equation can be used to relate velocity and slope.

$$v^2 = \frac{(2.21)}{n^2}(S)(r_H)^{4/3} \qquad 3.74$$

The resulting equation is

$$h = (1 + K_e)\frac{2.21}{2gn^2}(S)(r_H)^{4/3} + d_n \qquad 3.75$$

If the entrance head, h, is very large, d_n will be larger than the barrel diameter and the above equations are not applicable. Case 1 should be used for large values of h since the flow is under pressure. h should then be re-defined as the level of the headwater above the top of the culvert exit.

If h is only slightly larger than the culvert diameter, different values of d_n should be assumed (which will also set r_H) until equation 3.75 is satisfied.

The discharge will be

$$Q = Av \qquad 3.76$$

case 6: Exit Controlled; Long Culverts; Partially Submerged Entrance; Free Outfall

If the headwater level is below the top of the culvert, the flow may be in an unstable transition mode. Discharge should be calculated from cases 1 and 5. The smaller of the two discharges should be used.

case 7: Exit Controlled; Partially Submerged Entrance; Submerged Exit

If the difference between the tailwater and headwater levels is large, the backwater will not extend from the exit to the entrance. Case 4 should be used.

If the difference in tailwater and headwater levels is small, the culvert can be assumed to be flowing full. Case 1 should be used.

Example 3.17

Size a culvert which has the following characteristics:

slope = .01
length = 250 feet
capacity = 45 cfs
n = .013
entrance fluid level = 5 feet above barrel top
free exit

step 1: Assume a trial culvert size. To simplify the calculation of flow area and hydraulic radius, choose a square cross-section. In this example, select a square opening with 1 foot sides as the initial trial culvert.

step 2: Calculate the flow assuming case 2. With entrance control, the culvert will not flow full. The entrance will act like an orifice.

$$A = (1)(1) = 1 \text{ ft}^2$$

$$h = 5 + .5 = 5.5 \text{ feet}$$

$$C_d = .62 \text{ for square edged openings}$$

$$Q = (.62)(1)\sqrt{(2)(32.2)(5.5)} = 11.7 \text{ cfs}$$

Since 11.7 cfs is less than 45, the culvert is too small, and a larger culvert is necessary. Choose a square design with 2 foot sides.

$$A = (2)(2) = 4$$

$$h = 5 + 1 = 6$$

$$C_d = .62$$

$$Q = (.62)(4)\sqrt{(2)(32.2)(6)} = 48.7 \text{ cfs}$$

48.7 cfs exceeds the required capacity. So, a 2 foot square culvert should be sufficient if the entrance control assumption is valid.

step 3: Begin checking the entrance control assumption by calculating the maximum hydraulic radius. Because the flow is entrance controlled, contraction occurs and the upper surface of the culvert is not wetted. This will give a maximum hydraulic radius, and in turn, a maximum velocity.

$$r_H = A/p_w = 4/6 = .667 \text{ ft}$$

step 4: Calculate the maximum velocity using the Manning equation for open channel flow.

$$v = \frac{1.486}{n}(r_H)^{2/3}(S)^{1/2} = \frac{1.486}{.013}(.667)^{2/3}(.01)^{1/2}$$

$$= 8.72 \text{ ft/sec}$$

step 5: Calculate the normal depth.

$$d_n = \frac{Q}{(v)(\text{width})} = \frac{45}{(8.72)(2)} = 2.58$$

Since this normal depth is greater than the culvert size, the discharge will be full pipe flow under pressure. The entrance control assumption was, therefore, not valid for this size culvert. At this point, two things can be done. A larger culvert can be chosen if entrance control is desired. Or, the solution can continue by checking to see if the culvert has the required capacity as a pressure conduit.

step 6: Check the capacity as a pressure conduit. As explained in case 4, h is re-defined as the headwater height above the top of the culvert exit.

$$h = 5 + (250)(.01) = 7.5$$

step 7: Since the pipe is flowing full, the hydraulic radius is

$$r_H = 4/8 = .5$$

step 8: Now that r_H is known, the flow velocity can be found from equation 3.69. Assuming a square-edged entrance,

$$v = \sqrt{\frac{7.5}{\frac{(1 + .5)}{2(32.2)} + \frac{(.013)^2(250)}{(2.21)(.5)^{4/3}}}} = 10.24 \text{ ft/sec}$$

step 9: Check the capacity

$$Q = vA = (10.24)(4) = 40.96 \text{ cfs}$$

The culvert size is not acceptable since its discharge under the maximum head does not have a capacity of 45 cfs.

step 10: Repeat from step 2 and try a larger size. Assume a 2.5' x 2.5' square. Then,

$$A = 6.25 \qquad h = 6.25 \qquad C_d = .62$$

$$Q = 77 \qquad r_H = .833 \qquad v = 10.12$$

$$d_n = 1.78$$

step 11: Calculate the critical depth. For rectangular channels,

$$d_c = (Q^2/w^2g)^{.333} = 2.16 \qquad \text{(Equation 5.36)}$$

Since the normal depth is less than the critical depth, the flow is supercritical. The entrance control assumption was correct for the culvert. The culvert has sufficient capacity to carry 45 cfs.

B. Series Pipe Systems

A series pipe system has one or more diameters along its run. If Q or v are known in any part of the system, the friction loss can be easily found as the sum of the friction losses in the sections.

Figure 3.9

A Series Pipe System

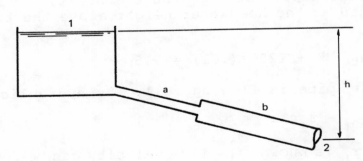

If both v and Q are unknown, a trial and error solution method is required. If the Darcy friction loss is to be used, the following procedure may be applied.

step 1: Using the Moody diagram with ε_a, ε_b, D_a, and D_b, find f_a and f_b for fully turbulent flow.

step 2: Write all of the velocities in terms of one unknown velocity.

$$v_a = v_a \qquad\qquad\qquad 3.77$$

$$v_b = (A_a/A_b)v_a \qquad\qquad 3.78$$

step 3: Write the friction loss in terms of the unknown velocity.

$$h_{f,total} = \frac{f_a L_a v_a^2}{2D_a g} + \frac{f_b L_b}{2D_b g}(A_a/A_b)^2 v_a^2 \qquad 3.79$$

$$= \frac{v_a^2}{2g}\left[\frac{f_a L_a}{D_a} + \frac{f_b L_b}{D_b}(A_a/A_b)^2\right] \qquad 3.80$$

step 4: Solve for the unknown velocity using Bernoulli's equation between points 1 and 2. Include the pipe friction, but ignore minor losses for convenience.

$$h = \frac{v_b^2}{2g} + h_f \qquad 3.81$$

$$= \frac{v_a^2}{2g}\left[(A_a/A_b)^2(1 + \frac{f_b L_b}{D}) + \frac{f_a L_a}{D_a}\right] \qquad 3.82$$

step 5: Using the values of v_a and v_b found from step 4, check the values of f_a and f_b. Repeat steps 3 and 4 using the new values of f_a and f_b if necessary.

If the Hazen-Williams coefficients are given for the pipe sections, the procedure for finding unknown velocities and flow quantitites is similar although considerably more difficult because the v^2 and $v^{1.85}$ terms cannot be combined. A first approximation, however, can be obtained by replacing $v^{1.85}$ with v^2 in the Hazen-Williams formula for friction loss. A trial and error method can then be used to find v.

C. Parallel Pipe Systems

A common method of increasing the capacity of an existing line is to install a second line parallel to the first. If that is done, the flow will divide in such a manner as to make the friction loss the same in both branches.

Figure 3.10

A Parallel Pipe System

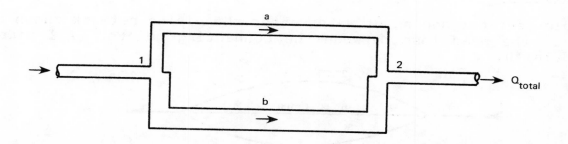

If the parallel system has only 2 branches, a simultaneous solution approach may be taken.

$$h_{f,a} = h_{f,b} = \frac{f_a L_a v_a^2}{2D_a g} = \frac{f_b L_b v_b^2}{2D_b g} \qquad 3.83$$

and

$$Q_a + Q_b = Q_c \qquad 3.84$$

$$\frac{1}{4}\pi(D_a^2 v_a + D_b^2 v_b) = Q_c \qquad 3.85$$

However, if the parallel system has 3 or more branches, it is easier to use the following method which can also be modified for use with the Hazen-Williams friction loss formula.

<u>step 1</u>: Write $h_f = \frac{fLv^2}{2Dg}$ for each branch.

<u>step 2</u>: Solve for v for each branch.

$$v = \sqrt{\frac{2Dg}{fL}h_f} \qquad 3.86$$

<u>step 3</u>: Solve for Q for each branch.

$$Q = Av = A\sqrt{\frac{2Dg}{fL}h_f} = K'\sqrt{h_f} \qquad 3.87$$

There will be a different value of K' for each branch.

<u>step 4</u>: $Q_{total} = Q_1 + Q_2 + Q_3 \qquad 3.88$

$$= (K_1' + K_2' + K_3')\sqrt{h_f} \qquad 3.89$$

Since Q_{total}, K_1', K_2', and K_3' are known, the friction loss can be solved for.

<u>step 5</u>: Check the values of f and repeat as necessary.

<u>Example 3.18</u>

3 cubic feet per second of water enter the piping network shown below. What is the head loss between the connecting points? All pipes are schedule 40.

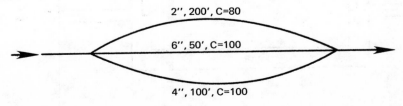

2", 200', C=80

6", 50', C=100

4", 100', C=100

The pipe dimensions are:

	2"	4"	6"
flow area	.0233	.0884	.2006
diameter	.1723	.3355	.5054

step 1: $h_f = \dfrac{(3.012)(v)^{1.85}L}{(C)^{1.85}(D)^{1.165}}$

step 2: $v = \dfrac{(.551)(C)(D)^{.63}}{L^{.54}}(h_f)^{.54}$ (equation 3.36)

$v_{2"} = \dfrac{(.551)(80)(.1723)^{.63}}{(200)^{.54}}(h_f)^{.54} = .833(h_f)^{.54}$

Similarly,

$v_{6"} = 4.335(h_f)^{.54}$

$v_{4"} = 2.303(h_f)^{.54}$

step 3: $Q_{2"} = (.0233)(.833)(h_f)^{.54} = .0194(h_f)^{.54}$

$Q_{6"} = (.2006)(4.335)(h_f)^{.54} = .8696(h_f)^{.54}$

$Q_{4"} = (.0884)(2.303)(h_f)^{.54} = .2036(h_f)^{.54}$

step 4: $3 = (.0194 + .8696 + .2036)(h_f)^{.54}$

$h_f = 6.49$ feet

◆

D. Reservoir Branching Systems

The 3-reservoir problem occurs occasionally. Its trial and error solution method is easily proceduralized. Although there are many possible choices for the unknown variable (pipe length, diameter, flow quantity, head, etc.), the three most common types of problems follow.

Figure 3.11

3-Reservoir System

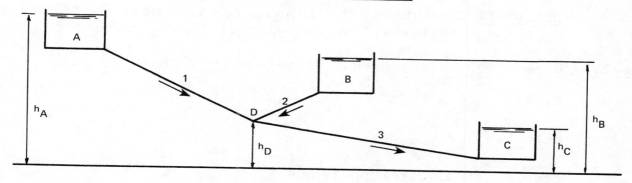

case 1: Given: all lengths, diameters, and elevations

Find: Q_1, Q_2, Q_3

Although an analytical solution method is possible, this problem is best solved by trial and error.

step 1: Assume Q_1 and use the Bernoulli equation to find the pressure at point D. Ignore minor losses and velocity head. Notice that the friction term depends on the assumed value of Q.

$$Q_1 = v_1 A_1$$

$$h_A = h_D + (p_D/\rho) + h_{f,1} \qquad 3.90$$

step 2: Once p_D is known, use it to find Q_2.

$$h_B = h_D + (p_D/\rho) + h_{f,2} \qquad 3.91$$

$$Q_2 = v_2 A_2$$

If $h_D + (p_D/\rho)$ is greater than h_B, flow will be into reservoir B. In that case, the friction term is negative, not positive.

step 3: Find Q_3.

$$h_C = h_D + (p_D/\rho) - h_{f,3} \qquad 3.92$$
$$Q_3 = v_3 A_3$$

step 4: Check that $Q_1 = Q_2 + Q_3$. If it does not, return to step 1. After two iterations, plot $(Q_1 - Q_2 - Q_3)$ versus Q. Estimate Q_1 by interpolation or extrapolation.

Alternative Solution Method: Assume a value of p_D and solve for the flow quantities. Repeat with different values of p_D.

case 2: Given: Q_1, all lengths, diameters, h_A, and h_B

Find: h_C

step 1: $v_1 = Q_1/A_1$

step 2: Solve for p_D from

$$h_A = h_D + (p_D/\rho) + h_{f,1} \qquad 3.93$$

step 3:   Solve for v_2 from

$$h_B = h_D + (p_D/\rho) + h_{f,2} \tag{3.94}$$

If $h_D + (p_D/\rho)$ is greater than h_B, the flow will be into reservoir B. In that case, the friction $h_{f,2}$ should be subtracted, not added.

step 4: $Q_2 = A_2 v_2$

step 5: $Q_3 = Q_1 \pm Q_2$

step 6: $v_3 = Q_3/A_3$

step 7:   Calculate $h_{f,3}$ from v_3, L_3, and D_3

step 8:   Find h_C from

$$h_C = h_D + (p_D/\rho) - h_{f,3} \tag{3.95}$$

case 3:   Given: All lengths, elevations, Q_1, and diameters D_1 and D_2
Find: D_3

steps 1-5: Repeat steps 1-5 from case 2.

step 6:   Find $h_{f,3}$ from

$$h_C = h_D + (p_D/\rho) - h_{f,3} \tag{3.96}$$

step 7:   Find D_3 from $h_{f,3}$

E. Pipe Networks

Network flows in multi-loop systems are best calculated by the Hardy Cross method when a hand solution is necessary. This is a systematic trial-and-error method which first assumes flows and then adds consecutive adjustments to the assumed flows. The Hardy Cross method is easy to apply. It is based on the following principles:

1. The flows entering a junction must equal the flows leaving the junction.

2. The algebraic sum of friction losses around any closed loop is zero.

If Q_a is the assumed flow in a pipe and Q_t is the true flow, the difference is δ, where

$$\delta = Q_t - Q_a \tag{3.97}$$

The true flow can be written in terms of the assumed flow and the correction:

$$Q_t = Q_a + \delta \qquad\qquad 3.98$$

The friction loss in the pipe has the form of $h_f = K'Q_t^n$ where $n = 2$ if the Darcy equation is used, and $n = 1.85$ if the Hazen-Williams equation is used.

- For Q in cfs, L in feet, and D in feet, the Hazen-Williams friction coefficient is

$$K' = \frac{(4.727)L}{D^{4.87}C^{1.85}} \qquad\qquad 3.99$$

- For Q in gpm, L in feet, and d in inches, the Hazen-Williams friction coefficient is

$$K' = \frac{(10.44)L}{C^{1.85}d^{4.87}} \qquad\qquad 3.100$$

- For Q in cfs, L in feet, and D in feet, the Darcy friction coefficient is

$$K' = \frac{(.0252)fL}{D^5} \qquad\qquad 3.101$$

  f is usually assumed to be the same (such as .02) in all parts of the network.

- For Q in gpm, L in feet, and D in feet, the Darcy friction coefficient is

$$K' = (1.251\ EE\text{-}5)\frac{fL}{D^5} \qquad\qquad 3.102$$

Combining and expanding as a series, the friction loss is

$$h_f = K'(Q_a + \delta)^n \simeq K'Q_a^n + nK'\delta Q_a^{n-1} + \cdots \qquad\qquad 3.103$$

Subsequent higher order terms can be omitted because it is assumed that the correction is small.

Around a loop in a network, the sum of the friction drops is zero. Therefore,

$$\Sigma h_f = \Sigma K'Q_a^n + \delta\Sigma nK'Q_a^{n-1} = 0 \qquad\qquad 3.104$$

δ has been taken outside of the summation because all branches in the loop have the same correction. (If n is the same for all pipes, it can be taken out of the summation also.)

Equation 3.104 can be solved for δ.

$$\delta = \frac{-\Sigma K'Q_a^n}{n\Sigma\left|K'Q_a^{n-1}\right|} = \frac{-\Sigma h_f}{n(\Sigma h_f/Q_a)} \qquad\qquad 3.105$$

To use this equation, you must first assume the flow directions as well as the flow rates. The numerator is the sum of head losses around the loop, taking signs into consideration. Because the denominator is a sum of the absolute values, δ must be applied in the same sense to each branch in the loop. If clockwise is assumed as the positive direction (an arbitrary decision), then δ is added to clockwise flows and subtracted from counterclockwise flows.

The application of the Hardy Cross method is as follows:

1.   Choose between the Darcy and Hazen-Williams friction loss equations. The Darcy equation results in an easier expression to evaluate because the exponent (n-1) is 1.

2.   Choose a positive direction (e.g., clockwise).

3.   Number all pipes in the network or identify all nodes.

4.   Divide the network into independent loops such that each branch is included in at least one loop.

5.   Calculate K' for each pipe in the network.

6.   Assume flow rates and directions. This may seem like a difficult step, but it is not. Most inaccurate first assumptions yield good results after several iterations.

7.   Calculate δ for each independent loop.

8.   Apply δ to each pipe in its loop using the previously mentioned sign convention.

9.   Return to step 7.

Example 3.19

Use a Moody friction factor of f = .02 to calculate the flow in each pipe in the network shown.

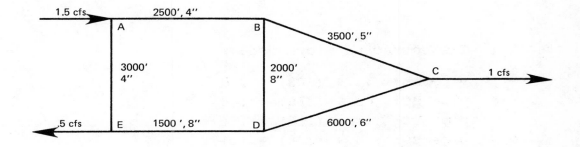

step 1: The Moody friction factor is given. So, the Darcy friction loss equation will be used.

step 2: Choose clockwise as the positive direction.

step 3: Use the identification system shown on the network.

step 4: Work with loops ABDE and BCD. Notice that loop ABCDE is not independent if the other two loops are used.

step 5: pipe AB: D = (4/12) = .3333

$$K' = \frac{(.0252)(.02)(L)}{D^5} = \frac{(.0252)(.02)(2500)}{(.3333)^5}$$

$$= 306.2$$

pipe BC: K' = 140.5
pipe DC: K' = 96.8
pipe BD: K' =   7.7
pipe ED: K' =   5.7
pipe AE: K' = 367.4

step 6: Assume the following directions and flows.

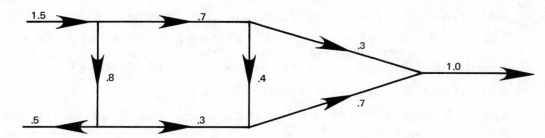

step 7: $\delta_{ABDE} = \dfrac{-[(306.2)(.7)^2+(7.7)(.4)^2-(5.7)(.3)^2-(367.4)(.8)^2]}{2[(306.2)(.7) + (7.7)(.4) + (5.7)(.3) + (367.4)(.8)]}$

$$= +.08$$

$\delta_{BCD} = \dfrac{-[(140.5)(.3)^2 - (96.8)(.7)^2 - (7.7)(.4)]}{2[(140.5)(.3) + (96.8)(.7) + (7.7)(.4)]}$

$$= +.16$$

step 8: The corrected flows are:

pipe AB: .7 + (.08) = .78
pipe BC: .3 + (.16) = .46
pipe DC: .7 - (.16) = .54
pipe BD: .4 + (.08) - (.16) = .32
pipe ED: .3 - (.08) = .22
pipe AE: .8 - (.08) = .72

<u>step 9</u>:   The procedure is now repeated using the corrected flows.

10. Flow Measuring Devices

The total energy in a fluid flow is the sum of pressure head, velocity head, and gravitational head.

$$h = \frac{p}{\rho} + \frac{v^2}{2g} + z$$

3.106

Change in gravitational head within a flow instrument is negligible. Therefore, if two of the three remaining variables (h, p, or v) are known, then the third can be found from subtraction. The flow measuring devices discussed in this section are capable of measuring total head (h) or pressure head (p).

A. Velocity Measurement

Velocity of a fluid stream is determined by measuring the difference between the static pressure and the stagnation pressure and then solving for the velocity head.

A piezometer tap may be used to measure the pressure head directly in feet of fluid.

$$h_p = p_o/\rho$$

3.107

Figure 3.12
Piezometer Tap

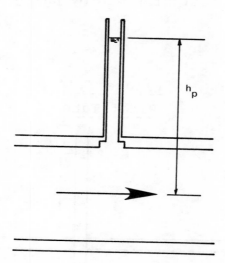

For liquids with pressures higher than the capability of the direct reading tap, a manometer may be used with a piezometer tap or with a static probe as shown in Figure 3.13.

Figure 3.13

Use of Manometers to Measure Static Pressure

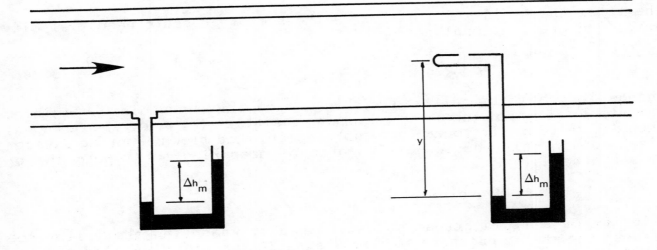

For either configuration of Figure 3.13, the static pressure is given by:

$$p_O = \rho_m \Delta h_m - \rho y \qquad\qquad 3.108$$

$$h_p = \frac{\rho_m \Delta h_m}{\rho} - y \qquad\qquad 3.109$$

Stagnation pressure, also known as total pressure or impact pressure, may be measured directly in feet of fluid by using a pitot tube as shown in Figure 3.14.

$$h_s = h_p + h_v \qquad\qquad 3.110$$

Figure 3.14

Pitot Tube

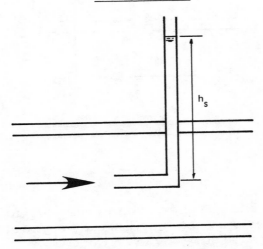

For high-pressure liquids, a manometer may be used to measure stagnation pressure:

$$p_s = \rho_m \Delta h_m - \rho y \qquad\qquad 3.111$$

$$\frac{p_s}{\rho} = \frac{p}{\rho} + \frac{v^2}{2g} \qquad\qquad 3.112$$

$$h_s = \frac{\rho_m \Delta h_m}{\rho} - y \qquad\qquad 3.113$$

Figure 3.15

Use of Manometer to Measure Total Pressure

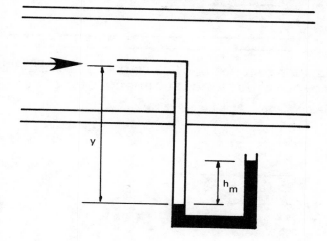

Using the results of the measurements above, the velocity head may be calculated from equation 3.114.

$$v = \sqrt{2g(h_s - h_p)} = \sqrt{2g(p_s - p_o)/\rho_o} \qquad\qquad 3.114$$

If Figures 3.13 or 3.15 apply, the velocity head (equation 3.110) is

$$\frac{v^2}{2g} = \left(\frac{\rho_m \Delta h_m}{\rho}\right)_s - \left(\frac{\rho_m \Delta h_m}{\rho}\right)_o \qquad\qquad 3.115$$

If the piezometer tap of Figure 3.12 and the pitot tube are placed at the same point, the velocity head in feet of fluid may be read directly.

$$\frac{v^2}{2g} = \Delta h \qquad\qquad 3.116$$

$$v = \sqrt{2g\,\Delta h} \qquad\qquad 3.117$$

Figure 3.16
Comparative Velocity Head Measurement

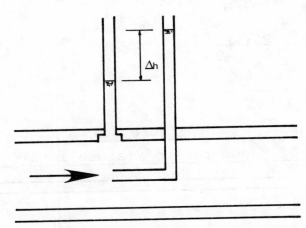

The instrumentation arrangement of Figure 3.16 may be combined into a single instrument to provide a measurement of velocity head as shown in Figure 3.17 and Figure 3.18.

$$\frac{v^2}{2g} = \frac{\Delta h (\rho_m - \rho)}{\rho} \qquad 3.118$$

$$v = \sqrt{\frac{2g (\rho_m - \rho) \Delta h}{\rho}} \qquad 3.119$$

Figure 3.17
Velocity Head Measurement

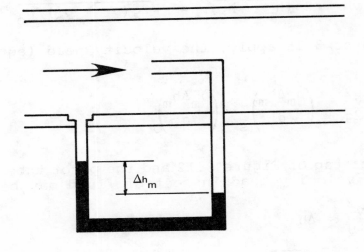

Figure 3.18
Velocity Head Measurement

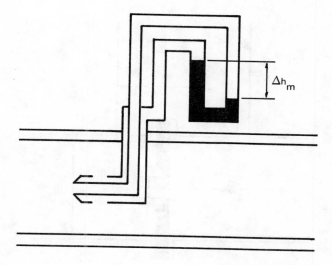

Example 3.20

50°F water is flowing through a pipe. A pitot-static gage registers a 3" deflection of mercury. What is the velocity within the pipe? (The density of mercury is 848.6 pcf.)

Using equation 3.119,

$$v = \sqrt{\frac{2(32.2)(848.6 - 62.4)(3/12)}{62.4}} = 14.24 \text{ fps}$$

◆

B. Flow Measurement

Using the same techniques described in the preceding section, the flow rate in a line may be determined by measuring the pressure drop across a restriction. Once the geometry of the restriction is known, the Bernoulli equation, along with empirically determined correction coefficients, may be applied to obtain an expression directly relating flow rate with pressure drop.

The simplest fluid flow measuring device is the orifice plate. This consists of a thin plate or diaphragm with a central hole through which the fluid flows.

Figure 3.19
Comparative Reading Orifice Plate

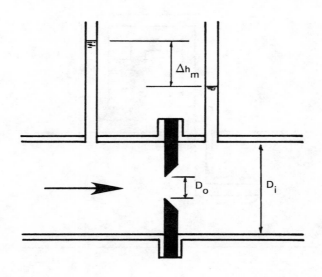

Figure 3.20
Direct Reading Orifice Plate

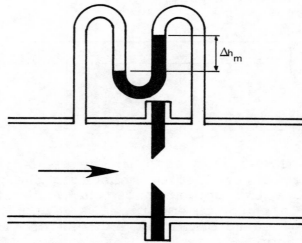

The governing orifice plate equations for liquid flow are given below.

$$Q = F_{va}C_dA_o\sqrt{\frac{2g(p_1 - p_2)}{\rho}} \qquad\qquad 3.120$$

$$= F_{va}C_dA_o\sqrt{\frac{2g(\rho_m - \rho)\Delta h_m}{\rho}} \qquad\qquad 3.121$$

F_{va} is the velocity of approach factor.

$$F_{va} = \sqrt{\frac{1}{1 - \left(\frac{C_c A_o}{A_i}\right)^2}} \qquad 3.122$$

The flow coefficient depends on the velocity of approach factor and the discharge coefficient. It may also be obtained from Figure 3.21.

$$C_f = F_{va} C_d \qquad 3.123$$

$$C_d = C_v C_c \qquad 3.124$$

Figure 3.21

Flow Coefficients For Orifice Plates

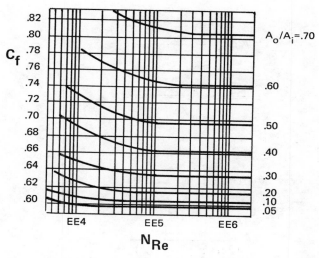

The area at the vena contracta is

$$A_{vc} = C_c A_o \qquad 3.125$$

$$v_{vc} = F_{va} C_v \sqrt{\frac{2g(p_1 - p_2)}{\rho}} \qquad 3.126$$

Operating on the same principles as the orifice plate, the venturi meter induces a smaller pressure drop. However, it is mechanically more complex.

Figure 3.22
Venturi Meter

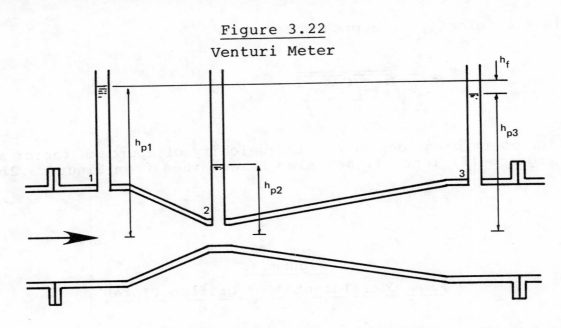

The governing equations are similar to those which apply to orifice plates.

$$Q = F_{va} C_d A_2 \sqrt{\frac{2g(p_1 - p_2)}{\rho}} \qquad 3.127$$

where

$$F_{va} = \frac{1}{\sqrt{1 - (A_2/A_1)^2}} \qquad 3.128$$

$$C_d = C_v C_c \qquad 3.129$$

C_c is usually 1.0 for venturi meters.

$$C_f = F_{va} C_d \qquad 3.130$$

$$v_2 = F_{va} \sqrt{\frac{2g(p_1 - p_2)}{\rho}} \qquad 3.131$$

Table 3.5
C_d for Venturi Meters

$2 < (A_1/A_2) < 3$	
C_d	N_{Re}
.94	6,000
.95	10,000
.96	20,000
.97	50,000
.98	200,000
.99	2,000,000

Example 3.21

150°F water is flowing in an 8" schedule 40 steel pipe at 2.23 cfs. If a 7 inch sharp edge orifice plate is bolted across the line, what manometer deflection in inches of Hg would be expected? (Mercury has a density of 848.6 pcf.)

	7" orifice	8" schedule 40
flow area	$.267$ ft^2	$.3474$ ft^2
diameter	$.583$ ft	$.6651$ ft

From table 3.4 for the orifice: $C_c = .63$, $C_d = .62$

Using equation 3.122, $F_{va} = 1.13$

From equation 3.127,

$$\Delta p = \left[\frac{2.23}{(1.13)(.62)(.267)}\right]^2\left[\frac{61.2}{2(32.2)}\right] = 135.06 \text{ psf}$$

150°F water has a density of 61.2 pcf. So,

$$h_m = \frac{135.06}{848.6 - 61.2} = 0.172 \text{ ft}$$ ◆

11. The Impulse/Momentum Principle

A force is required to cause a direction or velocity change in a flowing fluid. Conventions necessary to determine such a force are given here:

1. $\Delta v = v_2 - v_1$

2. A positive Δv indicates an increase. A negative Δv indicates a decrease.

3. F and x are positive to the right. F and y are positive upward.

4. F is the force on the fluid. The force on the walls or support has the same magnitude but opposite direction.

5. The fluid is assumed to flow from left to right.

The impulse momentum principle is expressed by equation 3.132.

$$F\Delta t = m\Delta v \qquad \text{or} \qquad F = (m/\Delta t)\Delta v = \dot{m}\Delta v \qquad 3.132$$

Since F is a vector, it may be broken into its components.

$$F_x = \dot{m}\Delta v_x \qquad \text{and} \qquad F_y = \dot{m}\Delta v_y \qquad 3.133$$

If the fluid flow is directed through an angle ϕ, then

$$\Delta v_x = v(\cos\phi - 1) \qquad \text{and} \qquad \Delta v_y = v\sin\phi \qquad 3.134$$

Application #1: Open Jet on Vertical Flat Plate

$$\Delta v_y = 0 \qquad\qquad\qquad\qquad\qquad\quad 3.135$$

$$\Delta v_x = -v \qquad\qquad\qquad\qquad\qquad 3.136$$

$$F_x = -\dot{m}v = -\dot{w}v/g = -Q\rho v/g \qquad 3.137$$

Figure 3.23

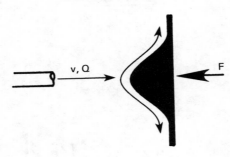

Application #2: Open Jet on Horizontal Flat Plate

At the plate, v has become $\sqrt{v_O^2 - 2gh}$

$$\Delta v_x = 0 \qquad\qquad\qquad\qquad\qquad\qquad\qquad 3.138$$

$$\Delta v_y = -\sqrt{v_O^2 - 2gh} \qquad\qquad\qquad\qquad\qquad 3.139$$

$$F = -\dot{m}\sqrt{v_O^2 - 2gh} = -(\dot{w}/g)\sqrt{v_O^2 - 2gh}$$

$$= (Q\rho/g)\sqrt{v_O^2 - 2gh} \qquad\qquad\qquad\qquad 3.140$$

Figure 3.24

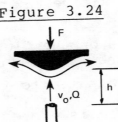

Application #3: Open Jet on Single Stationary Blade

v_2 may not be the same as v_1 if friction is present. If no information is given, assume that $v_2 = v_1$.

$$\Delta v_x = v_2\cos\phi - v_1 \qquad\qquad\qquad 3.141$$

$$\Delta v_y = v_2\sin\phi \qquad\qquad\qquad\qquad\quad 3.142$$

$$F_x = (Q\rho/g)(v_2\cos\phi - v_1) \qquad 3.143$$

$$F_y = (Q\rho/g)(v_2\sin\phi) \qquad\qquad\quad 3.144$$

Figure 3.25

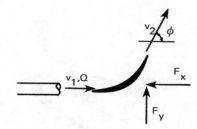

Application #4: Open Jet on Single Moving Blade

v_b is the blade velocity. Friction is neglected for simplicity. The discharge overtaking the moving blade is Q'.

$$Q' = (v - v_b)Q/v \qquad\qquad 3.145$$

$$F_x = (Q'\rho/g)(v - v_b)(\cos\phi - 1) \qquad\qquad 3.146$$

$$F_y = (Q'\rho/g)(v - v_b)\sin\phi \qquad\qquad 3.147$$

$$P = (F_x)v_b \qquad\qquad 3.148$$

$$P_{max} = Q'\rho v^2/2g \text{ (occurs when } v_b = \tfrac{1}{2}v \qquad\qquad 3.149$$
$$\text{and } \phi = 180\degree)$$

Figure 3.26

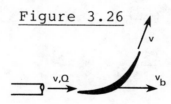

Application #5: Force on Confined Streams

Since the fluid is confined, the forces due to static pressures must be included along with the force from momentum changes. Using gage pressures and neglecting the fluid weight,

$$F_x = p_1A_1 - p_2A_2\cos\phi - (Q\rho/g)(v_2\cos\phi - v_1) \qquad\qquad 3.150$$

$$F_y = [p_2A_2 + (Q\rho v_2/g)]\sin\phi \qquad\qquad 3.151$$

Figure 3.27

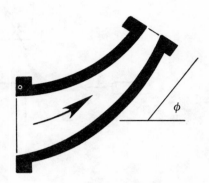

Example 3.22

60° water at 40 psig flowing at 8 ft/sec enters a 12" x 8" reducing elbow as shown and is turned 30°. (a) What is the resultant force on the water? (b) What other forces should be considered in the design of supports for the fitting?

(a) The total energy at point A is

$$\frac{(40)(144)}{(62.4)} + \frac{(8)^2}{(2)(32.2)} + 0 = 93.3 \text{ ft}$$

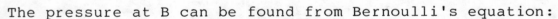

At point B, the velocity is

$$(8)(\frac{12}{8})^2 = 18 \text{ ft/sec}$$

The pressure at B can be found from Bernoulli's equation:

$$93.3 = \frac{p_B(144)}{62.4} + \frac{(18)^2}{(2)(32.2)} + \frac{26}{12}$$

So, p_B = 37.3 psig

$$Q = vA = (8)(\frac{1}{4})\pi(\frac{12}{12})^2 = 6.28 \text{ cfs}$$

From equation 3.150,

$$F_x = (40)(144)(\frac{1}{4})\pi(\frac{12}{12})^2 - (37.3)(144)(\frac{1}{4})\pi(\frac{8}{12})^2\cos30°$$

$$- (\frac{(6.28)(62.4)}{32.2})[(18)(\cos30°) - 8]$$

$$= 2808 \text{ lbf}$$

From equation 3.151

$$F_y = [(37.3)(144)(\frac{1}{4})\pi(\frac{8}{12})^2 + (\frac{(6.28)(62.4)(18)}{32.2})]\sin30°$$

$$= 1047 \text{ lbf}$$

The resultant force on the water is

$$R = \sqrt{(2808)^2 + (1047)^2} = 2997 \text{ lbf}$$

(b)   The support should be designed to also carry the weight of the water in the pipe and bend and the weight of the pipe and bend itself.

◆

Application #6: Water Hammer

Water hammer is another phenomenon that is momentum-related. Water hammer is an increase in pressure in a pipe caused by a sudden velocity decrease. The sudden velocity decrease will usually be caused by a valve closing.

Assuming the pipe material is inelastic, the time required for the water hammer shock wave to travel from a valve to the end of a pipe and back is given by

$$t = 2L/c$$

3.152

Most pipe materials, however, are elastic. This elasticity contributes to the velocity of the pressure wave.

$$t = 2L/a$$

3.153

where,

$$a = \sqrt{\frac{12}{\dfrac{\rho}{g}\left(\dfrac{1}{E_{water}} + \dfrac{d_i}{eE_{pipe}}\right)}}$$

3.154

The fluid pressure increase resulting from this shock wave is

$$\Delta p = \frac{\rho a \Delta v}{g}$$

3.155

Example 3.23

40°F water is flowing at 10 ft/sec through a 4" schedule 40 welded steel pipe. A check valve is suddenly closed. What increase in fluid pressure will occur?

Assume that the closing valve completely stops the flow. Therefore, Δv is 10 ft/sec.

Using equation 3.154, Tables 3.6 and 3.8,

$$a = \sqrt{\frac{12}{\dfrac{(62.4)}{(32.2)}\left(\dfrac{1}{294 \text{ EE3}} + \dfrac{4.026}{(30 \text{ EE6})(.237)}\right)}} = 4328 \text{ fps}$$

From equation 3.155,

$$\Delta p = \frac{(62.4)(4328)(10)}{32.2} = 83,872 \text{ psf}$$

This is an excessive pressure build up.

Table 3.6

Mechanical Properties of Water at Atmospheric Pressure

Temp °F	Density lb/ft³	Density slugs/ft³	Absolute Viscosity lb-sec/ft²	Kinematic viscosity ft²/sec	Surface tension lb/ft	Vapor pressure head ft	Bulk modulus of elasticity lb/in²
32	62.42	1.940	3.746EE-5	1.931EE-5	0.518EE-2	0.20	293EE3
40	62.43	1.940	3.229EE-5	1.664EE-5	0.514EE-2	0.28	294EE3
50	62.41	1.940	2.735EE-5	1.410EE-5	0.509EE-2	0.41	305EE3
60	62.37	1.938	2.359EE-5	1.217EE-5	0.504EE-2	0.59	311EE3
70	62.30	1.936	2.050EE-5	1.059EE-5	0.500EE-2	0.84	320EE3
80	62.22	1.934	1.799EE-5	0.930EE-5	0.492EE-2	1.17	322EE3
90	62.11	1.931	1.595EE-5	0.826EE-5	0.486EE-2	1.61	323EE3
100	62.00	1.927	1.424EE-5	0.739EE-5	0.480EE-2	2.19	327EE3
110	61.86	1.923	1.284EE-5	0.667EE-5	0.473EE-2	2.95	331EE3
120	61.71	1.918	1.168EE-5	0.609EE-5	0.465EE-2	3.91	333EE3
130	61.55	1.913	1.069EE-5	0.558EE-5	0.460EE-2	5.13	334EE3
140	61.38	1.908	0.981EE-5	0.514EE-5	0.454EE-2	6.67	330EE3
150	61.20	1.902	0.905EE-5	0.476EE-5	0.447EE-2	8.58	328EE3
160	61.00	1.896	0.838EE-5	0.442EE-5	0.441EE-2	10.95	326EE3
170	60.80	1.890	0.780EE-5	0.413EE-5	0.433EE-2	13.83	322EE3
180	60.58	1.883	0.726EE-5	0.385EE-5	0.426EE-2	17.33	313EE3
190	60.36	1.876	0.678EE-5	0.362EE-5	0.419EE-2	21.55	313EE3
200	60.12	1.868	0.637EE-5	0.341EE-5	0.412EE-2	26.59	308EE3
212	59.83	1.860	0.593EE-5	0.319EE-5	0.404EE-2	33.90	300EE3

Figure 3.28
Moody Friction Factor Chart

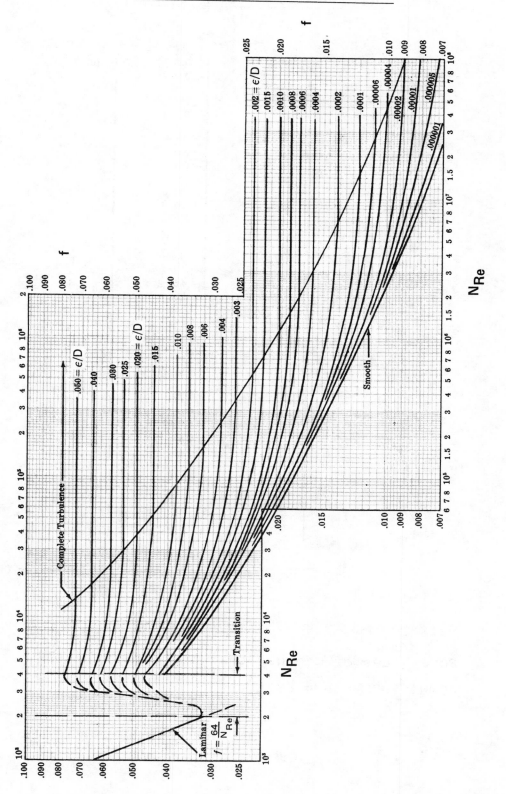

Figure 3.29

Hazen-Williams Nomograph
(C = 100)

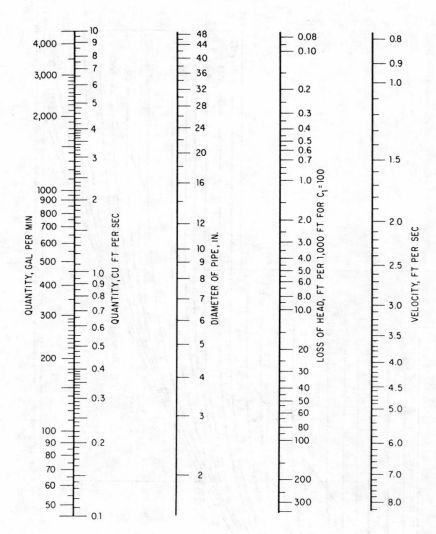

For values of C other than 100, multiply the nomograph values by $(\frac{100}{C})^{1.85}$

Table 3.7

Equivalent Length of Straight Pipe for Various Fittings (feet)

(Turbulent Flow Only, For Any Fluid)

FITTINGS			¼	⅜	½	¾	1	1¼	1½	2	2½	3	4	5	6	8	10	12	14	16	18	20	24
 REGULAR 90° ELL	SCREWED	STEEL	2.3	3.1	3.6	4.4	5.2	6.6	7.4	8.5	9.3	11.0	13.0										
		C.I.										9.0	11.0										
	FLANGED	STEEL			.92	1.2	1.6	2.1	2.4	3.1	3.6	4.4	5.9	7.3	8.9	12.0	14.0	17.0	18.0	21.0	23.0	25.0	30.0
		C.I.										3.6	4.8		7.2	9.8	12.0	15.0	17.0	19.0	22.0	24.0	28.0
 LONG RADIUS 90° ELL	SCREWED	STEEL	1.5	2.0	2.2	2.3	2.7	3.2	3.4	3.6	3.6	4.0	4.6										
		C.I.										3.3	3.7										
	FLANGED	STEEL			1.1	1.3	1.6	2.0	2.3	2.7	2.9	3.4	4.2	5.0	5.7	7.0	8.0	9.0	9.4	10.0	11.0	12.0	14.0
		C.I.										2.8	3.4		4.7	5.7	6.8	7.8	8.6	9.6	11.0	11.0	13.0
 REGULAR 45° ELL	SCREWED	STEEL	.34	.52	.71	.92	1.3	1.7	2.1	2.7	3.2	4.0	5.5										
		C.I.										3.3	4.5										
	FLANGED	STEEL			.45	.59	.81	1.1	1.3	1.7	2.0	2.6	3.5	4.5	5.6	7.7	9.0	11.0	13.0	15.0	16.0	18.0	22.0
		C.I.										2.1	2.9		4.5	6.3	8.1	9.7	12.0	13.0	15.0	17.0	20.0
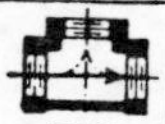 TEE-LINE FLOW	SCREWED	STEEL	.79	1.2	1.7	2.4	3.2	4.6	5.6	7.7	9.3	12.0	17.0										
		C.I.										9.9	14.0										
	FLANGED	STEEL			.69	.82	1.0	1.3	1.5	1.8	1.9	2.2	2.8	3.3	3.8	4.7	5.2	6.0	6.4	7.2	7.6	8.2	9.6
		C.I.										1.9	2.2		3.1	3.9	4.6	5.2	5.9	6.5	7.2	7.7	8.8
 TEE-BRANCH FLOW	SCREWED	STEEL	2.4	3.5	4.2	5.3	6.6	8.7	9.9	12.0	13.0	17.0	21.0										
		C.I.										14.0	17.0										
	FLANGED	STEEL			2.0	2.6	3.3	4.4	5.2	6.6	7.5	9.4	12.0	15.0	18.0	24.0	30.0	34.0	37.0	43.0	47.0	52.0	62.0
		C.I.										7.7	10.0		15.0	20.0	25.0	30.0	35.0	39.0	44.0	49.0	57.0
 180° RETURN BEND	REG. SCREWED	STEEL	2.3	3.1	3.6	4.4	5.2	6.6	7.4	8.5	9.3	11.0	13.0										
		C.I.										9.0	11.0										
	REG. FLANGED	STEEL			.92	1.2	1.6	2.1	2.4	3.1	3.6	4.4	5.9	7.3	8.9	12.0	14.0	17.0	18.0	21.0	23.0	25.0	30.0
		C.I.										3.6	4.8		7.2	9.8	12.0	15.0	17.0	19.0	22.0	24.0	28.0
	LONG RAD. FLANGED	STEEL			1.1	1.3	1.6	2.0	2.3	2.7	2.9	3.4	4.2	5.0	5.7	7.0	8.0	9.0	9.4	10.0	11.0	12.0	14.0
		C.I.										2.8	3.4		4.7	5.7	6.8	7.8	8.6	9.6	11.0	11.0	13.0
 GLOBE VALVE	SCREWED	STEEL	21.0	22.0	22.0	24.0	29.0	37.0	42.0	54.0	62.0	79.0	110.0										
		C.I.										65.0	86.0										
	FLANGED	STEEL			38.0	40.0	45.0	54.0	59.0	70.0	77.0	94.0	120.0	150.0	190.0	260.0	310.0	390.0					
		C.I.										77.	99.0		150.0	210.0	270.0	330.0					
 GATE VALVE	SCREWED	STEEL	.32	.45	.56	.67	.84	1.1	1.2	1.5	1.7	1.9	2.5										
		C.I.										1.6	2.0										
	FLANGED	STEEL							2.6	2.7		2.8	2.9	3.1	3.2	3.2	3.2	3.2	3.2	3.2	3.2	3.2	3.2
		C.I.										2.3	2.4		2.6	2.7	2.8	2.9	2.9	3.0	3.0	3.0	3.0
 ANGLE VALVE	SCREWED	STEEL	12.8	15.0	15.0	15.0	17.0	18.0	18.0	18.0	18.0	18.0	18.0										
		C.I.										15.0	15.0										
	FLANGED	STEEL			15.0	15.0	17.0	18.0	18.0	21.0	22.0	28.0	38.0	50.0	63.0	90.0	120.0	140.0	160.0	190.0	210.0	240.0	300.0
		C.I.										23.0	31.0		52.0	74.0	98.0	120.0	150.0	170.0	200.0	230.0	280.0
 SWING CHECK VALVE	SCREWED	STEEL	7.2	7.3	8.0	8.8	11.0	13.0	15.0	19.0	22.0	27.0	38.0										
		C.I.										22.0	31.0										
	FLANGED	STEEL			3.8	5.3	7.2	10.0	12.0	17.0	21.0	27.0	38.0	50.0	63.0	90.0	120.0	140.0					
		C.I.										22.0	31.0		52.0	74.0	98.0	120.0					
COUPLING OR UNION	SCREWED	STEEL	.14	.18	.21	.24	.29	.36	.39	.45	.47	.53	.65										
		C.I.										.44	.52										
BELL MOUTH INLET		STEEL	.04	.07	.10	.13	.18	.26	.31	.43	.52	.67	.95	1.3	1.6	2.3	2.9	3.5	4.0	4.7	5.3	6.1	7.6
		C.I.										.55	.77		1.3	1.9	2.4	3.0	3.6	4.3	5.0	5.7	7.0
SQUARE MOUTH INLET		STEEL	.44	.68	.96	1.3	1.8	2.6	3.1	4.3	5.2	6.7	9.5	13.0	16.0	23.0	29.0	35.0	40.0	47.0	53.0	61.0	76.0
		C.I.										5.5	7.7		13.0	19.0	24.0	30.0	36.0	43.0	50.0	57.0	70.0
RE-ENTRANT PIPE		STEEL	.88	1.4	1.9	2.6	3.6	5.1	6.2	8.5	10.0	13.0	19.0	25.0	32.0	45.0	58.0	70.0	80.0	95.0	110.0	120.0	150.0
		C.I.										11.0	15.0		26.0	37.0	49.0	61.0	73.0	86.0	100.0	110.0	140.0
SUDDEN ENLARGEMENT			$h = \dfrac{(V_1-V_2)^2}{2g}$ FEET OF FLUID; IF $V_2 = 0$ $\quad h = \dfrac{V_1^2}{2g}$ FEET OF FLUID																					

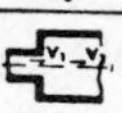

Table 3.8

Dimensions of Welded and Seamless Steel Pipe

Nominal Diameter Inches	Schedule	Outside Diameter Inches	Wall Thickness Inches	Internal Diameter Inches	Internal Area Square Inches	Internal Diameter Feet	Internal Area Square Feet
⅛	40 (S) 80 (X)	0.405	0.068 0.095	0.269 0.215	0.0568 0.0363	0.0224 0.0179	0.00039 0.00025
¼	40 (S) 80 (X)	0.540	0.088 0.119	0.364 0.302	0.1041 0.0716	0.0303 0.0252	0.00072 0.00050
⅜	40 (S) 80 (X)	0.675	0.091 0.126	0.493 0.423	0.1909 0.1405	0.0411 0.0353	0.00133 0.00098
½	40 (S) 80 (X) 160 (XX)	0.840	0.109 0.147 0.187 0.294	0.622 0.546 0.466 0.252	0.3039 0.2341 0.1706 0.0499	0.0518 0.0455 0.0388 0.0210	0.00211 0.00163 0.00118 0.00035
¾	40 (S) 80 (X) 160 (XX)	1.050	0.113 0.154 0.219 0.308	0.824 0.742 0.612 0.434	0.5333 0.4324 0.2942 0.1479	0.0687 0.0618 0.0510 0.0362	0.00370 0.00300 0.00204 0.00103
1	40 (S) 80 (X) 160 (XX)	1.315	0.133 0.179 0.250 0.358	1.049 0.957 0.815 0.599	0.8643 0.7193 0.5217 0.2818	0.0874 0.0798 0.0679 0.0499	0.00600 0.00500 0.00362 0.00196
1¼	40 (S) 80 (X) 160 (XX)	1.660	0.140 0.191 0.250 0.382	1.380 1.278 1.160 0.896	1.496 1.283 1.057 0.6305	0.1150 0.1065 0.0967 0.0747	0.01039 0.00890 0.00734 0.00438
1½	40 (S) 80 (X) 160 (XX)	1.900	0.145 0.200 0.281 0.400	1.610 1.500 1.338 1.100	2.036 1.767 1.406 0.9503	0.1342 0.1250 0.1115 0.0917	0.01414 0.01227 0.00976 0.00660
2	40 (S) 80 (X) 160 (XX)	2.375	0.154 0.218 0.344 0.436	2.067 1.939 1.687 1.503	3.356 2.953 2.235 1.774	0.1723 0.1616 0.1406 0.1253	0.02330 0.02051 0.01552 0.01232
2½	40 (S) 80 (X) 160 (XX)	2.875	0.203 0.276 0.375 0.552	2.469 2.323 2.125 1.771	4.788 4.238 3.547 2.464	0.2058 0.1936 0.1771 0.1476	0.03325 0.02943 0.02463 0.01711
3	40 (S) 80 (X) 160 (XX)	3.500	0.216 0.300 0.438 0.600	3.068 2.900 2.624 2.300	7.393 6.605 5.408 4.155	0.2557 0.2417 0.2187 0.1917	0.05134 0.04587 0.03755 0.02885
3½	40 (S) 80 (X)	4.000	0.226 0.318	3.548 3.364	9.887 8.888	0.2957 0.2803	0.06866 0.06172
4	40 (S) 80 (X) 120 160 (XX)	4.500	0.237 0.337 0.438 0.531 0.674	4.026 3.826 3.624 3.438 3.152	12.73 11.50 10.32 9.283 7.803	0.3355 0.3188 0.3020 0.2865 0.2627	0.08841 0.07984 0.07163 0.06447 0.05419

S = Wall thickness formerly designated 'standard weight'
X = Wall thickness formerly designated 'extra strong'
XX = Wall thickness formerly designated 'double extra strong'
Actual wall thicknesses may vary slightly.
Extracted from American Standard Wrought Steel and Wrought Iron Pipe (ASA B36.10-1959),
   The American Society of Mechanical Engineers.

Dimensions of Welded and Seamless Steel Pipe

Nominal Diameter Inches	Schedule	Outside Diameter Inches	Wall Thickness Inches	Internal Diameter Inches	Internal Area Square Inches	Internal Diameter Feet	Internal Area Square Feet
5	40 (S)	5.563	0.258	5.047	20.01	0.4206	0.1389
	80 (X)		0.375	4.813	18.19	0.4011	0.1263
	120		0.500	4.563	16.35	0.3803	0.1136
	160		0.625	4.313	14.61	0.3594	0.1015
	(XX)		0.750	4.063	12.97	0.3386	0.09004
6	40 (S)	6.625	0.280	6.065	28.89	0.5054	0.2006
	80 (X)		0.432	5.761	26.07	0.4801	0.1810
	120		0.562	5.501	23.77	0.4584	0.1650
	160		0.719	5.187	21.13	0.4323	0.1467
	(XX)		0.864	4.897	18.83	0.4081	0.1308
8	20	8.625	0.250	8.125	51.85	0.6771	0.3601
	30		0.277	8.071	51.16	0.6726	0.3553
	40 (S)		0.322	7.981	50.03	0.6651	0.3474
	60		0.406	7.813	47.94	0.6511	0.3329
	80 (X)		0.500	7.625	45.66	0.6354	0.3171
	100		0.594	7.437	43.44	0.6198	0.3017
	120		0.719	7.187	40.57	0.5989	0.2817
	140		0.812	7.001	38.50	0.5834	0.2673
	(XX)		0.875	6.875	37.12	0.5729	0.2578
	160		0.906	6.813	36.46	0.5678	0.2532
10	20	10.75	0.250	10.250	82.52	0.85417	0.5730
	30		0.307	10.136	80.69	0.84467	0.5604
	40 (S)		0.365	10.020	78.85	0.83500	0.5476
	60 (X)		0.500	9.750	74.66	0.8125	0.5185
	80		0.594	9.562	71.81	0.7968	0.4987
	100		0.719	9.312	68.11	0.7760	0.4730
	120		0.844	9.062	64.50	0.7552	0.4479
	140 (XX)		1.000	8.750	60.13	0.7292	0.4176
	160		1.125	8.500	56.75	0.7083	0.3941
12	20	12.75	0.250	12.250	117.86	1.0208	0.8185
	30		0.330	12.090	114.80	1.0075	0.7972
	(S)		0.375	12.000	113.10	1.0000	0.7854
	40		0.406	11.938	111.93	0.99483	0.7773
	(X)		0.500	11.750	108.43	0.97917	0.7530
	60		0.562	11.626	106.16	0.96883	0.7372
	80		0.688	11.374	101.61	0.94783	0.7056
	100		0.844	11.062	96.11	0.92183	0.6674
	120 (XX)		1.000	10.750	90.76	0.89583	0.6303
	140		1.125	10.500	86.59	0.87500	0.6013
	160		1.312	10.126	80.53	0.84383	0.5592
14 OD	10	14.00	0.250	13.500	143.14	1.1250	0.9940
	20		0.312	13.376	140.52	1.1147	0.9758
	30 (S)		0.375	13.250	137.89	1.1042	0.9575
	40		0.438	13.124	135.28	1.0937	0.9394
	(X)		0.500	13.000	132.67	1.0833	0.9213
	60		0.594	12.812	128.92	1.0677	0.8953
	80		0.750	12.500	122.72	1.0417	0.8522
	100		0.938	12.124	115.45	1.0104	0.8017
	120		1.094	11.812	109.58	0.98433	0.7610
	140		1.250	11.500	103.87	0.95833	0.7213
	160		1.406	11.188	98.31	0.93233	0.6827

Dimensions of Welded and Seamless Steel Pipe

Nominal Diameter	Schedule	Outside Diameter	Wall Thickness	Internal Diameter	Internal Area	Internal Diameter	Internal Area
Inches		Inches	Inches	Inches	Square Inches	Feet	Square Feet
16 OD	10	16.00	0.250	15.500	188.69	1.2917	1.3104
	20		0.312	15.376	185.69	1.2813	1.2895
	30 (S)		0.375	15.250	182.65	1.2708	1.2684
	40 (X)		0.500	15.000	176.72	1.2500	1.2272
	60		0.656	14.688	169.44	1.2240	1.1767
	80		0.844	14.312	160.88	1.1927	1.1172
	100		1.031	13.938	152.58	1.1615	1.0596
	120		1.219	13.562	144.46	1.1302	1.0032
	140		1.438	13.124	135.28	1.0937	0.9394
	160		1.594	12.812	128.92	1.0677	0.8953
18 OD	10	18.00	0.250	17.500	240.53	1.4583	1.6703
	20		0.312	17.376	237.13	1.4480	1.6467
	(S)		0.375	17.250	233.71	1.4375	1.6230
	30		0.438	17.124	230.00	1.4270	1.5993
	(X)		0.500	17.000	226.98	1.4167	1.5762
	40		0.562	16.876	223.68	1.4063	1.5533
	60		0.750	16.500	213.83	1.3750	1.4849
	80		0.938	16.124	204.19	1.3437	1.4180
	100		1.156	15.688	193.30	1.3073	1.3423
	120		1.375	15.250	182.65	1.2708	1.2684
	140		1.562	14.876	173.81	1.2397	1.2070
	160		1.781	14.438	163.72	1.2032	1.1370
20 OD	10	20.00	0.250	19.500	298.65	1.6250	2.0739
	20 (S)		0.375	19.250	291.04	1.6042	2.0211
	30 (X)		0.500	19.000	283.53	1.5833	1.9689
	40		0.594	18.812	277.95	1.5677	1.9302
	60		0.812	18.376	265.21	1.5313	1.8417
	80		1.031	17.938	252.72	1.4948	1.7550
	100		1.281	17.438	238.83	1.4532	1.6585
	120		1.500	17.000	226.98	1.4167	1.5762
	140		1.750	16.500	213.83	1.3750	1.4849
	160		1.969	16.062	202.62	1.3385	1.4071
24 OD	10	24.00	0.250	23.500	433.74	1.9583	3.0121
	20 (S)		0.375	23.250	424.56	1.9375	2.9483
	(X)		0.500	23.000	415.48	1.9167	2.8852
	30		0.562	22.876	411.01	1.9063	2.8542
	40		0.688	22.624	402.00	1.8853	2.7917
	60		0.969	22.062	382.28	1.8385	2.6547
	80		1.219	21.562	365.15	1.7802	2.5358
	100		1.531	20.938	344.32	1.7448	2.3911
	120		1.812	20.376	326.92	1.6980	2.2645
	140		2.062	19.876	310.28	1.6563	2.1547
	160		2.344	19.312	292.92	1.6093	2.0342
30 OD	10	30.00	0.312	29.376	677.76	2.4480	4.7067
	(S)		0.375	29.250	671.62	2.4375	4.6640
	20 (X)		0.500	29.000	660.52	2.4167	4.5869
	30		0.625	28.750	649.18	2.3958	4.5082

Table 3.9

Area, Wetted Perimeter and Hydraulic Radius of Partially Filled Circular Pipes

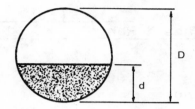

$\frac{d}{D}$	$\frac{area}{D^2}$	$\frac{wet.\ per.}{D}$	$\frac{hyd.\ rad.}{D}$	$\frac{d}{D}$	$\frac{area}{D^2}$	$\frac{wet.\ per.}{D}$	$\frac{hyd.\ rad.}{D}$
0.01	0.0013	0.2003	0.0066	0.51	0.4027	1.5908	0.2531
0.02	0.0037	0.2838	0.0132	0.52	0.4127	1.6108	0.2561
0.03	0.0069	0.3482	0.0197	0.53	0.4227	1.6308	0.2591
0.04	0.0105	0.4027	0.0262	0.54	0.4327	1.6509	0.2620
0.05	0.0147	0.4510	0.0326	0.55	0.4426	1.6710	0.2649
0.06	0.0192	0.4949	0.0389	0.56	0.4526	1.6911	0.2676
0.07	0.0242	0.5355	0.0451	0.57	0.4625	1.7113	0.2703
0.08	0.0294	0.5735	0.0513	0.58	0.4723	1.7315	0.2728
0.09	0.0350	0.6094	0.0574	0.59	0.4822	1.7518	0.2753
0.10	0.0409	0.6435	0.0635	0.60	0.4920	1.7722	0.2776
0.11	0.0470	0.6761	0.0695	0.61	0.5018	1.7926	0.2797
0.12	0.0534	0.7075	0.0754	0.62	0.5115	1.8132	0.2818
0.13	0.0600	0.7377	0.0813	0.63	0.5212	1.8338	0.2839
0.14	0.0688	0.7670	0.0871	0.64	0.5308	1.8546	0.2860
0.15	0.0739	0.7954	0.0929	0.65	0.5404	1.8755	0.2881
0.16	0.0811	0.8230	0.0986	0.66	0.5499	1.8965	0.2899
0.17	0.0885	0.8500	0.1042	0.67	0.5594	1.9177	0.2917
0.18	0.0961	0.8763	0.1097	0.68	0.5687	1.9391	0.2935
0.19	0.1039	0.9020	0.1152	0.69	0.5780	1.9606	0.2950
0.20	0.1118	0.9273	0.1206	0.70	0.5872	1.9823	0.2962
0.21	0.1199	0.9521	0.1259	0.71	0.5964	2.0042	0.2973
0.22	0.1281	0.9764	0.1312	0.72	0.6054	2.0264	0.2984
0.23	0.1365	1.0003	0.1364	0.73	0.6143	2.0488	0.2995
0.24	0.1449	1.0239	0.1416	0.74	0.6231	2.0714	0.3006
0.25	0.1535	1.0472	0.1466	0.75	0.6318	2.0944	0.3017
0.26	0.1623	1.0701	0.1516	0.76	0.6404	2.1176	0.3025
0.27	0.1711	1.0928	0.1566	0.77	0.6489	2.1412	0.3032
0.28	0.1800	1.1152	0.1614	0.78	0.6573	2.1652	0.3037
0.29	0.1890	1.1373	0.1662	0.79	0.6655	2.1895	0.3040
0.30	0.1982	1.1593	0.1709	0.80	0.6736	2.2143	0.3042
0.31	0.2074	1.1810	0.1755	0.81	0.6815	2.2395	0.3044
0.32	0.2167	1.2025	0.1801	0.82	0.6893	2.2653	0.3043
0.33	0.2260	1.2239	0.1848	0.83	0.6969	2.2916	0.3041
0.34	0.2355	1.2451	0.1891	0.84	0.7043	2.3186	0.3038
0.35	0.2450	1.2661	0.1935	0.85	0.7115	2.3462	0.3033
0.36	0.2546	1.2870	0.1978	0.86	0.7186	2.3746	0.3026
0.37	0.2642	1.3078	0.2020	0.87	0.7254	2.4038	0.3017
0.38	0.2739	1.3284	0.2061	0.88	0.7320	2.4341	0.3008
0.39	0.2836	1.3490	0.2102	0.89	0.7384	2.4655	0.2996
0.40	0.2934	1.3694	0.2142	0.90	0.7445	2.4981	0.2980
0.41	0.3032	1.3898	0.2181	0.91	0.7504	2.5322	0.2963
0.42	0.3130	1.4101	0.2220	0.92	0.7560	2.5681	0.2944
0.43	0.3229	1.4303	0.2257	0.93	0.7612	2.6061	0.2922
0.44	0.3328	1.4505	0.2294	0.94	0.7662	2.6467	0.2896
0.45	0.3428	1.4706	0.2331	0.95	0.7707	2.6906	0.2864
0.46	0.3527	1.4907	0.2366	0.96	0.7749	2.7389	0.2830
0.47	0.3627	1.5108	0.2400	0.97	0.7785	2.7934	0.2787
0.48	0.3727	1.5308	0.2434	0.98	0.7816	2.8578	0.2735
0.49	0.3827	1.5508	0.2467	0.99	0.7841	2.9412	0.2665
0.50	0.3927	1.5708	0.2500	1.00	0.7854	3.1416	0.2500

Table 3.10

Specific Roughness and Hazen-Williams Constants
for Various Pipe Materials

Type of pipe or surface	ε(ft) Range	Design	C Range	Clean	Design
STEEL					
welded and seamless	.0001-.0003	.0002	150-80	140	100
interior riveted, no					
projecting rivets				139	100
projecting girth rivets				130	100
projecting girth and					
horizontal rivets				115	100
vitrified, spiral-riveted,					
flow with lap				110	100
vitrified, sprial-riveted,					
flow against lap				100	90
corrugated				60	60
MINERAL					
concrete	.001-.01	.004	152-85	120	100
cement-asbestos			160-140	150	140
vitrified clays					110
brick sewer					100
IRON					
cast, plain	.0004-.002	.0008	150-80	130	100
cast, tar (asphalt)					
coated	.0002-.0006	.0004	145-50	130	100
cast, cement lined	.000008	.000008		150	140
cast, bituminous lined	.00008	.00008	160-130	148	140
cast, centrifugally spun	.00001	.00001			
galvanized, plain	.0002-.0008	.0005			
wrought, plain	.0001-.0003	.0002	150-80	130	100
MISCELLANEOUS					
fiber				150	140
copper and brass	.000005	.000005	150-120	140	130
wood stave	.0006-.003	.002	145-110	120	110
transite	.000008	.000008			
lead, tin, glass		.000005	150-120	140	130
plastic		.000005	150-120	140	130

Bibilography

1. Binder, R.C., Fluid Mechanics, 4th editon, Prentice Hall, Inc., New York, NY, 1962

2. Colt Industries, Hydraulic Handbook, 10th edition, Kansas City, KS, 1977

3. Crane Company, Technical Paper No. 410, Chicago, IL, 1976

4. De Nevers, Noel, Fluid Mechanics, Addison-Wesley, Reading, MA, 1970

5. Hydraulic Institute, Pipe Friction Manual, 3rd edition, Cleveland, OH, 1961

6. King, Reno C., Piping Handbook, 5th edition, McGraw-Hill Book Company, New York, NY, 1973

7. Linsley, Ray K., and Franzini, Joseph B., Water Resources Engineering, 2nd edition, McGraw-Hill Book Company, New York, NY, 1972

8. Olson, Reuben M., Essentials of Engineering Fluid Mechanics, International Textbook Company, Scranton, PA, 1964

9. Parker, Jerald D., Boggs, James H., and Blick, Edward F., Introduction to Fluid Mechanics and Heat Transfer, Addison-Wesley, Reading, MA, 1969

10. Peckworth, Howard F., Concrete Pipe Handbook, American Pipe Association, Chicago, IL, 1951

11. Streeter, Victor L., and Wylie, E. Benjamin, Fluid Mechanics, 6th edition, McGraw-Hill Book Company, New York, NY, 1975

12. Westaway, C.R., and Loomis, A.W., Cameron Hydraulic Data, 15th edition, Ingersoll-Rand Company, Woodcliff Lake, NJ, 1977

13. Vennard, John K., Elementary Fluid Mechanics, 2nd edition, John Wiley & Sons, Inc., New York, NY, 1947

Practice Problems: HYDRAULICS

Required

1. Three reservoirs (A, B, and C) are interconnected with a common junction at elevation 25 feet above some arbitrary reference point. The water surface levels for reservoirs A, B, and C are at elevations of 50, 40, and 22 feet respectively. The pipe from reservoir A to the junction is 800 feet of 3" steel pipe. The pipe from reservoir B to the junction is 500 feet of 10" steel pipe. The pipe from reservoir C to the junction is 1000 feet of 4" steel pipe. Is the flow into or out of reservoir B? Neglect minor losses and velocity heads. Assume f = .02.

2. A cast iron pipe has an outside diameter of 24" and a wall thickness of 0.75". Water is flowing at 6 fps. (a) If a valve is closed instantaneously, what will be the pressure created? (b) If the pipe is 500 feet long, over what length of time must be valve be closed to create a pressure equivalent to instantaneous closure?

3. Assume C = 100 and find the flow in each of the pipes in the distribution system shown below. An accuracy of ± 10 gpm is acceptable.

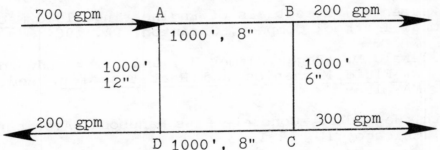

Optional

4. The distance between each labeled point on the distribution graph below is 1000 feet. Assume C = 100 for each pipe section. What is the pressure at all labeled points? If the pump receives 20 psig water, what horsepower is required?

point	pressure	elevation
A		200 ft
B		150
C	40 psig	300
D		150
E		200
F		150

5. Water is carried in a pipe which changes gradually from 6" at point A to 18" at point B. B is 15 feet higher than A. 5 cfs are flowing and the respective pressures at A and B are 10 psia and 7 psia. What is the direction of flow?

6. Points A and B are 3000 feet apart along a new 6" steel pipe. B is 60 feet above A. 750 gpm of 60°F water flow in the pipe. The flow direction is from A to B. What pressure must be maintained at A if the pressure at B is to be 50 psig?

7. A cylindrical tank 20 feet in diameter and 40 feet high has a 4" hole in the bottom with C_d = .98. How long will it take for the water level to drop from 40' to 20'?

8. A venturi meter with an 8" diameter throat is installed in a 12" diameter water line. Assume the venturi coefficient is equal to one. What is the flow in cubic feet per second if a mercury manometer registers a 4" differential?

9. What will be the measured pressure drop across a .2 foot diameter sharp-edged orifice installed in a 1 foot diameter pipe if 70°F water is flowing at 2 fps?

10. A pipe necks down from 24" at point A to 12" at point B. The discharge is 8 cfs in the direction of A to B. The pressure head at A is 20 feet. Assume no friction. Find the resultant force and its direction on the fluid if water is flowing.

NOTES

Hydraulic Machines

Nomenclature

bhp	brake horsepower	hp
c	specific heat	BTU/lb-°F
C_v	coefficient of velocity	-
d	diameter	inches
D	diameter	ft
ehp	electrical horsepower	hp
E	energy	ft-lb/lb
f	Darcy friction factor	-
fhp	friction horsepower	hp
g	acceleration due to gravity (32.2)	ft/sec²
h	static head	ft
H	dynamic head	ft
L	pipe length	ft
n	rotational speed	rpm
n_s	specific speed	rpm
n_{ss}	suction specific speed	rpm
NPSHA	net positive suction head available	ft
NPSHR	net positive suction head required	ft
p	pressure	psi
Q	flow quantity	gpm
S.G.	specific gravity	-
T	temperature	°F
v	velocity	ft/sec
w	weight	lb
whp	hydraulic ('water') horsepower	hp
z	height above datum	ft

Symbols

η	efficiency	-
ε	specific pipe roughness	ft
ρ	density	lb/ft³
β	blade angle	

Subscripts

a	atmospheric	p	pump or pressure	
A	added by pump	s	suction	
d	discharge	sd	static discharge	
f	friction	th	theoretical	
i	intake	ts	total static	
j	jet	T	turbine	
m	motor	v	velocity	
n	nozzle	v_p	vapor pressure	

1. Introduction

Pumps and turbines are the two types of hydraulic machines discussed in this chapter. Pumps convert mechanical energy into fluid energy. Turbines convert fluid energy into mechanical energy.

2. Pump Operation

Unless otherwise specified, pumps in this chapter are assumed to be centrifugal pumps. Liquid flowing into the suction side (the inlet) is captured by an impeller and thrown to the outside of the pump casing. Within the casing, the velocity imparted to the fluid by the impeller is converted into pressure energy.

Figure 4.1

A Centrifugal Pump

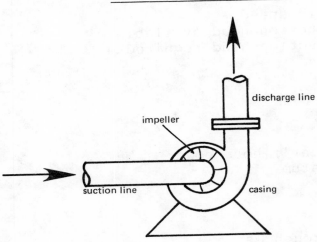

Two quantities are needed to select or evaluate a centrifugal pump: head added and flow rate. The head added by the pump (assuming pressures are in psf) is

$$h_A = \frac{p_d}{\rho} - \frac{p_i}{\rho} + \frac{v_d^2}{2g} - \frac{v_i^2}{2g} + z_d - z_i \qquad 4.1$$

The capacity or flow rate is governed by the impeller thickness. For a given impeller diameter, the wider the vanes, the greater the capacity of the pump. For a desired flow rate or a desired discharge head, there will be one optimum impeller design. The impeller that is best for developing a high discharge pressure will have different proportions than an impeller designed to produce a high flow rate. The quantitative index of this optimization is called specific speed (n_S). Specific speed is discussed in section 6 of this chapter.

3. Types of Pumps

Centrifugal pumps can be classified into three general categories according to the way the impeller imparts energy to the fluid. Each of these categories has its own range of specific speeds.

A. Centrifugal (radial) flow impellers impart energy primarily by centrifugal force. Liquid enters the impeller at the hub and flows radially to the outside of the casing. Single suction impellers have a specific speed less than 5000. Double suction impellers have a specific speed less than 6000.

Figure 4.2

Centrifugal (Radial) Flow Pump

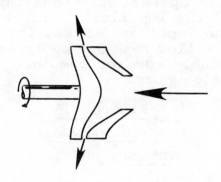

B. Mixed flow impellers impart energy partially by centrifugal force and partially by axial force, since the vanes are acting partly as an axial compressor. Liquid enters the impeller at the hub and flows both radially and axially to discharge. Specific speeds of mixed flow pumps range from 4200 to 9000.

Figure 4.3

Mixed Flow Pump

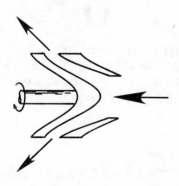

C. Axial flow impellers impart energy to the fluid by acting as axial flow compressors. Fluid enters and exits along the axis of rotation. Specific speed is greater than 9000.

Figure 4.4

Axial Flow Pump

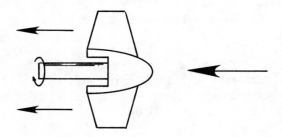

Radial flow and mixed flow centrifugal pumps may be designed for either single or double suction operation. In a single suction pump, fluid enters only one side of the impeller. In a double suction pump, fluid enters both sides of the impeller. Thus, for an impeller with a given specific speed, a greater flow rate can be expected from a double suction pump. In addition, a double suction pump has a lower NPSHR for a given flow than a single suction pump.

Figure 4.5

Radial Flow Pump

(Double Suction)

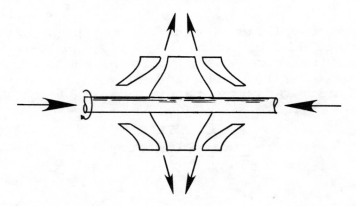

A multiple stage pump consists of two or more impellers within a single casing, such that the discharge of the first stage becomes the input of the second stage. In this manner, higher heads are achieved than would be possible with a single impeller.

4. Pump Head Terms

A. Friction head (h_f): The head required to overcome resistance to flow in the pipe, fittings, valves, entrances, and exits. (Refer to Chapter 3.)

$$h_f = \frac{fL_e v^2}{2Dg}$$

4.2

B. Velocity head (h_v): The head of a fluid as a result of its kinetic energy.

$$h_v = \frac{v^2}{2g}$$

4.3

C. Atmospheric head (h_a): Atmospheric pressure converted to feet of fluid being pumped. (The 2.31 constant is obtained by dividing 144 by 62.4.)

$$h_a = \frac{(p_a)(2.31)}{S.G.}$$

4.4

D. Pressure head (h_p): Pressure converted to feet of fluid being pumped.

$$h_p = \frac{(p)(2.31)}{S.G.}$$

4.5

E. Vapor pressure head (h_{vp}): Fluid vapor pressure converted to feet of fluid being pumped.

$$h_{vp} = \frac{(p_{vp})(2.31)}{S.G.}$$

4.6

F. Static suction head (h_s): The vertical distance in feet above the centerline of the inlet to the free level of the fluid source. If the free level of the fluid source is below the inlet, h_s will be negative. In this case, h_s is known as static suction lift.

Figure 4.6
Static Suction Head

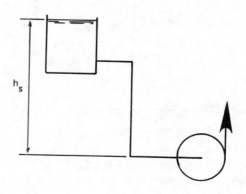

Figure 4.7
Static Suction Lift

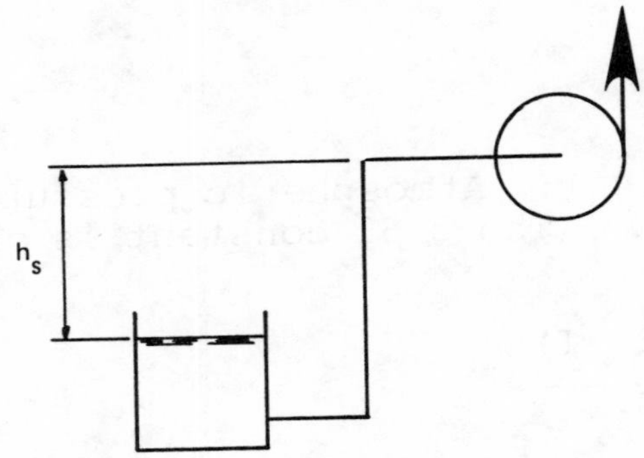

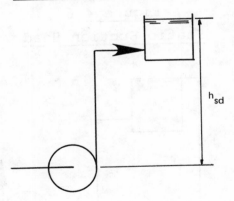

G. Total (dynamic) suction head (H_s): The static suction head minus the friction head in the suction line (i.e., the total energy of the fluid entering the impeller). If h_s is negative (i.e., the free level of the fluid source is below the inlet), then H_s will be negative. In this case, H_s is known as total (dynamic) suction lift.

$$H_s = h_s - h_{f(s)}$$ 4.7

H. Static discharge head (h_{sd}): The vertical distance in feet above the pump centerline to the free level of the discharge tank or point of free discharge.

Figure 4.8
Static Discharge Head

I. Total (dynamic) discharge head (H_d): The static discharge head plus the discharge velocity head plus the friction head in the discharge line (i.e., the total energy of the fluid leaving the pump).

$$H_d = h_{sd} + h_{vd} + h_{f(d)}$$ 4.8

$$= h_{sd} + \frac{v_d^2}{2g} + h_{f(d)}$$ 4.9

J. Total static head (h_{ts}): The vertical distance in feet between the free level of the supply and either the point of free discharge or the free level of the discharge tank.

Figure 4.9
Total Static Head

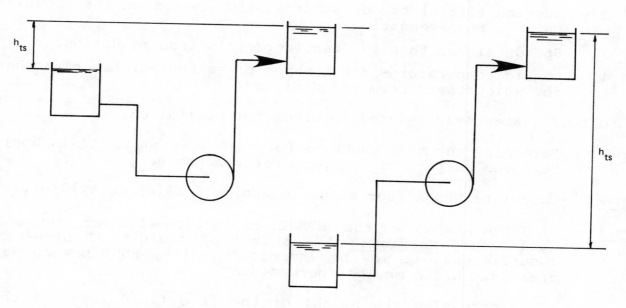

K. Total (dynamic) head (H): The total discharge head less the total suction head.

$$H = H_d - H_s \qquad\qquad 4.10$$

$$= h_{sd} - h_s + \frac{v_d^2}{2g} + h_{f(s)} + h_{f(d)} \qquad\qquad 4.11$$

5. Net Positive Suction Head and Cavitation

Liquid is not sucked into a pump. A positive head (normally atmospheric pressure) must push the liquid into the impeller (i.e., 'flood' the impeller). Net Positive Suction Head Required (NPSHR) is the minimum fluid energy required at the inlet by the pump for satisfactory operation. NPSHR is usually specified by the pump manufacturer. Net Positive Suction Head Available (NPSHA) is the actual fluid energy at the inlet.

$$NPSHA = h_a + h_s - h_{f(s)} - h_{vp} \qquad\qquad 4.12(a)$$

$$= h_{p(i)} + h_{v(i)} - h_{vp} \qquad\qquad 4.12(b)$$

If NPSHA is less than the NPSHR, the fluid will cavitate. Cavitation is the vaporization of fluid within the casing or suction line. If the fluid pressure is less than the vapor pressure, pockets of vapor will form. As vapor pockets reach the surface of the impeller, the local high fluid pressure will collapse them causing noise, vibration, and possible structural damage.

Cavitation may be caused by any of the following conditions:

1. Discharge heads far below the pump's calibrated head at peak efficiency.

2. Suction lift higher or suction head lower than the manufacturer's recommendation.

3. Speeds higher than the manufacturer's recommendation.

4. Liquid temperatures (thus vapor pressures) higher than that for which the system was designed.

The following steps may be used to check for cavitation:

step 1: Determine the minimum NPSHR for the given pump. This should be given as part of the pump performance data.

step 2: Calculate NPSHA from either equation 4.12(a) or 4.12(b).

step 3: If NPSHA is greater than NPSHR, cavitation will not occur. A good safety margin is 2 - 3 feet of fluid. If NPSHA is insufficient, it may be increased or the NPSHR may be decreased. NPSHA may be increased by:

a. increasing the height of the free fluid level of the supply tank,

b. reducing the distance and minor losses between the supply tank and the pump,

c. reducing the temperature of the fluid, or

d. pressurizing the supply tank.

NPSHR may be reduced by:

a. placing a throttling valve in the discharge line. (This will increase the total head thereby reducing the capacity of the pump and driving its operating point into a region of lower NPSHR.) Or, by

b. using a double suction pump.

Example 4.1

2 cfs of water are pumped from a feed tank mounted on a platform to an open reservoir through 6" schedule 40 ($\epsilon/D = .000293$) steel pipe. Determine the static suction head, total suction head, and NPSHA.

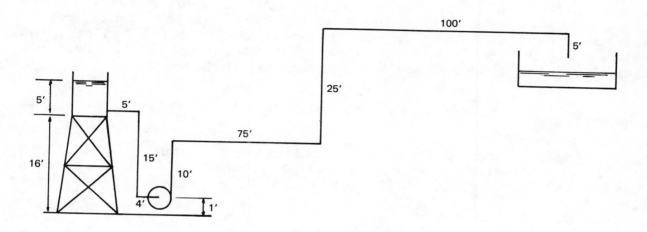

1. Assume 60°F and 14.7 psia. From equation 4.4,

$$h_a = \frac{(14.7)(2.31)}{(1)} = 33.96 \text{ ft}$$

2. For 6" schedule 40 steel pipe, D = .505 ft, A = .201 ft^2

3. $v = \frac{Q}{A} = \frac{2}{.201} = 9.95 \text{ ft/sec}$

4. The equivalent lengths of the pipe and flanged fittings are:

square entrance loss	1 x 16	= 16
90° long radius elbows	2 x 5.7	= 11.4
pipe run (5 + 15 + 4)		24
		51.4 ft

5. At 60°, the kinematic viscosity of water is 1.217 EE-5 ft^2/sec. The vapor pressure is .59 feet of water.

6. The Reynolds number is

$$N_{Re} = \frac{(.505)(9.95)}{1.217 \text{ EE-5}} = 4.13 \text{ EE5}$$

7. From the Moody friction factor chart, f = .0165, so

$$h_f = \frac{(.0165)(51.4)(9.95)^2}{(2)(.505)(32.2)} = 2.6 \text{ ft}$$

8. From equation 4.12(a),

$$NPSHA = 33.96 + 20 - 2.6 - .59 = 50.77 \text{ ft}$$

◆

6. Specific Speed (n_s):

Specific speed is a function of a pump's capacity, head, and rotational speed at peak efficiency. For a given pump and impeller configuration, the specific speed remains essentially constant over a range of flow rates and heads. Theoretically, specific speed is the rpm at which a homologous pump would have to turn in order to put out 1 gpm at 1 foot total head. (For double suction pumps, Q in equation 4.13 is divided by 2.) Use equation 4.37 for specific speeds of turbines.

$$n_s = \frac{n\sqrt{Q}}{H^{.75}} \qquad\qquad 4.13$$

Specific speed is used as a guide to selecting the most efficient pump type. Given a desired flow rate, pipeline geometry and motor speed, n_s is calculated from equation 4.13. The type of impeller is chosen from Table 4.1

Table 4.1
Specific Speed versus Impeller Types

Approximate Range of Specific Speed (rpm)	Impeller Type
500 - 1000	Radial Vane
2000 - 3000	Francis (Mixed) Vane
4000 - 7000	Mixed Flow
9000 and above	Axial Flow

Highest heads per stage are developed at low specific speeds. However, for best efficiency, a centrifugal pump's specific speed should be greater than 650 at its operating point. At low specific speeds, the impeller diameter is large with high mechanical friction and hydraulic losses. If the specific speed for a given set of conditions drops below 650, a multiple stage pump should be selected.

As the specific speed increases, the ratio of the impeller diameter to the inlet diameter decreases. As this ratio decreases, the pump is capable of developing less head. Best efficiencies are usually obtained from pumps with specific speeds between 1500 and 3000. At specific speeds of 10,000 or higher, the pump is suitable for high flow rates but low discharge heads.

7. Suction Specific Speed (n_{ss}):

If NPSHR is substituted for total head in the expression for specific speed, a formula for suction specific speed results.

Suction specific speed is an index of the suction characteristics of the impeller.

$$n_{ss} = \frac{n\sqrt{Q}}{NPSHR^{.75}} \qquad\qquad 4.14$$

Ideally, n_{ss} is 7900 for single suction pumps and 11,200 for double suction pumps. For double suction pumps, the value of Q in equation 4.14 is reduced by 50%.

Example 4.2

A pump is to discharge 150 gpm against a 300 foot total head at 3500 rpm. What type of pump should be selected?

1.   Calculate the specific speed from equation 4.13.

$$n_s = \frac{3500\sqrt{150}}{(300)^{.75}} = 594.7 \text{ rpm}$$

From Table 4.1, the pump should be a radial vane type. However, pumps with best efficiencies have n_s greater than 650. To increase the specific speed, the rotational speed may be increased or the total head may be decreased. Since 3600 rpm is about the maximum practical speed for centrifugal pumps, the better choice would be to divide the total head between two stages (or use two pumps in a series).

In a two stage system, the specific speed would be:

$$n_s = \frac{3500\sqrt{150}}{(150)^{.75}} = 1000$$

This is satisfactory for a radial vane pump.

8. Power and Efficiency

The work performed by a pump is a function of the total head and the weight of the liquid pumped in a given time. Pump output is measured in hydraulic horsepower. Relationships for finding the hydraulic horsepower are given in Table 4.2

Table 4.2
Hydraulic Horsepower Equations

	Q in gpm	$\dot{w}$ in lb/sec	$\dot{Q}$ in cfs
H is added head in feet	$\dfrac{HQ\,(S.G.)}{3960}$	$\dfrac{H\,\dot{w}}{550}$	$\dfrac{H\dot{Q}(S.G.)}{8.814}$
p is added head in psf	$\dfrac{pQ}{2.468\ EE5}$	$\dfrac{p\,\dot{w}}{(34320)(S.G.)}$	$\dfrac{p\dot{Q}}{550}$

The input horsepower delivered to the pump shaft is:

$$bhp = whp/\eta_p \qquad\qquad 4.15$$

The difference between hydraulic horsepower and brake horsepower is the power lost within the pump due to mechanical and hydraulic friction. This is referred to as heat horsepower and is determined from equation 4.16.

$$fhp = bhp - whp \qquad\qquad 4.16$$

Electrical power to the motor is

$$ehp = bhp/\eta_m \qquad\qquad 4.17$$

Overall efficiency is the pump efficiency multiplied by the motor efficiency:

$$\eta = (\eta_p)(\eta_m) = whp/ehp \qquad\qquad 4.18$$

Figure 4.10

Efficiency versus Specific Speed

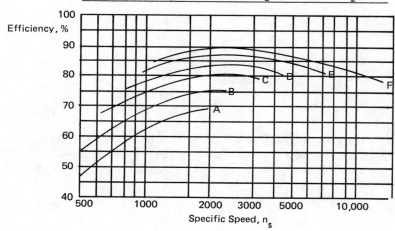

curve A: 100 gpm
curve B: 200 gpm
curve C: 500 gpm
curve D: 1000 gpm
curve E: 3000 gpm
curve F   10,000 gpm

Table 4.3

Standard Motor Sizes (BHP)

.5, .75, 1, 1.5, 2, 3, 5, 7.5
10, 15, 20, 25, 30, 40, 50, 60,
75, 100, 125, 150, 200, 250

Table 4.4

Standard Motor Speeds (rpm)

60 Cycle

Synchronous	Fully Loaded
3600	3500
1800	1770
1200	1170
900	870

Example 4.3

Recommend a motor size for the pump in example 4.1 if the friction loss in the discharge line is 12.00 feet.

The Bernoulli equation may be used to find the head added by the pump.

$$0 + 0 + 20 + H = 0 + \frac{(9.95)^2}{2(32.2)} + 30 + 2.6 + 12$$

$$H = 26.14 \text{ ft}$$

From Table 4.2,

$$whp = \frac{H\dot{Q}(S.G.)}{8.814} = \frac{(26.14)(2)(1)}{8.814} = 5.93 \text{ hp}$$

The flow rate is $(2)/(.002228) = 897.7$ gpm.

Assuming a motor speed of 1200 rpm, equation 4.13 gives the specific speed:

$$n_s = \frac{1200\sqrt{897.7}}{(26.14)^{.75}} = 3110 \text{ rpm}$$

From Figure 4.10, a pump efficiency of about 82% can be expected. So, the minimum motor horsepower would be 5.93/.82 = 7.23. From table 4.3, choose a 7.5 hp motor.

9. Affinity Laws - Centrifugal Pumps

Most parameters (impeller diameter, speed, and flow rate) determining a pump's performance can be varied. If the impeller diameter is held constant and the speed varied, the following ratios are maintained with no change of efficiency:

$$\frac{Q_2}{Q_1} = \frac{n_2}{n_1} \qquad\qquad 4.19$$

$$\frac{H_2}{H_1} = \left(\frac{n_2}{n_1}\right)^2 = \left(\frac{Q_2}{Q_1}\right)^2 \qquad\qquad 4.20$$

$$\frac{bhp_2}{bhp_1} = \left(\frac{n_2}{n_1}\right)^3 = \left(\frac{Q_2}{Q_1}\right)^3 \qquad\qquad 4.21$$

If the speed is held constant and the impeller size varied,

$$\frac{Q_2}{Q_1} = \frac{d_2}{d_1} \qquad\qquad 4.22$$

$$\frac{H_2}{H_1} = \left(\frac{d_2}{d_1}\right)^2 \qquad\qquad 4.23$$

$$\frac{bhp_2}{bhp_1} = \left(\frac{d_2}{d_1}\right)^3 \qquad\qquad 4.24$$

Example 4.4

A pump delivers 500 gpm against a total head of 200 feet operating at 1770 rpm. Changes have increased the total head to 375 feet. At what rpm should this pump be operated to achieve this new head at the same efficiency?

From equation 4.20,

$$n_2 = 1700\sqrt{375/200} = 2424$$

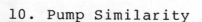

10. Pump Similarity

The performance of one pump can be used to predict the performance of a dynamically similar (homologous) pump. This can be done by using equations 4.25 through 4.30.

$$\frac{n_1 d_1}{\sqrt{H_1}} = \frac{n_2 d_2}{\sqrt{H_2}} \qquad\qquad 4.25$$

$$\frac{Q_1}{d_1^2 \sqrt{H_1}} = \frac{Q_2}{d_2^2 \sqrt{H_2}} \qquad\qquad 4.26$$

$$\frac{bhp_1}{\rho_1 d_1^2 H_1^{1.5}} = \frac{bhp_2}{\rho_2 d_2^2 H_2^{1.5}} \qquad\qquad 4.27$$

$$\frac{Q_1}{n_1 d_1^3} = \frac{Q_2}{n_2 d_2^3} \qquad\qquad 4.28$$

$$\frac{bhp_1}{\rho_1 n_1^3 d_1^5} \qquad \frac{bhp_2}{\rho_2 n_2^3 d_2^5} \qquad\qquad 4.29$$

$$\frac{n_1 \sqrt{Q_1}}{H_1^{.75}} = \frac{n_2 \sqrt{Q_2}}{H_2^{.75}} \qquad\qquad 4.30$$

Example 4.5

A 6" pump operating at 1770 rpm discharges 1500 gpm of cold water (S.G. = 1.0) against an 80 foot head at 80% efficiency. A homologous 8" pump operating at 1170 rpm is being considered as a replacement. What capacity and total head can be expected from the new pump? What would be the new power requirement?

From equation 4.25,

$$H_2 = [\frac{(8)(1170)}{(6)(1770)}]^2 (80) = 62.14 \text{ feet}$$

From equation 4.28,

$$Q_2 = \frac{(1170)(8)^3}{(1770)(6)^3}(1500) = 2350 \text{ gpm}$$

$$whp_2 = \frac{(2350.3)(62.14)(1.0)}{3960} = 36.88 \text{ hp}$$

$$bhp_2 = 36.88/.8 = 46.1 \text{ hp}$$

◆

11. Pump Performance Curves

Evaluating the performance of a pump is often simplified by examining a graphical representation of its operating characteristics. For a given impeller diameter and constant speed, the head added by a centrifugal pump will decrease as the flow rate increases. This is illustrated by Figure 4.11. Other operating characteristics also vary with the flow rate. These can be presented on individual graphs. However, since the independent variable (flow rate) is the same for all, common practice is to plot all characteristics together on a single graph. This is illustrated in Figure 4.12.

Figure 4.11

Head versus Flow Rate

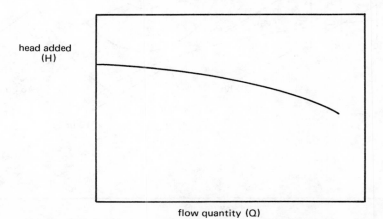

head added (H)

flow quantity (Q)

Figure 4.12

Pump Performance Curves

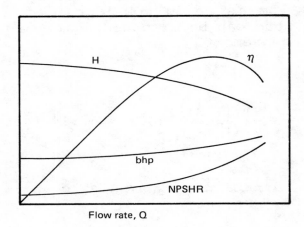

Flow rate, Q

Figures 4.11 and 4.12 are for a pump with a fixed impeller diameter and
rotational speed. The characteristics of a pump operated over a range
of speeds are illustrated in Figure 4.13. For maximum efficiency, the
operating point should fall along the dotted line.

Figure 4.13

Performance at Different Speeds

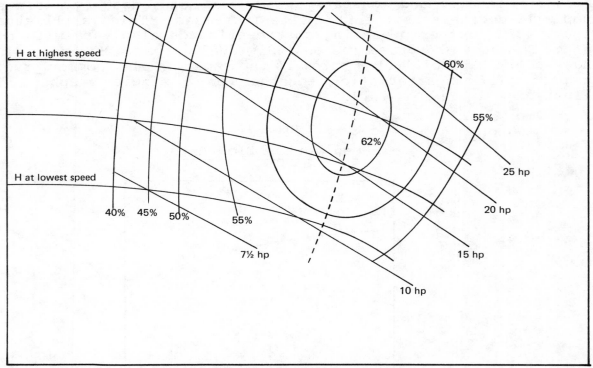

Flow rate, Q

Manufacturers' performance curves show pump performance at a limited number of calibration speeds. The desired operating point may be outside the range of the published curves. It is then necessary to estimate a speed at which the pump would give the required performance. This is done by using the affinity laws, as illustrated in example 4.6.

Example 4.6

A pump with the 1750 rpm performance curve shown below is required to pump 500 gpm at 425 foot total head. At what speed must this pump be driven to achieve the desired performance with no change in efficiency or impeller size?

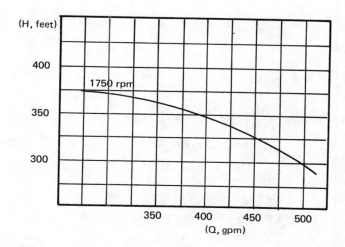

From equation 4.20, the quantity (H/Q^2) is constant for a pump with a given impeller size. In this case,

$$\frac{425}{(500)^2} = 1.7 \text{ EE-3}$$

In order to apply an affinity equation, it is necessary to know the operating point on the 1750 rpm curve. To find the operating point, choose random values of Q and solve for H such that $(H/Q^2) = 1.7$ EE-3.

Q	H
475	383
450	344
425	307
400	272

These four points are shown on the following graph. Notice that the intersection of the constant efficiency line and the original 1750 rpm curve is at 440 gpm.

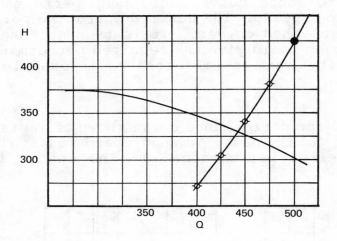

Then, from equation 4.19,

$$n_2 = 1750\left(\frac{500}{440}\right) = 1989 \text{ rpm}$$

◆

12. System Curves

A graph can also be made of the resistance to flow of the piping system. This resistance varies with the square of the flow rate since h_f varies with v^2 in the Darcy friction formula.

$$\frac{H_1}{Q_1^2} = \frac{H_2}{Q_2^2} \qquad\qquad 4.31$$

Equation 4.31 is illustrated by Figure 4.14 in which there is no static head (h_{ts}) to overcome.

Figure 4.14
System Performance Curve
(Dynamic Losses Only)

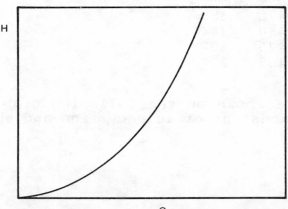

When a static head (h_{ts}) exists in a system, the loss curve is displaced upward an amount equal to the static head. This is illustrated in Figure 4.15.

Figure 4.15

System Performance Curve

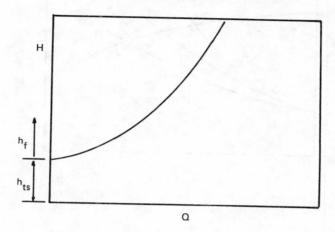

The intersection of the pump characteristic curve with the system curve defines the operating point as shown in Figure 4.16.

Figure 4.16

Operating Point

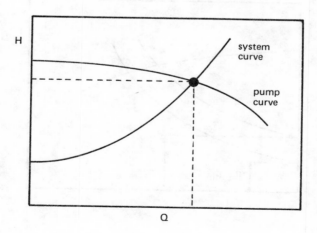

After a pump is installed, it may be desired to vary the pump's performance. If a valve is placed in the discharge line, the operating point may be moved along the performance curve by opening or closing the valve. This is illustrated in Figure 4.17. (A throttling valve should never be placed in the intake line since that would reduce NPSHA.)

Figure 4.17

Effect of Throttling the Discharge

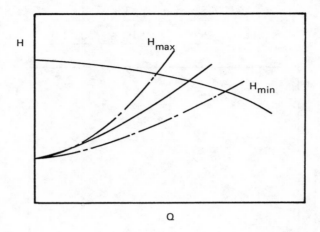

In most systems, the static head will vary as the feed tank is drained or as the discharge tank fills. The system head is then defined by a pair of parallel curves intercepting the performance curve. The two intercept points are the maximum and the minimum capacity requirements.

Figure 4.18

Extreme Operating Points

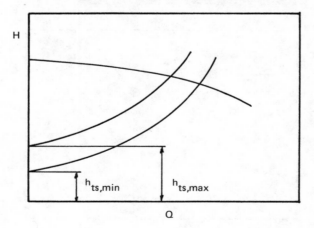

13. Pumps in Series or in Parallel

Parallel operation is obtained by having two pumps discharging into a common header. This type of connection is advantageous when the system demand varies greatly. A single pump providing total flow would have to operate far from its optimum efficiency at one point or other. With two pumps in parallel, one can be shut down during low demand. This allows the remaining pump to operate close to its optimum efficiency point.

Figure 4.19 illustrates that parallel operation increases the capacity of the system while maintaining the same total head.

Figure 4.19

Pumps Operating in Parallel

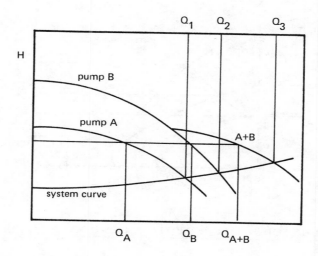

The performance curve for a set of pumps in parallel can be plotted by adding the capacities of the two pumps at various heads. Capacity does not increase at heads above the maximum head of the smaller pump. Furthermore, a second pump will operate only when its discharge head is greater than the discharge head of the pump already running.

When the parallel performance curve is plotted with the system head curve, the operating point is the intersection of the system curve with the A + B curve. With pump A operating alone, the capacity is given by Q_1. When pump B is added, the capacity increases to Q_3 with a slight increase in total head.

Series operation is achieved by having one pump discharge into the suction of the next. This arrangement is used primarily to increase the discharge head, although a small increase in capacity also results.

The performance curve for a set of pumps in series can be plotted by adding the heads of the two pumps at various capacities.

Figure 4.20
Pumps Operating in Series

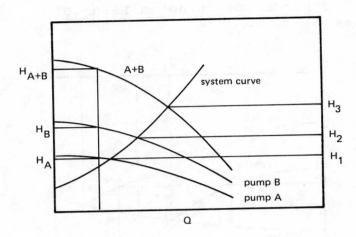

14. Special Considerations for Pumping Sewage

The primary consideration in choosing a pump to lift sewage is the pump's tendency to clog. Centrifugal pumps for sewage and liquids with large solids should always be of the single-suction type with non-clog, open impellers. Clogging can be minimized by limiting the number of impeller vanes to two or three, providing for large passageways, using a bar screen ahead of the pump, and choosing a single-suction pump. (Double suction pumps are prone to clogging because rags will catch and wrap around the shaft which extends through the impeller eye.)

Non-clog pumps are of heavy construction but are constructed for ease of cleaning and repair. Horizontal pumps usually have a split casing, one-half of which can be removed for maintenance. A hand-sized cleanout opening may also be built into the casing. Although designed for long life, a sewage pump should normally be used with a grit chamber for prolonged bearing life.

The solid-handling capacity of a pump may be given in terms of the largest sphere which can pass through it without clogging. For example, a wastewater pump with a 6" inlet should be able to pass a 4" sphere. Of course, the pump should also be capable of handling spheres with diameters slightly larger than the bar screen spacing.

15. Impulse Turbines

As shown in Figure 4.21, an impulse turbine converts the energy of a fluid stream into kinetic energy by use of a nozzle which directs the stream jet against the turbine blades. Impulse turbines are generally employed where the available head exceeds 800 feet.

Figure 4.21
A Simple Impulse Turbine

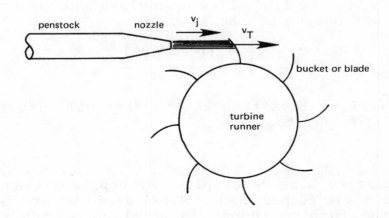

The total head available to an impulse turbine is given by equation 4.32. (p is the pressure of the fluid at the nozzle entrance.)

$$H = \frac{p}{\rho} + \frac{v^2}{2g} - h_n \qquad 4.32(a)$$

$$= h_s - \frac{fL_e v^2}{2Dg} - h_n \qquad 4.32(b)$$

The nozzle loss is

$$h_n = (\frac{p}{\rho} + \frac{v^2}{2g})(1 - C_v^2) \qquad 4.33$$

The velocity of the fluid jet is

$$v_j = C_v \sqrt{2gH} \qquad 4.34$$

Figure 4.22
Turbine Blade Geometry

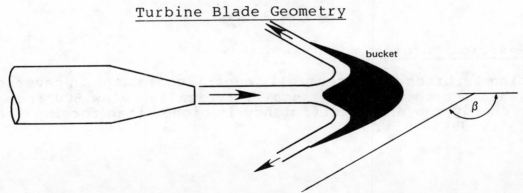

The energy transmitted by each pound of fluid to the turbine runner is given by equation 4.35.

$$E = \frac{v_T(v_j - v_T)}{g}(1 - \cos\beta) \qquad\qquad 4.35$$

Multiplying the energy by the fluid flow rate gives an expression for the theoretical horsepower output of the turbine.

$$bhp_{th} = \frac{Q\rho(v_j - v_T)v_T(1 - \cos\beta)}{(2.47\ EE5)g} \qquad\qquad 4.36$$

The actual output will be less than the theoretical output. Typical efficiencies range from 80% to 90%.

Example 4.7

A Pelton wheel impulse turbine developing 100 bhp is driven by a water stream from an 8" schedule 40 penstock. Total head (before nozzle loss) is 200 feet. If the turbine runner is rotating at 500 rpm and its efficiency is 80%, determine the diameter of the jet, the flow rate, and the pressure head at the nozzle entrance. (C_v = .95)

From equation 4.34, the jet velocity is

$$v_j = .95\sqrt{2(32.2)(200)} = 107.8 \text{ ft/sec}$$

Using Table 4.2, the flow rate is

$$Q = \frac{(8.814)(100)}{(200)(1)(.8)} = 5.51 \text{ cfs}$$

The jet area is

$$A_j = 5.51/107.8 = .051 \text{ ft}^2$$

The flow area of 8" schedule 40 pipe is .3474 ft^2

The velocity at the nozzle entrance is 5.51/.3474 = 15.89 fps.

The pressure head at the nozzle entrance is

$$200 - \frac{(15.89)^2}{2(32.2)} = 196 \text{ ft}$$

◆

16. Reaction Turbines

Reaction turbines are essentially centrifugal pumps in reverse. They are used when the total head is small, typically below 800 feet. However, their energy conversion efficiency is higher than for impulse turbines, typically 90% - 95%.

Reaction turbines are classified in the same way as centrifugal pumps, according to the manner in which the impeller extracts energy from the fluid. Each of these types is associated with a range of specific speeds.

1. Centrifugal radial flow and mixed flow turbines are designed to operate most efficiently under heads of 80' to 600' with specific speeds ranging from 10 to 110. Radial flow turbines have the lowest specific speeds. Best efficiencies are found in turbines with specific speeds between 40 and 60.

2. Axial flow (propeller) turbines operate with specific speeds between 100 and 200 rpm and have the best efficiencies between 120 and 160.

$$n_s = \frac{n\sqrt{bhp}}{H^{1.25}}$$

4.37

Since reaction turbines are centrifugal pumps in reverse, all of the affinity and similarity relationships (equations 4.19 through 4.30) may be used when comparing homologous turbines.

Example 4.8

A reaction turbine developes 500 bhp when running at 500 rpm. Flow through the turbine is 50 cfs. Water enters at 20 fps with a 100' pressure head. The turbine diameter is 24". Elevation of the turbine above tailwater level is 10'. Find the effective head and turbine efficiency.

The effective (total) fluid head is

$$H = 100 + \frac{(20)^2}{2(32.2)} + 10 = 116.2 \text{ ft}$$

From Table 4.2,

$$(bhp)_{in} = \frac{(116.2)(50)(1)}{8.814} = 659.2 \text{ hp}$$

$$n_T = \frac{500}{659.2} = .758$$

Bibliography

1. Binder, Raymond C., Fluid Mechanics, 4th edition, Prentice-Hall, Inc., Englewood Cliffs, NJ, 1962

2. Hicks, Tyler G., Pump Selection and Application, McGraw-Hill Book Company, New York, NY, 1957

3. King, Reno C., Piping Handbook, 5th edition, McGraw-Hill Book Company, New York, NY, 1973

4. Streeter, Victor L. and Wylie, E. Benjamin, Fluid Mechanics, 6th edition, McGraw-Hill Book Company, Scranton, PA, 1975

5. Worthington Pump International, Rotary and Centrifugal Pump Theory and Design, East Orange, NJ, 1971

Practice Problems: HYDRAULIC MACHINES

Required:

1. A sludge slurry with a specific gravity of 1.2 is pumped at the rate of 2000 gpm through an inlet of 12" and out an 8" discharge at the same level. The inlet gage reads 8" of mercury below atmospheric. The discharge gage reads 20 psig and is located 4 feet above the centerline of the pump outlet. If the pump efficiency is 85%, what is the input power?

2. A pump discharges water at 12 fps through a 6" line. The inlet is a section of 8" line. Suction is 5 psig below atmospheric. If the pump is 20 horsepower and is 70% efficient, what is the maximum height at which water at atmospheric pressure is available? Assume all friction losses add up to 10 feet of fluid.

3. Water flows from a source to a turbine, exiting 625 feet lower. The head loss is 58 feet due to friction, the flow rate is 1000 cfs, and the turbine efficiency is 89%. What is the output in kilowatts?

Optional:

4. A horizontal turbine reduces 100 cfs of water from 30 psig to 5 psi vacuum. Neglecting friction, what horsepower is generated?

5. A Francis hydraulic reaction turbine with 22" diameter blades runs at 610 rpm and developes 250 horsepower when 25 cfs of water flow through it. The pressure head at the turbine entrance is 92.5 feet. The elevation of the turbine above the tail water level is 5.26 feet. The inlet and outlet velocities are 12 fps. Find
   (a) The effective head
   (b) The turbine efficiency
   (c) The rpm at 225 feet effective head
   (d) The BHP at 225 feet effective head
   (e) The discharge in cfs at 225 feet effective head

6. Water (180°F, 80 psia) empties through 30 feet of 1½" pipe by a pump whose inlet and outlet are 20 feet below the surface of the water level when the tank is full. The pumping rate is 100 gpm and the NPSHR is 10 feet for that rate. If the inlet line contains two gate valves and two long radius elbows, and the discharge is into a 2 psig tank, when will the pump cavitate? Neglect entrance and exit losses.

7. What is the maximum suggested specific speed for a 2-stage pump adding 300 feet of head to water pulled through an inlet 10 feet below it?

8. Water at 500 psig will be used to drive a 250 horsepower turbine at 1750 rpm against a backpressure of 30 psig. What type of turbine would you suggest? If the 4" diameter jet discharging at 35 fps is deflected 80° by a single moving vane with velocity of 10 fps, what is the total force acting on the blade?

9. A 1750 rpm pump is normally splined to a ½ horsepower motor. What horsepower is required if the pump is to run at 2000 rpm?

10. The inlet of a centrifugal pump is 7 feet above a water surface level. The inlet is 12 feet of 2" pipe and contains one long-radius elbow and one check valve. The 2" outlet is 8 feet above the surface level and contains two long-radius elbows in its 80 feet of length. The discharge is 20 feet above the surface. The following pump curve data is available. What is the flow if water is at 70°F? Neglect the entrance loss.

gpm	head		gpm	head
0	110		50	93
10	108		60	87
20	105		70	79
30	102		80	66
40	98		90	50

NOTES

NOTES

OPEN CHANNEL FLOW

Nomenclature

A	area	ft²
b	weir width	ft
C	coefficient	-
d	depth	ft
d_H	hydraulic diameter	ft
D	pipe diameter	ft
E	specific energy	ft-lb/lb
f	Darcy friction factor	-
g	acceleration due to gravity (32.2)	ft/sec²
h	head	ft
H	head	ft
K	minor loss coefficient	-
L	channel length	ft
m	Bazen coefficient	-
n	Manning roughness coefficient	-
N	number of end contractions	-
p	pressure	lb/ft²
P	wetted perimeter, or weir height	ft, ft
Q	flow quantity	ft³/sec
r_H	hydraulic radius	ft
S	slope of energy line (energy gradient)	-
S_o	channel slope	-
v	velocity	ft/sec
w	channel width	ft
Y	weir height	ft
z	height above datum	

Symbols

ρ	density	lbm/ft³

Subscripts

b	brink
c	critical
e	equivalent
f	friction
H	hydraulic
s	spillway
t	total

1. Introduction

An open channel is a fluid passageway which allows part of the fluid to be exposed to the atmosphere. This type of channel includes natural waterways, canals, culverts, flumes, and pipes flowing under the influence of gravity (as opposed to pressure conduits which always flow full.) The discussion in this chapter is specifically limited to water flowing in open channels.

There are many difficulties in evaluating open channel flow. The unlimited geometric cross sections and variations in roughness have contributed to a small number of scientific observations upon which to estimate the required coefficients and exponents. Therefore, the analysis of open channel flow is more empirical and less exact than that of pressure conduit flow. This lack of precision, however, is more than offset by the percentage error in runoff calculations that generally precede the channel calculations.

Flow in open channels is almost always turbulent. However, within that category are many somewhat confusing categories of flow. Flow can be categorized on the basis of the channel material. Except for a short discussion of erodible canals, this chapter assumes the channel is non-erodible.

Flow can also be a function of time and location. If the flow quantity is invariant, it is said to be steady. If the flow cross section does not depend on the location along the channel, the flow is said to be uniform. Steady flow can be non-uniform, as in the case of a river with a varying cross section or on a steep slope. Furthermore, uniform channel construction does not ensure uniform flow, as will be seen in the case of hydraulic jumps.

Other types of flow are defined in the chapter glossary.

Due to the adhesion between the wetted surface of the channel and the water, the velocity will not be uniform across the area in flow. The velocity term used in this chapter is the mean velocity. The mean velocity, when multiplied by the flow area, gives the flow quantity.

$$Q = Av \qquad \text{5.1}$$

The location of the mean velocity depends on the distribution of velocities in the waterway, which is generally quite complex. The procedure for measuring the velocity of a channel (called stream gaging) involves measuring the average channel velocity at 10 or more places across the channel width. These sub-average velocities are then themselves averaged to give a grand average (mean) flow velocity.

The sub-average velocities are difficult to find. Some of the procedures used to find them are listed here:

1. Multiply the surface velocity (as measured by a surface float) by .9.
2. Measure the velocity at (.6)(depth).
3. Average the velocities at (.2)(depth) and (.8)(depth).

Figure 5.1

Distribution of Velocities in a Channel

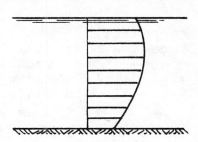

2. Definitions

Accelerated Flow: A form of varied flow in which the velocity is increasing and the depth is decreasing.

Apron: An underwater 'floor' constructed along the channel bottom to prevent scour. Aprons are almost always extensions of spillways and culverts.

Backwater: Water upstream from a dam or other obstruction which is deeper than it would normally be without the obstruction.

Backwater Curve: A plot of depth versus location along the channel containing backwater.

Check: A short section of built-up channel placed in a canal or irrigation ditch and provided with gates or flashboards to control flow or raise upstream level for diversion.

Colloidal State: A mixture of water and extremely fine sediment which will not easily settle out.

Conjugate Depth: The depth on either side of a hydraulic jump.

Contraction: A decrease in the width or depth of flow caused by the geometry of a weir, orifice, or obstruction.

Critical Flow: Flow at the critical depth and velocity. Critical flow minimizes the specific energy and maximizes discharge.

Critical Depth: The depth which minimizes the specific energy of flow.

Critical Slope: The slope which produces critical flow.

Critical Velocity: The velocity which minimizes specific energy. When water is moving at its critical velocity, a disturbance wave cannot move upstream since it moves at the critical velocity.

Downpull: A force on a gate, typically less at lower depths than at upper depths due to increased velocity when the gate is partially open.

Energy Gradient: The slope of the specific energy line (i.e., the sum of the potential and velocity heads.)

Flume: An open channel constructed above the earth's surface, usually supported on a trestle or piers.

Forebay: A reservoir holding water for use after it has been discharged from a dam.

Freeboard: The height of the channel side above the water level.

Gradient: See 'Slope.'

Headwall: Entrance to a culvert or sluiceway.

Hydraulic Gradient: Slope of the potential head relative to the channel bottom. Since the potential head is equal to the depth of the channel, the hydraulic gradient and channel bottom are parallel. (Static pressure is omitted from the hydraulic gradient since the pressure is atmospheric at all points on the surface.)

Hydraulic Jump: A spontaneous increase in flow depth from a velocity higher than critical to a velocity lower than critical.

Hydraulic Mean Depth: Same as 'hydraulic radius.'

Limit Slope: The smallest critical slope for a channel with a given shape and roughness.

Normal Depth: The depth of uniform flow. This is a unique depth of flow for any combination of channel conditions. Normal depth is found from the Chezy-Manning equation.

Overchute: A flume passing over a canal to carry floodwaters away without contaminating the canal water. An 'elevated culvert.'

q-curve: A plot of depth of flow versus quantity flowing for a channel with a constant specific energy.

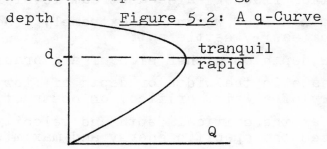

depth   Figure 5.2: A q-Curve

d_c — tranquil
        rapid

Q

Rapid Flow: Flow at less than the critical depth, as typically occurs on steep slopes.

Rating Curve: A plot of quantity flowing versus depth for a natural watercourse.

Reach: A section of channel.

Retarded Flow: A form of varied flow in which the velocity is decreasing and the depth increasing.

Sand Trap: A section constructed deeper than the rest of the channel to allow sediment to settle out.

Scour: Erosion at the exit of an open channel or toe of a spillway.

Settling Basin: A large, shallow basin through which water passes at low velocity, causing most of the suspended sediment to settle out.

Shooting Flow: See 'Rapid Flow.'

Sill: A submerged wall or weir.

Slope: The head loss per foot. For almost-level channels in uniform flow, the slope is equal to the tangent of the angle made by the channel bottom.

Stage: Same as 'depth.'

Standing Wave: A stationary wave caused by an obstruction in a watercourse. The wave cannot move (propagate) because the water is flowing at its critical speed.

Steady Flow: Flow which does not vary with time.

Stilling Basin: An excavated pool downstream from a spillway used to decrease tailwater depth and to ensure an energy-dissipating hydraulic jump.

Stream Gaging: A method of determining the velocity in an open channel.

Subcritical Flow: Flow at greater than the critical depth (less than the critical velocity.)

Supercritical Flow: Flow at less than the critical depth (greater than the critical velocity.)

Tail Race: An open waterway leading water out of a dam spillway back to a natural channel.

Tail Water: The water into which a a spillway or outfall discharges.

Tranquil Flow: Flow at greater than the critical depth.

Turnout: A pipe placed through a canal embankment to carry water from the canal for other uses.

Uniform Flow: Flow which has a constant depth, volume, and shape along its course.

Varied Flow: Flow that has a changing depth along the water course. The variation is with respect to location, not time.

Wasteway: A canal or pipe which returns excess irrigation water back to the main channel.

Wetted Perimeter: The length of the channel which has water contact. The air-water interface is not included in the wetted perimeter.

3. Parameters

The hydraulic radius is the ratio of area in flow to wetted perimeter.

$$r_H = \frac{A}{P} \qquad 5.2$$

For a circular channel flowing either full or half-full, the hydraulic radius is $(D/4)$. Hydraulic radii of other channel shapes is easily calculated from the basic definition. The hydraulic depth is the ratio of area in flow to the width of the channel at the fluid surface:

$$d_H = \frac{A}{w} \qquad 5.3$$

The slope, S, in open channel equations is the slope of the energy line. If the flow is uniform, the slope of the energy line will parallel the water surface and channel bottom. In general, the slope can be calculated as the energy loss per unit length of channel.

$$S = h_f/L \qquad 5.4$$

4. Governing Equations for Uniform Flow

Although it is of limited value, the incompressibility of the water allows the use of the continuity equation.

$$A_1 v_1 = A_2 v_2 \qquad 5.5$$

The most common equation used to calculate the flow velocity in open channels is the Chezy equation.

$$v = C\sqrt{r_H S} \qquad 5.6$$

Various equations for evaluating the coefficient C have been proposed. If the channel is small and very smooth (i.e., man-made with dimensions of a few feet), the Chezy formula can be used. f in equation 5.7 is dependent on the Reynolds number, and can be found from the Moody diagram in the usual manner.

$$C = \sqrt{8g/f} \qquad 5.7$$

If it is assumed that the channel is large, then the friction loss will not depend so much on the Reynolds number as on the channel roughness. The Manning formula is frequently used to evaluate the constant C. Notice that the value of C depends only on the channel roughness and geometry.

$$C = \frac{1.49}{n}(r_H)^{1/6} \qquad 5.8$$

n is the Manning roughness constant, and it is found in the appendix of this chapter. Putting equations 5.6 and 5.8 together produces the Chezy-Manning equation, applicable when the slope is less than .10.

$$v = \frac{1.49}{n}(r_H)^{2/3}\sqrt{S} \qquad 5.9$$

The Kutter coefficient has also seen widespread use, although it is much more cumbersome than the Manning coefficient to calculate. In the following equation, n is the same as for the Manning equation.

$$C = \frac{41.65 + \frac{.00281}{S} + \frac{1.811}{n}}{1 + (41.65 + \frac{.00281}{S})\frac{n}{\sqrt{r_H}}} \qquad 5.10$$

The Kutter formula has essentially been replaced by the Manning formula because of the former's complexity. There is also evidence that the Kutter equation is in error when S is very small (much smaller, however, than is usually encountered in normal design work.) Other than these drawbacks, the two give similar results.

The Bazin formula has been used extensively in France. It is given by equation 5.11.

$$C = \frac{157.6}{1 + m/\sqrt{r_H}}$$

5.11

Values of m are given in the following table.

Table 5.1
Bazin Coefficients (3:218)

type of surface	m
smooth cement	.109
planed wood	.109
brickwork	.290
rough planks	.290
rubble masonry	.833
smooth earth channels	1.540
ordinary earth channels	2.360
rough channels	3.170

Example 5.1

A rectangular channel on a .002 slope is constructed of finished concrete and is 8 feet wide. What is the uniform flow if water is at a depth of 5 feet? Evaluate C with both the Manning and Kutter equations.

The hydraulic radius is: $r_H = \frac{(8)(5)}{5+8+5} = 2.22$ ft

From table 5.6 the roughness coefficient for finished concrete is .012. The Manning coefficient is

$$C = \frac{1.49}{.012}(2.22)^{1/6} = 141.8$$

The discharge from equations 5.1 and 5.6 is

$$Q = (141.8)(8)(5)\sqrt{(2.22)(.002)} = 377.9 \text{ cfs}$$

The Kutter coefficient, as calculated from equation 5.10, is 144.0. This results in a flow of 383.8 cfs.

5. Energy and Friction Relationships

The Bernoulli equation can be written for two points along the bottom of an open channel experiencing uniform flow.

$$\frac{p_1}{\rho} + \frac{v_1^2}{2g} + z_1 = \frac{p_2}{\rho} + \frac{v_2^2}{2g} + z_2 + h_f$$

5.12

However, $(p/\rho) = d$. And since $d_1 = d_2$ and $v_1 = v_2$,

$$h_f = z_1 - z_2$$

5.13

For small slopes typical of almost all natural waterways, the channel length and horizontal run are essentially identical. Then, the hydraulic slope is

$$S = \frac{z_1 - z_2}{L} = h_f/L \qquad\qquad 5.14$$

Therefore, the friction loss in a length of channel is

$$h_f = LS \qquad\qquad 5.15$$

The friction loss can also be calculated from the Darcy equation using $D_e = 4r_H$, equation 5.7, and equation 5.8 to find f.

$$h_f = \frac{Ln^2 v^2}{2.21(r_H)^{4/3}} \qquad\qquad 5.16$$

Minor losses from obstructions, curves, and changes in velocity are calculated with loss coefficients as they are in a pressure-conduit flow.

$$h_{minor} = K\frac{v^2}{2g} \qquad\qquad 5.17$$

Little data is available, however, to evaluate K.

Example 5.2

In example 5.1, an open channel in normal flow had the following characteristics: S = .002, n = .012, v = 9.447 ft/sec, r_H = 2.22 ft. What is the energy loss per 1000 feet?

From equation 5.15, $h_f = (1000)(.002) = 2$ feet

From equation 5.16, $h_f = \dfrac{1000(.012)^2(9.447)^2}{2.21(2.22)^{4/3}} = 2$ feet

◆

6. Most Efficient Cross Section

The most efficient cross section (from an open channel standpoint) is the one which has maximum discharge for a given slope, area, and roughness. Wetted perimeter will be at a minimum (to minimize friction) when the flow is maximum.

Semicircular cross sections have the smallest wetted perimeter, and therefore the cross section with the highest efficiency is the semi-circle. Although such a shape can be constructed with concrete, it cannot be used with earth channels.

Rectangular channels are frequently used with wooden flumes. The most efficient rectangle is one which has a depth equal to one-half of the width.

For trapezoidal channels, the most efficient cross section will be one in which the depth is twice the hydraulic radius. The sides of such a trapezoid will be inclined at 30° from the horizontal, and the flow area is half a hexagon.

A semicircle with its center at the middle of the water surface can always be inscribed in a cross section with maximum efficiency.

Figure 5.3
Circles Inscribed in Efficient Channels

(a) circular   (b) rectangular   (c) trapezoidal

Example 5.3

A rubble masonry open channel is being designed to carry 500 cfs of water on a .0001 slope. Using n = .017, find the most efficient dimensions for a rectangular channel.

Let the depth and width be d and w respectively. For an efficient rectangle, $d = \frac{1}{2}w$. Therefore,

$$A = dw = \tfrac{1}{2}w^2$$

$$P = d + w + d = 2w$$

$$r_H = \tfrac{1}{2}w^2/2w = \tfrac{1}{4}w$$

From equation 5.1, Q = Av. Combining this with equation 5.9,

$$500 = (\tfrac{1}{2}w^2)(\frac{1.49}{.017})(\tfrac{1}{4}w)^{2/3}(.0001)^{\frac{1}{2}}$$

$$500 = (.1739)w^{8/3}$$

$$w = 19.82$$

So, $d = \frac{1}{2}w = 9.91$ feet.

◆

7. Circular Sections

Combining equations 5.1 and 5.9 gives

$$Q = vA = \frac{1.49}{n}(A)(r_H)^{2/3}\sqrt{S} \qquad\qquad 5.18$$

The area in flow also depends on the hydraulic radius. If the flow is known, the diameter of a round pipe flowing full is

$$D = 1.33(Qn/\sqrt{S})^{3/8} \qquad\qquad 5.19$$

If the round pipe is flowing half full, replace the 1.33 with 1.73.

Maximum discharge from a circular channel occurs at slightly less than full. Specifically, maximum flow occurs when d = .96D. Therefore, a non-pressurized circular channel without obstructions or backwater will not flow full. Maximum velocity will occur at .91D.

Table 5.2
Circular Channel Ratios
(Also see page 5-27)

d/D	Q/Q_{full}	v/v_{full}
.1	.02	.30
.2	.07	.48
.3	.14	.61
.4	.26	.71
.5	.41	.80
.6	.56	.88
.7	.72	.95
.8	.87	1.02
.9	.99	1.04
.95	1.02	1.03
1.00	1.00	1.00

Example 5.4

2.5 cfs of water are in uniform flow in a 20" sewer line (n = .015, S = .001). What are the depth and velocity?

If the pipe flowed full, it would carry Q_{full}.

$$D = 20/12 = 1.667 \text{ ft}$$

$$r_H = \tfrac{1}{4}D = \tfrac{1}{4}(1.667) = .417 \text{ ft}$$

$$Q_{full} = \tfrac{1}{4}\pi(1.667)^2(\tfrac{1.49}{.015})(.417)^{2/3}\sqrt{.001} = 3.83 \text{ cfs}$$

$$v_{full} = 3.83/\tfrac{1}{4}\pi(1.667)^2 = 1.75 \text{ ft/sec}$$

So, $Q/Q_{full} = 2.5/3.83 = .65$. From figure 5.20, (d/D) = .66 and $(v/v_{full}) = .92$. So,

$$v = (.92)(1.75) = 1.61 \text{ ft/sec}$$

$$d = (.66)(20) = 13.2 \text{ inches}$$

8. Analysis of Natural Watercourses

Natural watercourses do not always have uniform paths or cross sections. This complicates their analysis considerably. Frequently, analyzing the flow from a river is a case of 'doing the best you can.' Luckily, some types of problems can be solved with a reasonable amount of error.

As was seen in equation 5.16, the friction loss (and hence the hydraulic gradient) depends on the square of the roughness coefficient. Therefore, an attempt must be made to evaluate n as accurately as possible. If the channel consists of a river with overbank flood plains, it should be treated as parallel channels. The flow from each subdivision should be calculated independently, and the separate values added to obtain the total flow.

Figure 5.4
River with Flood Plain

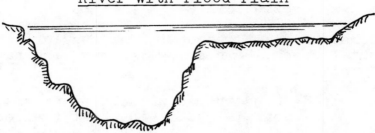

If the channel is divided by an island into two channels (figure 5.5), Q will usually be known. It may be necessary to calculate Q_1 and Q_2 in that case, or, if Q_1 and Q_2 are known, it may be necessary to find the slope.

Figure 5.5
Divided Channel

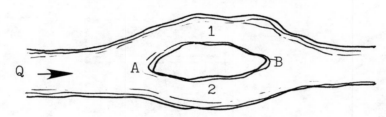

Since the drop $(z_B - z_A)$ between points A and B is the same regardless of flow path,

$$S_1 = \frac{z_B - z_A}{L_1} \qquad\qquad 5.20$$

$$S_2 = \frac{z_B - z_A}{L_2} \qquad\qquad 5.21$$

Once the slopes are known, Q_1 and Q_2 can be found from equation 5.18. The sum of Q_1 and Q_2 will probably not be the same as the given flow quantity, Q. In that case, Q should be prorated according to the ratios of Q_1 and Q_2 to $(Q_1 + Q_2)$.

If the lengths L_1 and L_2 are the same or almost so, the Chezy-Manning equation may be solved for the slope by writing equation 5.22:

$$Q = Q_1 + Q_2 = 1.49\left[\frac{A_1}{n_1}(r_{H,1})^{2/3} + \frac{A_2}{n_2}(r_{H,2})^{2/3}\right]\sqrt{S} \qquad 5.22$$

9. Flow Measurement with Weirs

A weir is an obstruction in an open channel over which flow occurs. Although a dam spillway is a specific type of weir, most weirs are designed for flow measurement. These weirs consist of a vertical flat plate with sharpened edges. Because of their construction, they are called sharp-crested weirs.

Sharp-crested weirs are most frequently rectangular, consisting of a straight, horizontal crest. However, weirs may also have trapezoidal and triangular openings.

If a rectangular weir is constructed with an opening width less than the channel width, the overfalling liquid sheet (called the nappe) decreases in width as it falls. This contraction of the nappe causes these weirs to be called contracted weirs, although it is the nappe that is actually contracted. If the opening of the weir extends the full channel width, the weir is called a suppressed weir, since the contractions are suppressed.

Figure 5.6

Contracted and Suppressed Weirs

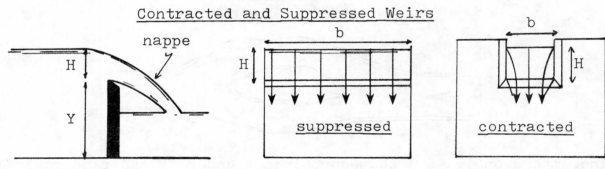

The derivation of the basic weir equation is not particularly difficult, but it is dependent on many simplifying assumptions. The basic weir equation (equation 5.23 or 5.24) is, therefore, an approximate result requiring correction by the inclusion of experimental coefficients.

If it is assumed that the contractions are suppressed, upstream velocity is uniform, flow is laminar over the crest, nappe pressure is zero, the nappe is fully ventilated, and viscosity, turbulence, and surface tension effects are negligible, then the following equation may be derived from the Bernoulli equation:

$$Q = \frac{2}{3}b\sqrt{2g}\left[\left(H + \frac{v_1^2}{2g}\right)^{3/2} - \left(\frac{v_1^2}{2g}\right)^{3/2}\right] \qquad 5.23$$

If v_1 is negligible, then

$$Q = \frac{2}{3}b\sqrt{2g}(H)^{3/2} \qquad 5.24$$

Equation 5.24 must be corrected for all of the assumptions made. This is done by introducing a coefficient, C_1, to account primarily for a non-uniform velocity distribution.

$$Q = \frac{2}{3}(C_1)b\sqrt{2g}(H)^{3/2} \qquad 5.25$$

A number of investigations have been done to evaluate C_1. Perhaps the most widely used is the coefficient formula developed by Rehbock:

$$C_1 = \left[.6035 + .0813(\tfrac{H}{Y}) + \frac{.000295}{Y}\right]\left[1 + \frac{.00361}{H}\right]^{3/2} \qquad 5.26$$

If the contractions are not suppressed (i.e., one or both sides do not extend to the channel sides) then the actual width, b, should be replaced with the effective width.

$$b_{effective} = b_{actual} - (.1)(N)(H) \qquad\qquad 5.27$$

N is one if one side is contracted, and N is two if there are two end contractions.

A submerged rectangular weir requires a more complex analysis, both because of the difficulty in measuring H and because the discharge depends on both the upstream and downstream depths. The following equation, however, may be used with little difficulty.

$$Q_{submerged} = Q_{free\ flow}\left[1 - (\frac{H_{downstream}}{H_{upstream}})^{3/2}\right]^{.385} \qquad 5.28$$

Equation 5.28 is used by first finding Q from equation 5.25 and then correcting it with the bracketed quantity.

Figure 5.7
Submerged Weir

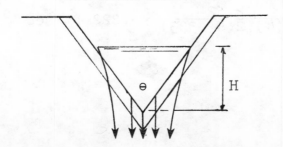

Triangular (V-notch) weirs should be used when small flow rates are to be measured. The flow over a triangular weir depends on the notch angle, θ.

$$Q = C_2(\tfrac{8}{15})\tan(\tfrac{1}{2}\theta)\sqrt{2g}(H)^{5/2} \qquad\qquad 5.29$$

$$C_2 = 2.48(H)^{2.48} \qquad\qquad 5.30$$

Figure 5.8
Triangular Weir

A trapezoidal weir is essentially a rectangular weir with a triangular weir on either wide. If the angle of the sides from the vertical is approximately 14° (i.e., 1:4 slope) the weir is known as a Cipoletti weir. The discharge from the triangular ends of a Cipoletti weir approximately make up for the contractions that reduce rectangular flow. Therefore, no correction is theoretically necessary. The discharge from a Cipoletti weir is given by equation 5.31.

$$Q = 3.367(b)(H)^{3/2} \qquad\qquad 5.31$$

Figure 5.9
Trapezoidal Weir

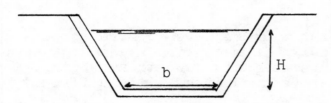

Equation 5.24 may also be used for broad-crested weirs (C = .5 to .57) and ogee spillways (C = .60 to .75.)

Example 5.5

A sharp-crested, rectangular weir with two contractions is $2\frac{1}{2}$ feet high and 4 feet long. A 4" head exists upstream from the weir. What is the velocity of approach?

$H = 4/12 = .333$ ft

From equation 5.27, N = 2 and the effective width is

$$b_{effective} = 4 - (.1)(2)(.333) = 3.93$$

The Rehbock coefficient (from equation 5.26) is

$$C_1 = (.6035 + .0813(\frac{.333}{2.5}) + \frac{.000295}{2.5})(1 + \frac{.00361}{.333})^{3/2}$$
$$= .624$$

From equation 5.24, the flow is

$$Q = \frac{2}{3}(.624)(3.93)\sqrt{(2)(32.2)}\ (.333)^{3/2}$$
$$= 2.52 \text{ cfs}$$
$$v = \frac{Q}{A} = \frac{2.52}{(4)(2.5+.333)} = .222 \text{ ft/sec}$$

10. Non-Uniform Flow

A. Critical Flow

If water is introduced down a path with a steep slope (as down a spillway) the effect of gravity will be to cause an increasing velocity. This velocity will be opposed by friction. Since the gravitational force is constant but friction varies as the square of velocity, these two forces eventually become equal. When they become equal, the velocity stops increasing, the depth stops decreasing, and the flow becomes uniform. Until they become equal, however, the flow is non-uniform (varied).

The total head was already shown to be given by the Bernoulli equation:

$$H_t = z + \frac{p}{\rho} + \frac{v^2}{2g} \qquad\qquad 5.32$$

Specific energy is the total head with respect to the channel bottom. In this case, $z=0$ and $(p/\rho) = d$.

$$E = d + \frac{v^2}{2g} \qquad\qquad 5.33$$

In uniform flow, total head decreases due to the frictional effects, but specific energy is constant. In non-uniform flow, total head decreases, but specific energy may increase or decrease.

Since $v = Q/A$, equation 5.33 can be written

$$E = d + \frac{Q^2}{2gA^2} \qquad\qquad 5.34$$

Since the area depends on the depth, fixing the channel shape and slope and assuming a depth will determine Q. This also will determine the specific energy, as illustrated in the following specific energy diagram.

Figure 5.10
Specific Energy Diagram

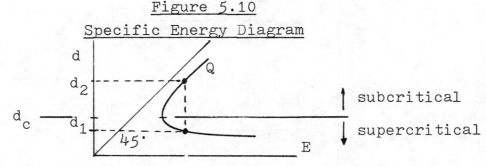

For a given flow rate, there are two different depths of flow that have the same energy - a high velocity with low depth and a low velocity with high depth. The former is called rapid (supercritical) flow; the latter is called tranquil (subcritical) flow.

There is one depth, the critical depth, which minimizes the energy of flow. Critical depth for a given flow depends on the shape of the channel.

If the channel is rectangular, the critical depth is two-thirds of the critical specific energy.

$$d_c = \frac{2}{3}E_c \qquad\qquad 5.35$$

The following equation can be used to calculate the critical depth:

$$d_c = \sqrt[3]{Q^2/g(w)^2} \qquad\qquad 5.36$$

Once the critical depth is known, the corresponding velocity and discharge are given by the following two equations:

$$v_c = \sqrt{gd_c} \qquad\qquad 5.37$$

$$Q_c = v_c w d_c = w\sqrt{g}(d_c)^{3/2} \qquad\qquad 5.38$$

The critical depth of a trapezoidal channel is

$$d_c = \frac{4zE - 3b + \sqrt{16z^2E^2 + 16zEb + 9b^2}}{10z} \qquad\qquad 5.39$$

$$z = e/D \qquad\qquad 5.40$$

If z=0, the trapezoidal section becomes a rectangle, and equation 5.39 reduces to 5.35. If b=0, the equation reduces to $d_c = (4/5)E$, the critical depth of a triangular channel.

Figure 5.11
Trapezoidal Channel

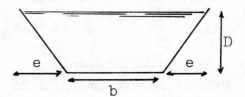

To find the discharge from a trapezoidal channel with flow at critical, first find d_c and then substitute it into the following equation:

$$Q = d(b+zd)\sqrt{2g(E-d)} \qquad\qquad 5.41$$

For non-rectangular shapes, the critical depth can be found by trial and error from the following equation in which b is the surface width.

$$\frac{Q^2}{g} = \frac{A^3}{b} \qquad\qquad 5.42$$

For any given discharge and cross section, there is a unique slope that will produce and maintain flow at critical depth. Once d_c is known, this critical slope can be found from the Chezy-Manning equation. Generally, the slope will not be critical, and flow will be supercritical (faster) or subcritical (slower).

If the slope changes abruptly from subcritical to supercritical, or from supercritical to subcritical as illustrated in figure 5.12, critical depth will occur at or near the abrupt change. The decrease in depth from subcritical to supercritical is known as a hydraulic drop. In a hydraulic drop, the critical depth occurs at the apex.

The occurrence of critical depth when the flow changes from supercritical to subcritical may be somewhat removed from the change point. A procedure for finding the exact point is given later in this chapter. The increase in depth from supercritical to subcritical is known as a hydraulic jump.

<u>Figure 5.12</u>

<u>Occurrence of Critical Depth</u>

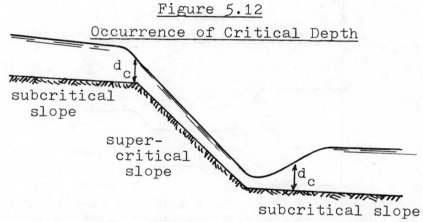

Critical depth also occurs at free outfall from a channel of mild slope. The occurrence is at the point of curvature inversion, just upstream from the brink. For mild slopes, the brink depth is

$$d_b = (.715)d_c$$

5.43

<u>Figure 5.13</u>

<u>Free Outfall</u>

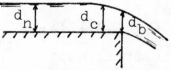

Critical flow occurs across a broad-crested weir. With no obstruction to hold the water, it falls from the normal depth to the critical depth, but it can fall no more than that because there is no source to increase the specific energy (to increase the velocity.) This is not a contradiction of the previous free outfall case where the brink depth is less than the critical depth. The flow curvatures in free outfall are a result of the constant gravitational acceleration.

Figure 5.14

Broad-Crested Weir

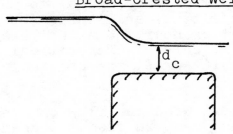

Critical depth also occurs when a channel bottom has been sufficently raised. A raised channel bottom is essentially a broad-crested weir.

Figure 5.15

Raised Channel Bottom

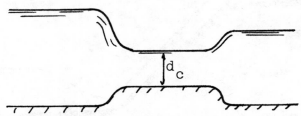

The critical depth is important, not because it minimizes the energy of flow, but because it maximizes the quantity flowing for a given cross section and slope. Critical flow is generally quite turbulent because of the large changes in energy that occur with small elevation and depth changes. Critical depth flow, then, is characterized by water surface undulations.

In all of the previous instances of critical depth, equation 5.36 may be used to calculate the actual value.

Example 5.6

At a particular point in an open rectangular channel (n=.013, S=.002, w = 10 feet) the flow is 250 cfs and the depth is 4.2 feet.

(a) Is the flow tranquil, normal, critical, or rapid?
(b) What is the normal depth?
(c) If the flow ends in a free outfall, what is the brink depth?

a) From equation 5.36, the critical depth is

$$d_c = \sqrt[3]{(250)^2/(32.2)(10)^2} = 2.69$$

Since the actual depth exceeds the critical depth, the flow is tranquil.

b) $$A = (d_n)(10)$$

$$P = 2d_n + 10$$

$$r_H = \frac{10d_n}{2d_n+10} = \frac{5d_n}{d_n+5}$$

From equation 5.18,

$$250 = (10)(d_n)(\frac{1.49}{.013})(\frac{5d_n}{d_n+5})^{2/3}\sqrt{.002}$$

By trial and error, d_n = 3.1 feet

c) d_b = .715(2.69) = 1.92 ft

B. Varied Flow Calculations

Accelerated flow occurs in any channel where the actual slope exceeds the friction loss per foot. That is,

$$S_o > h_f/L \qquad\qquad 5.44$$

Retarded flow occurs when

$$S_o < h_f/L \qquad\qquad 5.45$$

Consider the following figure.

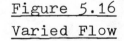

Figure 5.16
Varied Flow

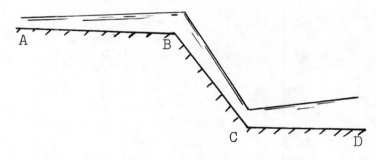

In sections AB and CD, the slopes are less than the energy gradient, so the flows are retarded. In section BC, the slope is greater than the energy gradient, and the velocity increases (i.e., the flow is accelerated.) If section BC were long enough, the friction loss would eventually become equal to the accelerating energy, and the flow would become uniform.

Cases of accelerated and retarded flow (except for hydraulic jump) can be evaluated from the following procedure which will give the distance between points of two known (or assumed) depths. Assuming that the friction losses are the same for varied flow as for uniform flow, the following equations are needed:

$$S = (\frac{nv_{ave}}{1.486(r_{H,ave})^{2/3}})^2 \qquad\qquad 5.46$$

$$v_{ave} = \frac{1.486}{n}(r_{H,ave})^{2/3}(S)^{\frac{1}{2}} \qquad\qquad 5.47$$

S is the slope of the energy gradient from equation 5.46, not the channel slope S_o. The usual method of finding the depth profile is to start at a point in the channel where d_2 and v_2 are known. Then, assume a depth d_1, find v_1 and S, and solve equation 5.48 for L.

$$L = \frac{(d_1 + \frac{v_1^2}{2g}) - (d_2 + \frac{v_2^2}{2g})}{S - S_o} = \frac{E_1 - E_2}{S - S_o} \qquad 5.48$$

Example 5.7

How far from the point in example 5.6 will the depth be 4 feet?

The difference between 4 feet and 4.2 feet is small, so a 1-step calculation will probably be sufficient.

$$d_1 = 4 \text{ feet}$$

$$v_1 = Q/A = \frac{250}{4(10)} = 6.25 \text{ ft/sec}$$

$$E_1 = 4 + \frac{(6.25)^2}{(2)(32.2)} = 4.607 \text{ ft}$$

$$r_H = \frac{(4)(10)}{4+10+4} = 2.22$$

$$d_2 = 4.2$$

$$v_2 = \frac{250}{(4.2)(10)} = 5.95$$

$$E_2 = 4.2 + \frac{(5.95)^2}{(2)(32.2)} = 4.75$$

$$r_H = \frac{(4.2)(10)}{4.2+10+4.2} = 2.28$$

$$v_{ave} = \tfrac{1}{2}(6.25 + 5.95) = 6.1$$

$$r_{H,ave} = \tfrac{1}{2}(2.22 + 2.28) = 2.25$$

From equation 5.46,

$$S = (\frac{(.013)(6.1)}{1.486(2.25)^{.667}})^2 = .000965$$

From equation 5.48,

$$L = \frac{4.607 - 4.75}{.000965-.002} = 138 \text{ ft}$$

C. Hydraulic Jump

If water is introduced at high (supercritical) velocity to a section of slow-moving (subcritical) flow (as in section C in figure 5.16) the velocity will be reduced through a hydraulic jump. A hydraulic jump is an abrupt rise in the water surface. The increase in depth is always from below the critical depth to above the critical depth.

Figure 5.17

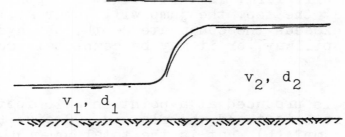

If the depths d_1 and d_2 are known, then the velocity v_1 can be found from

$$v_1^2 = \frac{gd_2}{2d_1}(d_1 + d_2) \qquad\qquad 5.49$$

d_1 and d_2 are known as conjugate depths, because they occur on either side of the jump. These depths are:

$$d_1 = -\tfrac{1}{2}d_2 + \sqrt{\frac{2v_2^2 d_2}{g} + \frac{d_2^2}{4}} \qquad\qquad 5.50$$

$$d_2 = -\tfrac{1}{2}d_1 + \sqrt{\frac{2v_1^2 d_1}{g} + \frac{d_1^2}{4}} \qquad\qquad 5.51$$

The specific energy lost in the jump is the energy lost per pound of water flowing.

$$\frac{\text{specific energy}}{\text{drop}} = (d_1 + \frac{v_1^2}{2g}) - (d_2 + \frac{v_2^2}{2g}) \quad (\frac{\text{ft-lbf}}{\text{lbm}}) \; 5.52$$

The location of the hydraulic jump can be found from the same procedure as was used with accelerating and retarded flow. In the case of an apron at the bottom of a dam spillway, the apron is usually insufficient to overcome friction, and the water depth will gradually increase.

Figure 5.18

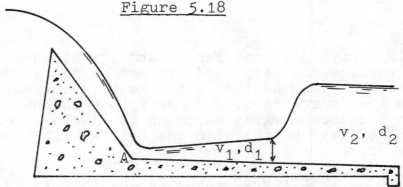

Computations start at the toe of the dam, point A, with a known velocity. Once d_1 is calculated from d_2 (equation 5.50) the distance between points A and B can be found from equation 5.48.

If the tailwater depth d_2 is less than critical, no hydraulic jump will occur. If the tailwater depth at the toe is less than the conjugate depth corresponding to d_2 but greater than the critical depth, flow will continue until the depth increases to d_1, and then a hydraulic jump will form. If the tailwater depth is equal to the conjugate depth at the toe, the jump will occur at the toe. If the tailwater depth exceeds the conjugate depth, the hydraulic jump may occur up on the spillway, or it may be completely submerged.

Example 5.8

A hydraulic jump is produced at a point in a 10 foot wide channel where the depth is 1 foot and the flow is 200 cfs. (a) What is the depth after the jump? (b) What is the total power dissipated?

a)
$$v_1 = Q/A = 200/(10)(1) = 20 \text{ ft/sec}$$

From equation 5.51,

$$d_2 = -\tfrac{1}{2}(1) + \sqrt{\frac{(2)(20)^2(1)}{32.2} + \frac{(1)^2}{4}} = 4.51$$

b) The weight flow is

$$(200)(62.4) = 12480 \text{ lb/sec}$$

The velocity after the jump is

$$v_2 = 200/(10)(4.51) = 4.43$$

From equation 5.52, the change in specific energy is

$$(1 + \frac{(20)^2}{2(32.2)}) - (4.51 + \frac{(4.43)^2}{2(32.2)}) = 2.4 \text{ feet}$$

The total power dissipated is

$$(12480)(2.4) = 29,952 \text{ ft-lb/sec}$$

◆

11. Application to Design

A. Spillways

A spillway is always designed for a capacity based on the dam's inflow hydrograph, turbine capacity, and storage capacity. Overflow spillways frequently have a cross section known as an ogee, which closely approximates the underside of a nappe from a sharp-crested weir. This cross section reduces cavitation which is likely to occur when the water surface breaks contact with the spillway at higher heads than were designed for.

Discharge from an overflow spillway is the same as for a weir:

$$Q = C_s b(H)^{3/2} \qquad \qquad 5.53$$

C_s is the spillway coefficient, which varies from about 3 to 4 for an ogee spillway. C_s is dependent on H, the head above the spillway top. C_s increases as H increases.

If the velocity of approach is significant, the flow quantity is

$$Q = C_s b(H + \frac{v^2}{2g})^{3/2}$$

5.54

Scour protection is usually needed at the toe of a spillway to protect the area exposed to a hydraulic jump. This protection usually takes the form of an extended horizontal or sloping apron. Other measures, however, are needed if the tailwater exhibits large variations in depth.

B. Sluiceways

A sluiceway carries water away from a dam or reservoir, and is usually constructed in such a manner as to allow withdrawal at various reservoir levels. Its intake is submerged, but almost never is the inlet level below the minimum reservoir level. Sluiceways are generally round or square in shape.

Analysis and design of sluiceways is the same as for culverts. Refer to chapter 3.

C. Culverts

Culvert design is presented in chapter 3.

D. Erodible Canals

Design of erodible channels is similar to that of concrete or pipe channels, except for the added considerations of maximum velocities and permissible side slopes.

The sides of the channel should not be slopes exceeding the natural angle of repose for the material used. Although there are other factors which determine the maximum permissible side slope, table 5.3 lists some guidelines.

Table 5.3
Recommended Side Slopes

Type of Channel	(horizontal:vertical)
firm rock	vertical to $\frac{1}{4}$:1
muck and peat soils	$\frac{1}{4}$:1
concrete lined stiff clay	$\frac{1}{2}$:1
fissured rock	$\frac{1}{2}$:1
firm earth with stone lining	1:1
firm earth, large channels	1:1
firm earth, small channels	$1\frac{1}{2}$:1
loose, sandy earth	2:1
sandy, porous loam	3:1

Maximum velocities that should be used with erodible channels are given in table 5.4.

Table 5.4
Maximum Velocities (fps)

Channel Material	Clear Water	Water with Colloidal Silt
fine colloidal sand	1.50	2.50
noncolloidal sandy loam	1.75	2.50
noncolloidal silt loam	2.00	3.00
ordinary firm loam	2.50	3.50
stiff colloidal clay	3.75	5.00
fine gravel	2.50	5.00
coarse gravel	4.00	6.00
cobbles and shingles	5.00	5.50
shales and hardpans	6.00	6.00

Seepage rates in unlined natural channels are given in table 5.5. These quantities are, of course, approximate since the actual seepage will depend on the head.

Table 5.5
Seepage Rates (ft^3/ft^2-day) (4:287)

Channel Material	Seepage Rate
clay loam	0.25 - 0.75
sandy loam	1.0 - 1.5
loose sandy soils	1.5 - 2.0
gravel soils	3.0 - 6.0

Bibliography

1. Binder, R.C., Fluid Mechanics, Prentice Hall, New York, NY 1943

2. Daugherty, R. L., Hydraulics, 4th ed., McGraw-Hill Book Company, New York, NY 1937

3. King, Horace W., and Wisler, Chester O., Hydraulics, 3rd ed., John Wiley and Sons, New York, NY 1933

4. Linsley, Ray K., and Franzini, Joseph B., Water Resources Engineering, 2nd ed., McGraw-Hill Book Company, New York, NY 1972

5. Vennard, John K., Elementary Fluid Mechanics, 2nd ed., John Wiley and Sons, New York, NY 1947

Table 5.6 (3:176 and 4:272)

Manning's and Kutter's n

Kind of pipe	Variation From	To	Design From	Use To
clean, uncoated cast iron	.011	.015	.013	.015
clean, coated cast iron	.010	.014	.012	.014
dirty, tuberculated cast iron	.015	.035		
riveted steel	.013	.017	.015	.017
lock-bar and welded	.010	.013	.012	.013
galvanized iron	.012	.017	.015	.017
brass and glass	.009	.013		
wood stave	.010	.014		
small diameter			.011	.012
large diameter			.012	.013
concrete	.010	.017		
with rough joints			.016	.017
dry mix, rough forms			.015	.016
wet mix, steel forms			.012	.014
very smooth, finished			.011	.012
vitrified sewer	.010	.017	.013	.015
common-clay drainage tile	.011	.017	.012	.014
asbestos			.011	
planed timber			.011	
canvas			.012	
unplaned timber			.014	
brick			.016	
rubble masonry			.017	
smooth earth			.018	
firm gravel			.023	
corrugated metal pipe			.022	
natural channels, good condition			.025	
natural channels with stones and weeds			.035	
very poor natural channels			.060	

Figure 5.19
Manning Nomograph

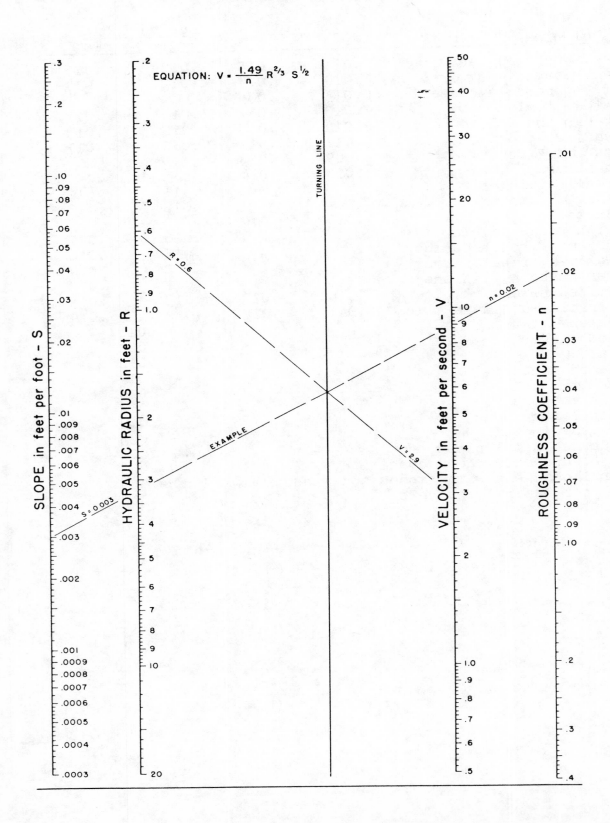

EQUATION: $V = \dfrac{1.49}{n} R^{2/3} S^{1/2}$

Figure 5.20
Circular Channel Ratios

Experiments have shown that n varies slightly with depth. This figure gives velocity and flow rate ratios for varying n (solid line) and constant n (broken line) assumptions.

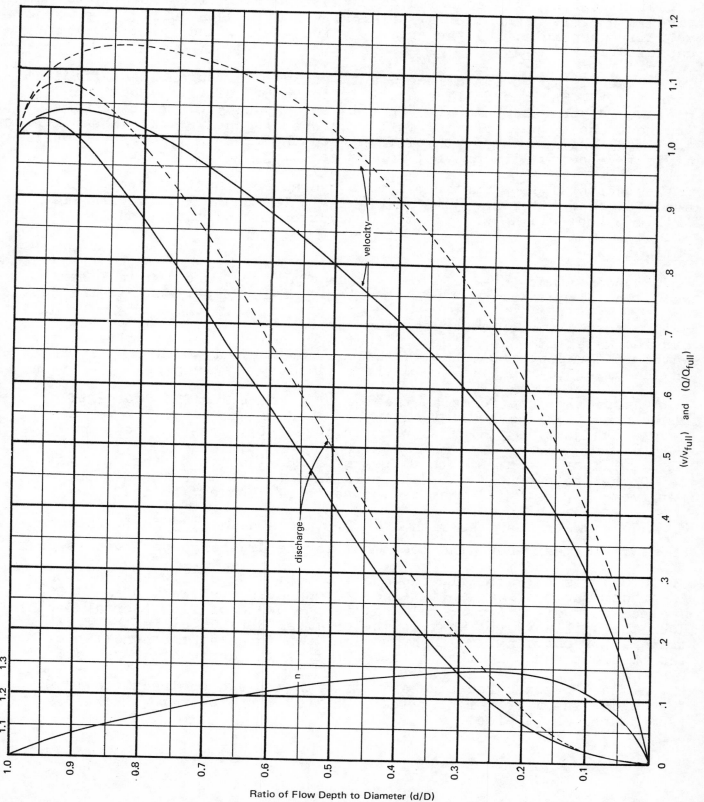

Ratio of Flow Depth to Diameter (d/D)

Practice Problems: OPEN CHANNEL FLOW

Required

1. 30 years ago, a 24 inch diameter pipe (n = .013) was installed on a .001 slope. Recent tests indicate that the full-flow capacity of the pipe is 6.0 cfs. Find the (a) original velocity when full, (b) present velocity when full, (c) the present value of 'n', and (d) the original capacity.

2. A sewer is to be installed on a 1% grade. Its roughness coefficient is n = .013. Maximum full capacity is to be 3.5 cfs. (a) What size pipe would you recommend? (b) What is the capacity of the pipe size you have chosen when flowing full? (c) What is the velocity when flowing full? (d) What is the depth of flow when the flow is .7 cfs? (e) What minimum velocity will prevent solids settling out in the sewer?

3. The depth of flow upstream from a hydraulic jump is 1 foot. The depth of flow after the jump is 2.4 feet. The channel is rectangular with a 5 foot width. What is the discharge rate of the channel?

Optional

4. A wooden flume (n = .012) of rectangular cross section is 2 feet wide. The flume carries 3 cfs of water on a 1% slope. What is the depth?

5. A 4 foot diameter concrete storm drain (n = .013, slope = .02) carries water at a depth of 1.5 feet. (a) What is the velocity of the water in the pipe? (b) What is the maximum velocity of water flowing in the pipe? (c) What is the maximum capacity of the pipe?

6. A hydraulic jump forms at the toe of a spillway. The depths of flow are .2 feet and 6 feet on either side of the jump. The velocity before the jump is 54.7 fps. What is the energy loss in the jump?

7. A spillway operates with 2 feet of head. The toe of the spillway is 40 feet below the top of the spillway. (a) What is the discharge per foot of crest? (b) What is the depth of flow at the toe?

8. An ogee weir operates with C_w = 3.5 and a head of 5 feet. The weir crest is 10 feet above the toe. What is the discharge per foot of crest?

9. 10,000 cfs of water flow down a 100 foot wide spillway placed on a 5% grade. The spillway surface has a roughness coeffient of n = .012. Neglect sidewall effects. (a) What is the depth of flow (normal depth) down the spillway? (b) What is the critical depth? (c) Is the flow tranquil or shooting? (d) A hydraulic jump forms at the junction of the 5% slope and a horizontal toe. What is the depth after the jump?

10. A sharp-crested rectangular weir with two end contractions is 5 feet wide. The weir height 6 feet. What is the flow rate if the head is measured as .43 feet?

NOTES

NOTES

HYDROLOGY

Nomenclature

A	area	ft^2
A_d	drainage area	acres
b	a constant	-
B	aquifer width	ft
C	rational runoff coefficient	-
C_p	pan coefficient	-
C_r	retardance coefficient	-
d	drawdown, or distance between stations	ft, or miles
E	evaporation	in/day
F	frequency of occurrence	-/yrs
I	rainfall intensity	in/hr
K	a constant	-
K_p	constant of permeability	$ft^3/day\text{-}ft^2$
L_c	centroidal stream length	miles
L_o	overland flow path length	ft
L_s	main stream length	miles
N	average precipitation per year, or time from peak to end of runoff	inches, or hours
P	precipitation over a short period	inches
Q	flow quantity	cfs
r	radial distance from well	ft
R	runoff	inches
s	slope of the hydraulic gradient	-
S_c	storage constant	-
t	time since pumping	hrs
t_c	storm duration (time of concentration)	minutes
t_e	overland flow time	minutes
t_p	time from start of storm to peak runoff	hrs
t_r	rain storm duration	hrs
T	transmissivity	$ft^3/day\text{-}ft$
u	a dimensionless variable	-
v	flow velocity	ft/sec
V	volume	ft^3
$W(u)$	well function	-
y	aquifer thickness after drawdown	ft
Y	original aquifer thickness	ft

Subscripts

o	at well
p	peak, or pan
r	at radius r
R	reservoir
u	unit
x	unknown

Important Conversions

Multiply	By	To Get
acre	43560	ft^2
acre-ft	43560	ft^3
acre-ft	325,851	gallons
acre-inches/hr	1.008	cfs
cubic feet	7.4805	gallons
cubic feet/sec	1.9834	acre-ft/day
cubic feet/sec	448.83	gpm
cubic feet/sec	.64632	MGD
darcy	1.062 EE-11	ft^2
gallon	.1337	ft^3
gallon	3.07 EE-6	acre-ft
gallon	1.547 EE-6	sec-ft-day
gallon/day	1.547 EE-6	ft^3/sec
gpm	1440	gallon/day
gpm	.002228	cfs
gpm	192.5	ft^3/day
horsepower	.7457	kw
horsepower	550	ft-lb/sec
hectare	2.471	acres
inch of runoff/sq. mile	53.3	acre-ft/sq. mile
inch of runoff/sq. mile	2.323 EE6	ft^3/sq. mile
Meinzer unit	1.00	gal/day-ft^2
MGD	1 EE6	gallon/day
second-ft-day (ft^3-day/sec)	86400	ft^3/day
square mile	640	acres
square mile	2.788 EE7	ft^2
square mile-inch	53.3	acre-ft
square mile-inch/day	26.88	cfs

1. Introduction

Engineering hydrology is a field which evaluates rainfall, stream flow, reservoir capacity, and subsurface water runoff in relationship to their effects on engineering projects. Although an important part of the study of hydrology is concerned with the collection of data, this chapter only treats the application of available data to capacity and related studies.

2. Definitions

Anabranch: The intertwining channels of a braided stream.

Anticlinal spring: A portion of an exposed aquifer (usually on a slope) between two impervious layers.

Aquiclude: An underground source of water with insufficient porosity to support any sufficient removal rate.

Aquifer: An underground source of water capable of supplying a well or other use.

Aquifuge: An underground geological formation which has no porosity or openings at all through which water can enter or be removed.

Artesian well: A spring in which water flows naturally out of the earth's surface due to pressure placed on the water by an impervious overburden and hydrostatic head.

Artesian formation: An aquifer in which the piezometric height is greater than the aquifer thickness.

Base flow: Runoff which percolates down to the water table and then discharges into a stream. Up to 2 years may elapse between precipitation and discharge.

Bifurcation ratio: The average number of streams feeding into the next side (order) waterway. The range is usually 2 to 4, with an average of around 3.5.

Blind drainage: Geographically large (with respect to the drainage basin) depressions which store water during a storm and, therefore, stop it from contributing to surface runoff.

Braided stream: A wide, shallow stream with many anabranches.

Capillary water: Water just above the water table which is drawn up out of an aquifer due to capillary action of the soil.

Cone of depression: The shape of the water table around a well during and immediately after use. The cone's apex differs from the original water table by the well's drawdown.

Confined water: Artesian water overlaid with an impervious layer, usually under pressure.

Connate water: Water, frequently saline, present in rock at its formation.

Depression storage: Initial storage of rain in small surface puddles.

Depth-Area-Duration analysis: A study made to determine the maximum amounts of rain within a given time period over a given area.

Dimple spring: A depression in the earth below the water table.

Drainage density: The total length of streams in a watershed divided by the drainage area.

Drawdown: The difference in water table level at a well head and far from it.

Dry weather flow: See 'base flow.'

Effluent stream: A stream which intersects the water table and receives groundwater. Effluent streams seldom go completely dry during the rainless periods.

Ephemeral stream: A stream which does dry during rainless periods.

Evaptotranspiration: Evaporation of water from a study area due to all sources including water, soil, snow, ice, vegetation, and transpiration.

Flowing well: A well which flows on its own accord to the surface. See also 'artesian well.'

Forebay: An area which recharges an aquifer.

Gravitational water: Water in transit downward through the earth.

Groundwater: Subsurface water flowing in an aquifer towards a stream. Groundwater is not water flowing on the ground - it is water flowing underground.

Hydrograph: A plot of discharge versus time for a stream or storm.

Hydrological cycle: The cycle experienced by water in its travel from the ocean, through evaporation and precipitation, percolation, runoff, and return to the ocean.

Hydrometeor: Any form of water falling from the sky.

Hygroscopic water: Moisture adhering in a thin film to soil grains.

Impervious layer: A geologic layer through which no water can pass.

Infiltration: The movement of water through the upper soil.

Influent stream: A stream above the water table. Influent streams may go dry during the rainless season.

Initial loss: The sum of interception and depression loss, but excluding blind drainage.

Interception: Rain which falls on vegitation and other impervious objects and which evaporates without contributing to runoff.

Interflow: Infiltrated subsurface water which travels to a stream without percolating down to the water level.

Juvenile water: Water formed chemically within the earth.

Lysimeter: A soil container used to observe and measure evaptotranspiration.

Meandering stream: A stream which flows in large loops, not in a straight line.

Meteoric water: See 'hydrometeor.'

Negative boundaries: A fault or similar geologic structure.

Net rain: Rain which contributes to surface runoff.

Overland flow: water which travels over the ground surface to a stream.

Pan: A container used to measure surface evaporation rates.

Perched spring: A localized saturated area which occurs above an impervious layer on a slope.

Percolation: The travel of water down through the soil to the water table.

Phreatic zone: The layer below the water table down to an impervious layer.

Phreatophytes: Trees with root systems which extend into the water table.

Piezometric level: The level to which water will rise in a pipe due to its own pressure.

Plat: A small plot of land.

Porosity: The ratio of pore volume to total formation volume.

Probable maximum rainfall: The rainfall corresponding to some given probability (e.g., 1 in 100 years).

Safe yield: The maximum rate of water withdrawal which is economically and ecologically feasible.

Seep: See 'spring.'

Sinuosity: The stream length divided by the valley length.

Specific yield: The ratio of water volume which will drain freely from a substance sample to the total volume. Specific yield is always less than porosity. Specific yield is also defined as the volume of water obtained by lowering one square foot of the water table by one foot.

Spring: A place where the earth surface and aquifer coincide.

Stream order: An artificial categorization of stream geneology. Small streams are 1st order, 2nd order streams are fed by 1st order order, 3rd order streams are fed by 2nd order streams, etc.

Subsurface runoff: See 'interflow.'

Surface detention: Rain water which collects as a film and runs off of the saturated surface during a storm.

Surface retention: That part of a storm which does not immediately appear as infiltration or surface runoff. Retention is made up of depression storage, interception, and evaporation.

Surface runoff: Water flow over the surface after a storm which reaches a stream.

Time of concentration: The time required for water to flow from the most distant point on a runoff area to the measurement or collection point.

Transpiration: The process in which plants give off internal moisture to the atmosphere.

Unit stream power: The product of velocity and slope, representing the rate of energy expenditure per pound of water.

Vaclose water: All water above the water table, including soil water, gravitational water, and capillary water.

Vadose zone: A zone above the water table containing both saturated and empty soil pores.

Water table: The top level of an aquifer, defined as the locus of points where the water pressure is equal to the atmospheric pressure.

Xerophytes: Drought-resistant plants, typically existing with root systems well above the water table.

Zone of aeration: See 'vadose zone.'

Zone of saturation: See 'phreatic zone.'

3. Precipitation

Although the word 'precipitation' encompasses all hydrometeoric forms, it is often applied only to rainfall in the liquid form. Precipitation data may be collected in a number of ways, but the open precipitation 'rain' gage is quite common.

If a rain measurement is lost or is not available, it can be estimated by one of the following procedures:

Method 1: Choose 3 stations evenly spaced around and close to the location which has missing data. If the normal annual precipitations at the 3 sites do not vary by more than 10% from the missing station's normal annual precipitation, the rainfall is estimated as the arithmetic mean of the 3 neighboring stations' precipitation for the period in question.

Method 2: If the difference is more than 10%, the Normal-Ratio Method can be used:

$$P_x = (\tfrac{1}{3})\left[(N_x/N_A)P_A + (N_x/N_B)P_B + (N_x/N_C)P_C\right]$$

Method 3: A method used by the U.S. National Weather Service for river forecasting is to use data from stations in the 4 nearest quadrants (North, South, East, and West of the unknown station) and to weight the data with the square of the distance between stations.

$$P_x = \frac{d_{A-x}^2 P_A + d_{B-x}^2 P_B + d_{C-x}^2 P_C + d_{D-x}^2 P_D}{d_{A-x}^2 + d_{B-x}^2 + d_{C-d}^2 + d_{D-x}^2}$$

The average precipitation over a specific area or time basis may be found from station data in several ways.

Method 1: If the stations are uniformly distributed over a flat site, their precipitations can be averaged. This also requires that the individual precipitation records not vary too much from the mean.

Method 2: The Thiessen method calculates the average by weighting station measurements by the area of the assumed basin for each station. These assumed basin areas are found by drawing dotted lines between all stations and bisecting these dotted lines with solid lines (which are extended outward until they connect with another solid line). The solid lines will form a polygon whose area is the assumed basin area.

Method 3: The most accurate method is the Isohyetal method. This method requires plotting lines of constant precipitation (isohyets) and weighting the isohyet values by the areas enclosed by the lines. Station data is used to draw isohyets, but not in the calculation of average rainfall.

Effective design of a structure will depend on the geographical location and degree of protection required. Once the location of a structure has been chosen, it is up to the engineer to design for the area's most probable maximum rainfall or probable maximum flood. Both of these require some judgement since the 'maximum' is not a deterministic number.

Rainfall intensity is the rate of precipitation per hour. Intensity will be low for most storms, but it will be high for some storms. These high intensity storms can be expected very infrequently - say every 20, 50, or 100 years. The average number of years between storms of a given magnitude is known as the design-storm frequency of occurrence.

The intensity of a storm may be calculated from the following general equation:

$$I = \frac{K'(F)^a}{(t_c+b)^c}$$

6.1

K', a, b, and c are constants which depend on the conditions and location.

The Steel formula is a simplification of the above equation:

$$I = \frac{K}{t_c+b}$$

6.2

K and b are dependent on the storm frequency and geographical location. t_c may be either the time of concentration (in runoff studies) or the rainfall duration. Values of K and b are found from the following figure and table:

Figure 6.1 (4:21-80)
Steel Formula Rainfall Regions

Table 6.1 (4:21-79)
Steel Formula Coefficients

Frequency in years	Coefficients	Region						
		1	2	3	4	5	6	7
2	K	206	140	106	70	70	68	32
	b	30	21	17	13	16	14	11
4	K	247	190	131	97	81	75	48
	b	29	25	19	16	13	12	12
10	K	300	230	170	111	111	122	60
	b	36	29	23	16	17	23	13
25	K	327	260	230	170	130	155	67
	b	33	32	30	27	17	26	10
50	K	315	350	250	187	187	160	65
	b	28	38	27	24	25	21	8
100	K	367	375	290	220	240	210	77
	b	33	36	31	28	29	26	10

Example 6.1

A storm has an intensity given by the following equation:

$$I = \frac{100}{t_c + 10}$$

15 minutes are required for runoff from the farthest corner of a 5 acre plat to reach a discharge culvert. What is the design intensity?

$$I = \frac{100}{15 + 10} = 4 \text{ inches/hour}$$

4. Subsurface Water

Subsurface water is a major source of all water used in the United States. In dry areas, it may be the only source of water used for domestic and irrigation uses. Subsurface zones are divided into two parts by the water table. The vadose zone exists above the water table, and pores in the vadose zone may be either empty or full. Below the water table is the phreatic zone, whose pores are always full.

Soil moisture content is measured in pounds per cubic foot. Soil moisture is usually determined by oven drying a sample of soil and measuring the weight loss. The moisture content may also be determined with a tensiometer, which measures the vapor pressure of the moisture in the soil.

Movement of water through an aquifer is given by equation 6.3 in which K_p is known as the coefficient of permeability (or 'hydraulic conductivity').

$$Q = (1.157 \text{ EE-5})K_p As \qquad 6.3$$

K_p in the United States is usually given in Meinzer units (gallons per day per square foot). For the sake of consistent nomenclature in this chapter, K_p is given in different units in table 6.2.

Table 6.2
Approximate Coefficients of Permeability, K_p

Material	ft^3/day-ft^2	gal/day-ft^2	darcys
Clay	1.3 EE-3	EE-2	EE-3
Sand	1.3 EE2	EE3	EE2
Gravel	1.3 EE4	EE5	EE4
Gravel/Sand	1.3 EE3	EE4	EE3
Sandstone	1.3 EE1	EE2	EE1
Dense shale & limestone	1.3 EE-1	1.0	EE-1
Quartzite & granite	1.3 EE-3	EE-2	EE-3

The transmissivity of flow from a saturated aquifer of thickness Y and width B is

$$T = K_p Y \qquad 6.4$$

But $\qquad A = BY = BT/K_p \qquad 6.5$

So, combining equation 6.3 with equation 6.5,

$$Q = (1.157 \text{ EE-5})BTs \qquad 6.6$$

A virgin aquifer with a well is shown in figure 6.2(a). Once pumping of the well starts, the water table will be lowered in the vicinity of the well, and the resulting water table surface is known as the cone of depression. The decrease in water level at the well is known as the drawdown, d_o. The drawdown at some distance r from the well is d_r.

If the drawdown is small with respect to the aquifer thickness, Y, and the well completely penetrates the aquifer, then the equilibrium (steady state) well discharge is given by the Druit equation, equation 6.7. The 86400 converts cubic feet per day to cubic feet per second.

$$Q = \frac{\pi K_p (y_1^2 - y_2^2)}{86400 \ln(r_1/r_2)} \qquad 6.7$$

Figure 6.2

(a)   Well Drawdown   (b)

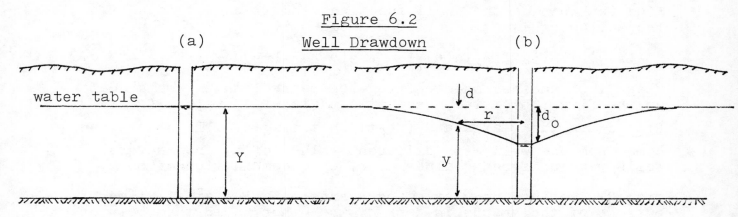

In equation 6.7, y_1 may be taken as the original aquifer depth, Y, if r_1 is the well's radius of influence.

For an artesian well fed by a confined aquifer of thickness Y, the discharge is

$$Q = \frac{2\pi K_p (y_1 - y_2)Y}{86400 \ln(r_1/r_2)}$$ 6.8

Example 6.2

A 9" diameter well is pumped at the rate of 50 gpm. The aquifer is 100 feet thick. The well sides cave in and are replaced with an 8" diameter tube. Assuming a 6 foot drawdown, what will be the steady flow from the new well? Assume the water table recovers its original thickness 2500 feet from the well.

$$(50)\text{gpm}(.002228)\frac{\text{cfs}}{\text{gpm}} = .1114 \text{ cfs}$$

From equation 6.7,

$$y_2 = 100 - 6 = 94 \text{ ft}$$
$$r_2 = 9/(2)(12) = .375 \text{ ft}$$

$$.1114 = \frac{\pi K_p((100)^2 - (94)^2)}{(86400) \ln(2500/.375)} \quad \text{or } K_p = 23.17 \text{ ft}^3/\text{day-ft}^2$$

For an 8" (r = .333 ft) pipe,

$$Q = \frac{\pi(23.17)((100)^2 - (94)^2)}{(86400) \ln(2500/.333)} = .110 \text{ cfs}$$

◆

When pumping first begins, the removed water also comes from the aquifer above the equilibrium cone of depression. Therefore, equation 6.7 cannot be used, and a non-equilibrium analysis is required. If the aquifer is relatively thick and permeable, and the removal rate Q is constant, then the Theis method may be used to find the drawdown d_r a distance r from the well.

step 1: Plot the 'Type Curve' of u versus W(u) on log-log paper. Plot u on the x axis and W(u) on the y axis. Choose values of u of .0001, .001, .01 and 1 as a start.

$W(u)$ is known as the 'well function' and it can be found from the following equation or from table 6.5:

$$W(u) \approx -.5772 - \ln(u) + u - \tfrac{1}{4}(u^2) + (\tfrac{1}{18})u^3 \cdots \qquad 6.9$$

step 2: On another sheet of log-log paper with the same scale as the first, plot the observation data of d_r versus (r^2/t). Plot d_r on the y axis and (r^2/t) on the x axis. This data can come from one well with different values of t, from several wells with different values of r, or a combination of both.

step 3: Keeping the y axes and x axes parallel at all times, put one of the graphs in back of the other, hold them both up to a light, and slide the papers around until the curves are superimposed. If no part of the curves appear to line up, go back to step 1 and extend the u versus $W(u)$ curve by choosing smaller or intermediate values of u.

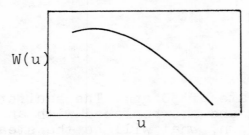
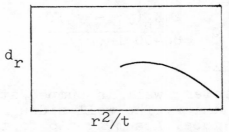

step 4: For a part of the curve where coincidence (agreement) is good, pick any single point (the 'match point') and read the corresponding values of u, $W(u)$, d_r, and r^2/t.

step 5: Using the match point values, calculate T and S_c from the following equations. S_c is the storage constant of the aquifer, essentially the same as the specific yield.

$$T = \frac{QW(u)}{4\pi d_r} \qquad 6.10$$

$$S_c = 4uT(t/r^2) \qquad 6.11$$

step 6: Once S_c and T are known, the drawdown can be found from

$$d_r = \frac{QW(u)}{4\pi T} \qquad 6.12$$

where $\quad u = \dfrac{r^2 S_c}{4Tt} \qquad 6.13$

If u is small (which it will be when t is large), T may be estimated by the modified Theis method.

step 1: For any two observations separated by a long period, calculate

$$T = \frac{Q \ln(t_2/t_1)}{4\pi(d_{r,2} - d_{r,1})} \qquad 6.14$$

Better accuracy can be obtained if d_2 and t_2 are extrapolated to the <u>right</u> from the straight part of the curve plotted in the next step.

<u>step 2</u>: On a semi-log graph, plot d (on the linear y-scale) versus t (on the log x-scale). Extend the straight line portion of the curve (when t is large) to the <u>left</u> and read t_o for d = 0.

<u>step 3</u>: Convert t_o to units of days.

<u>step 4</u>: Calculate $S_c = \dfrac{(2.25)Tt_o}{r^2}$

6.15

For adjacent wells which are close enough together to affect each other, the cones of depression will overlap. It is assumed that the actual drawdown is the sum of the individual drawdowns of all interfering wells.

The well efficiency is the ratio of the theoretical drawdown after some time period (e.g., 1000 minutes) to actual drawdown. An efficiency of .80 is usually acceptable.

Example 6.3

500 gallons per minute are drawn from a virgin well. The drawdown is measured at an observation well 100 feet away. The following data are collected. (a) Find the well's transmissivity and storage constant. (b) Estimate the drawdown after one year of steady pumping at the 500 gpm rate.

t (hours)	d (feet)
1	.6
2	1.4
3	2.4
4	2.9
5	3.3
6	4.0
8	5.2
10	6.2
12	7.5
18	9.1
24	10.5

<u>step 1</u>: Using values of W(u) from table 6.5, plot the type curve.

u	W(u)		u	W(u)
.001	6.33		.1	1.82
.005	4.73		.2	1.22
.01	4.04		.3	.91
.02	3.35		.5	.56
.03	2.96		1.0	.219
.05	2.48			

The resulting graph is shown on the following page.

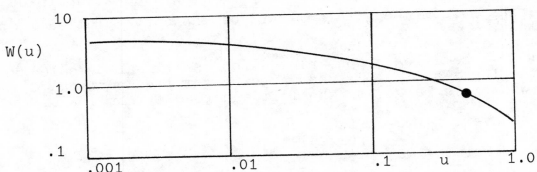

step 2: Plot d_r versus (r^2/t) on the same scale log-log paper.
For the first data point,   d = .6

$$\frac{r^2}{t} = \frac{(100)^2 ft^2 (24)hr/day}{1\ hr} = 240,000\ \frac{ft^2}{day}$$

Other values are given in the following table and plotted below.

r^2/t	d		r^2/t	d
240,000	.6		30,000	5.2
120,000	1.4		24,000	6.2
80,000	2.4		20,000	7.5
60,000	2.9		13,000	9.1
50,000	3.3		10,000	10.5
40,000	4.0			

steps 3 and 4: Aligning the two graphs and choosing an arbitrary
match point yields

   u = .46   W(u) = .6 r^2/t = 60,000   d = 2.8

step 5:   Q = (500)gpm(192.5)$\frac{ft^3}{day\text{-}gpm}$ = 96250 ft^3/day

$$T = \frac{(96250)(.6)}{4\pi(2.8)} = 1641.3\ \frac{ft^3}{day\text{-}ft}$$

$$S_c = \frac{(4)(.46)(1641.3)}{(60,000)} = .05$$

(b) From equation 6.13, u = $\frac{(100)^2(.05)}{(4)(1641.3)(365)}$ = 2.09 EE-4

From table 6.5, W(u) = 7.91, so equation 6.12 is solved for d.

$$d = \frac{(500)(24)(60)(7.91)}{(4\pi)(1641.3)(7.48)} = 36.9\ ft$$

Example 6.4

Assume that t is large enough to use the modified Theis equation and find T and S_c for the well in example 6.3.

step 1: For the 2 observations separated by a long period, choose

t	d
3	2.4
24	10.5

Then, from equation 6.14,

$$T = \frac{(96250)\,ft^3/day}{4\pi(10.5-2.4)}\,\ln(24/3) = 1966.3\ ft^3/day\text{-}ft$$

step 2: Plotting d versus t gives the following graph:

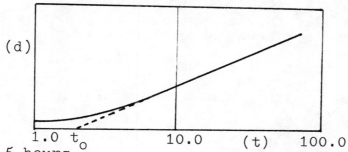

t_o = 2.5 hours

step 3: $t_o = (2.5)\ hr\ (\frac{1}{24})\ \frac{day}{hr} = .1042\ day$

step 4: From equation 6.15,

$$S_c = \frac{(2.25)(1966.3)(.1042)}{(100)^2} = .0461$$

◆

5. Total Surface Runoff

After a rain, water makes its way to a stream. A plot of stream discharge versus time is known as a hydrograph. Hydrograph periods may be very long (such as a year) down to very short (hours). A typical hydrograph is shown in figure 6.3.

Figure 6.3
A Stream Hydrograph

Q

rising
limb

falling limb
(recession)

t

In figure 6.3, the portion of the hydrograph to the left of the crest is known as the 'rising limb.' To the right of the crest, the curve is known as the 'recession.'

The stream discharge is assumed to consist of overland (surface) and groundwater (base) flow. Since culverts do not have to be designed to carry groundwater, a procedure called hydrograph separation or hydrograph analysis is necessary to separate surface and groundwater. This separation process is somewhat arbitrary.

step 1: Estimate the time from the crest to the cessation of significant direct runoff. An analytic estimate can be made if runoff and rainfall data are not available.

$$N = 6.59(A_d)^{.2} \qquad \text{6.16a}$$

N is typically the same for all storms in a river system, so it can be estimated by looking at several hydrographs.

step 2: Extend the discharge line existing prior to the storm forwards until it is under the crest. With a straight line, connect the end point with the new recession curve at a point N hours after the peak.

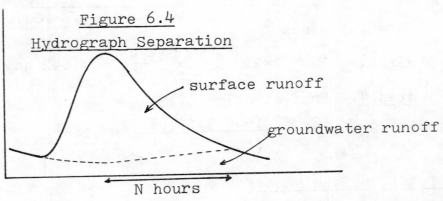

Figure 6.4

Hydrograph Separation

surface runoff

groundwater runoff

N hours

The total surface (direct) runoff can be calculated analytically from equation 6.16b.

$$\text{direct runoff} = \text{rainfall} - \text{basin recharge} - \text{ground water accretion} \qquad \text{6.16b}$$

Rainfall is easily determined from measuring station data, but the other terms in equation 6.16b are difficult to estimate. Therefore, the total runoff is commonly assumed to be a fraction of rainfall:

$$R = CP \qquad \text{6.17}$$

Typical values of C are given in table 6.4 . Equation 6.17 is best suited to areas where the imperviousness is large (i.e., large built-up or commercial areas).

6. Peak Runoff - The Rational Method

Although total runoff data is required for reservoir and dam design, the instantaneous peak runoff is needed to size culverts and sewers.

In the closed-form equations used to find peak runoff, it is usually assumed that the rainfall is applied to a surface at a constant rate. If this assumption is true and the surface is largely impervious, the runoff will eventually equal the rate of rainfall. The time between the start of rainfall and the start of steady runoff (which is also the time of peak flow) is known as the time of concentration, t_c. Typical values of t_c for plats less than 50 acres range from 5 to 30 minutes.

The rational formula is based on the assumptions given and has been in widespread use for small areas (i.e., less than 100 acres or so) despite its serious deficiencies for larger areas. Values of C are the same as for equation 6.17, and are found in table 6.4.

$$Q_p = CIA_d \qquad\qquad 6.18$$

Strictly speaking, Q_p is in acre-inches/hour, but it is typically taken as cubic feet per second since the conversion between these two units is 1.008. For a small drainage area, t_c is taken as the largest combination of overall flow time and channel time. Channel time is found by dividing the channel length by the (usually assumed) channel velocity. Assuming laminar flow, the overland flow time for small plats without defined channels can be found from the Izzard formula. Equation 6.19 should be used only if the product (IL_o) is less than 500.

$$t_e = t_c = \frac{(41)(b)(L_o)^{1/3}}{(CI)^{2/3}} \qquad\qquad 6.19$$

$$b = \frac{.0007(I) + C_r}{(s)^{1/3}} \qquad\qquad 6.20$$

The retardance coefficient, C_r, is given in table 6.3.

Table 6.3

Retardance Coefficient

Type of Surface	C_r
smooth asphalt	.007
concrete pavement	.012
tar & gravel pavement	.017
closely clipped lawn	.046
dense bluegrass turf	.060

The most important part of equation 6.18 is the rainfall intensity. Rainfall data can be compiled into intensity-duration-frequency curves similar to figure 6.5. The intensity used in equation 6.18 will depend on the time of concentration and the degree of protection desired. In design, 5-year stream curves are used for residential areas, 10 year frequencies for business sections, and 15 year frequencies for high-value districts where flooding will result in extensive damage.

Figure 6.5

Intensity-Duration-Frequency Curves

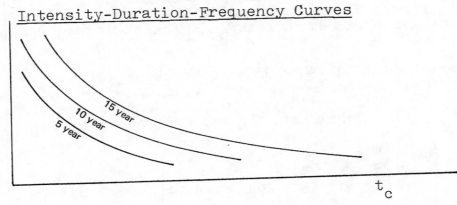

Example 6.5

A drainage area has the following characteristics:

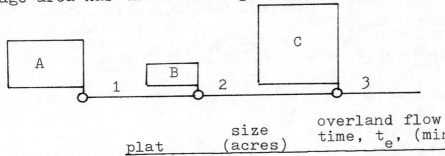

plat	size (acres)	overland flow time, t_e, (min)	C
A	10	20	.3
B	2	5	.7
C	15	25	.4

The rainfall intensity for the area is given by

$$I = \frac{115}{t_c + 15}$$

The manholes are 300 feet apart; the pipe slope is .009; and Manning's roughness is .015. What should be the pipe size in section 2? What is the maximum flow through section 3?

For plat A:

$$t_c = t_e = 20$$

$$I = \frac{115}{20 + 15} = 3.29 \text{ in/hr}$$

$$Q = (.3)(10)(3.29) = 9.87 \text{ cfs}$$

This value of Q (9.87) should be used to size line #1.

To find the flow time between manholes, a flow velocity is needed. Since the pipe diameter is unknown, it would have to be assumed to find the hydraulic radius and the flow velocity from the Chezy-Manning equation. It is just as easy to assume a flow velocity of 5 ft/sec.

Using a flow velocity of 5 ft/sec, the flow time between drainage inlets is

$$t = \frac{300}{5} = 60 \text{ seconds} = 1 \text{ minute}$$

For plat B:

At t=5, I = 5.75 in/hr. The run-off from 2 acres is (.7)(2)(5.75) or 8.05 cfs. However, at t = (20+1), the flow from plat A will reach the second manhole. At t = 21, I = 3.19 in/hr.

The sum of CA values is (.3)(10) + (.7)(2) = 4.4

Q = (4.4)(3.19) = 14.0 cfs

This value of Q should be used to design section 2 of the pipe. If the pipe is assumed to flow full, the required diameter is found from equation 5.19.

$$d = (1.33) \left[\frac{(14.0)(.015)}{\sqrt{.009}} \right]^{3/8} = 1.79 \text{ ft (round to 2.0 ft)}$$

For plat C:

Assuming 5 fps as the flow velocity in the pipe, the time from the start of the storm for water from plat A to reach the 3rd manhole is

20 + 1 + 1 = 22

Since 25 is larger than 22, the maximum runoff will occur 25 minutes after the start of the storm. The 22 minute datum is not used.

$$I = \frac{115}{25+15} = 2.875$$

The sum of CA values is (.3)(10) + (.7)(2) + (.4)(15) = 10.4

Q = (10.4)(2.875) = 29.9 cfs

7. Peak Runoff from the Unit Hydrograph

Consider a given drainage basin and the hydrograph from a storm of known duration. If the ordinates of the hydrograph are reduced (but not the duration) such that the area under the curve gives one inch of runoff, the result is a unit hydrograph. (Note that the storm duration and the hydrograph time base are not the same.)

Figure 6.6

Actual and Derived Unit Hydrographs

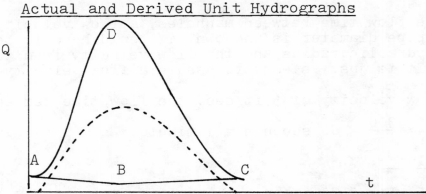

A unit hydrograph may be constructed from the runoff and rainfall data by using the following process.

step 1: Separate the runoff and groundwater using the procedure listed on page 6-14.

step 2: Measure the runoff area of the graph (area ABCD).

step 3: Divide the time axis from A to C into 10 or 20 sections. Divisions lengths are arbitrary, but 2 or 3 hours is a typical value.

step 4: For each division, subtract the base flow from the hydrograph reading and divide by the area ABCD.

step 5: Plot the data derived in step 4 versus elapsed time from point A (i.e., t = 0 at point A).

A unit hydrograph may be used to predict the runoff for storms which have durations differing as much as $\pm 25\%$ from the storm duration used to derive the unit hydrograph.

If a basin is ungaged so that no records are available to produce a unit hydrograph, important hydrograph parameters may be derived analytically (with some success). Knowing these parameters and recognizing that the total precipitation from a unit hydrograph must be one inch permits sketching a rough approximation to a unit hydrograph.

The Snyder synthetic hydrograph is shown in figure 6.7.

Figure 6.7

Snyder Synthetic Hydrograph

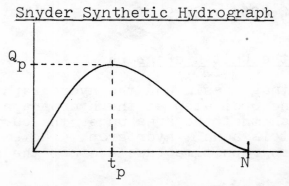

An estimate for the time to peak runoff for the Snyder hydrograph is

$$t_p = C_t(L_s L_c)^{.3} \tag{6.21}$$

C_t is a constant with wide variation, but it averages 2 for steep slopes. L_s is the main stream length from outlet to divide in miles. L_c is the distance in miles from the outlet to a point on the stream nearest the basin centroid. Equation 6.21 is valid for basin areas of 10 to 10,000 square miles.
The peak flow for the unit hydrograph is

$$Q_p = C_p A_d / t_p \tag{6.22}$$

C_p is a coefficient which depends on the geographical location, but values near .60 are typical.

The time base for the unit hydrograph is

$$N = 72 + 3t_p \tag{6.23}$$

Since the value of N can never be less than 72 hours, the Snyder hydrograph is not suitable for small basins which reach their peaks in a matter of hours.

Equations 6.22 and 6.23 work if the storm rain duration is

$$t_r = t_p/5.5 \tag{6.24}$$

If the duration is known to be otherwise, say t_r', the actual time to peak runoff that should be used in equations 6.22 and 6.23 is t_p'.

$$t_p' = t_p + \frac{t_r' - t_r}{4} \tag{6.25}$$

Example 6.6

After a 2-hour storm, a station downstream from a 45 square mile drainage basin measures 9400 cfs as a peak discharge and 3300 acre-feet as total runoff. Find the 2-hour unit hydrograph peak discharge. What would be the peak runoff and design flood volume if a 2-hour storm dropped $2\frac{1}{2}$ inches net precipitation?

1 inch of runoff from 45 square miles would be

$$(45) \text{ mile}^2 (1) \text{ inch} (53.3)\frac{\text{acre-ft}}{\text{sq. mile-in}} = 2399 \text{ acre-feet}$$

The runoff ratio is 3300/2399 = 1.38

The unit hydrograph peak discharge is 9400/1.38 = 6812 cfs.

For a $2\frac{1}{2}$" storm, peak runoff would be (2.5)(6812) = 17,030 cfs.
The design flood would be (2.5)(2399) = 5998 acre-feet

Example 6.7

A storm drops its significant rainfall in 6 hours on a 25 square mile basin. The resulting surface runoff is listed below. (a) Construct the unit hydrograph of this 6-hour storm. (b) Find the runoff at t=15 hours from a two storm system (both a 6 hours duration) if the first storm drops 2" net starting at t=0 and the second storm drops 5" net starting at t=12 hours.

Hours after rainfall starts	Runoff (cfs)	Hours after rainfall starts	Runoff (cfs)
0	0	21	600
3	400	24	400
6	1300	27	300
9	2500	30	200
12	1700	33	100
15	1200	36	0
18	800	TOTAL	9500

(a) The actual runoff is plotted below:

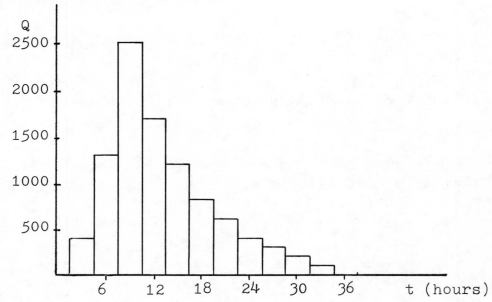

The total area under the curve is found by dividing the graph into squares and adding up the squares. This is equivalent to the following calculation:

$$\text{Volume} = (9500) \text{ cfs } (3) \text{ hours } (3600) \frac{\text{sec}}{\text{hr}} = 1.026 \text{ EE8 ft}^3 \text{ runoff}$$

The basin area is

$$(25) \text{ miles}^2 (5280)^2 \frac{\text{feet}}{\text{mile}} = 6.97 \text{ EE8 ft}^2$$

The net precipitation (after depression storage) was

$$\frac{(1.026 \text{ EE8}) \text{ ft}^3 (12) \text{ in/ft}}{(6.97 \text{ EE8}) \text{ ft}^2} = 1.767 \text{ inches}$$

The unit hydrograph has the same shape as the actual hydrograph, with all ordinates reduced by 1.767.

(b) The runoff at t=15 would be

$$2(\frac{1200}{1.767}) + 5(\frac{400}{1.767}) = 2490 \text{ cfs}$$ ◆

8. Reservoir Yield

The reservoir yield problem is an accounting problem (i.e., keeping track of what comes in and what goes out). The purpose of the reservoir yield analysis is to determine the proper size of a dam or reservoir, or to evaluate the ability of an existing dam to meet water demands. There are three basic methods of solving the reservoir yield problem.

Method #1: Tabular simulation is the easiest method to apply, although its precision is dependent on choosing time increments as small as possible. It is also necessary to make some assumption about the distribution of annual water inflows.

step 1: Determine the starting storage volume, V_n. If V_n is zero or considerably different than the average steady-state storage, a large number of iterations will be required before the simulation reaches its steady-state results.

step 2: For the next iteration (season of runoff and use), determine the inflow, discharge, evaporation, and seepage. Determine the starting storage volume for the next iteration (season) by solving equation 6.26.

$$V_{n+1} = V_n + (\text{inflow})_n - (\text{discharge})_n - (\text{seepage})_n$$
$$- (\text{evaporation})_n \qquad 6.26$$

Repeat step 2 as many times as necessary.

Reservoir seepage is generally very small compared to inflow and discharge. It is, therefore, neglected. Reservoir evaporation can be estimated from analytical relationships or by evaluating data from evaporation pans. Pan data may be extended to reservoir evaporation by the pan coefficient formula. In equation 6.27, the summation is taken over the number of days in a year.

$$E_R = \Sigma C_p E_p \qquad 6.27$$

Units of E are typically inches per day, but any other system of units may be used as long as the pan coefficient takes them into consideration. A typical value of C_p is .7.

Inflow may be taken as actual past history. However, since it is not very likely that history will repeat, the next method is mathematically more justifiable.

Method #2: This method is the same as method #1 except for its method of determining the inflow. Method #2 uses a Monte Carlo simulation which is dependent on enough historical data to establish a cumulative inflow distribution. A Monte Carlo simulation is quite suitable if long periods are to be simulated. If short periods are to be simulated, the simulation should be performed several times and the results averaged.

step 1: Tabulate or otherwise determine a frequency distribution of inflow quantities. This need not be a mathematical formula - a histogram is sufficient.

step 2: Form a cumulative distribution of inflow quantities.

step 3: Multiply the cumulative x axis (which runs from 0 to 1) by 100 or 1000, depending on the accuracy needed.

step 4: Generate random numbers between 0 and 100 (or 0 and 1000). Use of a random number table (table 6.6) is adequate for hand simulation.

step 5: Locate the inflow quantity corresponding to the random number from the cumulative distribution x axis.

Method #3: The non-sequential drought method is more complex than the above two methods, but it has the advantage of giving an estimate of the required reservoir size, rather than evaluating a trial size as the first two methods do.

Due to the absence of synthetic drought information, the first step in this method is to develop intensity-duration-frequency curves from stream flow records.

step 1: Choose a duration. Usually, the first duration used will be 7 days, although choosing 15 days will not introduce too much error.

step 2: Search the streamflow records to find the smallest flow during the duration chosen. (The first time through, for example, find the smallest discharge totaled over any 7 days.) The days do not have to be sequential.

step 3: Continue searching the discharge records to find the next smallest discharge over the number of days in the period. Continue searching and finding the smallest discharges (which gradually increase) until all of the days in the record have been used up. Do not use the same day more than once.

step 4: Give the values of smallest discharge order numbers. That is, M=1 is given to the smallest discharge, M=2 to the next smallest, etc.

step 5: For each observation, calculate the recurrence interval as

$$F = (n_y/M)$$

n_y is the number of years' worth of streamflow data that was searched to find the smallest discharges.

step 6: Plot the points as discharge
on the y axis versus F in years on the x axis. Draw a
reasonably continuous curve through the points.

step 7: Return to step 1 for the next duration. Repeat for
all of the following durations: 7, 15, 30, 60, 120, 183,
and 365 days.

Figure 6.8

Sample Family of Drought Curves

A synthetic drought can be constructed for any recurrence interval.
For example, if a 5-year drought is to be planned for, the discharges
Q_7, Q_{15}, Q_{30}, ..., Q_{365} are read from the appropriate curves for
F = 5 years.

The next step is to plot the mass diagram (also known as a Rippl
diagram) for the reservoir. This is a simultaneous plot of cumulative
demand and inflow. The storage requirements for the reservoir can be
found as the maximum separation of the inflow and demand lines.

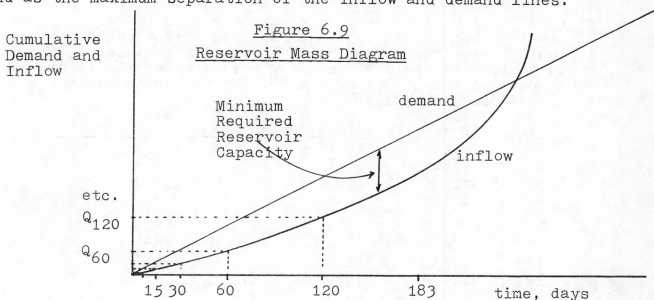

Figure 6.9

Reservoir Mass Diagram

Example 6.8

A well monitored stream has been observed for 50 years and has the following frequency distribution of total annual discharges.

discharge (units	frequency (years)	% of time
0 to .5	5	.10
.5 to 1.0	21	.42
1.0 to 1.5	17	.34
1.5 to 2.0	7	.14
	50	1.00 total

It is proposed to dam the stream and create a reservoir with a capacity of 1.8 units. The reservoir is to support a town which will draw 1.2 units per year. Simulate 10 years of reservoir operation assuming it starts with 1.5 units.

step 1: The frequency distribution was given.

steps 2 and 3: The cumulative distribution is

discharge	cumulative frequency	cum. freq. x 100
0 to .5	.10	10
.5 to 1.0	.52	52
1.0 to 1.5	.86	86
1.5 to 2.0	1.00	100

step 4: From table 6.6 , choose ten 2-digit numbers. Their choice is arbitrary, but they must come sequentially from a row or column. Use the first row for this simulation:

78, 46, 68, 33, 26, 96, 58, 98, 87, 27

step 5: For the first year, the random number is 78. Since 78 is greater than 52 but less than 86, the inflow is in the 1.0 to 1.5 unit range. The mid-point of this range is taken as the inflow which would be 1.25. The reservoir volume after the first year would be 1.5 + 1.25 - 1.20 = 1.55. The remaining years can be simulated just as easily.

year	starting volume	+ inflow	- usage	ending = volume	+ spill
1	1.5	1.25	1.2	1.55	
2	1.55	.75	1.2	1.1	
3	1.11	1.25	1.2	1.15	
4	1.15	.75	1.2	.7	
5	.7	.75	1.2	.25	
6	.25	1.75	1.2	.8	
7	.8	1.25	1.2	.85	
8	.85	1.75	1.2	1.4	
9	1.4	1.75	1.2	1.95	.15
10	1.8	.75	1.2	1.35	

No shortages were experienced; one spill was required.

Table 6.4
Rational Method Runoff Coefficients

Categorized by Surface

Forested	.05 - .2
Asphalt	.7 - .95
Brick	.7 - .85
Concrete	.8 - .95
Shingle roof	.75 - .95
Lawns, well drained (sandy soil)	
Up to 2% slope	.05 - .1
2% to 7% slope	.10 - .15
Over 7% slope	.15 - .2
Lawns, poor drainage (clay soil)	
Up to 2% slope	.13 - .17
2% to 7% slope	.18 - .22
Over 7% slope	.25 - .35
Driveways, walkways	.75 - .85

Categorized by Use

Farmland	.05 - .3
Pasture	.05 - .3
Unimproved	.1 - .3
Parks	.1 - .25
Cemetaries	.1 - .25
Railroad yard	.2 - .40
Playgrounds (except asphalt or concrete)	.2 - .35
Business districts	
neighborhood	.5 - .7
city (downtown)	.7 - .95
Residential	
single family	.3 - .5
multi-plexes, detached	.4 - .6
multi-plexes, attached	.6 - .75
suburban	.25 - .4
apartments, condominiums	.5 - .7
Industrial	
light	.5 - .8
heavy	.6 - .9

Table 6.5

W(u) versus u

u	1.0	2.0	3.0	4.0	5.0	6.0	7.0	8.0	9.0
1.0	0.219	0.049	0.013	0.0038	0.00114	0.00036	0.00012	0.000038	0.000012
EE-1	1.82	1.22	0.91	0.70	0.56	0.45	0.37	0.31	0.26
EE-2	4.04	3.35	2.96	2.68	2.48	2.30	2.15	2.03	1.92
EE-3	6.33	5.64	5.23	4.95	4.73	4.54	4.39	4.26	4.14
EE-4	8.63	7.94	7.53	7.25	7.02	6.84	6.69	6.55	6.44
EE-5	10.95	10.24	9.84	9.55	9.33	9.14	8.99	8.86	8.74
EE-6	13.24	12.55	12.14	11.85	11.63	11.45	11.29	11.16	11.04
EE-7	15.54	14.85	14.44	14.15	13.93	13.75	13.60	13.46	13.34
EE-8	17.84	17.15	16.74	16.46	16.23	16.05	15.90	15.76	15.65
EE-9	20.15	19.45	19.05	18.76	18.54	18.35	18.20	18.07	17.95
EE-10	22.45	21.76	21.35	21.06	20.84	20.66	20.50	20.37	20.25
EE-11	24.75	24.06	23.65	23.36	23.14	22.96	22.81	22.67	22.55
EE-12	27.05	26.36	25.95	25.66	25.44	25.26	25.11	24.97	24.86
EE-13	29.36	28.66	28.26	27.97	27.75	27.56	27.41	27.28	27.16
EE-14	31.66	30.97	30.56	30.27	30.05	29.87	29.71	29.58	29.46
EE-15	33.96	33.27	32.86	32.58	32.35	32.17	32.02	31.88	31.76

Table 6.6

Random Numbers

78466 83326	96589 88727	72655 49682	82338 28583	01522 11248
78722 47603	03477 29528	63956 01255	29840 32370	18032 82051
06401 87397	72898 32441	88861 71803	55626 77847	29925 76106
04754 14489	39420 94211	58042 43184	60977 74801	05931 73822
97118 06774	87743 60156	38037 16201	35137 54513	68023 34380
71923 49313	59713 95710	05975 64982	79253 93876	33707 84956
78870 77328	09637 67080	49168 75290	50175 34312	82593 76606
61208 17172	33187 92523	69895 28284	77956 45877	08044 58292
05033 24214	74232 33769	06304 54676	70026 41957	40112 66451
95983 13391	30369 51035	17042 11729	88647 70541	36026 23113
19946 55448	75049 24541	43007 11975	31797 05373	45893 25665
03580 67206	09635 84610	62611 86724	77411 99415	58901 86160
56823 49819	20283 22272	00114 92007	24369 00543	05417 92251
87633 31761	99865 31488	49947 06060	32083 47944	00449 06550
95152 10133	52693 22480	50336 49502	06296 76414	18358 05313
05639 24175	79438 92151	57602 03590	25465 54780	79098 73594
65927 55525	67270 22907	55097 63177	34119 94216	84861 10457
59005 29000	38395 80367	34112 41866	30170 84658	84441 03926
06626 42682	91522 45955	23263 09764	26824 82936	16813 13878
11306 02732	34189 04228	58541 72573	89071 58066	67159 29633
45143 56545	94617 42752	31209 14380	81477 36952	44934 97435
97612 87175	22613 84175	96413 83336	12408 89318	41713 90669
97035 62442	06940 45719	39918 60274	54353 54497	29789 82928
62498 00257	19179 06313	07900 46733	21413 63627	48734 92174
80306 19257	18690 54653	07263 19894	89909 76415	57246 02621
84114 84884	50129 68942	93264 72344	98794 16791	83861 32007
58437 88807	92141 88677	02864 02052	62843 21692	21373 29408
15702 53457	54258 47485	23399 71692	56806 70801	41548 94809
59966 41287	87001 26462	94000 28457	09469 80416	05897 87970
43641 05920	81346 02507	25349 93370	02064 62719	45740 62080
25501 50113	44600 87433	00683 79107	22315 42162	25516 98434
98294 08491	25251 26737	00071 45090	68628 64390	42684 94956
52582 89985	37863 60788	27412 47502	71577 13542	31077 13353
26510 83622	12546 00489	89304 15550	09482 07504	64588 92562
24755 71543	31667 83624	27085 65905	32386 30775	19689 41437
38399 88796	58856 18220	51056 04976	54062 49109	95563 48244
18889 87814	52232 58244	95206 05947	26622 01381	28744 38374
51774 89694	02654 63161	54622 31113	51160 29015	64730 07750
88375 37710	61619 69820	13131 90406	45206 06386	06398 68652
10416 70345	93307 87360	53452 61179	46845 91521	32430 74795

Bibliography

1. Hammer, Mark J., <u>Water and Waste-Water Technology</u>, John Wiley & Sons, New York, NY 1975

2. Linsley, Ray K., and Franzini, Joseph B., <u>Water Resources Engineering</u>, 2nd ed., McGraw-Hill Book Company, New York, NY 1972

3. Linsley, Ray K., Kohler, Mark A., and Paulhus, Joseph L. H., <u>Hydrology For Engineers</u>, 2nd ed., McGraw-Hill Book Company, New York, NY 1975

4. Merritt, Frederick S., <u>Standard Handbook for Civil Engineers</u>, 2nd ed., McGraw-Hill Book Company, New York, NY 1976

Practice Problems: HYDROLOGY

Required

1. Four 5-acre areas are served by a 1200 foot storm drain (n = .013 and slope = .005). Inlets to the storm drain are placed every 300 feet along the storm drain. The inlet time for each area served by an inlet is 15 minutes, and the run-off coefficient is .55. A storm to be used for design purposes has the following characteristics:

$$I = \frac{100}{t_c + 10}$$

I is in inches/hr
t is in minutes

What is the size of the last section of storm drain assuming that all flows are maximum?

2. An aquifer has a water table level of 100 feet below the ground surface. An 18 inch diameter well extends 200 feet into the aquifer, for a total depth of 300 feet. The aquifer transmissivity is 10,000 gallons/day-foot. The well's radius of influence is 900 feet with a 20 foot drawdown at the well. (a) What steady discharge is possible? (b) What horsepower is required to achieve the steady discharge?

3. A 2-hour storm over a 43 square mile area produced a flood volume of 3300 acre-feet with a peak discharge of 9300 cfs. (a) What is the unit hydrograph peak discharge? (b) If a 2-hour storm producing 2.5 inches of runoff is to be used to design a culvert, what is the design flood hydrograph volume? (c) What is the design discharge?

Optional

4. A .5 square mile drainage area has a suggested run-off coefficient of .6 and a time of concentration of 60 minutes. The drainage area is in Steel region #3, and a 10-year storm is to be used for design purposes. What is the run-off?

5. A well extends from the ground surface (elevation 383) through a gravel bed to a layer of bedrock at elevation 289 feet. The screened well is 1500 feet from a river whose surface level is 363 feet. The well is pumped by a 10" schedule 40 steel pipe which draws 120,000 gallons per day. The permeability of the well is 1600 gallons/day-ft^2. The pump discharges into a piping network whose friction head is 100 feet. What net horsepower is required for steady flow?

6. A measurement station on a stream recorded the following discharges from a 1.2 square mile drainage area.

hour	cfs	hour	cfs	hour	cfs
0	102	6	455	12	55
1	99	7	325	13	49
2	101	8	205	14	43
3	215	9	145		
4	507	10	100	15	38
5	625	11	70		

(a) Draw the actual hydrograph. (b) Draw the unit hydrograph. (c) Determine the time base (N) for direct run-off. (d) Separate the groundwater and surface water.

7. A watershed has an area of 100 square miles. The length of the main stream channel is 20 miles; and the distance to a point opposite the centroid is 11 miles. Assume that C_t = 1.8 and find (a) the time lag, (b) the duration of the synthetic hydrograph, (c) the peak discharge.

8. A reservoir has a total capacity of 7 units. At the beginning of a study, the reservoir contains 5.5 units. The monthly demand on the reservoir from a nearby city is .7 units. The monthly inflow to the reservoir is normally distributed with a mean of .9 units and standard deviation of .2 units. Simulate one year of reservoir operation.

9. Repeat problem 8 assuming that the monthly demand on the reservoir is normally distributed with a mean of .7 units and a standard deviation of .2 units.

10. A class A pan located near a reservoir shows an evaporation loss of .8 inches in one day. If the pan coefficient is .7, what is the approximate evaporation loss in the reservoir?

NOTES

WATER SUPPLY ENGINEERING

Nomenclature

A	area	ft²
AW	atomic weight	grams/gmole
B	width	ft
C	constant, or concentration	-, or mg/l
C_D	drag coefficient	-
d	particle diameter	ft
D	outside pipe diameter	ft
EW	equivalent weight	grams/gmole
f	change in charge	-
F_I	impact factor	-
g	acceleration due to gravity (32.2)	ft/sec²
h	height or depth	ft
H	height	ft
I	D.C. current	amps
K	rate constant	1/sec
K_{eq}	equilibrium constant	-
K_{sp}	solubility product	-
L	length	ft
LF	loading factor	-
MW	molecular weight	grams/gmole
N_{Re}	Reynolds number	-
p	pressure	psf
P	population, or force	1000's of people, or pounds
Q	flow	cfs
t	time	seconds
v	velocity	ft/sec
v*	surface loading	ft/sec
V	volume	ft³
w	pipe load	lb/ft
x	mole fraction	-
z	diagonal distance	ft

Symbols

ρ	density	lb/ft³
ν	kinematic viscosity	ft²/sec
μ	absolute viscosity	lb-sec/ft²
η	efficiency	-

Subscripts

f	flow through
i	incoming
o	outgoing
s	settling

1. Conversions

Multiply	By	To Obtain
Clark degrees	1	grains/Imperial gallon
cubic feet	7.481	gallons
cubic feet	28.32	liters
cubic feet/sec	.6463	million gallons/day
cubic feet/sec	448.8	gpm
feet	.3048	meters
gallons	3785	cubic centimeters
gallons	.1337	cubic feet
gallons	3.785	liter
gpm	.002228	cfs
grains	1.429 EE-4	pounds
grains/gallon	142.9	pounds/million gallons
grains/gallon	17.1	ppm
grains/gallon	17.1	mg/l
grains/gallon	1.2	Clark degrees
grams	.03527	ounces
grams	.002205	pounds
Imperial gallons	1.2	U.S. gallons
inches	2.540	centimeters
kilogram	2.205	pounds
kg/m^3	.06243	$pound/ft^3$
liter	1000	cubic centimeters
liter	.03532	cubic feet
liter	.2642	gallons
liter/sec	15.85	gpm
meters	3.281	ft
meters	39.37	inches
microns	.001	millimeters
MGD	1.547	ft^3/sec
mg/l	1.0	ppm
mg/l	.0583	grains/gallon
mg/l	8.345	pounds/million gallons
pounds	7000	grains
pounds	453.6	grams
ppm	.0583	grains/gallon
ppm	.07	Clark degrees

2. Definitions

acid: An acid is a compound containing hydrogen ions (H^+ or H_3O^+) in an
      aqueous solution. Acids have a sour taste, conduct electricity,
      turn blue litmus paper red, and have a pH between 0 and 7.
aeration: Mixing water with air, either by spraying water or bubbling
      air through it.
aerobic: requiring oxygen
air break: A means by which drinking water can be used for fire fighting
      without contaminating the drinking supply. Drinking water is
      freely discharged into the top of a tank which feeds the fire main.
algae: one-celled plant life
altitude valve: A valve which automatically opens to prevent overflow
      in storage tanks.

AMU: Abbreviation for atomic mass unit. One AMU is 1/12th the atomic weight of carbon.

anaerobic: not requiring oxygen

anion: negative ions that migrate to the positive electrode in an electroytic solution

atomic number: The number of protons in the nucleus of an atom.

atomic weight: Approximately the number of protons and neutrons in the nucleus of an element.

Avogadro's law: A gram-mole of any substance contains 6.023 EE23 molecules.

B. coli: 'bacteria coli.' See 'E. coil'

backfill: The soil that is used to cover a pipe put into a trench.

base: A compound containing hydroxide ions (OH^-) combined with alkali metals and alkali earths. Bases in aqueous solutions have a bitter taste, conduct electricity, turn red litmus paper blue, and have a pH between 7 and 14.

belt-line layout: See 'grid iron layout'

breakpoint chlorination: Application of chlorine which results in a minimum of chloramine residuals. No free chlorine residuals are produced unless the breakpoint is exceeded.

capita: person

carbonate hardness: hardness caused by bicarbonates

cation: a positive ion that migrates to the negative electrode in an electrolytic solution.

chloramine: compounds of chlorine and ammonia (e.g., NH_2Cl, $NHCL_2$, or NCL_3)

chlorine demand: The difference between applied chlorine and the chlorine residual. Chlorine demand is chlorine that has been reduced in chemical reactions and is no longer available for purification.

clear well: Storage in a water treatment plant which normally takes water from the filters.

coagulation: Floc formation as the result of adding coagulating chemicals. Coagulants destabilize (by reducing repulsive forces) suspended particles allowing them to agglomerate.

coliform: See 'colon bacilli'

colloid: a fine particle ranging in size from 1 to 500 millimicrons. colloids cause turbidity.

colon bacilli: bacterial residing normally in the intestinal tract. These are not necessarily dangerous if present in drinking water, but they are indicators of other possible pathogens.

combined residuals: compounds of an additive (such as chlorine) which have combined with something else. Chloramines are combined residuals.

compound: A homogeneous substance composed of two or more elements which can be decomposed by chemical changes only.

confirmed test: A second test used if the presumptive test for coliforms is positive.

cross connection: connecting fire and drinking water supplies together.

detention time: The average time spent by water in a settling basin.

detritus: See 'grit'

distillation: Salt removal in water by boiling and condensation.

domestic use: Water use by the public (home use)

double main system: Separate water mains for domestic and fire fighting use

E. coli: See 'escherichia coli'

element: A pure substance that cannot be decomposed by chemical means.

electrodialysis: A method of using induced currents and direction-selective membranes to remove dissolved salts from water.

electrolysis: The production of an oxidation-reduction reaction by a D.C. current.

enteric: intestinal

equivalent weight: The molecular weight divided by the change in charge or oxidation number.

escherichia coli: A common bacterium found in the digestive tract.

eutrophication: Aging of a lake due to plant growth and sedimentation.

facultative: able to live under different or changing conditions.

floc: agglomerated colloidal particles

flush hydrant: hydrants located in pits below street level.

free residuals: ions or compounds not combined or reduced. The presence of free residuals signifies excess dosage.

fungus: multi-cellular plant growth common to humid areas.

gravity distribution: A water supply which uses natural flow from an elevated tank or mountain reservoir to supply pressure.

grid-iron layout: A system of distribution pipes in which there are alternatve paths through which water can flow in case one path is disturbed.

grit: sand-like particles mixed with mud and other debris.

hard water: water containing bicarbonates of calcium and magnesium, as well as chlorides and sulphates.

hydrogen ion: positively charged combination of a proton and water molecule (H_3O)

hydrophilic: seeking or liking water

hydrophobic: disliking water

hydronium ion: See 'hydrogen ion'

ions: atoms which have lost or gained one or more electrons giving them a charge.

intrinsic water: Ultra-pure water with essentially no mineral or ion content. Typically used in electronic industries.

isotopes: Atoms of the same atomic number but with different atomic weights due to a variable number of neutrons.

mixture: A heterogeneous physical combination of two or more substances, each of which retains its identity and specific properties.

mole: A quantity of substance equal to its molecular weight in grams (gmole or gram-mole) or in pounds (pmole or pound-mole).

molecule: The smallest division of an element or compound.

molecular weight: The sum of the atomic weights of all atoms in the molecule.

non-pathogenic: not biologically harmful.

osmosis: The flow of a solvent through a permeable membrane separating two solutions of different concentrations.

oxidation: The loss of electrons.

oxidation number: An arbitrarily assigned number used in redox calculations used to balance molecular charges.

pathogenic: biologically harmful

permanent hardness: Hardness which cannot be removed by heating.

pH: A measure of a solution's hydrogen ion concentration (acidity).

pOH: A measure of a solution's hydroxyl ion concentration (alkalinity).

polished water: See 'intrinsic water'

post hydrant: a fire hydrant which rises from the sidewalk surface.

potable: drinkable

presumptive test: A 1st stage test which is inconclusive if positive, but is conclusive if negative.

protozoa: single-celled aquatic animals

radical: A charged group of atoms that combines as if it were a single element.

recarbonation: Addition of CO_2 to water to neutralize excess lime.

redox reaction: A reaction in which oxidation and reduction occur.

reduction: A gain in electrons.

residual: Compounds that are left over for later use after some of an additive has been combined or inactivated.

retention period: See 'detention time'

reverse osmosis: A process which uses pressure to force molecular motion of a solvent to flow against the ionization gradient.

salt: an ionic compound formed by direct union of elements, reactions between acids and bases, reactions of acids and salts, and reactions between different salts.

single main system: One main supplies potable and fire fighting water.

solution: A homogeneous mixture of solute and solvent.

stoichiometry: The study of how elements combine in predetermined quantities to form compounds.

superchlorination: Chlorination past the breakpoint.

syndets: Synthetic detergents containing phosphates.

temporary hardness: Hardness that can be removed by heating water.

thermocline: A layer of water in a lake or reservoir where temperature declines rapidly with depth.

turbidity: water clouded by colloids.

valence: The relative combining capacity of an atom or group of atoms compared with that of the standard hydrogen atom. Valence of an ion is the same as the ion's charge.

zeolite: A natural or synthetic resin which has an affinity for ions.

3. Chemistry Review

A. Valence

The charge of any element in its free state is zero. A charged condition will occur when the element has lost or gained one or more electrons. The valence of an ion is equal to its charge.

Common elements with one valence number are listed below:

cations			anions		
+1	+2	+3	-1	-2	+4
H	Mg	Al	F	O	C
Li	Ca	B	Cl	S	Si
Na	Sr		Br		
K	Ba		I		
Ag	Zn				
	Cd				

Some elements have two valence numbers. The lower valence is associated with the generic ending 'ous'. The higher valence is associated with the ending 'ic'.

+1/+2	+2/+3	+2/+4	+3/+5
Cu	Fe	Pb	As
Hg	Ni	Sn	Sb
	Co		Bi

The following elements have more than 2 valence numbers:

Cr: +2, +3, +6
Mn: +2, +3, +4, +6, +7
S: -2, +4, +6
N: -3, +1, +2, +3, +4, +5
P: -3, +3, +5

Common radicals are:

NH_4	Ammonium	+1	HCO_3	Bicarbonate	-1
ClO_3	Chlorate	-1	CO_3	Carbonate	-2
ClO_2	Chlorite	-1	SO_4	Sulfate	-2
ClO	Hypochlorite	-1	SO_3	Sulfite	-2
NO_3	Nitrate	-1	CrO_4	Chromate	-2
NO_2	Nitrite	-1	Cr_2O_7	Dichromate	-2
$C_2H_3O_2$	Acetate	-1	BO_3	Borate	-3
MnO_4	Permanganate	-1	PO_4	Phosphate	-3
OH	Hydroxide	-1	$Fe(CN)_6$	Ferricyanide	-3
			$Fe(CN)_6$	Ferrocyanide	-4

Compounds always form in such a manner as to obtain a neutral charge. For example, $MgNO_3$ would not be allowed because the magnesium has a charge of +2 and the nitrate ion has a charge of -1. However, $Mg(NO_3)_2$ would be allowed.

B. Chemical Reactions

During chemical changes, bonds between atoms are broken and new bonds are formed. Reactants are either converted to simpler products or are synthesized into more complex compounds. There are five common types of chemical reactions.

(1) <u>Direct combination or synthesis</u>. This is the simplest type of reaction where two elements or compounds combine directly to form a compound.

e.g., $2H_2 + O_2 \rightarrow 2H_2O$

$SO_2 + H_2O \rightarrow H_2SO_3$

(2) <u>Decomposition</u>. Bonds uniting a compound are disrupted by heat or other energy source to yield simpler compounds or elements.

e.g., $2HgO \rightarrow 2Hg + O_2$

$H_2CO_3 \rightarrow H_2O + CO_2$

(3) <u>Single displacements</u>. This type of reaction is identified by one element and one compound as reactants.

e.g., $2Na + 2H_2O \rightarrow 2NaOH + H_2$

$2KI + Cl_2 \rightarrow 2KCL + I_2$

(4) <u>Double decomposition</u>. These are reactions characterized by having two compounds as reactants and forming two new compounds.

e.g., $AgNO_3 + NaCl \rightarrow AgCl + NaNO_3$

$H_2SO_4 + ZnS \rightarrow H_2S + ZnSO_4$

(5) <u>Oxidation-Reduction</u> ('Redox'). These reactions involve oxidation of one substance and reduction of another. In the example below, calcium loses electrons and is oxidized; oxygen gains electrons and is reduced.

e.g., $2Ca + O_2 \rightarrow 2CaO$

Balancing chemical equations is largely a matter of deductive trial and error. The coefficients in front of each compound may be thought of as the number of molecules or moles taking part in the reaction. The number of atoms for each element must be equal on both sides of the equation. Total atomic weight must also be equal on both sides.

<u>Example 7.1</u>

Balance the following reaction equation:

$$Al + H_2SO_4 \rightarrow Al_2(SO_4)_3 + H_2$$

As written, the reaction is not balanced. For example, there is one aluminum on the left, but there are two on the right. The starting element in the balancing procedure is chosen somewhat arbitrarily.

<u>step 1</u>: Since there are two aluminums on the right, multiply (Al) by 2.

$$2Al + H_2SO_4 \rightarrow Al_2(SO_4)_3 + H_2$$

<u>step 2</u>: Since thare are three sulfate radicals (SO_4) on the right, multiply (H_2SO_4) by 3.

$$2Al + 3H_2SO_4 \rightarrow Al_2(SO_4)_3 + H_2$$

<u>step 3</u>: Now there are six hydrogens on the left, so multiply H_2 by 3 to finally balance the equation.

$$2Al + 3H_2SO_4 \rightarrow Al_2(SO_4)_3 + 3H_2$$

◆

C. Stoichiometry

Stoichiometric problems are known as 'weight and proportion' problems because their solutions use simple ratios to determine the weight of reactants required to produce some given amount of products. The procedure for solving these problems is essentially the same regardless of the complexity of the reaction.

    step 1: Write and balance the chemical equation.

    step 2: Calculate the molecular weight of each compound or element in the equation.

    step 3: Multiply the molecular weights by their respective coefficients and write the products under the formulas.

    step 4: Write the given weight data under the molecular weights calculated from step 3.

    step 5: Fill in missing information by using simple ratios.

Example 7.2

Caustic soda (NaOH) is made from sodium carbonate (Na_2CO_3) and slaked lime ($Ca(OH)_2$). How many pounds of caustic soda can be made from one ton of sodium carbonate?

The balanced chemical equation is

$$Na_2CO_3 + Ca(OH)_2 \longrightarrow 2NaOH + CaCO_3$$

MW's: 106   74   80   100
given: 2000   X

The simple ratio used is

$$\frac{NaOH}{Na_2CO_3} = \frac{80}{106} = \frac{X}{2000} \qquad \text{or } X = 1509 \text{ pounds}$$

◆

D. Equivalent Weights

The equivalent weight of a molecule which takes place in a chemical reaction is the molecular weight divided by the change in charge (or oxidation number) which is experienced by the molecule or ion.

Example 7.3

What are the equivalent weights of the following compounds?

    (a) Al in the reaction: $Al^{+++} + 3e^- \longrightarrow Al$

    (b) H_2SO_4 in the reaction: $H_2SO_4 + H_2O \longrightarrow 2H^+ + SO_4^{--} + H_2O$

    (c) NaOH in the reaction: $NaOH + H_2O \longrightarrow Na^+ + OH^- + H_2O$

(a) The atomic weight of aluminum is 27. Since the change in charge was three, the equivalent weight is $(27/3) = 9$.
(b) The molecular weight of sulfuric acid is 98.1. Since it went from neutral to ions with 2 charges each, the equivalent weight is $(98.1/2)$ or 49.05.

(c) Sodium hydroxide has a molecular weight of 40. The molecule went from neutral to a singly-charged state. So, EW = (40/1) = 40.

◆

E. Solutions of Solids in Liquids

Various methods exist for measuring the strengths of solutions. Some of them are:

(N)   Normality - the number of gram equivalent weights of solute per liter of solution. A solution is 'normal' if there is exactly one gram equivalent weight per liter.

(M)   Molarity - The number of gram moles of solute per liter of solution. A 'molar' solution contains one gram mole per liter of solution. Hydrated water adds to molecular weight.

(F)   Formality - The number of gram formula weights per liter of solution. Hydrated water does not add to formula weight.

(m)   Molality - The number of gram moles of solute per 1000 grams of solvent. A 'molal' solution contains 1 gmole per 1000 grams.

(x)   Mole fraction - The number of moles of solute divided by the number of moles of solvent and all solutes.

(mg/l) milligrams per liter - The number of milligrams per liter. Same as (ppm) for solutions of water.

(ppm) parts per million. The number of pounds (or grams) of solute per million pounds (or grams) of solution. Same as (mg/l) for solutions of water.

(meq/l) The number of milligram equivalent weights of solute per liter of solution. Calculated by multiplying normality by 1000.

Example 7.4

A solution is made by dissolving .353 grams of $Al_2(SO_4)_3$ in 730 grams of water. Assuming 100% ionization, what is the concentration expressed as normality, molarity, and (mg/l)?

The molecular weight of $Al_2(SO_4)_3$ is

$$2(26.98) + 3(32.06 + 4(16)) = 342.14$$

The equivalent weight is (342.14/6) = 57.02

The number of gram equivalent weights used is (.353/57.02) = 6.19 EE-3.

The number of liters of solution (same as the solvent volume if the small amount of solute is neglected) is .73.

$$N = \frac{6.19 \text{ EE-3}}{.73} = 8.48 \text{ EE-3}$$

The number of moles of solute used is (.353/342.14) = 1.03 EE-3.

$$M = \frac{1.03 \text{ EE-3}}{.73} = 1.41 \text{ EE-3}$$

The number of milligrams is $(.353/.001) = 353$.

$$(mg/l) = \frac{353}{.73} = 483.6$$

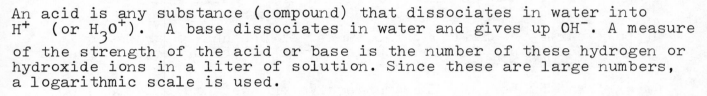

F. Solutions of Gases in Liquids

Henry's law states that the amount of gas dissolved in a liquid is proportional to the pressure of the gas. This applies separately to each gas to which the liquid is exposed.

Example 7.5

At 20°C and 760 mm Hg, one liter of water will absorb .043 grams of oxygen gas or .19 grams of nitrogen. Find the weight of oxygen and nitrogen in one liter of water exposed to air at 20°C and 760 mm Hg total pressure.

Partial pressure is volumetrically weighted. Air is 20.9% oxygen and the remainder is taken as nitrogen.

oxygen dissolved = $(.209)(.043) = .009$ g/l

nitrogen dissolved = $(.791)(.19) = .150$ g/l

G. Acids and Bases

An acid is any substance (compound) that dissociates in water into H^+ (or H_3O^+). A base dissociates in water and gives up OH^-. A measure of the strength of the acid or base is the number of these hydrogen or hydroxide ions in a liter of solution. Since these are large numbers, a logarithmic scale is used.

$$pH = -\log_{10}[H^+] \qquad\qquad 7.1$$

$$pOH = -\log_{10}[OH^-] \qquad\qquad 7.2$$

The following relationship always exists between pH and pOH:

$$pH + pOH = 14 \qquad\qquad 7.3$$

$[X]$ is defined as the ion concentration in moles per liter. The number of moles can be calculated from Avogadro's law by dividing the actual number of ions by 6.023 EE23. An easier method is to use equation 7.4.

$$[X] = (\text{fraction ionized})(\text{molarity}) \qquad\qquad 7.4$$

Example 7.6

Calculate the concentrations of H^+, OH^-, pH, and pOH in 4.2% ionized .010M ammonia solution prepared from ammonium hydroxide (NH_4OH).

$[OH^-] = (.042)(.010) = 4.2$ EE-4 moles/liter

$pOH = -\log(4.2$ EE-4$) = 3.38$

$pH = 14 - 3.38 = 10.62$

$[H^+] = 10^{-10.62} = 2.4$ EE-11 moles/liter

Since $(H^+ + OH^- \rightarrow H_2O)$, acids and bases neutralize each other by forming water. The volumes required for complete neutralization are:

$$\text{(vol. base)(normality of base)} = \text{(vol. acid)(normality of acid)} \qquad 7.5$$

If the concentrations are expressed as molarities,

$$\text{(vol. base)(molarity of base)(change in charge of base)} = \text{(vol. acid)(molarity of acid)(change in charge of acid)} \quad 7.6$$

Both equation 7.5 and 7.6 assume 100% ionization of the solute.

H. Reversible Reactions

Some reactions are capable of going in either direction. Such reactions are called reversible reactions and are characterized by the presence of all reactants and all products simultaneously. For example, the following reaction is reversible:

$$N_2 + 3H_2 \Longleftrightarrow 2NH_3 + 24,500 \text{ calories}$$

Le Chatelier's principle helps determine the direction the reaction will go when some property or condition is changed. That principle says that when a reversible reaction has resulted in an equilibrium condition and the reactants and products are stressed by changing the pressure, temperature or concentration, a new equilibrium will be formed in the direction which reduces the stress.

Consider the above reaction between nitrogen and hydrogen. When the reaction proceeds in the forward direction, heat is given off (an exothermic reaction). If the reaction proceeds in the reverse direction, heat is absorbed (endothermic reaction). Now, if the system is stressed by increasing the temperature, the reaction will proceed in the reverse direction because that direction will absorb heat and reduce the temperature.

For reactions that involve gases, the coefficients in front of the molecules may be taken as volumes. Thus, in the nitrogen-hydrogen reaction, 4 volumes combine to form 2 volumes. If the equilibrium system is stressed by increasing the pressure, then the forward reaction will occur because this direction reduces the volume and thereby reduces the pressure.

If the concentration of any substance is increased, the reaction proceeds in a direction away from the substance with the increased concentration.

If a catalyst is introduced, the equilibrium will not be changed. However, the reaction speed is increased, and equilibriums are reached more quickly. This reaction speed is known as the rate of reaction, reaction velocity, or rate constant.

The law of mass action says that the reaction speed is proportional to the concentrations of reactants. Given the following reaction,

$$A + B \Longleftrightarrow C + D$$

the rates of reaction may be calculated as the products of the reactants' concentrations.

$$v_{forward} = C_1[A][B] \qquad\qquad 7.7$$

$$v_{reverse} = C_2[C][D] \qquad\qquad 7.8$$

The constants, C, are needed to obtain the proper units. For reversible reactions, the equilibrium constant is defined by equation 7.9. This equilibrium constant is essentially independent of pressure, but will depend on temperature.

$$K_{eq} = \frac{[C][D]}{[A][B]} \qquad\qquad 7.9$$

If the reaction is

$$aA + bB \Longleftrightarrow cC + dD$$

then it is easy to see that

$$K_{eq} = \frac{[C]^c[D]^d}{[A]^a[B]^b} \qquad\qquad 7.10$$

The only special rule to be observed is that if any of A, B, C, or D are solids, then their concentrations are omitted in the calculation of the equilibrium constant.

Example 7.7

Acetic acid dissociates according to the following equation. What is the equilibrium constant if the ion concentrations are as given?

$$HC_2H_3O_2 + H_2O \Longleftrightarrow H_3O^+ + C_2H_3O_2^-$$

$$[HC_2H_3O_2] = .09866 \text{ moles/liter}$$
$$[H_2O] \qquad 55.5555 \text{ moles/liter}$$
$$[H_3O^+] \qquad .00134 \text{ moles/liter}$$
$$[C_2H_3O_2^-] \qquad .00134 \text{ moles/liter}$$

From equation 7.9,

$$K_{eq} = \frac{(.00134)(.00134)}{(.09866)(55.5555)} = 3.3 \text{ EE-7}$$

However, for weak aqueous solutions, the concentration of water is very large and essentially constant. Therefore, it is omitted. The new constant is known as the ionization constant to distinguish it from the equilibirum constant which does include the water concentration in its formula.

$$K_{ionization} = (K_{equilibrium})[H_2O] \qquad 7.11$$

Pure water is a very weak electrolyte and it ionizes only slightly by itself.

$$2H_2O \Longleftrightarrow H_3O^+ + OH^- \qquad 7.12$$

At equilibrium, the ion concentrations can be measured to be

$$[H_3O^+] = EE-7$$

$$[OH^-] = EE-7$$

The ionization constant for the ionization of pure water is

$$K_{ionization} = [H_3O^+][OH^-] \qquad 7.13$$
$$= (EE-7)(EE-7) = (EE-14)$$

Notice that the concentration of water was omitted because this was the ionization constant being calculated, not the equilibrium constant.

Taking logs of both sides of equation 7.13 will derive equation 7.3.

Example 7.8

A .1M acetic acid solution is 1.34% ionized. Find the

        (a) hydrogen ion concentration
        (b) acetate ion concentration
        (c) un-ionized acid concentration
        (d) ionization constant

(a) From equation 7.4, $[H_3O^+] = (.0134)(.1) = .00134$ moles/liter

(b) Since every hydrogen ion has a corresponding acetate ion (see example 7.7), the acetate ion concentration is also .00134 moles per liter.

(c) The concentration of un-ionized acid may be found from an equation similar to equation 7.4:

$$[HC_2H_3O_2] = \text{(fraction not ionized)(molarity)}$$
$$= (1 - .0134)(.1) = .09866 \text{ moles/liter}$$

The ionization constant is

$$K_{ionization} = \frac{(.00134)(.00134)}{(.09866)} = 1.82 \ EE-5$$

A faster way to find the ionization constant is to use the mass action equation:

$$K_{ionization} = \frac{(molarity)(fraction\ ionized)^2}{(1 - fraction\ ionized)} \qquad 7.14$$

Example 7.9

Find the concentration of the hydrogen ion for a .2M acetic acid solution if $K_{ionization} = 1.8$ EE-5

From equation 7.14 letting X be the fraction ionized,

$$1.8\ EE\text{-}5 = \frac{(.2)(X)^2}{1 - X}$$

If X is small, then (1 - X) = 1. Then, X = 9.49 EE-3.

From equation 7.4, the concentration is

$$[H_3O^+] = (9.49\ EE\text{-}3)(.2) = 1.9\ EE\text{-}3$$

◆

The common ion effect law is a form of Le Chatelier's law. This is a statement to the effect that if a salt containing a common ion is added to a weak acid or base solution, ionization will be repressed. This is a consequence of the need to have an unchanged ionization constant.

Example 7.10

What is the hydrogen ion concentration of a solution with .1 gmole of 80% ionized ammonium acetate in one liter of .1M acetic acid?

As before, $1.8\ EE\text{-}5 = \dfrac{[H_3O^+][C_2H_3O_2^-]}{[HC_2H_3O_2]}$

Let $X = [H_3O^+]$

Then $[HC_2H_3O_2] = .1 - X \approx .1$

But now $[C_2H_3O_2^-] = X + (.8)(.1) \approx (.8)(.1) = .08$

So, from the ionization constant equation,

$$1.8\ EE\text{-}5 = \frac{(X)(.08)}{.1} \qquad \text{or } X = 2.2\ EE\text{-}5$$

◆

I. Solubility Product

When an ionic solid (such as silver chloride) is dissolved in a solvent (such as water) it dissociates or ionizes:

$$AgCl \Longleftrightarrow Ag^+ + Cl^-$$

When the equation for the equilibrium constant is written, the term for solid components is omitted. When the equation for the ionization constant is written, the term for water concentration is also omitted. Thus, when an ionic solid is placed in water, the ionization constant will consist only of the ion concentrations. This ionization constant is known as the solubility product.

$$K_{sp} = [Ag^+][Cl^-]$$

As with the general case of ionization constants, the solubility product is essentially constant for slightly soluble solutes. Any time that the product of terms exceeds the standard value of the solubility product, solute will be precipitated out until the product of the remaining ion concentrations attains the standard value. If the product is less than the standard value, the solution is not saturated.

Table 7.1

SOLUBILITY PRODUCTS

Aluminum hydroxide @ 15°C	4	EE-13
@ 18°C	1.1	EE-15
@ 25°C	3.7	EE-15
Calcium carbonate @ 15°C	.99	EE-8
@ 25°C	.87	EE-8
Calcium fluoride @ 18°C	3.4	EE-11
@ 26°C	4.0	EE-11
Ferric hydroxide @ 18°C	1.1	EE-36
Magnesium hydroxide@ 18°C	1.2	EE-11

Example 7.11

What is the solubility product of lead sulfate ($PbSO_4$) if its solubility is 38 mg/l?

$$PbSO_4 \Longleftrightarrow Pb^{++} + SO_4^{--} \quad (in\ water)$$

The molecular weight of $PbSO_4$ is $(207.19 + 32.06 + 4(16)) = 303.25$

The number of moles of $PbSO_4$ in a liter of saturated solution is

$$.038/303.25 = 1.25\ EE-4$$

This is also the number of moles of Pb^{++} and SO_4^{--} that will form in the solution. Therefore,

$$K_{sp} = [Pb^{++}][SO_4^{--}] = (1.25\ EE-4)^2 = 1.56\ EE-8$$

◆

The method used in example 7.11 to find the solubility product works well with chromates (CRO_4^{--}), halides (F^-, CL^-, Br^-, I^-), sulfates (SO_4^{--}), and iodates (IO_3^-). However, sulfides (S^{--}), carbonates (CO_3^{--}) and phosphates (PO_4^{-2-}) and the salts of transition elements such as iron hydrolize and must be treated differently.

4. Qualities of Supply Water

A. Acidity and Alkalinity

Acidity is a measure of acids in solution. Acidity in surface water is caused by formation of carbonic acid from carbon dioxide in the air.

$$CO_2 + H_2O \longrightarrow H_2CO_3 \qquad\qquad\qquad 7.15$$

$$H_2CO_3 + H_2O \longrightarrow HCO_3^- + H_3O^+ \quad (pH > 4.5) \qquad 7.16$$

$$HCO_3^- + H_2O \longrightarrow CO_3^{--} + H_3O^+ \quad (pH > 8.3) \qquad 7.17$$

Measurement of acidity is done by titration with a standard basic measuring solution. Acidity in water is typically given in terms of the concentration of $CaCO_3$ (in mg/l) that would neutralize the acid.

$$\text{acidity (in terms of mg/l of } CaCO_3) = \frac{(\text{vol. titrant})(\text{titrant normality})(50{,}000)}{(\text{sample volume})}$$

$$7.18$$

Carbonic acid is very aggressive and must be neutralized to eliminate the cause of water pipe corrosion.

If the pH of water is greater than 4.5, carbonic acid ionizes to form bicarbonate (equation 7.16). If the pH is greater than 8.3, carbonate ions form which cause water hardness by combining with calcium (see equation 7.17).

Alkalinity is also measured in terms of mg/l of $CaCO_3$ by using an acidic titrant.

$$\text{alkalinity in terms of mg/l of } CaCO_3 = \frac{(\text{vol. titrant})(\text{titrant normality})(50{,}000)}{(\text{sample volume})}$$

$$7.19$$

Alkalinity and acidity of a titrated sample is determined from color changes in indicators added to the titrant. The following indicators are commonly used.

Table 7.2

Colorimetric Indicators

This titrant	turns from (basic)	to (less basic)	at pH
phenolphthalein	pink	clear	10.0 to 8.3
methyl orange	orange	pink	4.4 to 4.0
eriochrome black T	wine red	blue	10 to 7
bromocresol green/ with methyl red	gray	pink	4.8 to 4.6
thymol blue	yellow	red	2.8 to 1.2
methyl red	yellow	red	6.3 to 4.2
bromthymol blue	blue	yellow	7.6 to 6.0
cresol red	red	yellow	8.8 to 7.2
alizarin yellow	red	yellow	12.0 to 10.1

For strongly basic samples (pH > 8.3), titration with both phenolphthalein and methyl orange is used. The alkalinity is the sum total of the phenolphthalein alkalinity and the methyl orange alkalinity. 50 mg/l is an acceptable alkalinity for water used in boilers. (Note: Phenolphthalein alkalinity is not the same as carbonate alkalinity since CO_3^- has been converted to HCO_3^- but not neutralized. See equation 7.17. The carbonate alkalinity is twice the phenolphthalein alkalinity.)

Example 7.12

.02N sulfuric acid is used to titrate a 110 ml sample of water. 3.3 ml of titrant was needed to reach the phenolphthalein point, and 13.2 ml was needed to reach the methyl orange point. What are the total and phenolphthalein alkalinities?

From equation 7.19, the phenolphthalein alkalinity is

$$(3.3)(.02)(50,000)/(110) = 30 \text{ mg/l as } CaCO_3$$

The total alkalinity is

$$(13.2 + 3.3)(.02)(50,000)/(110) = 150 \text{ mg/l as } CaCO_3$$

The alkalinity of 150 mg/l is caused by carbonates (2 x 30 = 60 mg/l) and bicarbonates (150 - 60 = 90 mg/l)

B. Hardness

Water hardness is caused by doubly-charged (but not singly- or triply-charged) positive metallic ions such as calcium, magnesium, and iron. (Iron is not as common as the other two, however.) Hardness reacts with soap to reduce its cleansing effectiveness and to form scum on the water surface and ring around the bathtub.

Water containing bicarbonate (HCO_3^-) ions can be heated to precipitate a carbonate molecule. This hardness is known as temporary hardness or carbonate hardness.

$$Ca^{++} + 2HCO_3^- + heat \longrightarrow CaCO_3 + CO_2 + H_2O \qquad 7.20$$

$$Mg^{++} + 2HCO_3^- + heat \longrightarrow MgCO_3 + CO_2 + H_2O \qquad 7.21$$

The symbol ▒ means that the compound precipitates out until the water is saturated.

Remaining hardness due to sulfates, chlorides, and nitrates is known as permanent hardness or non-carbonate hardness because it cannot be removed by heating. Permanent hardness may be calculated numerically by causing precipitation, drying, and then weighing the precipitate.

$$Ca^{++} + SO_4^{--} + Na_2CO_3 \longrightarrow 2Na^+ + SO_4^{--} + CaCO_3 \qquad 7.22$$

$$Mg^{++} + 2Cl^- + 2NaOH \longrightarrow 2Na^+ + 2Cl^- + Mg(OH)_2 \qquad 7.23$$

Hardness may also be measured by the titration method using a titrant (complexione, versene, EDTA, or BDH) and an indicator (such as eriochrome black T). The standard hardness reagent used for titration has an equivalent hardness of 1 mg/l per ml used.

Example 7.13

A 75 ml water sample required 8.1 ml of EDTA. What is the hardness?

$$\text{hardness} = \frac{(8.1) \text{ ml } (1) \text{ mg/l}}{(75/1000)} = 108 \text{ mg/l}$$

◆

Hardness can be classified according to the following table.

Table 7.3
Hardness Classifications

Class	Type	Hardness
A	soft	below 60 mg/l
B	medium hard	60 - 120
C	hard	120 - 180
D	very hard	180 - 350
E	saline, brackish	above 350

Although high values of hardness are not organically dangerous, public acceptance of the water supply requires a hardness of less than 150 mg/l.

Example 7.14

Water is analyzed and is found to contain sodium (Na^+, 15 mg/l), magnesium (Mg^{++}, 70 mg/l), and calcium (Ca^{++}, 40 mg/l). What is the equivalent calcium hardness measured as $CaCO_3$?

Sodium is singly-charged, so it does not contribute to hardness. The approximate equivalent weights of the relevant compounds and elements are:

Mg: 12   Ca: 20 $CaCO_3$: 50

The equivalent hardness is

$$(70)(\tfrac{50}{12}) + (40)(\tfrac{50}{20}) = 391.6 \text{ mg/l as } CaCO_3$$

◆

C. Iron Content

Even in low concentrations, iron is objectionable because it stains bathroon fixtures, causes a brown bolor in laundered clothing, and affects taste. Water originally pumped from anaerobic sources may contain Fe^{++} ions which are invisible and soluble. When exposed to oxygen, insoluble Fe^{+++} ferric oxides form which give water the rust coloration.

Iron is measured optically by comparing the color of a sample with standard colors. The comparison may be made by eye or with a photoelectric colorimeter. Iron concentrations greater than .3 mg/l are undesireable.

D. Manganese Content

Manganese ions are similar in effect, detection, and measurement to iron ions. Manganous manganese (Mn^{++}) oxidizes to manganic manganese (Mn^{++++}) to give water a rust color. An undesireable concentration is .05 mg/l.

E. Fluoride Content

An optimum concentration of fluoride in the form of a fluoride ion, F^-, is between .8 mg/l for hot climates (80°F - 90°F average) to 1.2 mg/l for cold climates (50°F average). These amounts reduce the population cavity rate to a minimum without producing significant fluorosis (staining) of the teeth. The actual amount of fluoridation depends on the average outside temperature since the temperature affects the amount of water that is ingested by the population.

Fluoridation may be obtained by the following readily dissociating compounds:

Table 7.4

Fluoridation Chemicals

$(NH_4)_2SiF_6$

CaF_2

H_2SiF_6 (fluosilic acid)

NaF (sodium fluoride)

Na_2SiF_6 (sodium silicofluoride)

Fluoride content may be measured by colorimetric and electrical methods.

F. Chloride Content

Chlorine is used as a disinfectant for water, but its strong oxidation potential allows it also to be used to remove iron and manganese ions. Chlorine gas in water forms hypochlorous and hydrochloric acids:

$$Cl_2 + H_2O \underset{pH<4}{\overset{pH>4}{\rightleftharpoons}} HCl + HOCl \underset{pH<9}{\overset{pH>9}{\rightleftharpoons}} H^+ + OCl^- \qquad 7.24$$

Chlorine, hypochlorous acid, and hypochlorite ions are known as free chlorine residuals. Hypochlorous acid reacts with ammonia (if it is present) to form mono-, di-, and trichloramines. Chloramines are known as combined residuals. Chloramines are more stable than free residuals, but their disinfecting ability is weaker. Their action may extend for a considerable distance into the distribution system.

The amount of chlorine to be added depends on the organic and inorganic matter present in the water. However, most waters are effectively treated within 10 minutes if a free residual of .2 mg/l is maintained. Larger residual concentrations may cause objectionable odor and taste.

If the water contains phenol, it and the chlorine will form chlorophenol compounds which produce an objectionable taste. This may be stopped by adding ammonia to the water before chlorination.

Both free and combined residual chlorine may be tested for by color comparison. However, organic matter in waste water makes it necessary to use a test based on water conductivity. The color comparison test in supply water, however, is adequate.

G. Phosphorous Content

Orthophosphates ($H_2PO_4^-$, HPO_4^{--}, and PO_4^{---}) and polyphosphates (such as $Na_3(PO_3)_6$) result from the use of synthetic detergents ('syndets'). Phosphate content is more of a concern in waste water than in supply water. Excessive phosphate discharge contributes to aquatic plant growth and subsequent eutrofication.

Phosphates may be measured by a variety of means, including colorimetry and filtered precipitation analysis. Care should be taken not to confuse (mg/l) of phosphates with (mg/l) of phosphorus.

multiply	by	to obtain	
(mg/l) of phosphate	.326	(mg/l) of phosphorus	7.25

H. Nitrogen Content

Nitrogen is present in water in many forms, including organic (protein), ammonia, nitrite, nitrate, and gaseous ammonia. As with phosphates, nitrogen contamination is more of a problem with waste water than with supply water. Nitrogen pollution can promote algae growth. Ammonia is toxic to fishes.

Drinking water is typically tested only for nitrates. The following tests are used:

Table 7.5

to test for	use this method
ammonia	distillation
organic nitrogen	digestion with distillation
nitrate, nitrite	colorimetry

Gaseous nitrogen is of little concern since it is not normally metabolized by plants and it is of no danger to animal or human life.

I. Color

Color in domestic water is undesireable aesthetically, and it may dull the color of clothes and stain bathroom fixtures. Some industries (such as beverage production, dairy, food processing, paper manufacturing, and textile production) also have strict water color standards.

Water color is measured with a colorimeter or comparitively with tubes containing standard platinum/cobalt solutions. Color is graded on a scale of 0 (clear) to 70 color units.

J. Turbidity

Turbidity is a measure of the insoluble solids (soil, organics, and microorganisms) in water which impede light passage. Completely clean water measures 0 turbidity units (TU). 5 TU is noticeable to an average consumer, and this is an upper limit for drinking water. Muddy water exceeds 100 TU. A TU is equivalent to 1 mg/l of silica in suspension.

Turbidity is usually measured with a nephelometer, Jackson candle apparatus, or Baylis turbidimeter. Units may be NTU (nephelometer turbidity units) or JTU (Jackson turbidity units).

K. Suspended and Dissolved Solids

Solids present in a sample of drinking water may be divided into several categories, not all of which are mutually exclusive.

1. <u>Suspended solids</u>: Suspended solids, the same as 'filterable' solids, are measured by filtering a sample of water and weighing the residue.

2. <u>Dissolved solids</u>: Dissolved solids, same as non-filterable solids, is measured as the difference between total solids and suspended solids.

3. <u>Total solids</u>: Total solids is made up of suspended and dissolved solids. It is measured by drying a sample of water and weighing the residue.

4. <u>Volatile solids</u>: Volatile solids is the decrease in weight of total solids which have been ignited in an electric furnace.

5. <u>Fixed solids</u>: The fixed solids may be found as the difference between total solids and volatile solids.

6. <u>Settleable solids</u>: The volume (ml/l) of settleable solids is measured by allowing a sample to stand for one hour in a graduated conical container (Imhoff cone).

An upper limit of 500 mg/l of total solids is recommended.

L. Water-borne Diseases

Organisms that are present in water consist of bacteria, fungi, viruses, algae, protozoa, and multicellular animals. Not all of these are dangerous, but some are. Important organisms are listed in table 7.6.

Table 7.6
Water-Borne Organisms

this organism	causes this disease
BACTERIA	
Salmonella typhosa	typhoid fever
Vibrio comma	cholera
Shigella dysenteriae	dysentery
Escherichia coli	enteric problems
fecal streptococci	enteric problems
VIRUSES	
Poliomyelitis	polio
Infectious hepatitis	hepatitis
PROTOZOA	
Entamoeba histolytica	dysentery
PARASITES	
Schistosomiasis	flatworms
Bilharziasis	flatworms

5. Water Quality Standards

Minimum drinking water quality standards have been set by the Water Pollution Control Act of 1972 (source 1) and the Environmental Protection Agency in 1974 (source 2.) These minimum standards are given in table 7.7.

Table 7.7
Water Quality Standards

Contaminant/Quality	Permissible	Source
turbidity	1 TU	2
color	15 units	2
coliform organisms	1/100 ml	1
total dissolved solids	750 mg/l	1
arsenic	.05 mg/l	2
barium	1.0 mg/l	2
cadmium	.01 mg/l	2
chloride (Cl)	250 mg/l	2
chromium	.05 mg/l	2
copper	1.0 mg/l	2
cyanide	.005 mg/l	2
fluoride	see page 7-19	2
iron	.3 mg/l	2
lead	.05 mg/l	2
manganese	.05 mg/l	2
mercury	.002 mg/l	2
nitrate (NO_3)	10 mg/l	2
selenium	.01 mg/l	2
silver	.05 mg/l	2
sulfate (SO_4)	250 mg/l	2
zinc	5.0 mg/l	2

6. Water Demand

Water requirements come from a number of sources, including residential, commercial, industrial, and public consumers, as well unavoidable loss and waste. In project planning, a minimum of about 165 gallons per capita-day should be considered. This 165 gpcd is a total of all demands, as given in table 7.8. If large industries are present (such as canning, steel making, automobile production, electronics, etc.) then those industries' special needs must also be considered.

For ordinary domestic use, the water pressure should be 25 to 40 psi. A minimum of 60 psi at the fire hydrant is usually adequate, since that allows for up to 20 psi pressure drop in fire hoses. 75 psi and higher is common in commercial and industrial districts.

Table 7.8
Average Water Requirements (gpcd)

Residential	75 - 130
Commercial & Industrial	70 - 100
Public	10 - 20
Loss & Waste	10 - 20
Fire fighting	(See equation 7.27)
	165 - 270 TOTAL

Variations can be expected with the time of day and season. If the average daily demand is to be used to estimate peak demands or average fluctuations, then table 7.9 lists some multipliers. These multipliers are to be used against the 165 gpcd (or whatever other average is available.)

Table 7.9
Demand Multipliers

Consumption Time/Period	Multiplier
Winter	.80
Summer	1.30
Maximum daily	1.80
Maximum hourly	3.00
Early morning	.25 to .40
Noon	1.50 to 2.0

Water demand in gpcd must be multiplied by the population to obtain the total demand. Since a population changes, a supply system must be designed to handle demands through a reasonable time into the future. Five methods exist for estimating future demand:

1. Uniform growth rate
2. Constant percentage growth rate (e.g., 40% per decade)
3. Decreasing growth rate (e.g., 6% each decade)
4. Straight-line graphical extension
5. Comparison with neighboring cities

Economic aspects of the project dictate the number of years into the future which should be designed for. Table 7.10 lists some typical average lifetimes of water supply components (based on IRS depreciation guidelines.)

Table 7.10
Average Replacement Times of Components (years)

dams	
earthen, concrete	150
loose rock	60
steel	40
hydrants	50
water meters	30
pipes	
cast iron	
2" - 6"	50
8" - 12"	100
concrete	20
steel	30
pumps	20
reservoirs	75
tanks	
concrete	50
steel	40
wells	40

Since the maximum demand can be up to 3 times the average daily demand (table 7.9), the design rate should be approximately $2\frac{1}{2}$ to 3 times the average daily rate plus an allowance for fire fighting. If the water treatment plant's capacity is fixed, the distribution system should be able to handle the plant's capacity plus an allowance for fire fighting (which can be passed around the treatment plant.)

The requirements for fire fighting will vary between 500 gpm (a minimum) to 12,000 gpm for a single fire. Multiple fires will place a greater demand on the distribution system. A municipality must continue to serve its domestic, commercial, and industrial customers, however. The Insurance Services Office recommends that the fire system be able to operate with the remainder of the potable water system operating at the maximum daily rate, as taken over all 24 hour periods within the last three years.

Recommended fire flow in a neighborhood will depend on construction type, occupancy, and floor area. An estimate for a neighborhood can be found from equation 7.26 (taken from the Guide for Determination of Required Fire Flow, Insurance Services Office, 1974.)

$$Q = (.04)\ C\ \sqrt{A} \qquad\qquad 7.26$$

C is a constant which depends on construction: 1.5 for wood frame, 1.0 for ordinary construction, .8 for noncombustible construction, and .6 for fire resistant construction. A is the area (in square feet) of all stories in the building, except for basements. Special rules are used to find A for multi-story fire resistant structures, buildings with various fire loadings, or buildings with sprinkler systems. Q is rounded to the nearest 250 gpm, but it should not be less than 500 gpm or more than 8000 gpm for a single building.

In estimating the water requirements for fire fighting on a population basis, the American Insurance Association has recommended the following formula:

$$Q = 2.3\sqrt{P}(1 - .01\sqrt{P})$$

7.27

Most insurance requirements will be met if the flow rate can be maintained for T hours, where T is the flow rate in 1000's of gallons per minute, with a maximum of 10 hours. Mechanical components, such as power supplies, pumps, and treatment plants, should be able to operate at the fire flow for 2 to 5 days, depending on the availability of spares and expected out-of-service time.

7. Pipe Materials

Many types of pipes are available. Successful pipe materials used for distribution must have adequate strength to withstand external loads from backfill, traffic, and earth movement; high burst strength to withstand water pressure; smooth interior surfaces; corrosion resistant exteriors; and tight joints.

The following types of pipes are suitable for use in the water distribution system.

Table 7.11
Water Pipe Materials

Type	Comments
ductile and gray cast iron	Long life, strong, impervious. May be tar coated to resist exterior corrosion. Cement mortar may be used on the inside to reduce interior corrosion. Available in 2" to 54" standard sizes. Normally 18 feet lengths.
asbestos-cement	Immune to electrolysis and corrosion. Smooth interior surface. Available 4" to 36" diameter. Standard 13 feet lengths.
concrete	Durable, watertight, low maintenance, smooth interior surface. Diameters 16" to 144"; standard 8, 12, and 16 feet lengths.
steel	High strength, good yielding and shock resistance, but susceptible to corrosion. Exterior may be tarred, painted, or wrapped. Interior may have enamel or cement mortar lining. Smooth interior surface.
plastic	Chemically inert, corrosion resistant, smooth interior. PVC most popular. PVC available in ratings to 315 psi, diameters ½" to 16", and standard lengths of 20 feet and 39 feet.

8. Loads on Buried Pipes

If a pipe is buried (placed in an excavated trench and backfilled) it must support an external vertical load in addition to its internal pressure load. The magnitude of the load depends on the amount of backfill, type of soil, and type of pipe. For rigid pipes (concrete, cast iron, clay) which cannot deform and which are placed in narrow trenches (2 or 3 diameters) the load in pounds per foot of pipe is

$$w = C\rho B^2 \qquad\qquad 7.28$$

C and ρ are given in table 7.12. B is the trench width in feet at the top of the pipe. (A minimum trench width to allow working room is commonly estimated at $(4/3)$ times the pipe diameter plus 8".)

For a flexible pipe (steel, plastic, copper) placed in a narrow trench, the loading is

$$w = C\rho BD \qquad\qquad 7.29$$

D is the outside diameter of the pipe in feet.

Table 7.12 (3:319)

Backfill Material	Sand & Gravel	Saturated Topsoil	Clay	Saturated Clay
Density (pcf)	100	100	120	130
h/B		Values of C		
1	.84	.86	.88	.90
2	1.45	1.50	1.55	1.62
3	1.90	2.00	2.10	2.20
4	2.22	2.33	2.49	2.65
5	2.45	2.60	2.80	3.03
6	2.60	2.78	3.04	3.33
7	2.75	2.95	3.23	3.57
8	2.80	3.03	3.37	3.76
10	2.92	3.17	3.56	4.04
12	2.97	3.24	3.68	4.22

If a pipe is placed on undisturbed ground and covered with fill ('broad fill' or 'embankment fill') the load is

$$w = C_p \rho D^2 \qquad\qquad 7.30$$

Table 7.13 (3:320)

Values of C_p

(h/D)	Rigid pipe, rigid surface, noncohesive backfill	Flexible pipe, Average Conditions
1	1.2	1.1
2	2.8	2.6
3	4.7	4.0
4	6.7	5.4
6	11.0	8.2
8	16.0	11.0

Equation 7.30 is also the upper maximum for trenched pipes (equations 7.28 and 7.29) if B > 2D.

Figure 7.1
Backfilled Trenches

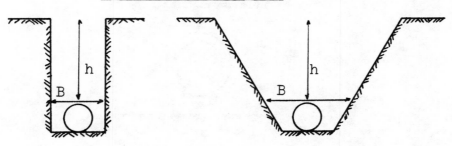

Boussinesq's equation may be used to calculate the load on a pipe due to a superimposed load, P, at the surface. This load should be added to the loadings calculated from equations 7.28, 7.29, and 7.30.

$$p = \frac{3h^3 P}{2\pi z^5} \qquad\qquad 7.31$$

$$w = Dp \qquad\qquad 7.32$$

If the pipe has less than 3 feet of cover, a multiplicative impact factor should also be used.

$$w = F_I Dp \qquad\qquad 7.33$$

Table 7.14
Impact Factors

cover depth, h	impact factor, F_I
less than 1 foot	1.3
1 to 2 feet	1.2
2 to 3 feet	1.1
more than 3 feet	1.0

Figure 7.2
External Loads on Buried Pipes

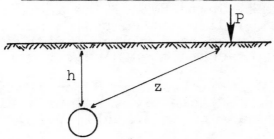

Concrete pipes are tested (ASTM specifications) in a 3-edge bearing mechanism, as shown in figure 7.3, to determine the crushing strength. Crushing strengths (taken as the load which produces a .01" crack) are given in pounds per foot of pipe per foot of inside diameter. There-fore, crushing strength is

crushing strength = (load per unit diameter)(diameter)   7.34

Figure 7.3

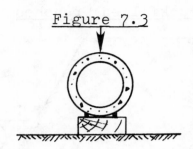

The weight supported by the pipe can be increased above the crushing strength by proper bedding. Different bedding methods and their load factors are given in figure 7.4. The allowable load is given by equation 7.35. The factor of safety varies from 1.2 to 1.5 for clay or unreinforced concrete. For reinforced concrete, the factor of safety is 1.0.

$$\text{allowable load} = \frac{(\text{Crushing strength})(\text{load factor})}{\text{factor of safety}} \qquad 7.35$$

Figure 7.4

Load Factors

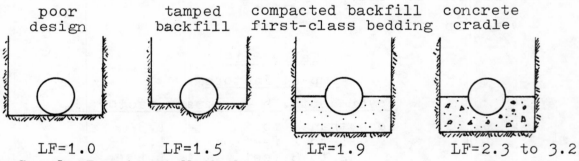

poor design	tamped backfill	compacted backfill first-class bedding	concrete cradle
LF=1.0	LF=1.5	LF=1.9	LF=2.3 to 3.2

9. Water Supply Treatment Methods

A. Plain Sedimentation (Clarification)

Water contaminated with sand, dirt, mud, etc., may be treated in a sedimentation basin or tank. Up to 80% of the incoming sediment may be removed in this manner. Sedimentation basins are usually concrete, rectangular or circular in plan, and equipped with scrapers or raking arms to periodically remove accumulated sludge.

Settlement of water-borne particles depends on the water temperature (which affects viscosity), particle size, and particle specific gravity. Typical specific gravities are given in table 7.15.

Table 7.15

Specific Gravities of Particles

discrete sand	2.65
flocculated mud	1.03

Settlement time can be calculated from the settling velocity, flow-through velocity, and depth of the tank. Tank depths are typically 10 to 15 feet. The settling velocities for spherical particles in 68°F water is given in figure 7.5. Of course, settling velocities will be much less than those shown in figure 7.5 because actual sediment particles are not spherical.

Figure 7.5
Settling Velocities
(spherical particles, 68°F water)

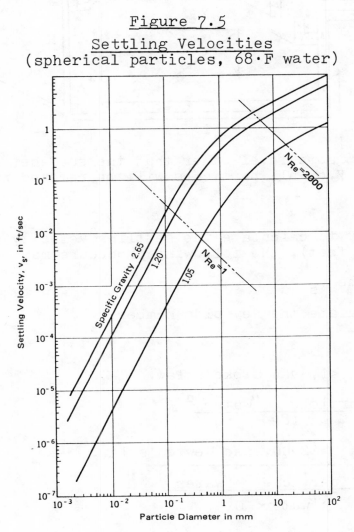

If it is assumed that the water velocity is a uniform v_f, then all particles with $v_s > v^*$ will be removed. v^* is known as the overflow rate (or 'surface loading', or 'critical velocity') and has typical values of 500 to 1000 gallons per day per square foot. B is the tank width, typically 30 to 40 feet, and L is the length, typically 100 to 200 feet. (For radial flow circular basins, a typical diameter is 100 feet.)

$$v^* = \frac{Q}{A_{surface}} = \frac{Q}{BL}$$

7.36

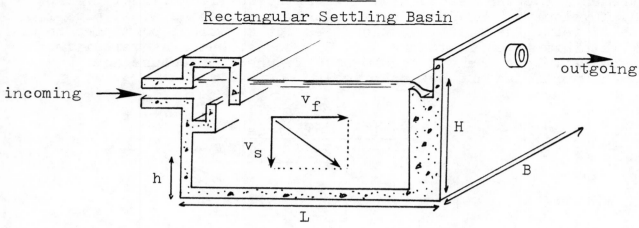

Figure 7.6
Rectangular Settling Basin

If water enters at some level other than the surface, such as at level h in figure 7.6, all particles will be removed which have

$$v_s > \frac{hQ}{HBL} \qquad 7.37$$

If it is necessary to calculate the settling velocity of a particle of diameter d (in feet), the following procedure may be used.

step 1: assume v_s

step 2: calculate the Reynolds number.

$$N_{Re} = \frac{v_s d}{\nu} \qquad 7.38$$

step 3: If $N_{Re} < 1$, use Stoke's law.

$$v_s = \frac{(\rho_{particle} - \rho_{water})d^2}{18\mu} \qquad 7.39$$

If $N_{Re} > 2000$, use Newton's law.

$$v_s = \sqrt{\frac{4g(\rho_{particle} - \rho_{water})d}{3(\rho_{water})C_D}} \qquad 7.40$$

Values of C_D are given in table 7.16.

Table 7.16
Approximate Drag Coefficients for Spheres

N_{Re}	C_D
2000	.4
10,000	.4
50,000	.5
100,000	.5
200,000	.4

The time spent by water in the basin is known as the detention (or retention) time (or period). The detention time is given by equation 7.41. A minimum time recommended is 4 hours, although periods from 1 to 10 hours are used.

$$t = \frac{\text{tank volume}}{Q} \qquad 7.41$$

The weir loading is the daily flow rate divided by the total effluent weir length, usually expressed in gallons/ft-day. A recommended maximum is 20,000 gpd/ft.

The basin efficiency is

$$\eta = \frac{(\text{flow through period})}{\text{detention time}} \qquad 7.42$$

The flow through period is found by using color dye in the basin.

B. Mixing and Flocculation

Chemicals may be added to obtain a desired water quality. These chemicals are added to the water in mixing basins. There are two types of mixing basins: complete mixing and plug flow mixing.

Complete mixing basins dispense the chemical immediately throughout the volume by using mixing paddles. If the volume of water being mixed is small, the tank is known as a flash (or 'quick') mixer. Quick mixing detention time is often less than 60 seconds.

The retention time required for complete mixing in a tank of volume V is given by equation 7.43.

$$t = \left(\frac{V}{Q}\right) = \frac{1}{K}\left(\frac{C_i}{C_o} - 1\right) \qquad 7.43$$

Plug flow mixing is accomplished by allowing the water and additive to flow through a long chamber at a uniform rate without mechanical agitation. The retention time in a plug flow mixer of length L is given by equation 7.44.

$$t = \frac{V}{Q} = \frac{L}{v} = \frac{1}{K}(\ln(C_i/C_o)) \qquad 7.44$$

If the mixer adds coagulant for the removal of colloidal sediment, it is known as a flocculator. Floc is a precipitate that forms when the coagulant allows the colloidal particles to agglomerate. Flocculation is enhanced by gentle agitation, but the floc disintegrates with violent agitation.

Common coagulants are aluminum sulfate ('alum', $Al_2(SO_4)_3 \cdot 18H_2O$), ferrous sulphate ('copperas', $FeSO_4 \cdot 7H_2O$), ferric chloride ($FeCl_3$), or chlorinated copperas (a mixture of ferrous sulphate and ferric chloride.) The most common coagulant is alum. The usual dosage is 10 to 40 mg/l (80 to 300 pounds per million gallons).

Alum reacts with alkalinity in the incoming water to form an aluminum hydroxide floc. If the water is not sufficiently alkaline, an auxiliary chemical (such as CaO (lime) or Na_2CO_3 (soda ash)) is used along with the alum. Lime is also used as an auxiliary chemical with $FeSO_4$ coagulant.

After flash mixing, a 20- to 30-minute period of gentle mixing is used to permit flocculation. During this period, the flow-through velocity should be between .5 and 1.5 ft/min. The peripheral speed of mixing paddles should range from .5 to 2.5 ft/sec. The flocculation is followed by sedimentation for 2 to 8 hours in a low-velocity basin.

C. Clarification with Flocculation

The flocculation-clarifier combines mixing, flocculation, and sedimentation into a single tank. These units are called 'solid contact units' and 'upflow tanks.' They are generally round in construction, with mixing and flocculation taking place near the central hub, and sedimentation occurring at the periphery. Flocculation-clarifiers are most suitable when combined with softening since the precipitated solids help seed the floc.

The following general guidelines may be used to design a flocculation-clarifier:

minimum flocculation & mixing time	30 minutes
minimum retention time	2 hours
maximum weir loading	10 gpm/ft
maximum upflow rate	1.0 gpm/ft^2
maximum sludge formation rate	5% of water flow

D. Filtration

Nonsettling floc may be removed by filtering. The rapid sand ('gravity') filter is the most common filter for this use. Rapid sand filters are essentially beds of granulated gravel and sand. Although the box depth may be 10 feet, the sand bed will be only about 2 feet deep. Since the water has been coagulant treated, it may be passed through the filter quickly - hence the name 'rapid'. Rapid filters are usually square or nearly square in design.

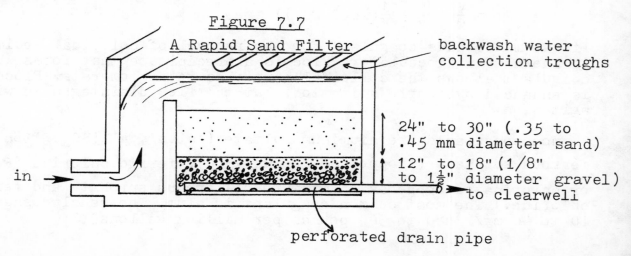

Figure 7.7
A Rapid Sand Filter

backwash water collection troughs

24" to 30" (.35 to .45 mm diameter sand)

12" to 18" (1/8" to 1½" diameter gravel)

to clearwell

in

perforated drain pipe

Optimum filter operation will occur when the top layer of sand is slightly more coarse than the rest of the sand. During backwashing, however, the finest sand rises to the top. Various designs using coal and garnet layers in conjunction with sand layers overcome this difficulty due to the differences in specific gravity.

Historically, the flow rate has been 2 gpm/ft^2 in rapid sand filter design, although some current filters operate at 8 gpm/ft^2. 4 gpm/ft^2 is a reasonable rate for modern designs.

A water treatment plant should have at least 3 filters so that 2 can be in operation when one is being cleaned. Typical total through-puts per filter range from 350 gpm for small plants to 3500 gpm for large plants.

Optimum design of a filtering system includes discharge into a clearwell. Clearwells are storage reservoirs with capacities of 30% to 60% of the daily output. Demand may be satisfied by the clearwell if one or more of the filters is serviced

The most common type of service needed by filters is backwashing. Filters require backwashing when the pores between sand particles clog up. Typically this occurs after 1 to 3 days of operation. Backwashing is done when the head loss through the filter bed reaches approximately 8 feet. Backwashing with filtered water expands the sand layer up to 50%, which dislodges the trapped material.

Water is pumped through the filter during backwashing from the bottom. The rate at which the water rises in the filter housing is usually 12 to 36 inches per minute. This rise time should not exceed the settling velocity of the smallest particle which is to be retained in the filter. Backwashing usually takes between 3 and 5 minutes. Water used for backwashing, which is collected in troughs for disposal, constitutes between 1% and 5% of the total processed water. The head loss after backwashing should be around 2 feet.

The actual amount of backwash water can be found from equation 7.45. Be sure to use consistent units.

$$\text{water volume} = \left(\frac{\text{backwash}}{\text{time}}\right)\left(\frac{\text{filter}}{\text{area}}\right)\left(\frac{\text{rise}}{\text{rate}}\right) \qquad 7.45$$

Slow sand filters are similar in design to rapid sand filters, except that the sand layer is thicker (24" to 48"), the gravel layer is thinner (6" to 12"), and the flow rate is much lower (.05 to .1 gpm/ft^2.) Slow sand filters are limited to low-turbidity applications not requiring chemical treatment. Cleaning is usually accomplished by removing a few inches of sand.

Other types of filters seeing limited use are:

- pressure filters: Similar to rapid sand filters except incoming water is pressurized. Filter rates of 2 to 4 gpm/ft^2. Not used in large installations.
- diatomaceous earth filters: Mostly used in swimming pools. 1 to 5 gpm/ft^2.
- microstrainers: Woven stainless steel fabric, usually mounted on a rotating drum.

E. Chlorination

Chlorination is used for disinfection and oxidation. As a disinfectant, it destroys bacteria and microorganisms. As an oxidant, it removes iron, manganese, and ammonia nitrogen.

Chlorine may be added as a gas or a solid. (If it is added to the water as a gas, it is stored as a liquid which vaporizes around -35°C.) Liquid chlorine is the predominant form since it is cheaper than hypochlorite solid $(Ca(OCl)_2)$. If chlorine liquid or gas is added to water, the following reaction occurs to form hypochlorous acid, which itself ionizes to a hypochlorite and hydrogen ions.

$$Cl_2 + H_2O \longrightarrow HCl + HOCl \underset{pH<7}{\overset{pH>8}{\rightleftharpoons}} H^+ + OCL^- + HCL \qquad 7.46$$

If calcium hypochlorite solid is added to water, the ionization follows immediately.

$$Ca(OCl)_2 + H_2O \longrightarrow Ca^{++} + 2OCl^- + H_2O \qquad 7.47$$

Chlorine existing in water as hypochlorous acid and hypochlorite ions is known as free available chlorine (free residuals). Chlorine in combination with ammonia is known as combined available chlorine (combined residuals.)

The average chlorine dose is in the 1 to 2 mg/l range. Minimum chlorine residuals for 70°F water are given in table 7.17. However, inactivated viruses (such as might be present in surface water) require a heavier chlorine concentration. Since treatment of water is by both free and combined residuals, ammonia may be added to the water to produce chloramines.

Excess chlorine may be removed with a reducing agent, usually called a 'dechlor.' Sulfur dioxide and sodium bisulfate are used in this manner. Aeration also reduces chlorine content, as does passing the water through an activated charcoal filter.

Table 7.17
Minimum Chlorine Residuals (mg/l)

pH Value	Free residuals after 10 minutes	Combined residuals after 60 minutes
6.0	.2	1.0
7.0	.2	1.5
8.0	.4	1.8
9.0	.8	3.0
10.0	.8	3.0

F. Fluoridation

Fluoridation may occur any time after filtering. Smaller utilities almost always choose liquid solution, volumetric feeding mechanism, with solutions being manually prepared. Larger utilities use gravimetric dry feeders with sodium silicofluoride or solution feeders with fluorsilic acid. The residual fluorine content has already been given (page 7-19). The characteristics and dose rates of common fluorine compounds are given in table 7.18.

Table 7.18
Dose Rates for Fluorine Compounds

Formula	H_2SiF_6	NaF	Na_2SiF_6
Form	liquid	solid	solid
Typical Purity	22-30%	90-98%	98-99%
Dose to obtain 1.0 mg/l (pounds per million gal.)	35.2 (with 30% purity)	18.8 (with 98% purity)	14.0 (with 98.5% purity)

Defluoridation (required if the fluoride exceeds 1.5 mg/l) may be achieved with alumina or bone char.

G. Iron Removal

Various methods exist for iron removal.

    1. Aeration, followed by sedimentation and filtration.

$$Fe^{++} + O_x \longrightarrow FeO_x \qquad\qquad 7.48$$

    2. Aeration, followed by chemical oxidation, sedimentation, and filtration. Chlorine or potassium permanganate may be used as an oxidizer.

    3. Manganese zeolite process: Manganese dioxide removes soluble iron ions.

    4. Lime water softening

H. Manganese Removal

Manganese is not easily removed by aeration alone. However, methods 2, 3, and 4 listed under 'Iron Removal' may be used successfully.

I. Water Softening

1. Lime and Soda Ash Softening

Water softening may be accomplished with lime and soda ash to precipitate calcium and magnesium ions from the solution. Lime treatment has added benefits of disinfection, iron removal, and clarification. Practical limits of precipitation softening are 30 mg/l of $CaCO_3$ and 10 mg/l of $Mg(OH)_2$ (as $CaCO_3$) because of intrinsic solubilities. Water softened with this method usually leaves the softening apparatus with a hardness of between 50 and 80.

Lime (CaO) is available as granular quicklime (90% CaO, 10% MgO) or hydrated lime (68% CaO, rest water). Both forms are slaked prior to use, which means that water is added to form a lime slurry in an exothermic reaction.

$$CaO + H_2O \longrightarrow Ca(OH)_2 + heat \qquad\qquad 7.49$$

Soda ash is usually available as 98% pure sodium carbonate (Na_2CO_3).

FIRST STAGE TREATMENT: In the first stage treatment, lime added to water reacts with free carbon dioxide to form calcium carbonate precipitate.

$$CO_2 + Ca(OH)_2 \longrightarrow CaCO_3\text{⫶} + H_2O \qquad\qquad 7.50$$

Next, the lime reacts with calcium bicarbonate.

$$Ca(HCO_3)_2 + Ca(OH)_2 \longrightarrow 2CaCO_3\text{⫶} + 2H_2O \qquad\qquad 7.51$$

Any magnesium hardness is also removed at this time.

$$Mg(HCO_3)_2 + Ca(OH)_2 \longrightarrow CaCO_3\text{⫶} + 2H_2O + MgCO_3 \qquad\qquad 7.52$$

To remove the soluble $MgCO_3$, the pH must be above 10.8. This is accomplished by adding an excess of approximately 35 mg/l of CaO or 50 mg/l of $Ca(OH)_2$.

$$MgCO_3 + Ca(OH)_2 \longrightarrow CaCO_3\text{⫶} + Mg(OH)_2\text{⫶} \qquad\qquad 7.53$$

The amount of lime required in the first stage softening is given by equation 7.54. Note that "alkalinity" includes both $Ca(HCO_3)_2$ and $Mg(HCO_3)_2$ contributions.

$$\begin{aligned}
\text{mg/l of pure CaO} &= \frac{56}{44}\binom{\text{mg/l of}}{CO_2} + \frac{56}{100}\binom{\text{mg/l of}}{\text{alkalinity}} & 7.54a \\
&= \frac{56}{44}\binom{\text{mg/l of}}{CO_2} + \frac{56}{40.1}\binom{\text{mg/l of}}{Ca^{++}} + \frac{56}{24.3}\binom{\text{mg/l of}}{Mg^{++}} & 7.54b
\end{aligned}$$

If slaked lime is used, substitute 74 in place of 56 in equation 7.54.

FIRST STAGE RECARBONATION: Lime added to precipitate hardness removes itself. This is good because any calcium that remains in the water has the potential for forming scale. Further stabilization can be achieved by recarbonation (treatment with carbon dioxide).

$$Ca(OH)_2 + CO_2 \longrightarrow CaCO_3\text{⫶} + H_2O \qquad\qquad 7.55$$

Excess recarbonation should be avoided. If the pH is allowed to drop below 9.5, then carbonate hardness reappears.

$$CaCO_3 + CO_2 + H_2O \longrightarrow Ca(HCO_3)_2 \qquad\qquad 7.56$$

At this time, any unsettled $Mg(OH)_2$ can be returned to a soluable state.

$$2Mg(OH)_2 + 2CO_2 \longrightarrow 2MgCO_3 + 2H_2O \qquad 7.57$$

The amount of carbon dioxide needed in the first stage of recarbonation is

$$\text{mg/l of } CO_2 = \frac{44}{74}\left(\text{mg/l of Ca(OH)}_2 \text{ excess}\right) + \frac{44}{58.3}\left(\text{mg/l of Mg(OH)}_2\right) \qquad 7.58$$

SECOND STAGE TREATMENT: The second stage treatment removes calcium noncarbonate hardness (sulfates and chlorides) which needs soda ash for precipitation.

$$CaSO_4 + Na_2CO_3 \longrightarrow CaCO_3{\ssstyle\vdots} + Na_2SO_4 \qquad 7.59$$

Magnesium noncarbonate hardness needs both lime and soda ash.

$$MgSO_4 + Ca(OH)_2 \longrightarrow Mg(OH)_2{\ssstyle\vdots} + CaSO_4 \qquad 7.60$$

$$CaSO_4 + Na_2CO_3 \longrightarrow CaCO_3{\ssstyle\vdots} + Na_2SO_4 \qquad 7.61$$

The amount of soda ash required is

$$\text{mg/l of } Na_2CO_3 = \frac{106}{100}(\text{mg/l of noncarbonate hardness as } CaCO_3) \qquad 7.62$$

Excess soda ash leaves sodium ions in the water. Luckily, noncarbonate hardness requiring soda ash is a small part of total hardness. Soda ash is also costly, so the actual dose may be slightly reduced from what is needed according to equation 7.62.

SECOND STAGE RECARBONATION: Second stage recarbonation is needed to remove $CaCO_3$. CO_2 is added until the pH is about 8.6, at which time no further precipitation will occur because

$$[Ca^{++}][CO_3^-] < K_{sp} \text{ of } CaCO_3$$

Sodium polyphosphate may be added at this time to inhibit crusting on filter sand and scale formation in pipes.

A split process may be used to reduce the amount of lime that is neutralized by recarbonation (and is wasted.) Excess lime is added in the first stage. This forces precipitation of magnesium in the first stage (instead of in the second stage as shown by equations 7.60 and 7.61.) Excess lime reacts with calcium hardness in the second stage. The amount of bypass depends on the allowable hardness of water leaving the plant. A typical split process is shown in figure 7.8.

Figure 7.8
A Split (Bypass) Process

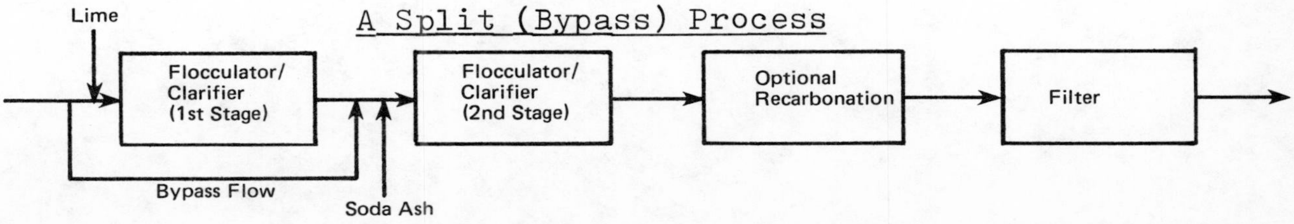

If hardness (as $CaCO_3$) and alkalinity (also as $CaCO_3$) are the same, then there are no SO_4^-, Cl^-, or NO_3^- ions present. (That is, there is no non-carbonate hardness.) If hardness is greater than alkalinity, however, then non-carbonate hardness is present. If hardness is less than alkalinity, then all hardness is bicarbonate hardness, and the extra HCO_3^- comes from other sources (such as $NaHCO_3$).

Example 7.15

How much slaked lime (90% pure), soda ash, and carbon dioxide are required to reduce the hardness of the water evaluated below to zero using the lime-soda ash process? Neglect the fact that this process cannot really produce zero hardness, and base your answer on stoichiometric considerations.

$$\text{total hardness: } 250 \text{ mg/l as } CaCO_3$$

$$\text{alkalinity: } 150 \text{ mg/l as } CaCO_3$$

$$\text{carbon dioxide: } 5 \text{ mg/l}$$

From equation 7.54, the pure slaked lime requirement is

$$CO_2: \quad (\tfrac{74}{44})(5) \quad = \quad 8.4$$

$$\text{alkalinity: } (\tfrac{74}{100})(150) = 111.0$$

$$\text{excess: } \quad = \underline{\ 50.0\ } \text{ (required to raise pH to 10.8)}$$
$$\text{TOTAL } 169.4 \text{ mg/l}$$

The actual requirement of 90% pure slaked lime is

$$169.4/.9 = 188.2 \text{ mg/l}$$

The noncarbonate hardess is $(250 - 150) = 100$ mg/l. The soda ash requirement is given by equation 7.62.

$$(\tfrac{106}{100})(100) = 106 \text{ mg/l}$$

The first stage recarbonation CO_2 requirement is given by equation 7.58. For the excess lime,

$$(\tfrac{44}{74})(50) \quad = 29.7 \text{ mg/l}$$

Additional CO_2 is required in the second stage recarbonation to reduce the pH to 8.6.

◆

2. Ion Exchange Method

In the ion exchange process (also known as 'zeolite process' or 'base exchange method') water is passed through a filter bed of exchange material. This exchange material is known as zeolite. Ions in the insoluble exchange material are displaced by ions in the water. When the exchange material is spent, it is regenerated with a rejuvenating solution such as sodium chloride (salt). Typical flow rates are 6 gpm per ft^2 of filter bed. The processed water will have a zero hardness. However, since there is no need for water with zero hardness, some water is usually bypassed around the unit.

There are 3 types of ion exchange materials. Greensand (glauconite) is a natural substance which is mined and treated with manganese dioxide. Siliceous-gel zeolite is an artificial solid used in small volume deionizer columns. Polystyrene resins are also synthetic. Polystyrene resins currently dominate the softening field.

During operation, the calcium and magnesium ions are removed according to the following reaction in which R is the zeolite anion.

$$\begin{Bmatrix} Ca \\ Mg \end{Bmatrix} \begin{Bmatrix} (HCO_3)_2 \\ SO_4 \\ Cl_2 \end{Bmatrix} + Na_2R \longrightarrow Na_2 \begin{Bmatrix} (HCO_3)_2 \\ SO_4 \\ Cl_2 \end{Bmatrix} + \begin{Bmatrix} Ca \\ Mg \end{Bmatrix} R \qquad 7.63$$

The resulting sodium compounds are soluble.

Typical characteristics of an ion exchange unit are:

exchange capacity:	3000 grains hardness/ft^3 zeolite for natural; 5000 - 30,000 (20,000 typical) for synthetic.
flow rate:	2 to 6 gpm/ft^3
backwash flow:	5 to 6 gpm/ft^2
salt dosage:	5 to 20 pounds/ft^3. Alternatively, .3 to .5 pound of salt per 1000 grains of hardness removed
brine contact time:	25 to 45 minutes

Example 7.16

A municipal plant uses water with a total hardness of 200 mg/l. The designed discharge hardness is 50 mg/l. If an ion exchange unit is used, what is the bypass factor?

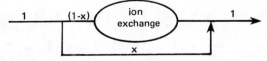

Since the water passing through the ion exchange unit is reduced to zero hardness,

$$(1-x)0 + x(200) = 50$$

$$x = .25 = \text{bypass factor}$$

J. Turbidity Removal

Coagulants may be based on aluminum (e.g., aluminum sulfate, sodium aluminate, potash alum, or ammonia alum) or iron (e.g., ferric sulfate, ferrous sulfate, chlorinated copperas, or ferric chloride.) If significant hydrolysis of iron and aluminum salts is ignored, the relationships in this section may be used to calculate the approximate stoichiometric quantities.

The most-used coagulant is aluminum sulfate $(Al_2(SO_4)_3 \cdot nH_2O)$. Filter alum is about 17% soluble material. The hydrolysis of the aluminum ion is complex. Assuming that the aluminum floc is $Al(OH)_3$ and the water pH is near neutral, then 1 mg/l of alum with a molecular weight of 600 removes by flocculation the following quantities:

.5 mg/l $(CaCO_3)$ of natural alkalinity

.39 mg/l of 95% hydrated lime $(Ca(OH)_2)$

.33 mg/l of 85% quicklime (CaO)

.53 mg/l of soda ash (Na_2CO_3)

If the alum has a molecular weight which is different than 600 (due to the variation in the number of waters of hydration), multiply the above quantities by (600/actual molecular weight).

Typical doses of alum are 5 to 50 mg/l, depending on turbidity. Alum flocculation is effective within pH limits of 5.5 to 8.0.

Ferrous sulfate $(FeSO_4 \cdot 7H_2O)$, also known as copperas, reacts with lime $(Ca(OH)_2)$ to flocculate ferric hydroxide $(Fe(OH)_3)$. This is an effective method of clarifying turbid waters at higher pH, as in lime softening. 1 mg/l of ferrous sulfate with a molecular weight of 278 will react with .27 mg/l of lime.

Ferric sulfate $(Fe_2(SO_4)_3)$ reacts with natural alkalinity or lime to creat floc. 1 mg/l of ferric sulfate will react with

1.22 mg/l of $Ca(HCO_3)_2$

.56 mg/l of $Ca(OH)_2$

.62 mg/l of natural alkalinity (as $CaCO_3$)

Ferric sulfate may be used for color removal at low pH; at high pH, it is useful for iron and manganese removal, as well as a coagulant with precipitation softening.

The total amount of sludge produced from coagulation and flocculation is:

$$\text{sludge solids (pounds per million gallons)} = 8.34\left[\frac{\text{alum dose (mg/l)}}{4} + \text{turbidity in JTU}\right] \quad 7.64$$

K. Taste and Odor Control

Some of the methods of treating water for taste and odor removal is:

    aeration: Suitable for groundwater with odors cause by dissolved gases. Not effective if odors are non-volatile.

    oxidation: Chlorine or potassium permanganate may be used to destroy odor.

    activated carbon: Useful in odor and taste control.

L. Demineralization/Desalination

If dissolved salts are to be removed, then one of the following methods must be used.

    distillation: The water is vaporized leaving the salt behind. The vapor is reclaimed by condensation.

    electrodialysis: Positive and negative ions flow through selective membranes under the influence of an induced electrical current.

    ion exchange: This is the same process described for water softening

    reverse osmosis: This is the cheapest method of demineralization. In operation, a thin membrane of cellulose acetate plastic separates two salt solutions of different concentration. Although ions would normally flow through the membrane into the solution with the lower concentration, the migration direction can be reversed by applying pressure to the low concentration fluid. Typical reverse osmosis units operate at 400 psi and produce about 2 gallons per day of fresh water for each quare foot of surface.

10. Typical Municipal Systems

The processes employed in treating incoming water will depend on the characteristics of the water. However, some sequences work better than others due to the physical and chemical nature of the processes. Listed in this section are some typical sequences. Not present in the lists are the usual system hardware items such as intake screens, pumps, pipes, hydrants, reservoirs, and holding basins.

A. For Well Ground Water (typically cleaner than surface water)

    sequence #1: intake
                     chlorination
                     fluoridation

sequence #2: intake
              aeration
              oxidation (chlorine or potassium permanganate)
              settling
              filtering
              chlorination
              fluoridation

sequence #3: intake
              aeration
              lime addition
              soda ash addition
              rapid mix
              flocculation
              settling
              recarbination
              filtering
              chlorination
              fluoridation

B. For Lake or Surface Water (typically turbid, and carrying odor and color).

sequence #1: intake
              chlorination
              coagulation
              rapid mixing
              flocculation
              optional chlorination
              addition of activated carbon
              settling
              addition of activated carbon
              filtering
              chlorination
              fluoridation

C. For River Surface Water (very turbid)

sequence #1: intake
              presedimentation (holding basin)
              chlorination
              coagulation
              rapid mix
              flocculation
              settling
              coagulant
              rapid mix
              flocculation
              addition of activated carbon
              settling
              addition of activated carbon
              filtering
              chlorination
              fluoridation

D. For Hard Water

sequence #1: intake
(bypass to second flocculator)
lime addition
alum addition
rapid mixing
flocculation
sedimentation
oxidation (chlorine or potassium permanganate)
second flocculation
sedimentation
filtering
fluoridation
chlorination

sequence #2: intake
presedimentation (in a basin)
chlorination
mixing
addition of activated carbon
lime addition
alum addition
flocculation
sedimentation
addition of activated carbon
mixing
filtering
fluoridation
chlorination
soda ash addition

sequence #3: intake
lime addition
alum addition
addition of activated carbon
mixing
flocculation
chlorination
sedimentation
recarbonation
filtering
(bypass to discharge)
zeolite treatment

Bibliography

1. De Nevers, Noel, _Fluid Mechanics_, Addison-Wesley Publishing Company, Reading, MA 1970

2. Hammer, Mark J., _Water and Waste-Water Technology_, John Wiley & Sons, Inc., New York, NY 1975.

3. Linsley, Ray K., and Franzini, Joseph B., _Water Resources Engineering_, 2nd ed., McGraw-Hill Book Company, New York, NY 1972.

4. Mahan, Bruce H., _University Chemistry_, Addison-Wesley Publishing Company, Inc., Reading, MA, 1966

5. Merritt, Frederick S., _Standard Handbook for Civil Engineers_, 2nd ed., McGraw-Hill Book Company, New York, NY 1976

6. Munro, Lloyd A., _Chemistry in Engineering_, Prentice-Hall, Englewood Cliffs, NJ 1964

7. Pauling, Linus, _College Chemistry_, 3rd ed., W. H. Freeman and Company, San Francisco, CA 1964.

8. Sienki, M. J., _Stoichiometry and Structure_, Part 1, W. A. Benjamin, Inc., New York, NY 1964.

Table 7.19
Atomic Weights of Elements Referred to Carbon (12)

Element	Symbol	Atomic Weight	Element	Symbol	Atomic Weight
Actinium	Ac	(227)	Mercury	Hg	200.59
Aluminum	Al	26.9815	Molybdenum	Mo	95.94
Americium	Am	(243)	Neodymium	Nd	144.24
Antimony	Sb	121.75	Neon	Ne	20.183
Argon	Ar	39.948	Neptunium	Np	(237)
Arsenic	As	74.9216	Nickel	Ni	58.71
Astatine	At	(210)	Niobium	Nb	92.906
Barium	Ba	137.34	Nitrogen	N	14.0067
Berkelium	Bk	(249)	Osmium	Os	190.2
Beryllium	Be	9.0122	Oxygen	O	15.9994
Bismuth	Bi	208.980	Palladium	Pd	106.4
Boron	B	10.811	Phosphorus	P	30.9738
Bromine	Br	79.909	Platinum	Pt	195.09
Cadmium	Cd	112.40	Plutonium	Pu	(242)
Calcium	Ca	40.08	Polonium	Po	(210)
Californium	Cf	(251)	Potassium	K	39.102
Carbon	C	12.01115	Praseodymium	Pr	140.907
Cerium	Ce	140.12	Promethium	Pm	(145)
Cesium	Cs	132.905	Protactinium	Pa	(231)
Chlorine	Cl	35.453	Radium	Ra	(226)
Chromium	Cr	51.996	Radon	Rn	(222)
Cobalt	Co	58.9332	Rhenium	Re	186.2
Copper	Cu	63.54	Rhodium	Rh	102.905
Curium	Cm	(247)	Rubidium	Rb	85.47
Dysprosium	Dy	162.50	Ruthenium	Ru	101.07
Einsteinium	Es	(254)	Samarium	Sm	150.35
Erbium	Er	167.26	Scandium	Sc	44.956
Europium	Eu	151.96	Selenium	Se	78.96
Fermium	Fm	(253)	Silicon	Si	28.086
Fluorine	F	18.9984	Silver	Ag	107.870
Francium	Fr	(223)	Sodium	Na	22.9898
Gadolinium	Gd	157.25	Strontium	Sr	87.62
Gallium	Ga	69.72	Sulfur	S	32.064
Germanium	Ge	72.59	Tantalum	Ta	180.948
Gold	Au	196.967	Technetium	Tc	(99)
Hafnium	Hf	178.49	Tellurium	Te	127.60
Helium	He	4.0026	Terbium	Tb	158.924
Holmium	Ho	164.930	Thallium	Tl	204.37
Hydrogen	H	1.00797	Thorium	Th	232.038
Indium	In	114.82	Thulium	Tm	168.934
Iodine	I	126.9044	Tin	Sn	118.69
Iridium	Ir	192.2	Titanium	Ti	47.90
Iron	Fe	55.847	Tungsten	W	183.85
Krypton	Kr	83.80	Uranium	U	238.03
Lanthanum	La	138.91	Vanadium	V	50.942
Lead	Pb	207.19	Xenon	Xe	131.30
Lithium	Li	6.939	Ytterbium	Yb	173.04
Lutetium	Lu	174.97	Yttrium	Y	88.905
Magnesium	Mg	24.312	Zinc	Zn	65.37
Manganese	Mn	54.9380	Zirconium	Zr	91.22
Mendelevium	Md	(256)			

Table 7.20 (2:9)
Inorganic Chemicals Used in Water Treatment

Chemical Name	Formula	Use	Molecular Weight	Equivalent Weight
Activated carbon	C	Taste and odor control	12.0	----
Aluminum sulfate (filter alum)	$Al_2(SO_4)_3 \cdot 14.3H_2O$	Coagulation	600	100
Aluminum hydroxide	$Al(OH)_3$	(Hypothetical combination)	78.0	26.0
Ammonia	NH_3	Chloramine disinfection	17.0	----
Ammonium fluosilicate	$(NH_4)_2SiF_6$	Fluoridation	178	----
Ammonium sulfate	$(NH_4)_2SO_4$	Coagulation	132	66.1
Calcium bicarbonate	$Ca(HCO_3)_2$	(Hypothetical combination)	162	81.0
Calcium carbonate	$CaCO_3$	Corrosion control	100	50.0
Calcium fluoride	CaF_2	Fluoridation	78.1	----
Calcium hydroxide	$Ca(OH)_2$	Softening	74.1	37.0
Calcium hypochlorite	$Ca(ClO)_2 \cdot 2H_2O$	Disinfection	179	----
Calcium oxide (lime)	CaO	Softening	56.1	28.0
Carbon dioxide	CO_2	Recarbonation	44.0	22.0
Chlorine	Cl_2	Disinfection	71.0	----
Chlorine dioxide	ClO_2	Taste and odor control	67.0	----
Copper sulfate	$CuSO_4$	Algae control	160	79.8
Ferric chloride	$FeCl_3$	Coagulation	162	54.1
Ferric hydroxide	$Fe(OH)_3$	(Hypothetical combination)	107	35.6
Ferric sulfate	$Fe_2(SO_4)_3$	Coagulation	400	66.7
Ferrous sulfate (copperas)	$FeSO_4 \cdot 7H_2O$	Coagulation	278	139
Fluosilicic acid	H_2SiF_6	Fluoridation	144	----
Hydrochloric acid	HCl	pH adjustment	36.5	36.5
Magnesium hydroxide	$Mg(OH)_2$	Defluoridation	58.3	29.2
Oxygen	O_2	Aeration	32.0	16.0
Potassium permanganate	$KMnO_4$	Oxidation	158	----
Sodium aluminate	$NaAlO_2$	Coagulation	82.0	----
Sodium bicarbonate (baking soda)	$NaHCO_3$	pH adjustment	84.0	84.0
Sodium carbonate (soda ash)	Na_2CO_3	Softening	106	53.0
Sodium chloride (common salt)	NaCl	Ion-exchange regeneration	58.4	58.4
Sodium fluoride	NaF	Fluoridation	42.0	----
Sodium hexametaphosphate	$(NaPO_3)_n$	Corrosion control	----	----
Sodium hydroxide	NaOH	pH adjustment	40.0	40.0
Sodium hypochlorite	NaClO	Disinfection	74.4	----
Sodium silicate	Na_4SiO_4	Coagulation aid	184	----
Sodium fluosilicate	Na_2SiF_6	Fluoridation	188	----
Sodium thiosulfate	$Na_2S_2O_3$	Dechlorination	158	----
Sulfur dioxide	SO_2	Dechlorination	64.1	----
Sulfuric acid	H_2SO_4	pH adjustment	98.1	49.0
Water	H_2O	----	18.0	----

Table 7.21

CONVERSIONS FROM mg/l AS A SUBSTANCE TO mg/l AS CACO$_3$

Multiply the mg/l of the substances listed below by the corresponding factors to obtain mg/l as CaCO$_3$. For example, 70 mg/l of Mg^{++} would be (70)(4.10) = 287 mg/l as CaCO$_3$.

Substance	Factor	Substance	Factor
Al^{+++}	5.56	MgCO$_3$	1.19
AlCl$_3$	1.13	Mg(HCO$_3$)$_2$	.68
Al(OH)$_3$	1.92	MgO	2.48
Ba^{++}	.73	Mg(OH)$_2$	1.71
Ba(OH)$_2$	.59	Mg(NO$_3$)$_2$	.67
BaSO$_4$	.43	MgSO$_4$	.83
Ca^{++}	2.50	NaCl	.85
CaCl$_2$	.90	Na$_2$CO$_3$	.94
CaCO$_3$	1.00	NaHCO$_3$	.60
Ca(HCO$_3$)$_2$	.62	NaNO$_3$	.59
CaO	1.79	NaOH	1.25
Ca(OH)$_2$	1.35	Na$_2$SO$_4$	.70
CaSO$_4$	.74	NH$_3$	2.94
Cl$^-$	1.41	NH$_4^+$	2.78
CO$_3^{--}$	1.67	NH$_4$OH	1.43
CO$_2$	2.27	(NH$_4$)$_2$SO$_4$	.76
Cu^{++}	1.57	NO$_3^-$	.81
CuSO$_4$	.63	OH$^-$	2.94
FeCl$_3$	.93	PO$_4^{---}$	1.58
FeSO$_4$	.66	SO$_4^{--}$	1.04
HCO$_3^-$	.82	Zn^{++}	1.54
K$^+$	1.28		
KCl	.67		
K$_2$CO$_3$	.72		
Mg^{++}	4.10		
MgCl$_2$	1.05		

Practice Problems: WATER SUPPLY ENGINEERING

Required

1. A water treatment plant has four rapid sand filters, each of which has a capacity of 600,000 gallons per day. Each filter is backwashed once a day for eight minutes. (a) Determine the inside dimensions of the sand filter. (b) What percentage of the filtered water is used for backwashing?

2. Design a circular, mechanically cleaned primary clarifier using the following specifications:

   flow rate: 2.8 million gallons/day
   detention period: 2 hours
   surface loading: 700 gallons/ft^2-day

If the initial flow rate is only 1.1 million gallons per day, what are the surface loading and average detention period?

3. A town's water supply is to be taken from a river with the following quality characteristics:

   turbidity: varies between 20 and 100 units
   total hardness: less than 60 mg/l (as $CaCO_3$)
   coliform count: varies between 200 and 1000 per 100 ml

The town has a design population of 15,000 people and an average consumption of 110 gpcd. (a) What rate (gpm) should the distribution be designed to carry? (b) What total filter area would you recommend? (c) Is softening required? (d) If 2 mg/l of chlorine is required to obtain the necessary chlorine residual, how many pounds per 24 hours of chlorine are required?

Optional

4. A town's water supply has the following hypothetical ion concentrations. (a) What is the total hardness in mg/l (as $CaCO_3$)? (b) How much lime ($Ca(OH)_2$) and soda ash are required to remove the calcium hardness?

Ca^{++}	80.2 mg/l	CO_3^{--}	0
Na^+	46.0 mg/l	Mg^{2+}	24.3 mg/l
NO_3^-	0	Fl^-	0
Cl^-	85.9 mg/l	SO_4^{--}	125 mg/l
CO_2	19 mg/l	Fe^{++}	1.0 mg/l
Al^{+++}	0.5 mg/l	HCO_3^-	185 mg/l

5. The following concentrations of inorganic compounds are found during a routine analysis of a city's water supply.

$Ca(HCO_3)_2$ = 137 mg/l (as $CaCO_3$)

CO_2 = 0 mg/l

$MgSO_4$ = 72 mg/l (as $CaCO_3$)

problem 5, continued

(a) How many pounds of lime ($Ca(OH)_2$) and soda ash (Na_2CO_3) are required to soften one million gallons of this water to 100 mg/l if 30 mg/l excess lime is required for a complete reaction? (b) How many pounds of salt would be required if a zeolite process is used with the following characteristics:

> exchange capacity: 10,000 grains hardness/ft^3
> salt requirement: .5 pound/1000 grains hardness removed

6. A water treatment plant has five square rapid sand filters, each of which has a capacity of one million gallons per day. (a) What are the recommended dimensions for the filters? (b) If each filter is backwashed each day for 5 minutes, what percentage of the plant's filtered water is used for backwashing?

7. Water from an underground aquifer is to be reduced from 245 mg/l hardness to 80 mg/l hardness by the zeolite process. (a) Draw a line schematic of the process used to accomplish this reduction. (b) What is the time between regenerations of the softener if the exchanger has the following characteristics:

> flow volume: 20,000 gallons per day
> exchanger resin volume: 2 cubic feet
> resin exchange capacity: 20,000 grains per cubic foot

8. A 12" standard strength clay sewer pipe is to be installed under a backfill of 11 feet of saturated topsoil which has a density of 120 pounds per cubic foot. The pipe strength is 1,500 pounds per foot. Design a bedding using a factor of safety of 1.5.

9. A settling tank has an overflow rate of 100,000 gal/ft^2-day. Water carrying grit of various sizes is introduced. The grit has the following distribution of settling velocities:

settling velocity (fpm)	weight fraction remaining
10.0	.54
5.0	.45
2.0	.35
1.0	.20
.75	.10
.50	.03

What is the percentage by weight of the grit removed?

10. What is the settling velocity of a spherical sand particle which has a specific gravity of 2.6 and a diameter of 1 millimeter?

NOTES

WASTE-WATER ENGINEERING

Nomenclature

A	area	ft^2
b	y-intercept	various
BOD	biological (biochemical) oxygen demand	mg/l
C	concentration	mg/l
COD	chemical oxygen demand	mg/l
d	stream flow depth, or particle diameter	ft, or ft
D	oxygen deficit	mg/l
DO	dissolved oxygen	mg/l
f	Darcy friction factor	-
F	effective number of passes	-
g	acceleration due to gravity (32.2)	ft/sec^2
K_D	deoxygenation coefficient	-/days
K_R	reoxygenation coefficient	-/days
K_t	oxygen transfer coefficient	-/hrs
L	loading	for BOD: $lb/1000\ ft^3$-day
m	slope of line	-
MLSS	mixed liquor suspended solids	mg/l
N	normality	gew/l
P	population	1000's of people
Q	flow quantity	gallons/day
Q'	flow quantity	cfs
R	ratio	-
ROT	rate of oxygen transfer	mg/l-hr
s	sludge suspended solids	decimal
SA	sludge age	days
SG	specific gravity	-
SS	suspended solids	mg/l
SVI	sludge volume index	
t	time	days
T	temperature	·C
v	velocity	ft/sec
V	volume	ml
W	sludge removal rate	lb/day (dry)
x	a fraction	decimal

Symbols

η	efficiency	-
β	oxygen saturation coefficient	-

Subscripts

A	aeration
AS	activated sludge
BOD	biological oxygen demand
c	critical
d	discharge
e	equivalent
f	final

F-M	food to microorganism
H	hydraulic
i	initial
ML	mixed liquor
o	immediately after mixing
p	particle, or primary
R	recirculation, recirculated, or return
RS	return sludge
req	required at discharge
s	standard 5-day, or secondary
sat	saturated
ss	suspended solids
t	at time t in days
T	at temperature T in °C
u	ultimate carbonaceous
w	wastewater

1. Conversions

Multiply	By	To Obtain
acre-feet	43.56	1000's of cubic feet
cubic feet	7.48	gallons
cubic feet/sec (cfs)	.6463	MGD
cubic feet/sec (cfs)	448.8	gpm
gallons	.1337	cubic feet
gallons/day (gpd)	1.547 EE-6	cfs
gallons/min (gpm)	.002228	cfs
gallons/acre-day (gad)	2.296 EE-5	gallons/day-ft^2
gallons/ft^2-day (gpd/ft^2)	.04356	million gallons/acre-day
million gallons/acre-day (mgad)	22.96	gpd/ft^2
million gallons/day (MGD)	1.547	cfs
milligrams (mg)	2.205 EE-6	pounds
milligrams/liter (mg/l)	8.345	pounds/million gallons
meters	3.281	feet
millimeters/meter	.012	in/foot
m^3/m^2-day	24.54	gallons/ft^2-day
m^3/m-day	80.52	gallons/ft-day
miles per hour (mph)	1.466	ft/sec
pounds	4.536 EE5	milligrams
pounds/acre-ft-day	.02296	lbs/1000 ft^3-day
pounds/1000 ft^3-day	43.56	lbs/acre-ft-day
pounds/1000 ft^3-day	133.7	lbs/million gal-day
pounds/million gallons	.1198	mg/l
pounds/million gallons-day	.00748	lbs/1000 ft^3-day

2. Definitions

activated sludge: Solids from aerated settling tanks which are rich in
        bacteria.
aerated lagoon: A holding basin into which air is mechanically intro-
        duced to speed up aerobic decomposition.
appurtenance: A thing which belongs with (or is designed to complement)
        something else. For example, a manhole is a sewer appurtenance.
bioactivation process: A process using sedimentation, trickling filter,
        and secondary sedimentation before adding activated sludge.
        Aeration and final sedimentation are the follow up processes.
biosorption process: A process which mixes raw sewage and sludge which
        has been pre-aerated in a separate tank.
biota: The flora and fauna of a region, process, or tank.
branch sewer: A sewer off the main sewer.
bulking: See 'sludge bulking'
carbonaceous demand: Oxygen demand due to biological activity in a
        water sample
chemical precipitation: Causing suspended solids to settle out by
        adding coagulating chemicals
clean-out: A pipe through which snakes can be pushed to unplug a sewer.
combined system: A system using a single sewer for domestic waste and
        storm water
comminutor: A device which cuts solid waste into small pieces.
complete mixing: Mixing accomplished by mechanical means (stirring).
cunette: A small channel in the invert of a large combined sewer for
        dry weather flow.
deoxygenation: The act of removing dissolved oxygen from water.
dewatering: Removal of excess moisture from sludge waste.
digestion: Conversion of sludge solids to gas.
dilution disposal: Relying on a large water volume (lake or stream)
        to dilute waste to an acceptable concentration.
domestic waste: Waste which originates from households.
effluent: That which flows out of a process.
elutriation: A counter-current sludge washing process used to remove
        dissolved salts.
first-stage demand: See 'carbonaceous demand'
floatation: Adding chemicals or bubbling air through waste to get solids
        to float to the top as scum.
force main: A sewer line which is pressurized.
humus: A greyish brown sludge consisting of relatively large particle
        biological debris, as the material sloughed off from a trickling
        filter.
infiltration: Ground water which enters sewer pipes through cracks and
        joints.
influent: Flow entering a process
inverted siphon: A sewer line which drops below the hydraulic gradient.
Kraus process: Mixing raw sewage, activated sludge, and material from
        sludge digesters.
lamp holes: Sewer inspection holes large enough to lower a lamp into
        but too small for a man.
lateral: A sewer line which goes off at right angles to another.
main: A large sewer from which all other branches originate.
malodorous: Offensive smelling.

mesophilic bacteria: Bacteria growing between 10 and 40°C, with an
        optimum temperature of 37°C. 40°C is, therefore, the upper
        limit for most wastewater processes.
Mohlman index: Same as the 'sludge volume index'
nitrogenous demand: Oxygen demand from nitrogen-consuming bacteria.
outfall: The pipe which discharges completely treated wastewater
        into a lake, stream, or ocean.
partial treatment: Primary treatment only.
post-chlorination: Addition of chlorine after all other processes
        have been completed.
pre-chlorination: Addition of chlorine prior to sedimentation to help
        control odors and to aid in grease removal.
putrefaction: Anaerobic decomposition of organic matter with accompanying
        foul odors.
refractory: Dissolved organic materials which are biologically resistant
        and difficult to remove.
regulator: A device or weir which deflects large volume flows into a
        special high-capacity sewer.
sag pipe: See 'inverted siphon'
second stage demand: See 'nitrogenous demand'
seed: The activated sludge initially taken from the secondary settling
        tank and returned to the aeration tank to start the activated
        sludge process.
separate system: Separate sewers for domestic and storm waste water.
septic: Produced by putrefaction.
sludge bulking: Failure of suspended solids to completely settle out.
split chlorination: Addition of chlorine prior to sedimentation and
        after final processing
submain: See 'branch'
supernatant: The clarified liquid floating on top of a digesting sludge
        layer.
thermophilic bacteria: Bacteria which thrive in the 45°C to 75°C range
        (optimum near 55°C).
volatile solid: Solid material in a water sample or in sludge which can
        be burned or vaporized at high temperature.
wet well: A short-term storage tank containing a pump or pump entrance,
        and into which the raw influent is brought.
zooglea: The gelatinous film of aerobic organisms which cover the
        rocks in a trickling filter.

3. Wastewater Quality Characteristics

A. Dissolved Oxygen: Fish and most aquatic life require oxygen.
The biological decomposition of organic solids is also dependent on
oxygen. If the dissolved oxygen content of water is less than the
saturated values (given in table 8.13) there is good reason to believe
that the water is organically polluted. Other reasons for measuring
the dissolved oxygen concentration are for aerobic treatment monitoring,
aeration process monitoring, BOD testing, and pipe corrosion studies.

Dissolved oxygen is frequently evaluated by the iodometric method (azide
modification). A water sample is taken and biological activity is
stopped; solids are then flocculated by the addition of an inhibitor
(such as copper sulfate-sulfuric acid). The sample is then mixed with
a manganese sulfate and alkali-iodide-azide reagent. If no oxygen is

present, the manganous ion reacts with the hydroxide ion to form a white precipitate.

$$Mn^{++} + 2OH^- \longrightarrow Mn(OH)_2 \qquad\qquad 8.1$$

If oxygen is present, some of the Mn^{++} is oxidized to Mn^{++++} which reacts to form a brownish oxide.

$$2Mn(OH)_2 + O_2 \longrightarrow 2Mn^{++} + 4OH^- + O_2$$
$$2MnO_2 + 2H_2O \qquad\qquad 8.2$$

After thorough mixing and settling, sulfuric acid is added. The iodide ion in the reagent is oxidized.

$$MnO_2 + 2I^- + 4H^+ \longrightarrow Mn^{++} + I_2 + 2H_2O \qquad\qquad 8.3$$

The free iodine concentration is equivalent to the original dissolved oxygen content. The amount of iodine is determined by titration with thiosulfate. The iodine solution is originally yellow, and the titration endpoint (clear solution) cannot accurately be determined. Therefore, starch is added which turns blue in the presence of free iodine.

Electrode methods are also used to measure dissolved oxygen.

Example 8.1

A 200 ml sample of inhibited water is placed in a 300 ml BOD water bottle. 2 ml each of manganese sulfate solution and 2 ml of alkali-iodide-azide reagent are added. After stoppering, the bottle is repeatedly inverted to mix the solutions. A brownish tint appearing in the water, indicates the presence of oxygen. After allowing sufficient time for precipitation, 2 ml of concentrated sulfuric acid is added, and the stoppered bottle is again inverted for mixing. The mixing continues until the liquid color is a uniform yellow. 203 ml of the solution is titrated with .0250 N sodium thiosulfate to a pale straw color, and 2 ml of a soluble starch solution is added. If 7 ml of titrant are added before the blue color first starts to disappear, what is the dissolved oxygen concentration?

With the volumes and concentrations listed, 1 ml of titrant corresponds to 1 mg/l of dissolved oxygen. Therefore, the dissolved oxygen concentration is 7 mg/l. ◆

 B. Biological Oxygen Demand: When they oxidize organic waste material in water, biological organisms remove oxygen from the water. Therefore, oxygen use is an indication of the organic waste content. The biological oxygen demand (BOD) of a biologically active sample (no seed required) is given by equation 8.4:

$$(BOD)_s = \dfrac{\dfrac{(DO)_i - (DO)_f}{V_{sample}}}{V_{sample} + V_{dilution}} \qquad\qquad 8.4$$

BOD is determined by adding a measured amount of wastewater (which supplies the organic material) to a measured amount of dilution water (which reduces toxicity and supplies dissolved oxygen). An oxygen use curve similar to figure 8.1 will result. (More than one identical sample must be prepared in order to determine initial and final concentrations of dissolved oxygen.)

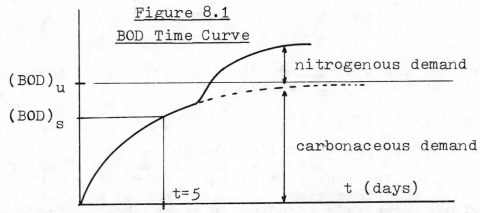

Figure 8.1
BOD Time Curve

The standard BOD test typically calls for a 5-day incubation period at 20°C. The BOD at any time can be found from equation 8.5.

$$(BOD)_t = (BOD)_u(1 - 10^{-K_D t})$$ 8.5

K_D is the deoxygenation rate constant, typically taken as .1. The ultimate BOD cannot be found from long term studies due to the effect of nitrogen-consuming bacteria in the sample. However, if K_D is .1, the ultimate BOD can be found from equation 8.6.

$$(BOD)_u = 1.47(BOD)_s$$ 8.6

The rate constant K_D can be experimentally determined by evaluating BOD for various incubation periods. Referring to figure 8.2, K_D is given by equation 8.7.

$$K_D = 2.61\left(\frac{m}{b}\right)$$ 8.7

Figure 8.2

The Rate Constant Graph

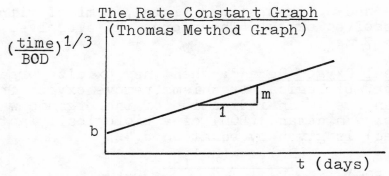

(Thomas Method Graph)

K_D for other temperatures can be found from equation 8.8. (The 1.047 constant is often quoted in literature. Recent research suggests 1.135 for 4°C to 20°C, and 1.056 for 20°C to 30°C.)

$$K_{D,T} = (1.047)^{T-20} K_{D,20°C}$$ 8.8

The variation in BOD with temperature is given by equation 8.9.

$$(BOD)_T = (BOD)_{20 \cdot C} (.02T + .6) \qquad 8.9$$

Table 8.1
Typical Values of K_D

treatment plant effluents: .05 - .10
highly polluted shallow streams: .25

Example 8.2

Ten 5-ml samples of wastewater are placed in 300 ml BOD bottles. Half of the bottles are titrated immediately with an average initial concentration of dissolved oxygen of 7.9 mg/l. The remaining bottles are incubated for 5 days, after which the average dissolved oxygen is determined to be 4.5 mg/l. What is the standard BOD and ultimate carbonaceous BOD assuming K = 0.13?

From equation 8.4: $(BOD)_s = \dfrac{(7.9 - 4.5)}{\dfrac{5}{300}} = 204$ mg/l

From equation 8.5, the ultimate BOD is

$$(BOD)_u = \frac{204}{1-10^{(-.13)(5)}} = 263 \text{ mg/l}$$

◆

If a sample of industrial wastewater is taken, it will probably lack sufficient microorganisms to metabolize the organic matter. In such a case, seed organisms must be added. The BOD for seeded experiments is found by measuring dissolved oxygen in the seeded sample after 15 minutes $(DO)_i$ and after 5 days $(DO)_f$ as well as the dissolved oxygen of the seed material itself after 15 minutes $(DO^*)_i$ and after 5 days $(DO^*)_f$.

$$(BOD)_s = \frac{(DO)_i - (DO)_f - x[(DO^*)_i - (DO^*)_f]}{\dfrac{V_{sample}}{V_{sample} + V_{dilution}}} \qquad 8.10$$

$$x = \frac{\text{volume of seed added to sample}}{\text{volume of seed used to find DO*}} \qquad 8.11$$

The BOD of domestic waste is typically taken as .17 - .20 pounds per capita-day, excluding industrial wastes. This makes it possible to calculate the population equivalent of any BOD loading.

$$P_e = \frac{(BOD)(Q)(8.345 \text{ EE-9})}{(.17)} \quad \text{(in 1000's of people)} \quad 8.12$$

Values of BOD for various industrial effluents are given in table 8.2.

Table 8.2

Typical BODs of Industrial Effluents

Industry/Type of Waste	BOD	COD
canning		
corn	19.5 lb/ton corn	
tomatoes	8.4 lb/ton tomatoes	
dairy milk processing	1150 lb/ton raw milk	1900 mg/l
	1000 mg/l	
beer brewing	1.2 lb/barrel beer	
commercial laundry	1250 lb/1000 pounds dry	2400 mg/l
	700 mg/l	
slaughterhouse (meat pack-ing)	7.7 lb/animal	2100 mg/l
	1400 mg/l	
papermill	121 lb/ton pulp	
synthetic textile	1500 mg/l	3300 mg/l
chlorophenolic manufacturing	4300 mg/l	5400 mg/l
milk bottling	230 mg/l	420 mg/l
cheese production	3200 mg/l	5600 mg/l
candy production	1600 mg/l	3000 mg/l

BOD of 100 mg/l is considered a weak effluent; BOD of 200-250 is considered a medium strength effluent; above 350 mg/l the effluent is considered strong.

C. Relative Stability

The relative stability test is much easier than the BOD test to perform, although it is much less accurate. The relative stability of an effluent is defined as the percent of initial BOD that has been satisfied. The test consists of taking a sample of effluent and adding a small amount of methylene blue dye. When all oxygen has been removed from the water, anaerobic bacteria start to remove the dye. The time for the color to start degrading is known as the 'stabilization time' or 'decoloration time.'

The relative stability can be found from the stabilization time by using table 8.3.

Table 8.3

Relative Stability (at 20°C)

stabilization time (days)	relative stability %	stabilization time (days)	relative stability %
½	11	8	84
1	21	9	87
1½	30	10	90
2	37	11	92
2½	44	12	94
3	50	13	95
4	60	14	96
5	68	16	97
6	75	18	98
7	80	20	99

Example 8.3

A sample of treatment plant effluent begins to clarify after 13 days.
What percent of the original BOD remains unsatisfied?

From table 8.3, the relative stability is 95%. Therefore, only 5%
of the initial BOD remains unsatisfied.

◆

D. Chemical Oxygen Demand: Unlike BOD which is a measure of oxygen
removed by biological organisms, chemical oxygen demand (COD) is a
measure of maximum oxidizable substances. Therefore, COD is an excellent
measure of effluent 'strength'.

COD is calculated by first adding a solution containing excessive
dichromate ions to the sample and then heating the sample in the presence
of silver sulfate catalyst in sulfuric acid.

$$K_2Cr_2O_7 \longrightarrow 2K^+ + Cr_2O_7^{--} \qquad \text{(in water)} \qquad 8.13$$

$$H_2SO_4 \longrightarrow 2H^+ + SO_4^{--} \qquad \text{(in water)} \qquad 8.14$$

$$Cr_2O_7^{--} + 14\ H^+ + 6e^- \xrightarrow[Ag^+]{heat} 2Cr^{+++} + 7H_2O \qquad 8.15$$

Equation 8.15 shows that each dichromate ion has the oxidizing power of
3 oxygen atoms (since 6 electrons are transferred and the oxidation
number of oxygen is -2.) When organic materials (such as sugars, carbo-
hydrates, cellulose, fats, organic acids, and hydrocarbons) are present,
the dichromate ion oxidizes these also to form carbon dioxide. For
example, the oxidation for glucose sugar is given by equation 8.16.

$$C_6H_{12}O_6 + 4Cr_2O_7^{--} + 32H^+ \longrightarrow 6CO_2 + 8Cr^{+++} + 22H_2O \qquad 8.16$$

Example 8.4

A solution containing 160 mg/l of glucose is oxidized by dichromate ions
according to equation 8.16. What is the chemical oxygen demand?

The molecular weights are:
$$180 \quad + 4(216) + 32(1) \longrightarrow 6(44) + 8(52) + 22(18)$$

The number of dichromate ions needed to oxidize one mole of glucose
solution is
$$\frac{4}{180} = \frac{X}{160} \qquad \text{or } X = 3.56 \text{ ions/mole}$$

Each dichromate ion has the oxidation power of 3 oxygen atoms, so the
COD is $(3.56)(3)(16) = 170.9$ mg/l ◆

After the actual oxidation process, there will be dichromate ions
remaining (because an excessive dichromate solution was used.) These
ions are back titrated using ferrous-ammonium sulfate and ferroin
indicator.
$$6Fe^{++} + Cr_2O_7^{--} + 14H^+ \longrightarrow 6Fe^{+++} + 2Cr^{+++} + 7H_2O \qquad 8.17$$

The net dichromate usage is the measure of COD.

$$COD = \frac{8000}{V_{sample}}[(V_{dichromate\ solution})(N_{dichromate})-(V_{titrant})(N_{titrant})]$$

8.18

The same procedure is repeated using distilled water as a control to compensate for any possible organic matter in the reagents. The true COD is given by equation 8.19.

$$(COD)_{true} = (COD)_{sample} - (COD)_{control}$$

8.19

E. Chlorine Demand

Chlorination destroys bacteria, hydrogen sulfide, and other smelly substances by oxidation. For example, hydrogen sulfide is oxidized according to equation 8.20.

$$H_2S + 4H_2O + 4Cl_2 \longrightarrow H_2SO_4 + 8HCl$$

8.20

The chlorine demand is the amount of chlorine (or its chloramine or hypochlorite equivalent) required to give a .5 mg/l residual after 15 minutes of contact time. 15 minutes is the recommended contact and mixing time prior to discharge since this period will kill nearly all pathogenic bacteria in the water. Typical doses for wastewater effluent are given in table 8.4.

Table 8.4

Typical Chlorine Doses

Final Process	dose (mg/l)
no treatment (straight discharge)	10-30
secondary sand filter	2-6
secondary activated sludge	2-8
secondary trickling filter	3-15
primary sedimentation	5-25

F. Grease: Greases are organic substances including fats, vegetable and mineral oils, waxes, fatty acids from soaps, and other hydrocarbons. Grease's low solubility causes adhesion problems in pipes and tanks, reduces contact area during various filtering processes, and produces sludge which is difficult to dispose of.

Grease content may be found by the Soxhlet extraction method. Essentially, this method isolates the grease in a flask from which a solvent (usually petroleum ether) is completely evaporated. The grease content is the increase in flask weight divided by the original water sample volume.

G. Volatile Acids: Volatile acids (acetic, propionic, and butyric) occur in anaeorobically digested sludge. These acids can be used to indicate the completion of a sludge digestion process. Column chromatography is used, and the acid content is given in mg/l (as acetic acid.)

4. Wastewater Quality Standards

Applicable standards have been set by both the Water Pollution Control Act of 1972 and the Environmental Protection Agency. General standards set by the Water Pollution Control Act are given in table 8.5. These standards must be met by facilities that receive federal funding.

Table 8.5
Surface Water Standards

Water Use	Minimum DO	Maximum Dissolved Solids	Maximum Coliforms per 100 ml
domestic use (food preparation)	6	*	none
water contact recreation	4 - 5	*	1000 total ave. 200 fecal ave. Not more than 10% exceeding 200 (400 fecal)
fisheries	4 - 6	*	5000 ave
industrial supply	3 - 5	750 to 1500 mg/l *	-
agricultural irrigation	3 - 5	750 to 1500 mg/l *	-
shellfish harvesting	4 - 6	*	70 total ave. Not more than 10% exceeding 230

* No floating solids or settling solids that form deposits.

The EPA's standards for secondary treatment are given in terms of 5-day BOD, suspended solids, coliform count, and pH.

Table 8.6
Secondary Effluent Standards

Quality	Average Over	Discharge Maximum
BOD (5-day)	30 days	30 mg/l
	7 days	45 mg/l
	30 days	15% of incoming BOD
Suspended solids	30 days	30 mg/l
	7 days	45 mg/l
	30 days	15% of incoming SS
Fecal coliforms(**)	30	200 per 100 ml
	7	400 per 100 ml
pH	at all times	within 6 to 9

** A geometric mean is used, not arithmetic.

5. Design Flow Quantity

Approximately 2/3 of a community's domestic and industrial water use
will return as wastewater. This water is discharged into the sewer
systems, which may be separate or the same as storm drains. Therefore,
the nature of the return system must be known before sizing can occur.

Sanitary sewer sizing may be based on an average of 100-125 gpcd, or
it may be assumed that the wastewater quantity will be the same as the
water supply quantity. There will be variations with time in the flow,
although the variations are not as pronounced as they are for water
supply.

Table 8.7
Variations in Wastewater Flow
Based on the average daily flow

description	when/where	variation
daily peak	10 - 12 am (residential)	225%
	constant during day (commercial)	150%
	12 noon at the outfall	150%
daily minimum	4 - 5 am	40%
seasonal peak	late summer	125%
seasonal minimum	winter's end	90%
seasonal average	May, June	100%
maximum peaks	in laterals	300%
	treatment plant influent	200%

Several codes specify a design loading of 400 gpcd (laterals and submains)
and 250 gpcd (main, trunk, and outfall). Both of these include the effect
of infiltration. Infiltration due to cracks and poor joints is limited
by many municipal codes to 500 gallons per day per mile of pipe per
inch of diameter. Modern piping materials and joints should be able to
reduce this quantity to 200 gpd/inch-mile. Infiltration may also be
roughly estimated at 3% - 5% of the peak hourly domestic rate, or as
10% of the average rate.

6. Collection Systems

A. Storm Drains and Inlets: Curb inlets to storm drains should be
placed no more than 600 feet apart, and a limit of 300 feet is advisable.
Of course, inlets are required at all low points where pondage could
occur. A common practice is to install 3 inlets in a sag vertical curve -
one at the lowest point and one on each side with an elevation of .2 feet
above the center inlet. Opening may be of the covered grate type or the
curb inlet type.

The capacity of a curb-type opening which diverts 100% of gutter flow is
given by equation 8.21. A typical curb depression is 5 inches.

$$Q' = (.7)\left(\begin{array}{c}\text{curb} \\ \text{opening} \\ \text{length, ft}\end{array}\right)\left(\begin{array}{c}\text{inlet flow} \\ \text{depth, ft}\end{array} + \begin{array}{c}\text{curb inlet} \\ \text{depression, ft}\end{array}\right)^{3/2} \quad 8.21$$

Grate inlets accepting flows less than .4 feet deep have a capacity
given by equation 8.22. The bars should be parallel to the flow and at
least 18 inches long. Equation 8.22 should also be used for combined

curb-grate inlets.

$$Q' = 3 \binom{\text{grate}}{\text{perimeter, ft}} \binom{\text{inlet flow}}{\text{depth, ft}}^{3/2}$$ 8.22

   B. Manholes: Manholes should be provided at junctions, and at changes in elevation, size, diameter, or slope of sewers. If the sewer is too small for a man to enter, manholes should be placed every 500 feet to allow for cleaning. A maximum recommended spacing is every 700 feet.

   C. Pipes: Concrete pipe is commonly used for storm sewers. It is sized on the assumption that capacity will be exceeded every 10 (or 15, etc.) years. Circular pipe is used in most applications, although special shapes (arch, egg, elliptical, etc.) are available at extra cost. Concrete pipe in diameters up to 24" are usually not reinforced, and are available in standard 3 and 4 foot lengths. Reinforced pipe in diameters ranging from 12" to 144" is available in lengths from 4 to 12 feet.

Sewer pipes are constructed from clay tile, concrete, cast iron, asbestos-cement, and plastic-lined concrete. Clay tile is the least susceptible to sulfuric acid corrosion (see equation 8.20). Clay is typically used for diameters less than 36". Clay pipe is available in standard diameters of 4" and 6" in 2 foot lengths; 8", 10", 12", 15", 18", 21", and 24" diameters in $2\frac{1}{2}$ foot lengths; and 27", 30", 33", and 36" in 3 foot lengths.

Two strengths of clay pipe are available. The standard strength is suitable for pipes less than 12" in diameter for any depth of cover if the '4/3D + 8"' trench width rule is observed. Double strength pipe is recommended for large pipe deeply trenched. No less than 3 feet of cover should be used for any pipe.

Sewer velocities should not be less than 2 feet/sec when flowing full. This is a self-cleansing velocity. Velocities greater than 15 ft/sec require special provisions to protect against erosion and momentum effects. Table 8.8 gives minimum slopes to achieve a 2 ft/sec flow. These slopes will give a 1 ft/sec flow (minimum self-cleansing) when the flow depth is (1/6) diameter.

Table 8.8
Self Cleansing Slopes

Pipe diameter (inches)	slope (ft/ft)
8	.0033
10	.0025
12	.0019
15	.0014
18	.0011
21	.00092
24	.00077
27	.00066
30	.00057
36	.00045

7. Dilution Purification

Dilution purification (also known as 'self purification') refers to discharge of partially treated sewage into a body of water such as a stream or river. If the body is large and is adequately oxygenated, the sewage's BOD may be satisfied without putrefaction. Other conditions which must be monitored besides BOD are oxygen content and suspended solids.

Equation 8.23 may be used to calculate the final concentration of BOD, oxygen, and sediment when the two flows are mixed. Dilution requirements may be expressed in terms of ratios (e.g., 23 stream volumes per discharge volume) or absolute flow quantities (e.g., 4 to 7 cfs per 1000 population).

$$C_1 Q_1 + C_2 Q_2 = C_f(Q_1 + Q_2) \qquad 8.23$$

Example 8.5

Wastewater (DO = .9 mg/l, 6 MGD) is discharged into a 50°F stream flowing at 40 cfs. Assuming the stream is saturated with oxygen, what is the oxygen content of the stream immediately after mixing?

From table 8.13, the saturated oxygen content at 50°F (10°C) is 11.3 mg/l.

$$(6) \text{ MGD } (1.547)\frac{\text{cfs}}{\text{MGD}} = 9.28 \text{ cfs}$$

$$C_f = \frac{(.9)(9.28) + (11.3)(40)}{(9.28) + (40)} = 9.34 \text{ mg/l}$$

◆

The oxygen deficit is the difference between actual and saturated oxygen concentration. Since reoxygenation and deoxygenation of a polluted river occur simultaneously, an oxygen deficit will occur only if the reoxygenation rate is less than the deoxygenation rate. If the oxygen content goes to zero, aerobic decomposition and putrefaction will occur.

The oxygen deficit at any time t is given by the Streeter-Phelps equation:

$$D_t = DO_{sat} - DO_t = \frac{K_D(BOD)_u}{K_R - K_D}(10^{-K_D t} - 10^{-K_R t}) + D_o(10^{-K_R t}) \quad 8.24$$

D_t is the dissolved oxygen deficit, t is in days, and BOD_u is the ultimate carbonaceous BOD of the stream immediately after mixing. K_D and K_R are the deoxygenation and reoxygenation rate constants respectively, and D_o is the dissolved oxygen deficit immediately after mixing.

K_R may be approximated by equation 8.25 if field test data is not available.

$$K_{R,20°C} \approx 3.3(v)/(d^{1.33}) \qquad 8.25$$

K_R for different temperatures is given by equation 8.26. Typical values of K_R are given in table 8.9.

$$K_{R,T} = (1.016)^{T-20} K_{R,20°C} \qquad 8.26$$

Table 8.9 (2:571)

Typical Reoxygenation Constants (base 10)

'white water'	.5 and above
swiftly flowing	.3 to .5
large streams	.15 to .3
large lakes	.10 to .15
sluggish streams	.10 to .15
small ponds	.05 to .10

Equations 8.5 and 8.24 may be plotted simultaneously as shown in figure 8.3. The plot of equation 8.24 is known as the 'oxygen sag curve'. The difference between the two curves is the effect of reoxygenation.

Figure 8.3
The Oxygen Sag Curve

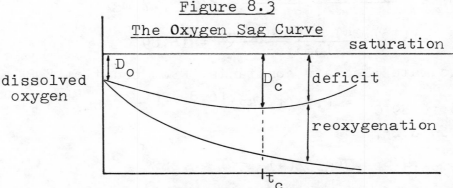

The time to the minimum or critical point on the sag curve is given by equation 8.27.

$$t_c = \frac{1}{K_R - K_D} \log_{10}\left[\left(\frac{K_D(BOD)_u - K_R D_o + K_D D_o}{K_D(BOD)_u}\right)\left(\frac{K_R}{K_D}\right)\right] \qquad 8.27$$

The ratio of (K_R/K_D) is known as the self-purification coefficient.

The critical oxygen deficit is given by equation 8.28.

$$D_c = \left(\frac{K_D(BOD)_u}{K_R}\right)(10)^{-K_D t_c} \qquad 8.28$$

Values of D_c are usually established by law or contract. However, fishes need a minimum of 4 or 5 mg/l of dissolved oxygen.

Knowing t_c and the stream flow velocity will enable you to find the location where the oxygen level is the lowest.

Example 8.6

A treatment plant discharge has the following characteristics:

        15 cfs
        45 mg/l BOD (5 day, 20°C)
        2.9 mg/l DO
        24°C
        $K_{D,20°C}$ = .1 per day (when mixed with river water)

The outfall is located in a river with the following characteristics:

.55 ft/sec velocity
4.0 feet average depth
120 cfs
4 mg/l BOD (5 day, 20°C)
8.3 mg/l DO
16°C

Determine the distance downstream where the oxygen level is minimum, and predict if the river can support fish life at that point.

step 1: Find the river conditions immediately after mixing. Use equation 8.23 three times.

$$BOD = \frac{(15)(45) + (120)(4)}{15 + 120} = 8.56 \text{ mg/l}$$

$$DO = \frac{(15)(2.9) + (120)(8.3)}{135} = 7.7 \text{ mg/l}$$

$$T = \frac{(15)(24) + (120)(16)}{135} = 16.89°C$$

step 2: Calculate the rate constants. From equation 8.8,

$$K_{D,16.89°C} = (.1)(1.047)^{16.89-20} = .0867$$

From equation 8.25,

$$K_{R,20°C} \approx 3.3(.55/(4)^{1.33}) = .287$$

From equation 8.26,

$$K_{R,16.89°C} = (.287)(1.016)^{16.89-20} = .275$$

step 3: Estimate $(BOD)_u$. Using equation 8.5,

$$(BOD)_u = \frac{8.56}{(1-(10)^{-.0867(5)})} = 13.56$$

step 4: Calculate D_o. From table 8.13, the saturated oxygen concentration at 16.89°C is approximately 9.7 mg/l. So, $D_o = 9.7-7.7 = 2.0$.

step 5: Calculate t_c from equation 8.27.

$$t_c = \frac{1}{.275-.0867}\log[(\frac{.0867(13.56)-.275(2)+.0867(2)}{.0867(13.56)}(\frac{.275}{.0867})]$$
$$= 1.77 \text{ days}$$

step 6: The distance downstream is

$$\frac{(1.77)\text{days}(.55)\text{ft/sec}(86400)\text{sec/day}}{(5280)\text{ft/mile}} = 15.9 \text{ miles}$$

step 7: The critical oxygen deficit is found from equation 8.28.
$$D_c = \frac{(.0867)(13.56)}{.275}(10)^{-.0867(1.77)} = 3.0$$

<u>step 8</u>: If the temperature 15.9 miles downstream is 16°C, the saturated oxygen content is 10 mg/l. Since the critical deficit is 3, the minimum oxygen content is 7 mg/l. This is adequate for fish life.

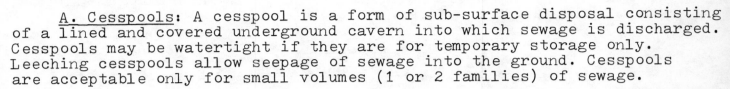

8. Small Volume Disposal

   <u>A. Cesspools</u>: A cesspool is a form of sub-surface disposal consisting of a lined and covered underground cavern into which sewage is discharged. Cesspools may be watertight if they are for temporary storage only. Leeching cesspools allow seepage of sewage into the ground. Cesspools are acceptable only for small volumes (1 or 2 families) of sewage.

   <u>B. Septic Tank</u>: A septic tank is a simple tank which allows both sedimentation and digestion to occur. Typical detention times at 8 to 24 hours. Effluent is discharged into underground tile fields which allow the water to percolate into the soil. Only 30-50% of the suspended solids are removed by septic tanks. The remaining solids eventually clog the tank and must be mechanically removed.

<u>Figure 8.4</u>

<u>Typical Septic Tank</u>

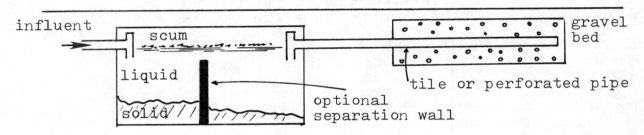

Typical design parameters of a domestic septic tank are given below:

Minimum capacity below flow line	300 gal for 5 persons or less 500 gal preferred. No garbage disposals. Add 30 gal per additional person.
Plan aspect ratio:	1:2
Minimum depth below flow line	3 - 4 feet
Minimum freeboard	1 foot
Tank burial depth	1 - 2 feet
Tile length	30 feet per person
Maximum tile run length	60 feet
Minimum tile depth	15 inches (30" preferred)
Lateral line spacing	6 feet
Gravel bed size	4 inch radius around tile, 12" below
Minimum soil layer below tile bed	10 feet

Municipal septic tanks should be designed to hold 12-24 hours of flow plus stored sludge. A general rule is to allow at least 25 gallons per person served by the tank.

C. Imhoff Tank: An Imhoff tank is similar to a septic tank in that sedimentation and sludge digestion both occur. However, these two processes occur in different parts of the tank, and Imhoff vessels are larger in capacity than simple domestic septic tanks. Wastewater enters the tank at the top where sedimentation occurs. The sediment slides down the sloped inner baffles.

Figure 8.5

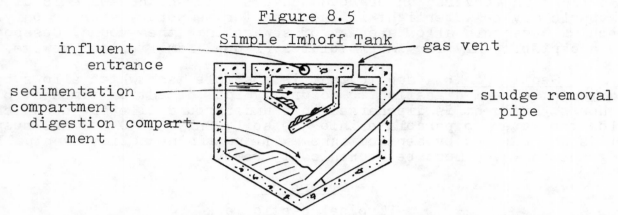

Simple Imhoff Tank

One of the inner baffles extends past the other so that gas produced in the digester chamber will not enter the sedimentation chamber.

Imhoff tanks are usually very large (i.e., 2 stories high) and have been used in the past where the loading is between 250,000 gpd and 1,000,000 gpd. Imhoff tanks can remove up to 60% of the suspended solids during the 1 to 2 hour retention time. They are more efficient than septic tanks, but require very frequent (up to hourly) attention and they are much more costly.

Typical technical characteristics for an Imhoff tank are given below.

sludge chamber capacity	2.5 ft^3/person
slope of inner baffles	2 vertical:1 horizontal
total depth	15 feet minimum
gas vent area	15%-25% of top area
fall-through slot width	8" minimum
baffle overhang	8"
sludge pipe diameter	8" minimum
distance from slot to sludge	18" minimum

9. Wastewater Processes

A. Preliminary Preparation

1. Screens: Coarse screens with openings 2 inches or larger should preceed pumps to prevent clogging. Screenings usually consist of paper, wood, and rags. Medium screens ($\frac{1}{2}$" to $1\frac{1}{2}$" openings) and fine screens (1/16" to 1/8") are also used to relieve the load on grit chambers and sedimentation basins. Screens are cleaned by automatic scraping arms. Screen capacities and head losses are specified by the manufacturers.

2. **Grit Chambers**: Grit is an abrasive that wears pumps, clogs pipes, and accumulates in excessive volumes. A grit chamber (also known as 'grit clarifier' or 'detritus tank') slows the wastewater down to approximately 1 ft/sec or less. This velocity allows the grit to settle out but moves the organic matter through. The grit may be manually or mechanically removed with buckets or a screw conveyor. A ½ to 1 minute detention time is typical.

Typical design standards for grit chambers are listed below:

grit removal rate	2 to 5 ft^3/million gallons
grit size	.2 mm or larger
grit specific gravity	above 2
depth	3 to 4 feet
length	60 feet
detention time	30 to 60 seconds

Scouring of the particles which have already settled will be prevented if the velocity is <u>kept below</u> that given in equation 8.29.

$$v = 2.2 \sqrt{\frac{gd_p}{f}(SG_p - 1)} \qquad\qquad 8.29$$

d_p and SG are the particle diameter and specific gravity respectively. f is the grit chamber friction factor.

3. **Skimming Tanks**: If the sewage has excessive grease or oil, a basin with a 10 minute detention time will allow the grease to rise to the surface. An aerating device below will help to coagulate and float grease to the surface. Approximately .1 cubic feet of air per gallon of wastewater should be used for this purpose. Surface grease can be mechanically removed by skimming troughs.

If the skimming tank is enclosed and the air evacuated to about 9" of mercury, rising air bubbles in the sewage will expand and help float the grease upwards without the need for mechanical aeration.

4. **Shredders**: Shredders (also called 'comminutors') cut waste solids to approximately ¼" in size. They reduce the amount of screenings which must be disposed of.

B. **Primary Treatment**

1. **Plain Sedimentation**: Plain sedimentation basins are described in chapter 7. Design characteristics for wastewater treatment are given below:

BOD reduction	20% - 40%
total solids reduction	35% - 65%
bacteria reduction	50% - 60%
organic content of settled solids	50% - 75%
specific gravity of settled solids	1.2 or less
typical settling velocity	above 4 feet/hr
plan shape	rectangular or circular
basin depth	6 to 10 feet (8 typical)
basin width	10 to 50 feet

CONTINUED NEXT PAGE

detention time	1.5 - 2 hours
circular diameter	30 - 150 feet
flow through velocity	.005 ft/sec
flow through time	at least 30% of detention time
overflow rate	400 to 2000 gpd/ft^2 (600-1000 typical)
bottom	slight slope (8%) towards hopper
inlet	baffled for uniform velocity
scum removal	mechanical or manual
weir loading	10,000 - 20,000 gpd/ft
DESIGN PARAMETERS	overflow rate and depth
GOOD DESIGN STANDARD	2 tanks in parallel

2. Chemical Sedimentation: Chemical flocculation ('clarification') is similar to that described in chapter 7 except that the coagulant doses are greater. Typically, the most economical coagulant used is ferric chloride. Lime and sulfuric acid many be used to adjust the pH for proper coagulation. Chemical precipitation is used when the stream into which the outfall discharges is running low, when there is a large increase in sewage flow, and generally when plain sedimentation is insufficient.

Typical design and operation characteristics are given below.

rapid mix period	1 to 2 minutes
flocculation period	10 to 30 minutes
coagulating retention period	1 to 2 hours
suspended solids reduction	60 to 80%
BOD reduction	60 to 85%
maximum blade peripheral speed	5 ft/sec
DESIGN PARAMETERS	overflow rate and depth

C. Secondary Treatment

Secondary treatment became mandatory for all publically owned water treatment plants as of July 1977 under the Federal Water Pollution Control Act ammendments of 1972.

1. Trickling Filters: Trickling filters (also known as 'biological beds') consist of a bed of 2" to 5" rocks up to 9 feet thick (6 feet typical) over which influent is sprayed. The biological and microbial slime growth attached to the rocks purify the wastewater as it trickles through the rocks. The water is introduced into the filter by rotating arms which move by virtue of the spray reaction. The clarified water is collected by an underdrain system. Some water may be returned to the filter for a longer contact time.

On the average, one acre of standard filter is needed for each 20,000 people served. Trickling filters can remove 70% to 90% of the suspended solids, 65% to 95% of the BOD, and 70% to 95% of the bacteria. Most of the reduction occurs in the first few feet of bed, and organisms in the lower part of the bed may be in a near-starvation condition. The bed will periodically slough off (unload) parts of its slime coating, and sedimentation after filtering is necessary.

High rate filters are now in use by most modern facilities. The higher hydraulic loading flushes the rockpile and inhibits excess biological growth. High rate filters may be only 3 to 4 feet deep. The high rate is possible because much of the filter discharge is recirculated.

The hydraulic loading of a trickling filter is the through-put divided by the plan area. Typical values of hydraulic loading are 25 to 100 gpd/ft^2 for standard filters, and up to 1000 gpd/ft^2 for high-rate filters.

$$L_H = \frac{Q_w + Q_R}{A_{filter}} = \frac{Q_w + R_R(Q_w)}{A_{filter}} = \frac{Q_w(1 + R_R)}{A_{filter}} \qquad 8.30$$

The recirculation ratio is given by equation 8.31. It may be as high as 3 for high rate filters, although it is zero for standard low-rate filters.

$$R_R = Q_R/Q_w = \frac{L_H A_{filter}}{Q_w} - 1 \qquad 8.31$$

The BOD loading (same as 'organic loading') is calculated without considering any recirculated flow. BOD loading is essentially the BOD of the applied wastewater divided by the filter volume.

$$L_{BOD} = \frac{(Q_w)(BOD_i)(8.345 \ EE\text{-}3)}{filter \ volume \ in \ ft^3} \qquad 8.32$$

BOD loading is given in pounds per 1000 cubic feet per day. Typical values are 5 to 25 lb/1000 cubic feet-day (low rate) and 30 to 90 lb/1000 cubic feet-day (high rate.)

The efficiency of a trickling filter is

$$\eta = \frac{(BOD)_i - (BOD)_f}{(BOD)_i} \qquad 8.33$$

If it is assumed that the biological layer and hydraulic loading are uniform, the water is at 20°C, and the filter is single-stage rock followed by a settling tank, then the following equation developed by the National Research Council may be used to calculate the BOD removal efficiency of the filter/clarifier combination.

$$\eta = \frac{1}{1 + .0561\sqrt{L_{BOD}/F}} \qquad 8.34$$

If the incoming and required BOD are known, the approximate BOD loading required is given by equation 8.35.

$$L_{BOD} = 317.7\left(\frac{BOD_{req}}{BOD_i - BOD_{req}}\right)^2 \qquad 8.35$$

The BOD loading for a high rate, single stage filter with recirculation is given by equation 8.37, where F is the effective number of passes through the filter.

$$F = \frac{1 + R_R}{(1 + .1(R_R))^2} \qquad 8.36$$

$$L_{BOD} = (317.7)F(\frac{BOD_{req}}{BOD_i - BOD_{req}})^2 \qquad 8.37$$

There are a number of ways to recirculate water back to the filter.
Some methods are shown in figure 8.6.

Figure 8.6
One-Stage Recirculation Methods

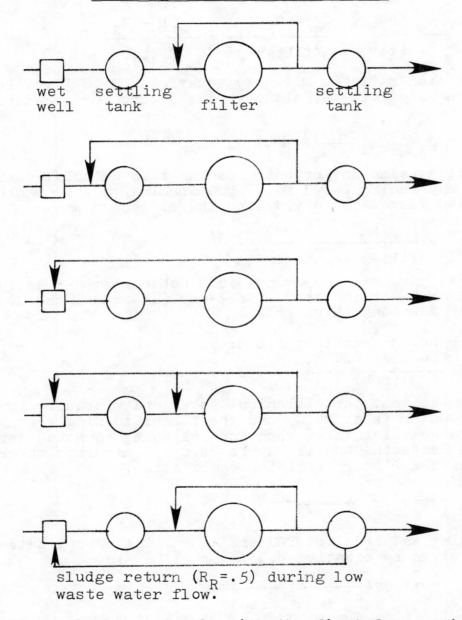

wet well · settling tank · filter · settling tank

sludge return (R_R=.5) during low
waste water flow.

Equation 8.37 should be used only with the first four recirculation
schemes shown in figure 8.6.

If even more BOD and solids removal is needed, two filters can be
connected in series to form a 2-stage filter system. A 2-stage filter
system is shown in figure 8.7(a) with an optional intermediate settling
tank.

Typical 2-stage performance is given below:

BOD loading	45 to 70 lb/1000 ft^3-day
hydraulic loading	.16 to .48 gpm/ft^2
recirculation ratio	.5 to 4.0
final effluent BOD	30 mg/l

<u>Figure 8.7</u>

<u>2-Stage Filter Systems</u>

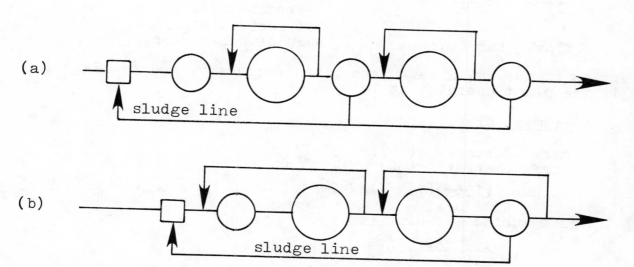

The approximate BOD loading that is required to accomplish a given reduction in BOD is given by equation 8.38.

$$L_{BOD} = (317.7)F_{2nd \atop stage} \left(\frac{BOD_{out,\ 1st}}{BOD_{in,\ 1st}}\right)^2 \left(\frac{BOD_{out,2nd}}{BOD_{in,2nd} - BOD_{out,2nd}}\right)^2$$

8.38

The efficiency of a second-stage filter is considerably less than that of a first-stage filter because much of the biological food has been removed. This lowered efficiency can be considered as an increase in BOD loading, as given by equation 8.39.

$$L_{BOD,\ adjusted \atop 2nd\ stage} = \frac{L_{BOD,\ actual\ second\ stage}}{[1 - first\ stage\ efficiency]^2}$$

8.39

The actual second stage load is calculated from equation 8.32 using the incoming BOD from the intermediate clarifier. Equation 8.34 may then be used to find the 2nd stage efficiency.

The overall BOD efficiency of the 2-stage system is:

$$\eta_{overall} = 1 - (1-\eta_{settling \atop basin})(1-\eta_{1st \atop stage})(1-\eta_{2nd \atop stage})$$

8.40

For temperatures other than 20°C, equation 8.41 may be used.

$$\eta_T = \eta_{20} \cdot c (1.035)^{T-20}$$

8.41

Example 8.7

A trickling filter plant is to process 1.4 MGD of domestic waste (170 mg/l BOD) with the following units:

primary clarifier	50 feet diameter
	7 feet wet depth
	peripheral weir
trickling filter	90 feet diameter
	7 feet wet depth
	50% recirculation
final clarifier	same dimensions as primary

Determine if the units have been sized correctly, and estimate the final BOD if the plant operates at 16°C.

Primary Clarifier

circumference: $\pi(50) = 157$ ft
surface area: $\frac{1}{4}\pi(50)^2 = 1963$ ft^2
volume: $(1963)(7) = 13740$ ft^3

The surface loading is

$$\frac{1.4 \text{ EE6}}{1963} = 713 \text{ gpd/ft}^2 \text{ (ok)}$$

The retention time is

$$t = \frac{V}{Q} = \frac{(13740)\text{ft}^3 (24) \text{ hr/day}}{(1.4 \text{ EE6}) \text{ gal/day} (.1337) \text{ ft}^3/\text{gal}} = 1.76$$

1.76 is ok.

The weir loading is $\frac{1.4 \text{ EE6}}{157} = 8917 \text{ gpd/ft}$ (ok)

Assume a 30% reduction in BOD at 20°C. The effluent BOD is

$$BOD = (.7)(170) = 119 \text{ mg/l}$$

Trickling Filter

area: $\frac{1}{4}\pi(90)^2 = 6362$ ft^2
volume: $(6362)(7) = 44534$ ft^3

From equation 8.30, the hydraulic load on the filter is

$$\frac{(1.5)(1.4 \text{ EE6})}{6362} = 330 \text{ gpd/ft}^2 \text{ (ok for high-rate filter)}$$

The BOD loading is

$$\frac{(1.4) \text{ MGD} (119) \text{ mg/l} (8.345) \frac{\text{lb-l}}{\text{MG-mg}} (1000) \text{ ft}^3/1000 \text{ ft}^3}{(44534) \text{ ft}^3}$$

$$= 31.2 \text{ lb/day-1000 ft}^3 \text{ (ok for high-rate filter)}$$

From equation 8.36,

$$F = \frac{1 + .5}{(1 + .1(.5))^2} = 1.36$$

From equation 8.34, the approximate BOD removal efficiency is

$$\eta = \frac{1}{1 + .0561\sqrt{31.2/1.36}} = .788 \text{ at } 20°C$$

The final clarifier effluent BOD is

$$(1 - .788)(119) = 25.2 \text{ mg/l}$$

Final Settling Tank

Same area, volume, surface loading, weir loading, and retention time as the primary clarifier. The BOD removal effect is included in equation 8.34.

16°C Performance

The 20°C overall efficiency is $\frac{170 - 25.2}{170} = .852$

From equation 8.41, the efficiency at 16°C is

$$\eta_{16°C} = .852(1.035)^{16-20} = .742$$

$$BOD_{16°C} = (1 - .742)170 = 43.9 \text{ mg/l}$$

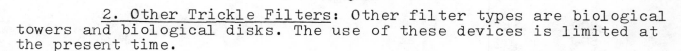

2. Other Trickle Filters: Other filter types are biological towers and biological disks. The use of these devices is limited at the present time.

3. Intermittent Sand Filters: For small populations, an intermittant sand filter may be employed. Because of the lower flow rate, the filter area per person is higher than for a trickling filter. Roughly one acre is needed for a population of 1000. The filter is constructed as a sand bed 2 to 3 feet deep over a 6" to 12" gravel bed. Application rates are usually 2 to $2\frac{1}{2}$ gpd/ft².

The filter is alternately exposed to water from a settling tank and to air. Straining and aerobic decomposition cleans the water. If the water is applied continuously as a final process from a secondary treatment plant, the filter is known as a 'polishing filter.' The water rate of a polishing filter may be as high as 10 gpd/ft². Up to 99% of the BOD can be satisfied in an intermittant sand filter.

4. Stabilization Ponds: A stabilization pond (also known as an 'oxidation pond') holds partially treated water for 3 to 6 weeks (up to 6 months in cold weather) at a depth of 2 to 6 feet (4 typical.) Aquatic plants, weeds, algae, and microorganisms are used to stabilize the organic matter. The algae gives off oxygen from growth in sunlight. The oxygen is used by microorganisms to digest organic matter. The microorganisms give off CO_2, ammonia, and phosphates which the algae use.

Areas required are large, up to one acre per 50 pounds BOD per day (about one acre per 300 people) for warm southern states. In cold northern states, twice this area may be required.

 5. Aerated Lagoons: If a stabilization pond has air added mechanically, it is known as an aerated lagoon. Such a lagoon is typically deeper and has a shorter detention time (4 to 10 days) than a stabilization pond. With floating aerators, one acre can support 500 to 1000 pounds of BOD per day (sewage from 1500 to 3000 people). In cold climates, twice this area is required.

 6. Activated Sludge Processes: Sludge produced during the oxidation process has an extremely high concentration of active aerobic bacteria. For this reason, partially oxidized sludge is called 'activated sludge.' Purification of raw sewage can be speeded up considerably if the raw sewage is mixed with ('seeded') activated sludge. The mixture of raw sewage and activated sludge is known as 'mixed liquor' (ML). The biological systems in the mixed liquor are known as 'mixed liquor suspended solids' (MLSS).

In operation, an activated sludge plant takes raw water and allows it to settle. The settled effluent is mixed with activated sludge in the ratio of 1 part sludge per 3 or 4 parts effluent. Mechanical aeration is used with a detention time of 6 to 8 hours. The effluent is next settled for 1 to 2 hours in a second sedimentation tank, chlorinated, and then discharged. Settled sludge from this last tank supplies the continuous seed for the activation.

Activated sludge processes are highly efficient, with the following average characteristics for a conventional system.

BOD reduction	90% to 95%
BOD loading	.25 to 1 lb/lb MLSS
sludge/sewage ratio	20% to 30%
maximum aeration chamber volume	5000 ft^3
aeration chamber depth	10 to 15 feet
aeration chamber width	20 ft
aeration period	4 to 8 hours
air rate	$\frac{1}{2}$ to 2 ft^3/gal raw sewage
	1500 ft^3/lb BOD reduction
minimum dissolved oxygen	2 mg/l
biological mass density	1000 to 4000 mg/l
actual dissolved oxygen	5 to 100 mg/l
sedimentation basin depth	15 ft
sedimentation basin detention time	2 hours
basin overflow rate	400 to 2000 gpd/ft^2
	(1000 typical)
% sludge returned	25% to 35%
frequency of sludge transfer	once each hour
activated sludge volume index	50 to 150
weir loading	10,000 gpd/ft
maximum tank volume	2500 ft^3

The rate of oxygen transfer (ROT) from the air to the mixed miquor during aeration is given by equation 8.42.

$$ROT = K_t D \qquad 8.42$$

K_t is a transfer coefficient which depends on the equipment and waste characteristics. The oxygen deficit, D, is given by equation 8.43.

$$D = \beta(DO)_{saturated} - (DO)_{actual\ ML} \qquad 8.43$$

β is the water's oxygen saturation coefficient, usually .8 or .9.

Example 8.8

20°C wastewater (β=.9) is processed in an aerating system with an oxygen transfer coefficient of 2.7 per hour. If the wastewater dissolved oxygen is 3 mg/l, what is the rate of oxygen transfer?

At 20°C, the saturated oxygen content is 9.2 mg/l. So, the oxygen deficit is (.9)(9.2) - 3 = 5.28 mg/l. From equation 8.42,

$$ROT = (2.7)(5.28) = 14.3 \text{ mg/l}$$

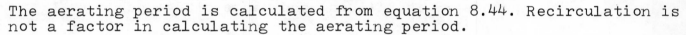

The aerating period is calculated from equation 8.44. Recirculation is not a factor in calculating the aerating period.

$$t_A = \frac{(\text{aeration tank volume, gallons})}{Q_w} \qquad 8.44$$

The BOD loading on the aeration tank is

$$L_{BOD} = \frac{(Q_w)(BOD_w)(.0624)}{(\text{aeration tank volume, gallons})} \qquad 8.45$$

The food-to-microorganism ratio is defined by equation 8.46.

$$R_{F-M} = \frac{(Q_w)(BOD_w)}{(MLSS)(\text{aeration tank volume, gallons})} \qquad 8.46$$

The rate of return sludge is

$$R_{RS} = Q_{RS}/Q_w \qquad 8.47$$

The BOD efficiency of an activated sludge process is

$$BOD = \frac{BOD_w - BOD_{after\ settling}}{BOD_w} \qquad 8.48$$

Sludge bulking refers to a condition in which the sludge will not settle out. Since the solids do not settle, they leave the sedimentation tank and cause problems in subsequent processes. The sludge volume index (also known as the Mohlman index) can be calculated by taking one liter of mixed liquor and measuring the volume of settled solids after 30 minutes. Then, the sludge volume index (SVI) is

$$SVI = \frac{(\text{settled volume in ml})(1000)}{MLSS} \qquad 8.49$$

If SVI is above 150, the sludge is bulking. Remedies include the addition of lime, chlorine, more aeration, and the reduction in MLSS.

SVI is related to the concentration of suspended solids in the activated sludge by equation 8.50.

$$SS_{AS} = \frac{1,000,000}{SVI} \qquad 8.50$$

The theoretical quantity of return sludge required can be calculated from the SVI test results:

$$Q_R/Q_W = \frac{(\text{settled volume in ml/l})}{(1000 - \text{settled volume in ml/l})} \qquad 8.51$$

Another important parameter is sludge age. Although the water passes through the system in a matter of hours, the sludge is recycled continuously and has an average stay much longer in duration. Two measures of sludge age are used: age of the suspended solids and age of BOD. Sludge age (SA_{SS}) is typically 3 to 5 days.

$$SA_{SS} = \frac{\text{pounds MLSS in aerating basin}}{\text{pounds SS in effluent and waste sludge per day}} \qquad 8.52$$

$$SA_{BOD} = 1/R_{F-M} = \frac{\text{pounds MLSS in aerating basin}}{\text{pounds BOD applies to basin per day}} \qquad 8.53$$

Example 8.9

One liter of liquid is taken from an aerating lagoon near its discharge point. After settling for 30 minutes in a one liter graduated cylinder, 250 ml of solids have settled out. A second water sample is taken and the suspended solids concentration determined to be 2300 mg/l. Find the SVI and percentage of required return sludge.

From equation 8.49, $SVI = \dfrac{(250)(1000)}{2300} = 109$

From equation 8.51, $Q_R/Q_W = \dfrac{250}{1000 - 250} = .33$ or 33%

◆

Various methods of aeration are used, each having its own characteristic ranges of operating parameters. These methods are discussed below.

   a. Extended Aeration: Small flows can be treated with the extended aeration method. This method uses mechanical floating or fixed sub-surface aerators to oxygenate the mixed liquor for 24 to 36 hours. There is no primary clarification and there are generally no sludge wasting facilities. Sludge is allowed to accumulate at the bottom of the lagoon for several months. Then, the system is shut down and the lagoon is pumped out.

Sedimentation basins are sized very small with low overflow rates (200 to 600 gpd/ft^2) and they have long retention times. All sludge is returned to the aerating basin.

Typical operating characteristics of the extented aeration method are:

MLSS: 1000 to 10,000 mg/l
L_{BOD}: 10 to 30 lb/day-1000 ft^3
R_{F-M}: .05 to .2
t_A: 20 to 30 hours
R_R: 100%
η_{BOD}: 85% to 95%

Figure 8.8

Extended Aeration

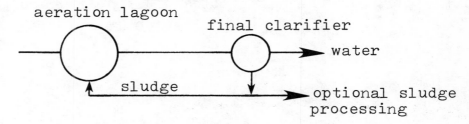

aeration lagoon

final clarifier

water

sludge

optional sludge processing

b. Conventional Aeration: In this method, (also known as 'plug flow') the influent is taken from a primary clarifier and then aerated. The amount of aeration may be decreased (i.e., 'tapered aeration') as the wastewater travels through the circuitous route since the BOD also decreases along the route.

Figure 8.9

Conventional Aeration

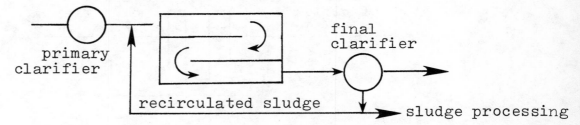

primary clarifier

final clarifier

recirculated sludge

sludge processing

Typical operating characteristics of the conventional aeration method are:

L_{BOD}: 30 to 40 lb/1000ft^3-day
R_{F-M}: .2 to .5
t_A: 6 to 7$\frac{1}{2}$ hours
R_R: 30%
η_{BOD}: 90% to 95%

c. Step Flow: In step flow, aeration is constant along the length of the aeration path, but influent is introduced at various points along the path.

Figure 8.10

Step Aeration

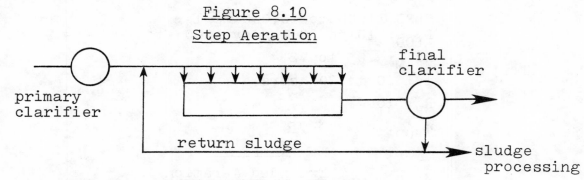

primary clarifier

final clarifier

return sludge

sludge processing

Typical operating characteristics of step flow are:

L_{BOD}: 30 to 50 lb/1000 ft^3-day

R_{F-M}: .2 to .5

t_A: 5 to 7 hours

R_R: 50%

η_{BOD}: 85 to 95%

d. Complete Mix: Waste is added uniformly and the mixed liquor is removed uniformly over the length of the tank. This method is often used for industrial waste processing.

Figure 8.11

Complete Mixing

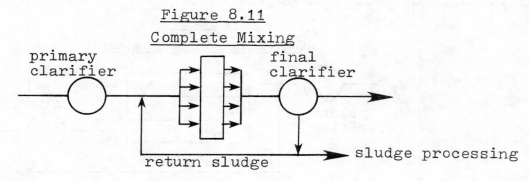

primary clarifier

final clarifier

return sludge

sludge processing

e. Contact Stabilization: Units for the contact stabilization ('biosorption') method are typically factory-built and erected on a site. They are compact, but not as economical or efficient as a regular plant. The units may be constructed as concentric compartmentalized cylinders. This method is designed primarily to handle colloidal wastes.

In this method, the aeration tank is called a 'contact tank.' The stabilization tank takes the sludge from the clarifier and aerates it also. Less time and space is required for this process because the sludge stabilization is done when the sludge is still concentrated.

Typical operating characteristics are given below.

Figure 8.12

Contact Stabilization

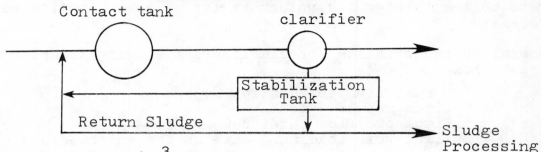

L_{BOD}: 30 to 50 lb/ft^3-day

R_{F-M}: .2 to .5

t_A: 6 to 9 hours

R_R: 100%

η_{BOD}: 85% to 90%

f. High Rate Aeration: This method uses mechanical mixing along with aeration to decrease the aeration period and to increase the BOD load per unit volume. Typical operating characteristics are:

L_{BOD}: above 80 lb/1000 ft^3-day

R_{F-M}: .5 to 1.0

t_A: 2.5 to 3.5 hours

R_R: 100%

η_{BOD}: 80 - 85%

g. High Purity Oxygen Aeration: This method requires the use of bottled or manufactured oxygen which is introduced to closed aerating tanks in place of atmospheric air. Mechanical mixing is needed to take full advantage of the oxygen. Typical operating charac-teristics are given below.

L_{BOD}: above 120 lb/1000 ft^3-day

R_{F-M}: .6 to 1.5

t_A: 1.0 to 3.0 hours

R_R: 50%

η_{BOD}: 90 to 95%

7. Intermediate Clarifiers: Sedimentation tanks located between trickling filter stages (see figure 8.7) or between a filter and subsequent aeration are known as intermediate clarifiers. Recom-mended standards are given below.

maximum overflow rate: 1000 gpd/ft^2
minimum water depth: 7 feet
maximum weir loading: 10,000 gpd/ft (plants 1 MGD or less)
15,000 gpd/ft (plants over 1 MGD)

   8. Final Clarifiers: Final sedimentation in secondary treatment is done in final clarifiers. The purpose of final clarifiers is
to collect sloughed off filter material (trickling filter processes)
or to collect sludge and return it for aeration (activated sludge
processes).

General characteristics for clarifiers following trickling filters are
given below.

    BOD removal: See equation 8.34
    minimum depth: 7 feet
    maximum overflow rate: 800 gpd/ft^2
    maximum weir loading: Same as for intermediate clarifiers, but
                      lower preferred.

If the final clarifier follows an activated sludge process, the sludge
should be removed rapidly from the entire bottom of the clarifier.
Characteristics of clarifiers following an activated sludge process are
given in table 8.10.

Table 8.10 (1:363)

Final Clarifiers for Activated Sludge Processes

Type of Aeration	Design Flow (MGD)	Minimum Detention time (hr)	Maximum Overflow rate (gpd/ft^2)
Conventional, high rate, and step	< .5	3.0	600
	.5 to 1.5	2.5	700
	> 1.5	2.0	800
Contact stabilization	< .5	3.6	500
	.5 to 1.5	3.0	600
	> 1.5	2.5	700
Extended aeration	< .05	4.0	300
	.05 to .15	3.6	300
	> .15	3.0	600

D. Advanced Tertiary Treatment

   1. Suspended Solids: Suspended solids are removed by microstrainers or polishing filter beds.

   2. Phosphorus Removal: Phosphorus may be removed by chemical
precipitation. Aluminum and iron coagulants, as well as lime, are
effective in removing phosphates. Although the actual coagulating
process is not completely understood, the following is probably a
typical reaction.

$$Al_2(SO_4)_3 \cdot nH_2O + 2PO_4^{--} \longrightarrow 2AlPO_4 + 3SO_4^{--} + nH_2O \qquad 8.54$$

Similarly, iron coagulants combine to form $FePO_4$. Much larger doses
than stoichiometric quantities are usually needed because of the complex
hydrolysis that occurs.

Lime may also be used to remove phosphorus:

$$5Ca^{++} + 4OH^- + 3HPO_4^{--} \longrightarrow Ca(OH)(PO_4)_3 + 3H_2O \qquad 8.55$$

3. Nitrogen Removal: In the ammonia stripping ('air stripping') method, lime is added to water to increase its pH to above 10. The water is then passed through a packed tower into which air is blown. The air (at the rate of approximately 400 ft^3/gallon) strips the ammonia gas out of the water. Recarbonation follows to remove the excess lime.

$$NH_4 + OH^- \underset{pH<10}{\overset{pH>11}{\rightleftharpoons}} NH_4OH \xrightarrow{air} H_2O + NH_3 \qquad 8.56$$

In the nitrification and denitrification process, bacteria oxidize ammonium ions to nitrate and nitrite in an aeration tank kept at low BOD.

$$NH_3^+ \xrightarrow{\frac{bacteria}{oxygen}} NO_2^- + NO_3^- \qquad 8.57$$

Following sedimentation and sludge recirculation, the effluent is treated to convert nitrates and nitrites to nitrogen gas. Methanol supplies the energy required by the denitrification bacteria.

Figure 8.13
Nitrification and Denitrification Process

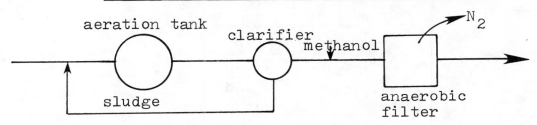

Ammonia may be removed by breakpoint chlorination. The relationships are not totally understood, and the following equation is not balanced. The stoichiometric requirements for Cl_2:N range from 8:1 to 10:1 by weight.

$$NH_3 + HOCL \longrightarrow N_2(gas) + N_2O(gas) + NO_2^- + NO_3^- + Cl^- \qquad 8.58$$

Other methods of nitrogen/ammonia removal include anion ion exchange and algae ponds.

4. Inorganic Salt Removal: Ions from inorganic salts may be economically removed by electrodialysis and reverse osmosis.

5. Dissolved Solids Removal: The so-called 'refractory' substances are dissolved organic solids that are biologically resistant. They may be removed by filtering through activated carbon.

10. Sludge Disposal

A. Sludge Quantities

Sludge removed from sedimentation basins is 90% to 99% moisture, and it must be disposed of. With primary treatment only, about .1 pounds of dried sludge can be expected per capita day. With secondary treatment, the total sludge load will be about .2 pounds per capita day (when dried.) This amounts to approximately 2 quarts of sludge per 100 gallons of wastewater processed.

The dry weight of solids from primary settling basins can be found from equation 8.59.

$$W_p = (\text{decrease in SS})(Q)(8.34 \text{ EE-6}) \qquad 8.59$$

The dry weight of solids from secondary aeration lagoons and biological filters can be found from equation 8.60.

$$W_s = K(\text{BOD})_{incoming}(Q)(8.34 \text{ EE-6}) \qquad 8.60$$

K in equation 8.60 depends on the food-to-microorganism ratio, as given in table 8.11. K is the fraction of BOD that appears as excess biological solids. For trickling filters and extended aeration, K ranges from .2 to .33. For conventional and step aeration, K ranges from .33 to .42. K is known as the cell yield or yield coefficient.

Table 8.11 (1:411)

K Values for Equation 8.60

R_{F-M}	K
.05	.2
.07	.21
.1	.24
.15	.28
.2	.33
.3	.37
.4	.4
.5	.43

Assuming the sludge specific gravity is near 1, the actual volume of wet sludge with a solids concentration, s, can be found from the dry weight by using equation 8.61.

$$\frac{\text{gallons of}}{\text{sludge/day}} = \frac{\text{dried weight per day, lbs}}{(s)(8.34)} \qquad 8.61$$

Typical values of s are given in table 8.12.

Table 8.12

Total Sludge Solids, s

Source or Type	s
primary settling tank sludge	.06 to .08
primary settling tank sludge mixed with filter sludge	.04 to .06
primary settline tank sludge mixed with activated sludge from aeration lagoons	.03 to .04
excess activated sludge	.005 to .02
filter backwashing water	.01 to .1
softening sludge	.02 to .15

The actual sludge specific gravity of the wet sludge is given by equation 8.62.

$$\frac{1}{(SG)_{total}} = \frac{\text{fraction moisture}}{1.0} + \frac{\text{fraction solids}}{(SG)_{solids}} \qquad 8.62$$

The volume of sludge can be found from equation 8.63.

$$V = \frac{W}{(SG)_{total}(62.4)} \quad \text{in } ft^3/day \qquad 8.63$$

Example 8.10

A trickling filter plant processes domestic waste with the following characteristics: 190 mg/l BOD, 230 mg/l SS, and 4,000,000 gpd.
(a) What is the wet sludge volume from the primary sedimentation tank and trickling filter? Assume the combined sludge solids content is 5%.
(b) What is the approximate weight of dry solids produced per person-day?

step 1: Find the weight of the dry solids obtained from the primary settling basin. From equation 8.59,

$$W_p = (.5)(230)(4 \text{ EE6})(8.34 \text{ EE-6}) = 3836 \text{ pounds/day}$$

Assume a 30% BOD reduction, so the BOD leaving the basin is $(.7)(190) = 133$ mg/l

step 2: Pick a K value of approximately .25. Then, the weight of dry solids from the filter is given by equation 8.60.

$$W_s = (.25)(133)(4 \text{ EE6})(8.34 \text{ EE-6}) = 1109 \text{ pounds/day}$$

step 3: The wet sludge volume can be found from equation 8.61.

$$\frac{(3836 + 1109)}{(.05)(8.34)} = 11860 \text{ gallons/day}$$

step 4: The equivalent population is given by equation 8.12.

$$P_e = \frac{(190)(8.34)(4)}{.17} = 37280 \text{ people}$$

step 5: The per capita dried solids rate is

$$\frac{3836 + 1109}{37280} = .13 \frac{\text{lb dry solids}}{\text{person-day}}$$

◆

B. Sludge Thickening

Since the volume of wet sludge is inversely proportional to its solids content (equation 8.61), thickening of sludge is desireable. Thickening is required to at least 4% solids if dewatering is to be feasible. Gravity thickening employs a stirred sedimentation tank into which sludge is fed. A doubling of solids content is usually possible with a gravity thickener.

Thickening may also be accomplished by the dissolved air flotation method. Air is bubbled through a tank containing sludge. The solid particles adhere to the air bubbles, float to the surface, and are skimmed away. The skimmed scum has a solids content of approximately 4%. Up to 85% of the total solids may be recovered, although chemical flocculants may be used to increase this to 95%. 2 to 4 pounds of solids are obtained each hour for each square foot of surface area.

C. Sludge Dewatering

Once the sludge has been thickened, it may be digested or dewatered prior to disposal. At least four methods of dewatering are available: vacuum filtration, pressure filtration, centrifugation, and drying beds of sand or gravel.

The most common method of dewatering is vacuum filtration in a rotary drum filter. Suction is applied from within the drum to attract solids to the filter and to extract moisture. The dried cake is scraped off of the drum in the discharge section of the device. Chemical flocculants are used to collect fine particles on the filter drum. A final solids content of 20% to 25% is attained. This is sufficient for sanitary landfill. A solids content of 30% is needed, however, for direct incineration.

For drying beds, allow 1.25 ft^2 per capita.

D. Digestion

Much of the organic material in sludge is easily digested by anaerobic microbes. Solids which are capable of being digested are known as volatile solids. Digestion of volatile solids results in methane, carbon dioxide, and hydrogen sulfide gases.

1. Anaerobic Digestion: If the digestion takes place in the absense of oxygen, it is known as anaerobic digestion. Two types of bacteria are involved: acid forming and acid splitting. The pH must be kept above 6.5 for the methane producing bacteria to function.

In a single-stage, floating-cover digester (figure 8.14) raw sludge is brought into the tank at the cover and top of the dome. The contents of the digester stratify into 4 layers: scum on top, supernatant next, a layer of actively digesting sludge, and a bottom layer of concentrated sludge. Some sludge may be withdrawn, heated, and returned to keep the temperature up.

Supernatant is removed along the periphery of the digester and returned to the inlet of the processing plant. Digested sludge is removed from the bottom and is then dewatered. Gas is removed from the gas dome and is burned. The heat from the burning methane is used to warm the sludge that is withdrawn.

Figure 8.14
Floating Cover Digester

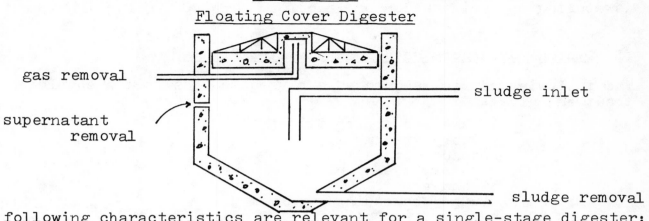

gas removal

supernatant removal

sludge inlet

sludge removal

The following characteristics are relevant for a single-stage digester:

> sludge loading: .2 lb of volatile solids/ft^3 tank-day
> optimum temperature: 95 to 98°F
> optimum pH: 6.7 to 7.8 (7.0 to 7.1 preferred)
> gas production: 7 to 10 ft^3/pound volatile solids added
>
> .5 to 1.0 ft^3/capita day
> gas composition: 65% methane, 35% CO_2
> retention time: 30 to 90 days (conventional)
> 15 to 25 days (high rate)
> final moisture content: 90% to 95%
> gas heat content: 600 BTU/ft^3
> depth: 15 to 20 feet

A single stage digester performs the functions of digestion, gravity thickening, and storage in one tank. In a 2-stage process, two digesters in series are used. Heating and mechanical mixing occur in the first digester. Since the sludge is continually mixed, it will not settle. Settling and further digestion occur in the unheated second tank.

Historically, rules of thumb were used to size digesters. For single-stage, heated digesters taking primary waste from a trickling filter, 3 to 4 cubic feet of digester volume per equivalent capita is required. For primary and secondary waste, 6 cubic feet are required.

If the required information is available, equation 8.64 may be used to size the digester.

$$\text{digester volume (gallons)} = \tfrac{1}{2}(Q_{\substack{raw \\ sludge}} + Q_{\substack{digested \\ sludge}})t_{digestion} + (Q_{\substack{digested \\ sludge}})t_{storage} \qquad 8.64$$

2. <u>Aerobic Digestion</u>: Long term aeration will also destroy volatile solids. Construction of an aerobic lagoon is similar to the aerobic lagoons discussed in this chapter. Typical characteristics are:

> sizing: 2 to 3 ft^3/capita
> aeration period: 10 to 20 days
> solids loading: .01 to .02 lb/ft^3-day

    E. Residual Disposal: Sludge may be disposed of after dewatering by incineration or landfill.

11. Typical Sequences Used in Wastewater Plants

The following partial sequences are used to construct a complete treatment plant:

        I   intake and preconditioning
        P   primary treatment
        S   secondary treatment
        T   tertiary treatment
        D   discharge
        SP   sludge processing
        SD   sludge disposal

I: Intake and Preconditioning

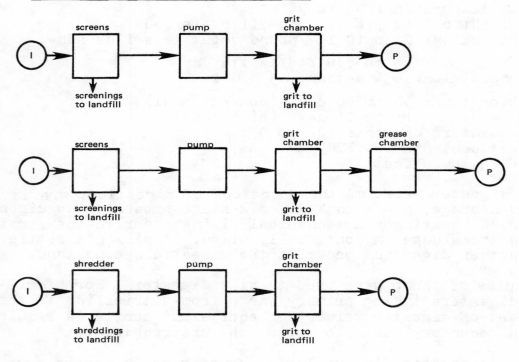

P: Primary Treatment

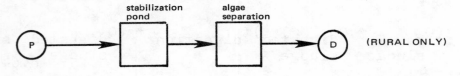

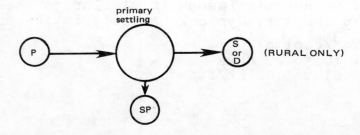

Primary Treatment, Continued

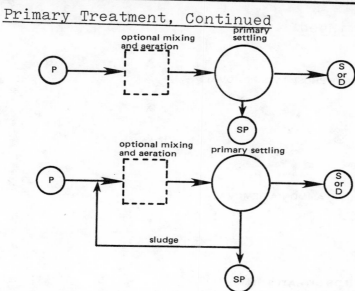

S: Secondary Treatment

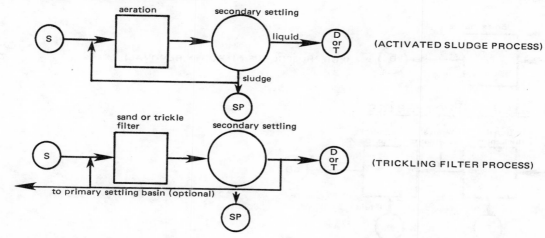

(ACTIVATED SLUDGE PROCESS)

(TRICKLING FILTER PROCESS)

T: Tertiary Treatment

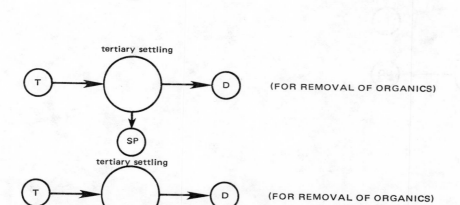

(FOR REMOVAL OF ORGANICS)

(FOR REMOVAL OF ORGANICS)

(FOR REMOVAL OF ORGANICS)

Tertiary Treatment, Continued

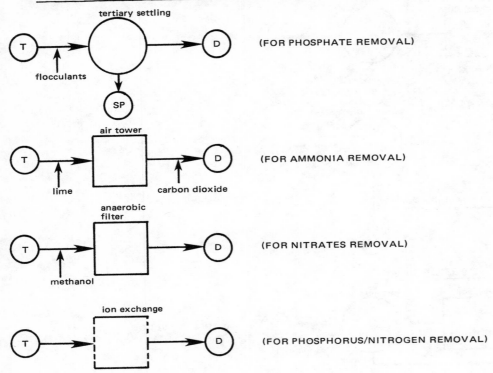

SP: Sludge Processing

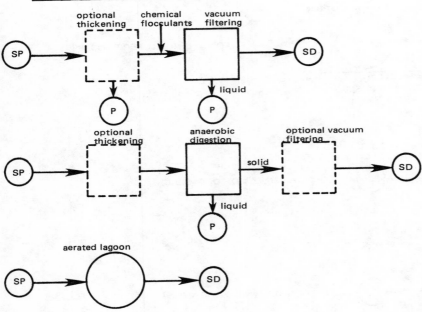

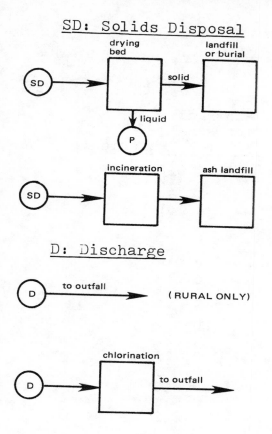

SD: Solids Disposal

D: Discharge

(RURAL ONLY)

Bibliography

1. Hammer, Mark J., _Water and Waste-Water Technology_, John Wiley and Sons, Inc., New York, NY 1975

2. Linsley, Ray K., and Franzini, Joseph B., _Water Resources Engineering_, 2nd ed., New York, NY 1972

3. Merritt, Frederick S., _Standard Handbook for Civil Engineers_, 2nd ed., McGraw Hill Book Company, New York, NY 1976

4. Munro, Lloyd A., _Chemistry in Engineering_, Prentice-Hall, Inc., Englewood Cliffs, NJ 1964

5. O'Rourke, Charles E., _General Engineering Handbook_, 2nd ed., McGraw-Hill Book Company, Inc., New York, NY 1940

Table 8.13 (1:16)

Saturated Oxygen Concentrations

(1 Atmosphere)

Temperature °C	Dissolved Oxygen, mg/l	Subtract for each 100 mg/l Chloride
0	14.6	0.017
1	14.2	0.016
2	13.8	0.015
3	13.5	0.015
4	13.1	0.014
5	12.8	0.014
6	12.5	0.014
7	12.2	0.013
8	11.9	0.013
9	11.6	0.012
10	11.3	0.012
11	11.1	0.011
12	10.8	0.011
13	10.6	0.011
14	10.4	0.010
15	10.2	0.010
16	10.0	0.010
17	9.7	0.010
18	9.5	0.009
19	9.4	0.009
20	9.2	0.009
21	9.0	0.009
22	8.8	0.008
23	8.7	0.008
24	8.5	0.008
25	8.4	0.008
26	8.2	0.008
27	8.1	0.008
28	7.9	0.008
29	7.8	0.008
30	7.6	0.008

Note: Unless told otherwise, assume a zero
      chloride concentration.

Practice Problems: WASTE WATER ENGINEERING

Required:

1. It is estimated that the BOD of raw sewage received at a treatment plant you are designing will be 300 mg/l from a population of 20,000. A single-stage, high-rate filter is to be used to reduce the plant effluent to 50 mg/l. 30% of the raw sewage BOD is removed by settling. Recirculation is from the filter effluent to the primary settling influent. Use '10-State Standards'.

   a. What is the design flow rate?
   b. If a round filter is used, what should be its depth and
      diameter?
   c. What volume should be recirculated?
   d. Draw a flow diagram of the process.
   e. What is the overall plant efficiency?

2. The average waste-water flow from a community of 20,000 is 125 gpcd. The 5-day, 20°C BOD is 250 mg/l. The suspended solids content is 300 mg/l. The final plant effluent is to be 50 mg/l BOD through use of settling tanks and trickling filters. The settling tanks are to have a surface settling rate of 1000 gpd/ft². The trickling filters are to be 6 feet deep, used without recirculation.

   a. Design circular settling tanks for this plant.
   b. Estimate the BOD removal in the settling tanks.
   c. Size the trickling filters.

3. A small town of 10,000 discharges its wastes directly into a stream which has the following characteristics:

minimum flow rate:	120 cfs
minimum dissolved oxygen: (at 15°C)	7.5 mg/l
velocity:	3 mph
temperature:	15°C
reaeration coefficient of stream and wastes:	.2 @ 20°C
BOD reaction coefficient of stream and wastes:	.1 @ 20°C

The town's waste consists of the following:

	volume	BOD @ 20°C	Temperature
domestic	122 gpcd	.191 lb/cd	64°F
infiltration	116,000 gpd		51°F
industrial #1	180,000 gpd	800 mg/l	95°F
industrial #2	76,000 gpd	1700 mg/l	84°F

   a. What is the domestic waste BOD in mg/l?
   b. What is the overall waste BOD in mg/l just before discharge
      into the stream?
   c. What is the temperature of the waste just before discharge
      into the stream?
   d. What is the theoretical minimum dissolved oxygen concentration
      in the stream?
   e. How far downstream would you expect the minimum dissolved
      oxygen concentration to occur?

Optional

4. A sewage treatment plant is being designed to handle both domestic and industrial waste with the following characteristics:

industrial #1:   1.3 MGD; 1100 mg/l BOD
industrial #2:   1.0 MGD; 500 mg/l BOD
domestic: 100 gpcd; .18 lb/cd BOD

The city has a population of 20,000; and, an expansion factor of 15% is to be used.

a. What is the design population equivalent for the plant?
b. What is the plant's organic loading?
c. What is the plant's hydraulic loading?

5. Sewage from a city of 40,000 has an average daily flow of 4.4 MGD. An analysis of the raw sewage shows the following characteristics:

pH   7.8
suspended solids   180 mg/l
5-day BOD   160 mg/l
COD   800 mg/l
total solids   900 mg/l
volatile solids   320 mg/l
settleable solids   8 mg/l

a. What are the dimensions of a circular sedimentation basin that would remove about 30% of the BOD?
b. How many basins are required?
c. What are the dimensions of square sedimentation basins that would remove about 30% of the BOD?
d. How many square basins would be required?
e. Design a single-stage trickling filter and final sedimentation basin (circular) which would produce, in combination with the primary unit, a final effluent having a 5-day BOD of not more than 20 mg/l.

6. A cheese factory discharges 35,000 gpd with the following effluent characteristics: 10,000 gpd with 1000 mg/l BOD; 25,000 gpd with 250 mg/l BOD. (a) Assuming an average depth of 4 feet, what lagoon size is required for stabilization? (b) What is the detention time?

7. An activated sludge plant processes 10 MGD of influent with 240 mg/l BOD and 225 mg/l of suspended solids. The discharge from the final clarifier contains 15 mg/l BOD and 20 mg/l suspended solids. The following assumptions may be used:

- Primary clarification removes 60% of the suspended solids and 35% of the BOD.
- Sludge has a specific gravity of 1.02 and consists of 6% solids by weight.
- The cell yield (BOD conversion to biological solids) in the aeration basin in 60%.
- The final clarifier does not reduce BOD.

(a) What is the daily weight of sludge produced? (b) Assuming the sludge is completely dried prior to disposal, determine the daily sludge volume.

NOTES

Soils

Nomenclature

a	area	cm²
A	area	in² or cm²
C_u	uniformity coefficient	-
C_z	coefficient of curvature	-
CBR	California bearing ratio	-
D	diameter	millimeters
e	void ratio	-
F	percent passing through the sieve	-
G_H	hydraulic gradient	-
h	head	cm
I_c	compression index	-
I_d	density index	-
I_g	group index	-
I_l	liquid index	-
I_p	plasticity index	-
k	coefficient of permeability	cm/sec
L	flow path length	cm
LS	linear shrinkage	-
n	porosity	-
p	pressure	psi
P	load	pounds
PPS	percent pore space	-
Q	flow quantity	cm³/sec
R	overconsolidation ratio, or Hveem's resistance	-,-
s	degree of saturation	-
S	strength	psi
SG	specific gravity	-
SL	shrinkage limit	-
SR	shrinkage ratio	-
t	time	seconds
v	velocity	cm/sec
V	volume	cm³
vs	volumetric skrinkage	-
w	water content	-
W	weight	grams
w_l	liquid limit	-
w_p	plastic limit	-
x	gravimetric fraction	-

Symbols

ρ	density	g/cm³ or lb/ft³
σ	normal stress	psi
ϕ	angle of internal friction	degrees
τ	shear stress	psi
ϵ	strain	-
θ	angle of principal stress plane	

Subscripts

A	axial
c	compressive
d	oven dried
eq	equilibrium
f	final
g	air
i	the ith component, or initial
n	unconfined
o	consolidated
R	radial
s	soil
sat	saturated
t	total
u	ultimate
us	ultimate shear
v	void
w	water
z	zero air voids

1. Conversions

Multiply	By	To Obtain
centimeters	.3937	inch
centimeters squared (cm^2)	.155	square inches (in^2)
cubic yards	27	cubic feet
cubic yards	202.2	gallons
dynes	EE-5	newtons
cubic feet (ft^3)	7.48	gallons
cubic feet (ft^3)	.03704	cubic yards
gallons	.1337	cubic feet (ft^3)
gallons	4.95 EE-3	cubic yards
gallons of water	8.33	pounds of water
grams/cubic centimeter (g/cm^3)	62.428	lb/cubic foot
inches	2.54	centimeters
square inches	6.4516	square centimeters
kilogram	2.20462	pounds
newtons	.22481	pounds
newtons	EE5	dynes
pascals	.145 EE-3	psi
pounds	.4536	kilograms
pounds	453.59	grams
pounds	4.448	newtons
pounds per square inch (psi)	144	psf
pounds per square inch (psi)	6.894 EE3	pascals
pounds per cubic feet (lb/ft^3)	.01602	$gram/cm^3$

2. Definitions

admixture: Material added to soil to increase its workability, strength, or imperviousness, or to lower its freezing point.

adsorption: Absorption characterized by a higher concentration of water at the surface of the solid than throughout.

adsorbed water: Water held near the surface of a material by electro-chemical forces.

aggregate: A mixture of various soil components (e.g., sand, gravel, and silt).

bentonite: A volcanic clay which exhibits extremely large volume changes with moisture content changes.

caisson: An air- and water-tight chamber used to work or excavate below the water level.

catena: A group of different soils which frequently occur together.

cation: Positively charged ion.

dilatancy: Property of increasing in volume when changing shape.

fine: Combined silt and clay.

friable: Easily crumbled.

frost susceptibility: Susceptible to having water continually drawn up from the water table by capillary action, forming ice crystals below the surface (but above the frost line).

gap graded: Having large particles and small particles, but no medium-sized particles.

glacial till: Soil resulting from a receding glacier, consisting of mixed clay, sand, gravel, and boulders.

gumbo: Silty soil that becomes soapy, sticky, or waxy when wet.

horizons: Dividing lines between layers of soil with different colors or compositions.

loess: A yellowish-brown loam.

normally loaded soil: A soil which has never been loaded to a greater extent than at present.

pycnometer: A stoppered flask with graduations.

pedology: The study of the soil constituting the upper 4 or 5 feet of the earth's crust.

rock flour: Fine grained, rounded quartz grains with little plasticity.

stratum: Layer

thixotropic: Gradually increasing in strength as absorbed water distributes itself through the soil.

till: See 'glacial till'.

3. Soil Types

Soil is an aggregate of loose mineral and organic particles that can be manipulated or excavated without blasting. This definition distinguishes soil from rock, which exhibits strong and permanent cohesive forces between the mineral particles.

Proper calculations for foundations and retaining walls require that the nature of the soil be known. This can be done either quantitatively or qualitatively. The primary components of any soil are gravel, sand, silt, and clay. Organic material may also be present in surface samples. If the soil is a mixture of two or more of these components, the soil is given the name of constituent which has the greatest influence on its mechanical behavior (e.g., silty clay, or sandy loam).

Particle size limits for defining gravel and sand have been suggested by ASTM, and these are given in table 9.4 later in this chapter. Whereas sand and gravel are classified as coarse grained soils, inorganic silt and clay are classified as fine grained soils. The clay-silt distinction cannot be made on the basis of size alone. Silt possesses little plasticity and cohesion. Clay, on the other hand, is very plastic and cohesive.

Clays may be distiguished from silts by using the following simple tests:

Dry Strength Test: Mold a small brick of soil and allow it to air dry. Break the brick and place a small (1/8") fragment between thumb and finger. A silt fragment will break easily whereas clay will not.

Dilatancy Test: Mix a small sample with water to form a thick slurry. When the sample is squeezed, water will flow back into a silty sample quickly. The return rate will be much lower for clay.

Plasticity Test: Roll a moist soil sample into a thin (1/8") thread. As the thread dries, silt will be weak and friable, but clay will be tough.

Dispersion Test: Dispense a sample of soil in water. Measure the time for the particles to settle. Sand will settle in 30 to 60 seconds. Silt will settle in 15 to 60 minutes, and clay will remain in suspension for a long time.

Organic matter may also be present in soil, and this presence may have a significant effect on the mechanical properties of the soil. Organic material is classified into organic silt and organic clay. Generally, the greater the organic content, the darker will be the soil color.

Indexing of the soil is necessary in order to apply some of the quantitative property relationships contained in this chapter. Indexing is accomplished by performing various classification tests on the soil. Index properties are of two types: grain properties and aggregate properties.

Soil grain properties include particle size distribution, density, and mineral composition. Density is found from a hydrometer test. Particle distribution is found from a sieve test for coarse soils (ASTM D422 or AASHO T88) and by a dispersion test for fine soils.

Table 9.1

SIEVE SIZES

This size seive	has this size openings	This size sieve	has this size openings
4"	100 mm	no. 16	1.18 mm
3"	75	20	0.850
2"	50	30	0.600
1½"	37.5	40	0.425
1"	25	50	0.300
¾"	19	60	0.250
½"	12.5	70	0.212
3/8"	9.5	100	0.150
no. 4	4.75	140	0.106
8	2.36	200	0.075
10	2.00		

The results of particle size distribution tests are graphed as a particle-size distribution, figure 9.1.

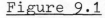

Figure 9.1

Particle-Size Distribution

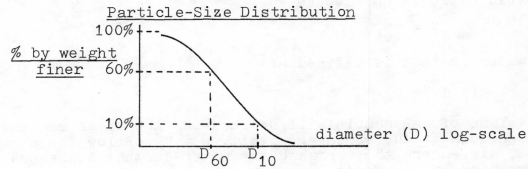

The effective grain size is defined as the diameter for which only 10% of the particles are finer (D_{10}). The 'Hazen' uniformity coefficient is given by equation 9.1.

$$C_u = D_{60}/D_{10} \qquad\qquad 9.1$$

The coefficient of curvature is

$$C_z = (D_{30})^2/(D_{10})(D_{60}) \qquad\qquad 9.2$$

Table 9.2

Soil Index Coefficients

Soil	C_u	C_z
gravel	>4	1-3
fine sand	5-10	1-3
coarse sand	4-6	
mixture of silty sand & gravel	15-300	
mixture of clay, sand, silt, and gravel	25-1000	

The aggregate index properties are essentially weight-volume relationships. In any sample of soil, there will be some air-filled voids, water-filled voids, and solid material. The percentages of these constituents (by both volume and weight) are used to calculate the aggregate properties.

Figure 9.2

Soil Sample Constituents

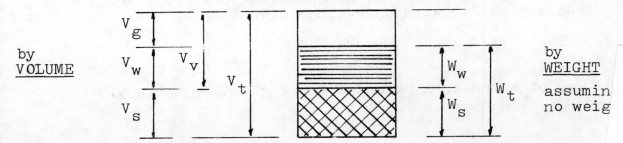

The porosity is defined as

$$n = V_v/V_t \qquad 9.3$$

The void ratio is defined as

$$e = V_v/V_s \qquad 9.4$$

The water content is defined as

$$w = W_w/W_s \qquad 9.5$$

The volume of the sample will decrease as the water content is reduced down to the shrinkage limit, w = SL. Below the shrinkage limit, air enters the voids and the water content decreases are not accompanied by decreases in volume.

The degree of saturation is defined as

$$s = V_w/V_v \qquad 9.6$$

The soil density (or unit weight) is

$$\rho = W_t/V_t \qquad 9.7$$

The dry density is

$$\rho_d = W_s/V_t \qquad 9.8$$

If the water content is known, the dry density (also known as the bulk density) of a moist sample can be found from equation 9.9.

$$\rho_d = \frac{W_t}{(1+w)V_t} = \frac{\rho}{(1+w)} \quad \frac{W_t}{V_t(1+w)} \qquad 9.9$$

The compacted density (zero air voids density) is

$$\rho_z = \frac{W_s}{V_w + V_s} \qquad 9.10$$

The density of the solid constituents is

$$\rho_s = W_s/V_s \qquad 9.11$$

The percent pore space is defined as

$$PPS = V_v/V_t = 1 - (V_s/V_t) = 1 - (\rho_d/\rho_s) \qquad 9.12$$

The specific gravity of the solid constituents is given by equation 9.13. The specific gravity of sand is approximately 2.65, and for clay it ranges from 2.5 to 2.9 with an average around 2.7.

$$SG_s = \rho_s/\rho_w \qquad 9.13$$

Typical values of these soil parameters are given in table 9.3.

Table 9.3 (5:13)

Typical Soil Indexes

Description	n	e	w_{sat}	ρ_d	ρ_{sat}
Sand, loose and uniform	.46	.85	.32	90	118
Sand, dense and uniform	.34	.51	.19	109	130
Sand, loose and mixed	.40	.67	.25	99	124
Sand, dense and mixed	.30	.43	.16	116	135
Glacial clay, soft	.55	1.20	.45	76	110
Glacial clay, stiff	.37	.60	.22	106	129

The density index (also known as 'relative density') is a measure of the tendency or ability to compact during loading. The density index is equal to one for a very dense soil; it is equal to 0 for a very loose soil.

$$I_d = \frac{e_{max} - e}{e_{max} - e_{min}} \qquad\qquad 9.14$$

There are many inter-relationships between the above soil indexes and parameters, but is is just as easy to work with the basic definitions as it is to rely on a table containing all possible combinations of variables.

Example 9.1

What is the degree of saturation for a sand sample with the following characteristics? SG = 2.65 (sand only), ρ = 115 lb/ft^3, w = .17.

In these problems, it is always best to keep track of the various weight and volume phases on a sketch.

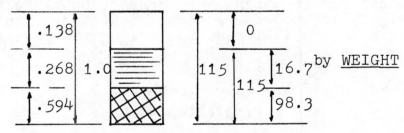

From equation 9.5, $W_w = wW_s = (.17)(W_s)$. But since $W_w + W_s = 115$,

$$(.17)W_s + W_s = 115 \quad \text{or} \quad W_s = 98.3 \quad \text{and} \quad W_w = 16.7$$

The solids volume is given by equation 9.11:

$$V_s = W_s/\rho_s = W_s/(SG)_s \rho_w = \frac{98.3}{(2.65)(62.4)} = .594 \text{ ft}^3$$

Similarly, the water volume is

$$V_w = \frac{16.7}{62.4} = .268$$

The air volume must be 1 - .594 - .268 = .138

The degree of saturation is

$$s = \frac{.268}{.268 + .138} = .66$$

◆

Example 9.2

Borrow soil is used to fill a 100,000 cubic yard depression. The borrow soil has the following characteristics: density = 96.0 lb/ft^3; water content = 8%; specific gravity of the solids = 2.66. The final in-place dry density should be 112.0 lb/ft^3 and the final water content should be 13% (dry basis).

(a) How many cubic yards of borrow soil are needed? (b) Assuming no evaporation loss, how many pounds of water are needed to achieve 13% moisture? (c) What will be the density of the in-place fill after a long rain?

The first step in borrow problems is to draw the phase diagrams for both the borrow and compacted fill soils. Use subscript B for borrow soil and F for fill soil, and work with 1 cubic foot of fill material.

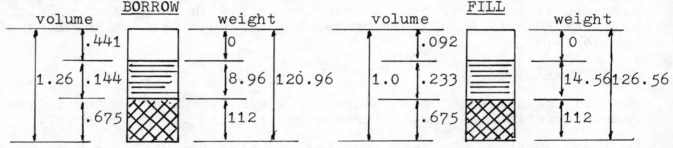

step 1: The air has no weight. Dry density precludes water. The soil content (weight) is the same at both locations. (That is, getting 112 pounds of soil in the fill requires taking 112 pounds of borrow soil.)

$$W_{sB} = W_{sF} = 112$$

step 2: The weight of the water in the fill is

$$W_{wF} = wW_s = (.13)(112) = 14.56$$

step 3: Total weight and density of the fill are

$$W_{tF} = 112 + 14.56 = 126.56$$

$$\rho_F = 126.56/1 = 126.56$$

step 4: The volume of the soil in the fill is

$$V_{sF} = \frac{112}{(2.66)(62.4)} = .675$$

step 5: The volume of the water in the fill is

$$V_{wF} = \frac{14.56}{62.4} = .233$$

step 6: The air volume in the fill is

$$1 - .233 - .675 = .092$$

step 7: The weight of the water in the borrow soil is

$$W_{wB} = wW_{sB} = (.08)(112) = 8.96$$

step 8: The total weight of the borrow soil is

$$W_{tB} = 112 + 8.96 = 120.96$$

step 9: The total volume of the borrow soil is

$$V_{tB} = \frac{120.96}{96} = 1.26$$

step 10: The volume of solids in the borrow soil is

$$V_{sB} = V_{sF} = .675$$

step 11: The volume of water in the borrow soil is

$$V_{wB} = \frac{8.96}{62.4} = .144$$

step 12: The air volume in the borrow soil is

$$1.26 - .144 - .675 = .441$$

step 13 (a): $V_{required,B} = (\frac{1.26}{1})(100,000) = 126,000$ cubic yards

(b): The actual moisture in the compacted borrow soil is

$$(126,000)(27)(8.96) = 3.05 \text{ EE7 pounds}$$

The required moisture in the fill soil is

$$(100,000)(27)(14.56) = 3.93 \text{ EE7 pounds}$$

The required additional moisture is
$(3.93 \text{ EE7} - 3.05 \text{ EE7}) = 8.8 \text{ EE6 pounds}$

(c): The density is $(.233+.092)62.4 + 112 = 132.3 \text{ lb/ft}^3$

◆

General soil classifications have been established by a number of organizations. The most common single index used in classification is particle size. The various classification schemes are presented in table 9.4. The actual classification of a soil will depend on the percentages of each constituent. For example, a soil might be classified as "4% gravel, 45% sand, 15% silt, and 36% clay (MIT)."

One method of giving qualitative descriptions to the soil is by using the USDA triangular chart (figure 9.3). This classification ignores the presence of any gravel, although the adjective 'stony' may be used in conjunction with the chart classifications.

Table 9.4

Soil Classification by Particle Size

System	date	gravel	sand	silt	clay
		sizes (mm)			
Bureau of Soils	1890	1-100	.05-1	.005-.05	<.005
Atterberg	1905	2-100	.2-2	.002-.2	<.002
MIT	1931	2-100	.06-2	.002-.06	<.002
USDA	1938	2-100	.05-2	.002-.05	<.002
Unified	1953	4.75-75	.075-4.75	<.075	
ASTM	1967	4.75-75	.075-4.75	<.075	
AASHTO	1970	2-75	.075-2	.002-.075	.001-.002

Figure 9.3

USDA Triangle Chart

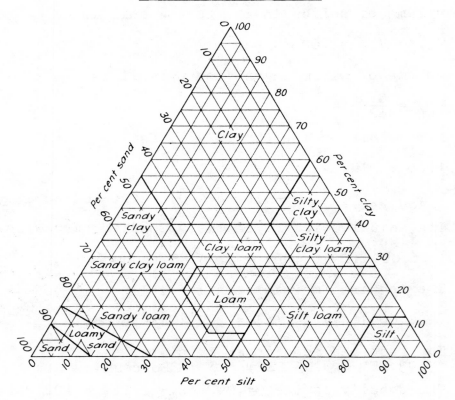

Since the qualitative description obtained from the USDA classification chart does not necessarily reflect the mechanical properties of the soil, other systems have been developed. The American Association of State Highway Officials (AASHTO) has developed a system based on the sieve analysis, liquid limit, and plasticity index.

Soils excellent for roadway subgrade construction are classified as A-1. Highly organic soils not suitable for roadway subgrade construction as classified as A-8. Subgroup classifications are also used, as well as group indexes for fine grained soil. The group index of a fine grained soil is given by equation 9.15.

$$I_g = (F_{200}-35)[.2+.005(w_1-40)] + .01(F_{200}-15)(I_p-10) \quad 9.15$$

F_{200} is the percentage of soil that passes through a #200 sieve. The AASHTO classification system is given in table 9.5.

Table 9.5

AASHTO Soil Classification System

Classification procedure: Using the test data, proceed from left to right in the chart. The correct group will be found by process of elimination. The first group from the left consistent with the test data is the correct classification. The A-7 group is subdivided into A-7-5 or A-7-6 depending on the plastic limit. For w_p less than 30, the classification is A-7-6. For w_p greater than or equal to 30, it is A-7-5. NP means non-plastic.

	Granular materials. (35% or less passing No. 200 sieve)								Silt-clay materials (More than 35% passing No. 200 sieve)			
	A-1		A-3		A-2							A-7
	A-1-a	A-1-b		A-2-4	A-2-5	A-2-6	A-2-7	A-4	A-5	A-6	A-7-5 / A-7-6
Sieve analysis: % passing											
No. 10	50 max										
No. 40	30 max	50 max	51 min								
No. 200	15 max	25 max	10 max	35 max	35 max	35 max	35 max	36 min	36 min	36 min	36 min
Characteristics of fraction passing No. 40:											
Liquid limit				40 max	41 min	40 max	41 min	40 max	41 min	40 max	41 min
Plasticity index	6 max		NP	10 max	10 max	11 min	11 min	10 max	10 max	11 min	11 min
Usual types of significant constituents	Stone fragments gravel and sand		Fine sand	-----Silty or clayey gravel and sand------				------Silty soils------		----Clayey soils----	
General subgrade rating	Excellent to good							Fair to poor			

Example 9.3

Determine the AASHTO classification of an inorganic soil with the following characteristics:

soil size	%
<.002	.19
.002-.005	.12
.005-.05	.36
.05-.075	.04
.075-2.0	.29
>2.0	0

$w_l = .53$

$I_p = .22$

$F_{200} = .04+.36+.12+.19 = .71$

From table 9.5, the classification is A-7-5 or A-7-6. Since the plastic limit is (.53-.22) = .31, the classification is A-7-5. The group index is

$$I_g = (71-35)[.2+.005(53-40)] + .01(71-15)(22-10) = 16.26 \text{ (say 16)}$$

So, the soil would be classified as A-7-5(16).

◆

ASTM standards (D-2487) also provide a method of classifying soils based on qualitative descriptions. Coarse-grained soils are divided into two categories: gravelly soils (symbol G) and sandy soils (symbol S). Sands and gravels are further subdivided into 5 subcategories:

symbol W: well-graded, fairly clean
symbol C: well-graded, with excellent clay binder
symbol P: poorly graded, fairly clean
symbol M: coarse materials with fines, not in preceding 3 groups

Fine-grained soils are divided into three categories: inorganic silty and very fine sandy soils (symbol M), inorganic clays (symbol C), and organic silts and clays (symbol O). These three are subdivided into two subcategories:

symbol L: low compressibilities (w_l 50 or less)

symbol H: high compressibilities (w_l greater than 50)

4. Soil Testing and Mechanical Properties

There are many tests for soil properties. Some of the more common tests are listed in this section.

A. Specific Gravity Tests (ASTM D854; AASHTO T100)

A sample of soil with known weight is placed in a calibrated pycnometer. The bottle is filled with water (or kerosene) and weighed. The bottle is then emptied, cleaned, and dried. It is filled with water only and weighed. The specific gravity is

$$SG = \frac{W_{dry\ sample\ alone}}{W_{dry\ sample\ alone} + W_{filled\ with\ water\ only} - W_{filled\ with\ water\ and\ dry\ soil\ sample}}$$

9.16

The specific gravity of a soil containing various percentages of minerals with different specific gravities is given by equation 9.17.

$$SG_{composite} = \frac{1}{\Sigma x_i/SG_i}$$

9.17

B. Moisture-Density Relationships (Standard Proctor Test) (ASTM D698; AASHTO T-99)

A soil sample is taken and compacted in 3 layers in a $1/30$ ft^3 cylinder by a specific number of hammer blows. The actual density is then given by equation 9.7. The dry density of the sample can be found from equation 9.9. This procedure is repeated for various water contents, and a graph similar to figure 9.4 is obtained.

Figure 9.4
Proctor Test Results

saturation density (ρ_z)

ρ_d^* is known as the maximum dry density, or density at 100% compaction.

w^* is known as the optimum water content.

Based on the standard proctor test, the following results can be expected.

Table 9.6
Typical Proctor Test Results

Soil Type	Maximum Density (lb/ft^3)	Optimum Water Content
clay	90 - 105	.2 - .3
silty clay	100 - 115	.15 - .25
sandy clay	110 - 135	.08 - .15

Example 9.4

A soil is tested according to ASTM D698 with a 1/30 ft^3 mold. The following data was collected.

test no.	sample net weight	water content
1	4.28 lb	7.3%
2	4.52	9.7
3	4.60	11.0
4	4.55	12.8
5	4.50	14.4

If .032 cubic feet of compacted soil tested at a construction site weighed 3.87 pounds wet and 3.74 pounds dry, what is the percent of compaction?

The actual density of sample number 1 is $(4.28)(30) = 128.4$ lb/ft^3. From equation 9.9, the dry density is

$$\rho_d = \frac{100(128.4)}{100+7.3} = 119.7$$

The following table is constructed for all 5 tests.

test no.	dry density
1	119.7
2	123.6
3	124.3
4	121.0
5	118.0

If this data is graphed, the following figure results. It appears that the peak is near point 3, so take the maximum density to be 124.3 lb/ft^3. The sample density was $(3.74/.032) = 116.9$, so the percentage of compaction was $(116.9/124.3) = .94$.

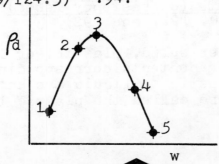

C. Modified Proctor Test (ASTM D1557; AASHTO T180)

This test is similar to the standard Proctor test except that the soil is compacted in 5 layers with a heavier hammer falling a greater distance. The result is a denser soil which is more representative of compaction densities available from modern equipment. Table 9.6 may be used by adding 10 to 20 lb/ft^3 to the densities and taking 3 to 10% from the moisture contents.

D. In-Place Density Test (ASTM D1556 or D2167; AASHTO T147)

This test, also known as the 'field density test', starts by compacting soil in the field and digging a 3" to 5" deep hole with smooth sides. All soil taken from the hole is saved and weighed before the water content can change. The hole volume is determined by filling with sand (T147 and D1556) or a water-filled rubber balloon (D2167). The required densities are given by equations 9.7 and 9.9.

E. Unconfined Compressive Strength Test

A cylinder of soil with height $1\frac{1}{2}$ to 2 times the diameter is loaded to compressive failure. Failure of elastic soils is taken as a 20% strain. The unconfined compressive strength is given by equation 9.18. The ultimate shear strength is taken as one half of the unconfined compressive strength.

$$S_{nc} = P/A \quad . \qquad\qquad 9.18$$

F. Sensitivity Tests

Clay will become softer as it is worked, and clay soils may turn into viscous liquids during construction. This tendency is determined by measuring the ultimate strength of two unconfined samples, one of which has been packed and extruded.

$$\text{Sensitivity} = \frac{S_{nc, \text{ undisturbed}}}{S_{nc, \text{ remolded}}} \qquad\qquad 9.19$$

Table 9.7

Sensitivity Classifications

Sensitivity	Class
1 - 8	natural clays
4 - 8	sensitive
8 - 15	extra sensitive
> 15	quick

G. Atterberg Limit Tests (Consistency Tests) (ASTM D423, D424, and D427; AASHTO T89, T90, T91, and T92)

Clay soils may be either solid, plastic, or liquid depending on the water content. The water contents corresponding to the transitions from solid to plastic or plastic to liquid are known as the Atterberg limits. These transitions are called the plastic limit (w_p) and liquid limit (w_l) respectively.

The difference between the liquid and plastic limits is known as the plasticity index.

$$I_p = w_l - w_p \qquad\qquad 9.20$$

Atterberg limits vary with the clay content, type of clay, and the ions (cations) contained in the clay.

The liquidity index of a clay soil is

$$I_l = \frac{w - w_p}{I_p} \qquad\qquad 9.21$$

The Atterberg liquid limit is found by taking a soil sample and placing it in a shallow container. The sample is parted in half with a special grooving tool. The container is dropped 25 times. At the liquid limit, the sample will have rejoined for a length of $\frac{1}{2}$".

The plastic limit test consists of rolling a soil sample into a 1/8" thread. The sample will crumble when it is at the plastic limit when rolled to that diameter.

Example 9.5

A clay has the following Atterberg limits: liquid limit = 60%; plastic limit = 40%; shrinkage limit = 25%. The clay shrinks from 15 cubic centimeters to 9.57 cubic centimeters when the moisture content is decreased from the liquid limit to the shrinkage limit. What is the clay's specific gravity (dry)?

The water reduction was (15-9.57) = 5.43 cubic centimeters. Since 1 cubic centimeter weighs 1 gram, the weight loss was 5.43 grams. The percentage weight loss (dry basis) is 60% = 25% = 35%. Therefore, the solid weight is

$$W_s = \frac{\Delta W_w}{\Delta w} = \frac{5.43}{.35} = 15.5$$

The water volume at the shrinkage limit is

$$V_w = (.25)(15.5) = 3.875$$

The volume of solid at the shrinkage limit is (9.57 - 3.875) = 5.695. The density of the solid is

$$\rho = \frac{15.5}{5.695} = 2.72 \text{ g/cm}^3 \qquad SG = 2.72$$

H. Permeability Tests

Permeability of a soil is a measure of continuous voids. A permeable material permits a significant flow of water. The flow of water through a permeable soil is given by equation 9.22.

$$v = kG_H \qquad\qquad 9.22$$

$$Q = Av \qquad\qquad 9.23$$

For loose filter sands, k is given by equation 9.24.

$$k \approx 100 \, (D_{10})^2 \qquad\qquad 9.24$$

Actual numerical values need to be calculated from permeability tests using constant- or falling-head permeators (figure 9.5). For constant head tests, k can be found from equation 9.25.

$$k = \frac{VL}{hAt} \qquad\qquad 9.25$$

For falling-head tests, k can be found from equation 9.26.

$$k = \frac{aL}{At} \ln(h_i/h_f) \qquad\qquad 9.26$$

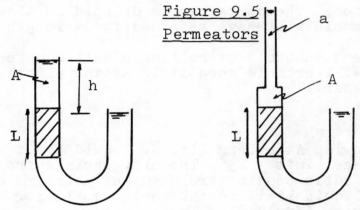

Figure 9.5
Permeators

Permeability may also be found from well pump-down tests (see chapter 6.)

Example 9.6

The permeability of a semi-impervious soil was determined in a falling-head permeator whose head decreased from 100 to 40 cm in 5 minutes. The body diamter was 13 cm; the standpipe diameter was .3 cm; and the sample length was 8 cm. What was the permeability of the soil?

From equation 9.26:

$$k = \frac{\frac{1}{4}\pi(.3)^2(8)}{\frac{1}{4}\pi(13)^2(5)(60)} \ln(100/40) = 1.3 \ EE{-}5 \text{ cm/sec}$$

◆

Table 9.8

Typical Permeabilities, cm/sec

clean gravel	5 - 10
clean sand (coarse)	.4 to 3
(medium)	.05 - .15
(fine)	.004 - .02
silty sand & gravel	EE-5 to .01
silty sand	EE-5 to EE-4
sandy clay	EE-6 to EE-5
silty clay	EE-6
clay	EE-7
colloidal clay	EE-9

I. Consolidation Tests

Consolidation tests (also known as 'confined compression' and 'oedometer' tests) start with a disc of soil (usually clay) confined by a metal ring. The faces of the disc are covered with porous plates. The disc sandwich is loaded in a water tank. The testing time is very long, since the water out-seepage is very slow. The load (P) versus the void ratio is plotted as an 'e-log p'curve (figure 9.7).

Figure 9.6
Consolidation Test

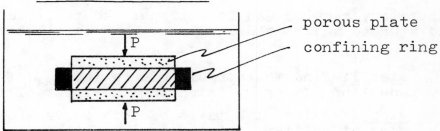

porous plate

confining ring

Figure 9.7 shows an e-log p curve for a soil sample from which the load has been removed at two places (m and n) allowing the clay to recover as much as it could.

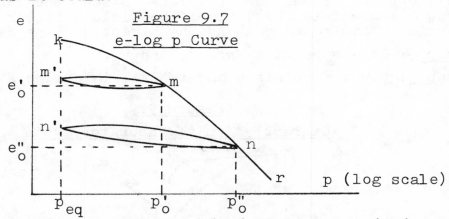

Figure 9.7
e-log p Curve

The line segment k-r is known as the virgin branch or virgin consolidation line. It is this line that would result if the load had not been removed. Lines m-m' and n-n' are rebound curves. Lines m'-m and n'-n are known as reloading curves.

Notice that point m' can only be achieved by loading the soil to a pressure of p_o' and then removing the pressure. This clay is said to have been preloaded or overconsolidated. Although the pressure of the clay is essentially the same as when it started, its void ratio has been reduced. The overconsolidation ratio is defined by equation 9.27.

$$R_o = p_o'/p_{eq}$$

9.27

The overconsolidation pressure, p_o, can be estimated by eye as a point slightly above the point of maximum curvature. Graphical means are also used. (See figure 9.9.)

The shape of the e-log p curve will depend on the degree of previous overconsolidation, as shown in figure 9.8.

Figure 9.8

Consolidation of Various Soils

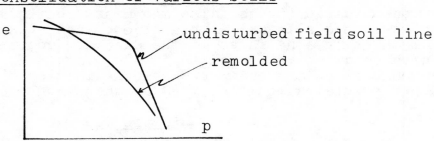

Laboratory results can be extrapolated to field results by comparing the undisturbed and remolded curves. This is illustrated in the following procedure.

step 1: Find the normal void ratio, e_o, of the field soil from equation 9.4 or by extending the front of the curve to the left.

step 2: Extend the tails of both curves downward until they intersect at point f. Point f is approximately at $.4e_o$.

step 3: Find the overconsolidation pressure, p_o, by visual or graphical means. (Draw a horizontal line from and a tangent line to the point of maximum curvature. Bisect the resulting angle. Draw a tangent to the tail of the field soil line. The intersection of this tangent and the bisection line defines p_o.)

step 4: Locate point a from e_o and p_o.

step 5: Connect points a and f. Call this line k.

Figure 9.9

Extrapolation of Laboratory Data

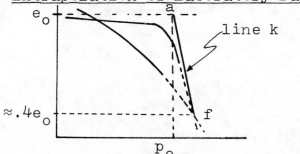

The line k is the data that should be used to predict consolidation of the soil under various loadings. The compressive index is given by equation 9.28 where e_1 and p_1 correspond to any single point on line k.

$$I_c = \frac{e_o - e_1}{\log_{10}(p_1/p_o)}$$

9.28

If the clay is soft and near its liquid limit, the compression index may be approximated by equation 9.29. In equation 9.29, w_1 is a whole number, not a decimal percentage.

$$I_c \approx .009(w_1 - 10)$$

9.29

J. Triaxial Stress Tests (R- and S-tests)

In a triaxial stress test, a cylindrical sample is loaded on both ends and all around its surface. Usually the radial stress (σ_R) is kept constant and the axial stress (σ_A) is varied. The normal and shear stresses on a plane of any angle can be found from the combined stress equations. (Consider compression positive.)

$$\sigma_\theta = \tfrac{1}{2}(\sigma_A + \sigma_R) + \tfrac{1}{2}(\sigma_A - \sigma_R)\cos 2\theta \qquad 9.30$$

$$\tau_\theta = +\tfrac{1}{2}(\sigma_A - \sigma_R)\sin 2\theta \qquad 9.31$$

These equations represent points on Mohr's circle, which can be easily constructed once σ_A and σ_R are known. (Care must be taken when plotting this graph. The sample is usually exposed to a pressure p_R over all of its surface, including the ends. Thus, p_R is equal to σ_R. The pressure applied to the ends, p_A, is in addition to radial pressure. Therefore, $\sigma_A = p_R + p_A$.) Test results are shown in figure 9.10 for two different samples which were both tested to failure. The ultimate shear strength can be read directly from the y axis.

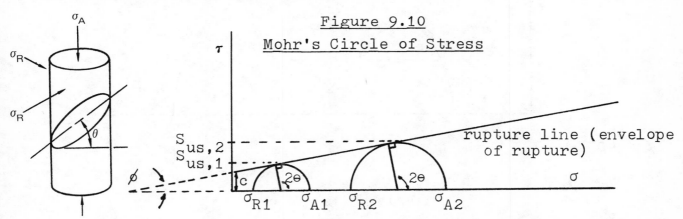

Figure 9.10
Mohr's Circle of Stress

The equation for the rupture line is given by Coulomb's equation:

$$\tau = c + \sigma(\tan\phi) \qquad 9.32$$

The plane of failure (plane of maximum principal stress) is

$$\theta = 45° + \tfrac{1}{2}\phi \qquad 9.33$$

For dry or drained sands and gravels, the intercept, c, is zero. Therefore, it is possible to draw the rupture line with only one test. Representative values of ϕ for sands and silts are given in table 9.9. ϕ is known as the angle of internal friction.

Presence of water in the pores of a sample will not affect these results much if the triaxial test is conducted in such a manner as to allow pore water pressure to dissipate (i.e., pore water to flow freely). Such triaxial tests are known as 'S-tests'. If the test is performed quickly so that the pore pressure does not have a chance to dissipate, the test is known as an 'R-test'. In such a case, much of the axial load may be carried by the pore moisture.

Table 9.9
Typical Values of ϕ

	loose	dense
dry uniform sand, round grains	27.5°	34°
dry uniform sand, angular grains	33°	45°
sandy gravel	35°	50°
silty sand	27-33°	30-35°
inorganic silt	27-30	30-34

Example 9.7

A sample of dry sand is taken and a triaxial test performed. The added axial stress causing failure was 5.43 tons/ft^2 when the radial stress was 1.5 tons/ft^2. What is the angle of internal friction? What is the angle of the failure plane?

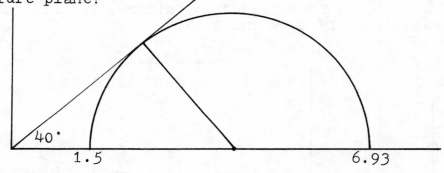

40° 1.5 6.93

$\phi = 40°$

$\theta = 45° + \frac{1}{2}(40) = 65°$

For any given radial pressure, a stress-strain curve may be plotted. This is illustrated in figure 9.11. The strain is volumetric strain due to the axial load only. The stress is the difference between the axial and radial stresses. The ultimate compressive stress (S_{uc}) may be read directly from the chart. S_{uc} is usually taken as the stress difference for which the strain is 20%.

Figure 9.11
Stress Strain Curves

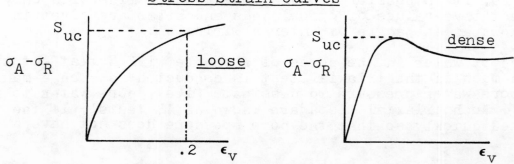

The initial slope of the line is known as the elastic modulus.

K. California Bearing Ratio Test: Shearing Resistance (ASTM D1883)

The California Bearing Ratio (CBR) test consists of measuring the load required to cause a standard-sized (3 square inches) plunger to penetrate a water-saturated soil specimen at a specific rate (.05 inches/minute). Depths of .1 or .2 inches are common, but .3, .4, and .5 inches are also used as the test points. Specifically, CBR is the relative load (in psi) required to force the piston into the sample. The word 'relative' is needed because the actual load is compared to a standard load derived from a sample of crushed stone. The ratio is multiplied by 100 and the percent omitted. Prior to testing, the sample is pre-loaded with a surcharge equal to the weight of any permanent structure (e.g., asphalt roadway) above it.

The resulting data will be in the form of inches of penetration versus load, as measured every .025 inch. This data can be plotted as shown in figure 9.12, curve A. If the plot is concave upward (curve B), the steepest slope is extended downward to the x axis. This point is taken as the zero penetration point and all penetration values adjusted accordingly.

Figure 9.12

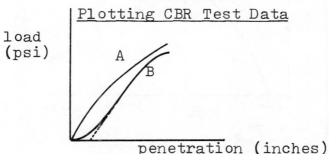

Plotting CBR Test Data

load (psi)

penetration (inches)

Standard loads for crushed stone are given in table 9.10. For a plunger of 3 square inches, the CBR is the ratio of the load for a .1 inch penetration divided by 1000 psi. The CBR for .2 inches should also be calculated. The test should be done over if $CBR_{.2} > CBR_{.1}$. If the results are similar, use $CBR_{.2}$.

$$CBR = \frac{\text{actual load (psi)}}{\text{standard load (psi)}} (100)$$

9.34

Table 9.10

Standard CBR Loads

Inches of Penetration	Standard Load
.1	1000 psi
.2	1500
.3	1900
.4	2300
.5	2600

Example 9.8

The following load data is collected for a 3 square inch plunger test. The loads are in addition to a 5 pound surcharge weight and a 10 pound preload. What is the CBR?

penetration (inches)	load (psi)
.025	20
.050	130
.075	230
.100	320
.125	380
.150	470
.175	530
.200	600
.250	700
.300	830

Upon graphing the data, it is apparent that a .02 inch correction is required. Therefore, the .1" load is interpolated from the graph for a .12 inch loading.

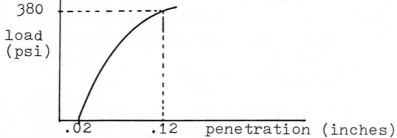

$$CBR_{.1} = \frac{(380)(100)}{1000} = 38$$

$$CBR_{.2} = \frac{(645)(100)}{1500} = 43$$

Since $CBR_{.2}$ is greater than $CBR_{.1}$, the test should be rerun.

◆

L. Plate Bearing Value Test: The Subgrade Modulus (ASTM D1195)

A standard diameter round steel plate of 1" thickness or greater is set over soil on a bed consisting of fine sand and/or plaster of paris. Smaller diameter plates are placed on top of the bottom plate to ensure rigidity. After the plate is seated by a quick but temporary load, it is loaded to a deflection of about .04 inches. This load is maintained until the deflection rate decreases to .01 inch/minute. Then the load is released. The deflection prior to loading, the final deflection, and the deflection each minute are recorded. This process is repeated 10 times for initial deflections of .04", .2", and .4".

For each repetition of each load, the deflection ('end-point deflection') is found for which the deflection rate is exactly .001 inch/minute. The loads are then corrected for dead weights of jacks, plates, etc. The deflection correction is found by plotting the deflection for the 5th repetition of each of the 3 loads. This deflection is added to all deflection data.

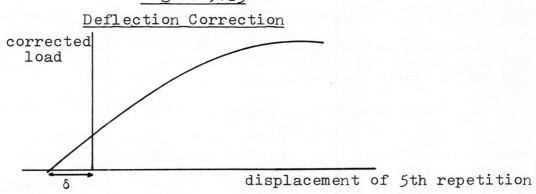

Figure 9.13
Deflection Correction

The corrected load versus the corrected deflection is graphed for the 10th repetition of each load. The bearing value is the interpolated load which would produce a deflection of .5 inches.

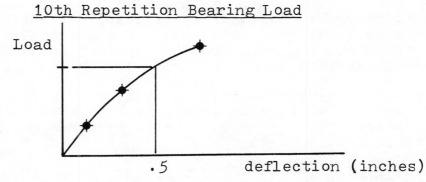

Figure 9.14
10th Repetition Bearing Load

Figure 9.14 may also be used to find the subgrade modulus, which is the slope of the line (in psi/inch) in the loading range encountered by the soil.

The remaining data may be used to plot deflection versus repetition characteristics on semi-log paper.

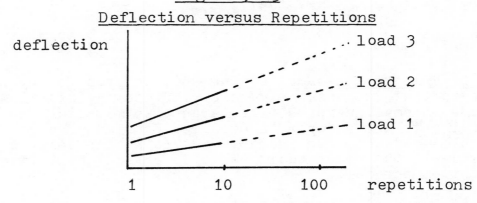

Figure 9.15
Deflection versus Repetitions

M. Field Moisture Equivalent (FME) Test (ASTM D424, AASHTO T93)

The FME test determines the saturated moisture content. FME is the minimum moisture content for which a smooth soil sample will absorb no slowly-applied water.

N. Centrifuge Moisture Equivalent (CME) Test (ASTM D425, AASHTO T94)

In this test, saturated soil is turned in a centrifuge for one hour under a force equal to 1000 g's. CME is the resulting moisture content. Permeable sands and silts have low (less than 12) CME's, whereas high values (20 or above) indicate semi-permeable clays.

O. Hveem's Resistance Value Test: The R-value (AASHTO T173, T174, and T175; California Test Method 301)

The term 'resistance', as it is used in this chapter and in chapter 16, refers to the ability of a soil to resist lateral deformation when a vertical load acts upon it. When displacement does occur, the soil moves out and away from the applied load. This movement must be kept within limits which depend on the expected loading.

Measuring the R-value of a soil is done by means of a stabilometer test. The R-value will range from zero (the resistance of water) to 100 (the approximate resistance of steel). R-values of soil and aggregate usually range from 5 to 85.

The R-value is determined using soil samples which are compacted as they would be during normal construction. They are tested as near to saturation as possible to give the lowest expected R-value. Thus, the R-value represents the worst possible state the soil might attain during use.

The procedure also takes into account the fact that some soils are expansive. When a compacted soil expands due to the absorption of water, the R-value also decreases. The steps in this test procedure account for this by lowering the R-value.

The following steps constitute the procedure for determining the R-value.

step 1: Prepare the samples: The soil is prepared by breaking all lumps so that it passes a #4 sieve. Since the stabilometer test is made on small samples, all large-size material should be removed according to the accepted procedure. Four samples are measured out (weighed) and water is added to saturate the soil. A fifth sample is used without water addition to determine the initial moisture content of the soil.

step 2: Compact the samples: The soil is compacted in a mechanical kneading compactor which duplicates field compaction conditions.

step 3: Determine the exudation pressure: In this step, the soil is placed in a specially-instrumented cylinder and compressed by a

plunger. The pressure is recorded that causes the desired
response on the instrumentation. The pressure should be more than
100 psi but less than 800 psi, although some very expansive soils
will give a pressure of less than 100 psi. Heavy clay materials
that exude along with the water are indications of very poor soil.
An R-value of less than 5 is specified for such soils.

step 4: Determine the expansive pressure. The soil is placed
into another cylinder with a lightly-loaded plunger. The plunger
is held against the top of the sample by a spring. Water is
allowed to percolate through the sample, and this causes the soil
to expand. The expansive pressure is calculated from the increase
in height and the plunger's retaining spring rate.

$$p_{expansive} = k_{spring}(\text{change in height}) \qquad 9.35$$

step 5: Determine the stabilometer R-value: The sample of soil
is placed in a stabilometer. This triaxial stress testing device
consists of a rubber sleeve in a metal cylinder with a liquid
separating them. As a vertical load is applied to the soil in the
cylinder, lateral deformation causes a horizontal pressure to be
transferred to the liquid. A pressure gage is used to measure the
liquid pressure. The stabilometer also has a crank handle to apply
horizontal pressure to the soil. This crank handle turns once
for each 0.1 inch of vertical displacement.

During the test, the soil sample is loaded vertically to 160 psi,
and the horizontal pressure, p_h, is noted. Then, the vertical load
is removed and the soil loaded by the crank handle until the horizontal pressure is 100 psi. The number of turns, D, of the displacement handle (crank) is noted.

The R-value is given by equation 9.36.

$$R = 100 - \frac{100}{\frac{2.5}{D}(\frac{160}{p_h} - 1) + 1} \qquad 9.36$$

step 6: The R-value determined in step 5 is modified by use of
the exudation pressure. The R-value determined in step 5 is also
modified by use of the expansion pressure. The lower of these two
modified values is taken as the design R-value.

P. Approximating the R-value Using Soil Classification

Rough estimates of the R-value can be made using simple soil classification tests. The soil type can be found from a sieve analysis, hydrometer test, and figure 9.3.

After the soil has been classified (e.g., sandy clay, etc.) figure 9.16
may be used to determine an approximate R-value. It will be seen that
each soil type encompasses a certain R-value range. The curves representing the various soils are stylized (approximate) frequency distributions.

Figure 9.16 (2:17)

R-Values of Various Soils

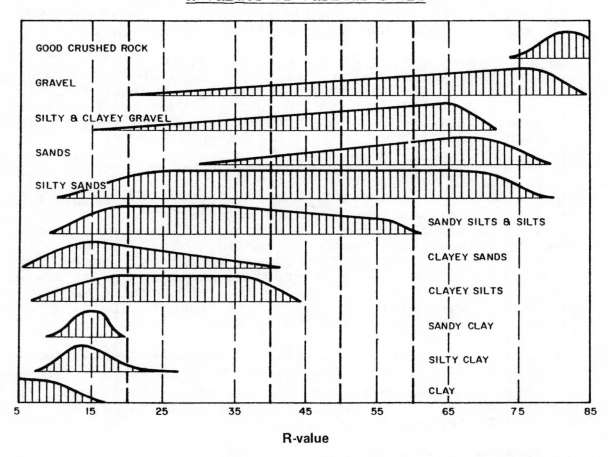

For fine-grained materials, the upper-tail (high R-value) represents a lower plasticity; the lower tail represents solids with higher plasticity.

The curve for coarse-grained materials are affected in the same manner. Lower tails represent materials with either more clay or clay with a higher activity. In coarse-grained materials with little or no clay, the lower tail represents hard, smooth-surfaced and uniform sized material. The upper tail represents rough-surfaced material with a distribution of sizes.

Q. Approximating the R-value Using CBR Data

Figure 9.17 may be used to approximate the R-value from California Bearing Ratio data. The nomograph is <u>not</u> applicable when <u>all</u> three of the following conditions are met:

1. 75% of the soil passes the #4 sieve
2. More than 8% passes the #200 sieve
3. The product of the plasticity index (in percent) and the percent passing the #200 sieve exceeds 200

Example 9.9

What is the approximate R-value for a soil having the following characteristics?

CBR: 4
Swell test: 4%
% passing the #200 sieve: 67%
% passing the #4 sieve: 100%
plasticity index: 10

This soil does not meet the first condition above, so the nomograph <u>may</u> be used. The solution is illustrated, yielding an R-value of approximately 26.

Bibliography

1. The Asphalt Institute, <u>Soils Manual</u>, 2nd ed., College Park, New York, 1963

2. County Engineers Association of California, <u>Flexible Pavement</u>, 3rd ed., Sacramento, California 1979

3. Sowers, George F., <u>Introductory Soil Mechanics and Foundations</u>, 4th ed., Macmillan Publishing Company, Inc., New York, New York, 1979

4. Merritt, Frederick, S., <u>Standard Handbook for Civil Engineers</u>, 2nd ed., McGraw-Hill Book Company, New York, New York, 1976

5. Peck, Ralph B., Hanson, Walter E., and Thornburn, Thomas H., <u>Foundation Engineering</u>, 2nd ed., John Wiley and Sons, New York, New York, 1974

Figure 9.17

CBR-to-R Nomograph

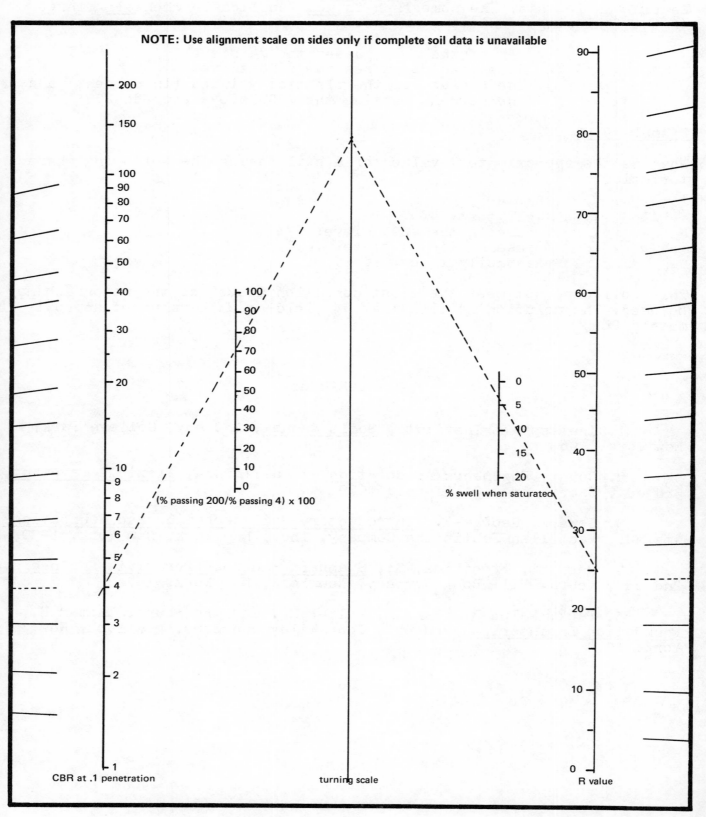

NOTE: Use alignment scale on sides only if complete soil data is unavailable

(% passing 200/% passing 4) x 100

% swell when saturated

CBR at .1 penetration

turning scale

R value

Practice Problems: SOILS

Required

1. A sample of moist soil was taken and found to have the following characteristics:

volume: .01456 m^3 (as sampled)

mass: 25.74 kg (as sampled)

22.10 kg (after oven drying)

specific gravity of solids: 2.69

Find the density, unit weight, void ratio, porosity, and degree of saturation for the soil.

2. A proctor test was performed on a soil which has a specific gravity for its solids of 2.71. For the data below, (a) plot the moisture-dry density curve. (b) Find the maximum density and optimum moisture. (c) What range of moisture is permitted if a contractor must achieve 90% compaction?

Water content %	actual density pcf	Water content %	actual density pcf
10	98	20	129
13	106	22	128
16	119	25	123
18	125		

3. For the soil in problem #2, how many gallons of water per cubic yard of soil need to be added to reach the maximum density if the soil is originally at 10% water content (dry basis)?

Optional

4. A triaxial shear test is performed on a sand sample. Failure occurred when the normal stress was 6260 psf and the shear stress was 4175 psf. What is the angle of internal friction? What are the principal stresses?

5. The results of a sieve test below give the percentage passing through the sieve. (a) Plot the particle size distribution. (b) Calculate uniformity coefficient. (c) Calculate the coefficient of curvature.

Sieve	% finer by weight
$\frac{1}{2}$"	52
#4	37
#10	32
#20	23
#40	11
#60	7
#100	4

6. A permeability test is conducted with a sample of soil which is 60 mm in diameter and 120 mm long. The head is kept constant at 225 mm. The flow is 1.5 ml in 6.5 minutes. What is the coefficient of permeability in units of meter/year?

7. A consolidation test is performed on a soil with the following results.

pressure (psf)	e	pressure	e
250	.755	8350	.724
520	.754	16,700	.704
1040	.753	33,400	.684
2090	.750	8350	.691
4180	.740	250	.710

(a) Graph the curve of stress versus void ratio. (Hint: use log or semi-log paper.) (b) What is the compression index? (c) If the initial pressure on the soil layer is 1400 psf, how much stress can a 10 foot thickness of this soil carry without settling more than .50 inches?

8. A sample of soil has the following characteristics:

% passing #40 screen: 95

% passing #200 screen: 57

liquid limit of 40 fraction: 37

plasticity index of 40 fraction: 18

Use the AASHTO system to classify this soil. Include the GI number.

9. A sample of sand has a relative density of 40% with a solids specific gravity of 2.65. The minimum void ratio is .45; the maximum is .97. (a) What is the density of this sand in a saturated condition? (b) If the sand is compacted to a relative density of 65%, what will be the decrease in thickness of a 4 foot thick layer?

10. Specifications on a job require a fill using borrow soil to be compacted to 95% of its standard Proctor maximum dry density. Tests indicate that this maximum is 124.0 pcf (dry) with 12% moisture. The borrow material has a void ratio of .60 and a solid specific gravity of 2.65. What is the minimum volume of borrow soil required to fill 1.0 cubic foot?

NOTES

NOTES

Foundations and Retaining Walls

Nomenclature

a_v	coefficient of volume compressibility	ft^2/lb
A	area	ft^2
B	width	ft
c	cohesion	lbf/ft^2
C	multiplicative correction factor	-
C_v	coefficient of consolidation	ft^2/sec
d	thickness	ft
D	depth	ft
e	void ratio	-
$e*$	eccentricity	ft
F	factor of safety	-
h	depth	ft
H	soil layer thickness	ft
I_c	compression index	-
k^c	permeability coefficient, or a constant	ft/sec, -
k_o	coefficient of earth pressure at rest	-
L	length	ft
M	moment	ft-lb
N	compressive multiplicative factor, or number of blows	-, -
p	pressure	lb/ft^2
q	uniform surcharge magnitude	lb/ft or lb/ft^2
PS	pile spacing	ft
r	distance	ft
R	force (resistance)	lb/ft of wall
S	strength	lb/ft^2
$S*$	settlement	ft
SG	specific gravity	-
t	time	various
T_v	time factor	-
U_z	percent of total consolidation	-
w	water content	-
w_l	liquid limit	-
W	weight	lb
z	distance (depth)	ft

Symbols

ϕ	angle of internal friction	degrees
δ	angle of wall friction	degrees
σ	normal stress	lb/ft^2
α	reduction factor	-
ρ	density	lb/ft^3
ρ'	submerged density	lb/ft^3

Subscripts

a	arc, or allowable
A	active, or axial
b	below mudline
c	compressive
f	footing
g	gross
h	horizontal component
H	heel
i	the ith component
n	unconfined
P	passive
R	radial
s	shear, or sliding
T	tail
v	vertical or vertical component
w	water

1. Conversions

Multiply	By	To Obtain
kips (kilo-pounds)	1000	pounds
pounds	5 EE-4	tons
pounds	.001	kips
pounds/square foot	5 EE-4	$tons/ft^2$
pounds/square inch	3.47 EE-6	$tons/ft^2$
tons	2000	pounds
tons/square foot	2000	$pounds/ft^2$
tons/square foot	2.88 EE5	$pounds/inch^2$

2. Definitions

abutment: A retaining wall which also supports a vertical load.
active pressure: Pressure causing a wall to move away from the soil.
batter pile: A pile inclined from the vertical.
bell: An enlarged section at the base of a pile or pier used as an anchor.
berm: A narrow shelf or ledge.
cased hole: An excavation whose sides are lined or sheeted.
dead load: An inert, inactive load, primarily due to the structure's own weight.
dredge level: See 'mud line'.
freeze (of piles): A large increase in the ultimate capacity (and required driving energy) of a pile after it has been driven some distance already.
grillage: A footing or part of a footing consisting of horizontally laid timbers or steel beams.
lagging: Heavy planking used to construct walls in excavations and braced cuts.
live load: The weight of all non-permanent objects in a structure, including people and furniture. Live load does not include seismic or wind loading.

mud line: The lower surface of an excavation or braced cut.

passive pressure: A pressure acting to counteract active pressure.

pier shaft: The part of a pier structure which is supported by the pier foundation.

ranger: See 'wale'.

rip rap: Pieces of broken stone weighing 15 to 150 pounds each, used to protect the sides of waterways from erosion.

sheeted pit: See 'cased hole'.

slickenside: A polished surface (plane) in stiff clay which is a potential slip plane.

soldier pile: An upright pile used to hold lagging.

stringer: See 'wale'.

surcharge: A surface loading in addition to the soil load behind a retaining wall. The surcharge may be due to temporary equipment or a permanent roadway.

wale: A horizontal brace used to hold timbers in place against the sides of an excavation, or to transmit the braced loads to the lagging.

3. Footings

A. General Considerations: A footing is an enlargement at the base of a load-supporting column designed to transmit forces to the subsoil. The area of the footing will depend on the load and the soil characteristics. The following types of footings are used.

spread footing: A footing used to support a single column. This is also known as an 'individual column footing' and 'isolated footing'.

continuous footing: A long footing supporting a continuous wall. Also known as 'wall footing'.

combined footing: A footing carrying more than one column.

cantilever footing: A combined footing that supports a column and an exterior wall (or column).

Figure 10.1

Types of Footings

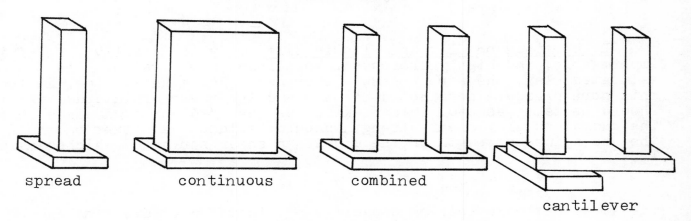

spread   continuous   combined

cantilever

Footings should be designed according to the following general considerations:

1. The footing should be located below the frost line and below the level which is affected by moisture content changes.

2. Footings need not be any lower than the highest-adequate stratum.

3. The centroid of the footing should coincide with the centroid of the applied load.

4. Allowable soil pressures (table 10.1) should not be exceeded.

5. Below-grade footings should be equipped with a drainage system.

6. Footings should be used on fill only if the original soil can support the combined load of fill and footing load.

7. Footings on fill over loose sand should be densified with piles.

8. If possible, footings should be placed in excavations made in compacted fill. They should not be put in place prior to compaction.

9. If the surcharges on either side of the footing differ by more than half of the unconfined compressive strength, the footing should be designed as a retaining wall.

10. Size footings to the nearest 3" above or equal to the theoretical size.

Table 10.1

Maximum Allowable Soil Bearing Pressures*
(Also see page 10.33)

crystalline bedrock, sound	100 tons/ft^2
layered rock, sound	40
sedimentary rock, sound	15
compacted sand or gravel	6-10
loose gravel	4
compact coarse sand	4
loose coarse sand	3
loose sand/gravel mixtures	3
compact fine sand	3
confined coarse sand (wet)	3
loose fine sand	2
confined fine sand (wet)	2
stiff clay	4
medium stiff clay	2
soft clay	1

B. Footings on Clay and Plastic Silt: Clay is normally soft, fairly impermeable, and highly preloaded. When loads are first applied to saturated clay, the pore pressure increases. For a short time, at least, this pore pressure does not dissipate and the angle of internal friction should be taken as $\phi=0$. This is known as the '$\phi=0$' or 'undrained' case. The undrained clay shear strength should be taken as either one half of the unconfined strength or the undrained shear strength, whichever is less.

$$c = S_s = \tfrac{1}{2}S_{nc} \qquad\qquad 10.1$$

The gross ultimate bearing capacity, p_g, is the pressure that can be supported by a footing. The net ultimate bearing capacity, p_{net}, is the

*As in the definition of p_a, the term 'allowable' implies that a factor of safety has already been applied.

load pressure in excess of the surcharge pressure. Both depend on the bearing capacity factor, N_c, and the shearing resistance, S_s. A safety factor of 3 (based on p_{net}) should be used for average conditions. Exceptional loadings and improbable combinations of snow, wind, and seismic forces may be allowed to reduce this safety factor to 2.

$$p_g = cN_c + \rho D_f \qquad\qquad 10.2$$

$$p_{net} = p_g - \rho D_f = cN_c = \frac{\text{bearing capacity}}{\text{area}} \qquad 10.3$$

$$p_a = (1/3)p_{net} = (1/3)cN_c = (1/6)N_c S_{nc} \qquad 10.4$$

Strength tests for S_{nc} are usually performed at every 6 inches of depth. The value of S_{nc} to be used is the strength at depth B below the footing.

<u>Figure 10.2</u>
<u>A Spread Footing</u>

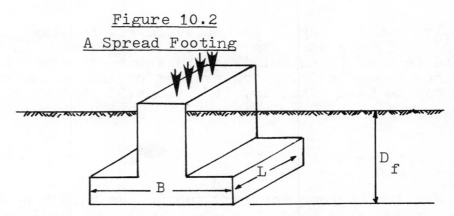

B is the width of a square footing, short side of a rectangular footing, or diameter of a round footing.

<u>Figure 10.3a (4:271)</u>

<u>Bearing Capacity Factor</u>
($\phi = 0$)

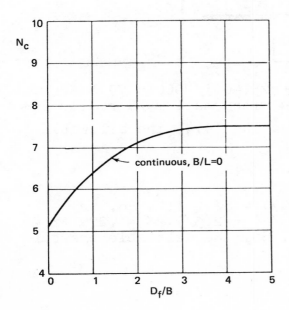

N_c Multipliers for Other
Values of B/L

B/L	Multiplier
1 (square)	1.25
.5	1.12
.2	1.05
1 (circular)	1.20

Example 10.1

In-place, unconfined compression tests of two clays are performed with the results listed below. What are the unconfined strengths to be used with $2\frac{1}{2}$ foot wide footings to be placed 2 feet deep?

depth (ft)	S_{nc} (tons/ft^2)	
	sample #1	sample #2
$\frac{1}{2}$	.58	.59
1	.61	.68
$1\frac{1}{2}$	.65	.57
2	.67	.67
$2\frac{1}{2}$	.70	.69
3	.75	.64
$3\frac{1}{2}$	.78	.75
4	.79	.69
$4\frac{1}{2}$ ·	.83	.73
5	.87	.68

The strength of clay #1 increases fairly regularly. At a depth of $(2+2\frac{1}{2})$ = $4\frac{1}{2}$, the strength is .83 tons/ft^2. That is the strength that should be used in calculating the shear strength. For clay #2, the readings are erratic, so an average of the values of 4, $4\frac{1}{2}$, and 5 feet may be used. That is, $(1/3)(.69+.73+.68)$ = .70.

◆

Example 10.2

An individual column footing carries an 83,800 pound dead load and a 75,400 pound live load. The unconfined compressive strength of the supporting clay is .84 tons/ft^2 and its density is 115 lb/ft^3. The footing is covered by a 6" basement slab. The footing is 2 feet thick. What size footing is required?

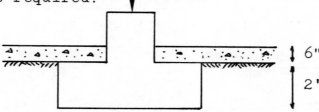

The total load on the column is $(83,800 + 75,400)/2000$ = 79.6 tons. From figure 10.3a, N_c for square footings with small D_f/B ratios is approximately $(5.5)(1.25)$ = 6.9. From equation 10.4, the allowable pressure is approximately

$$p_a = \frac{(.84)(6.9)}{6} = .97 \text{ tons/ft}^2$$

The approximate area required is $(79.6/.97)$ = 82.1 ft^2. So, try a 9'3" square footing (85.6 ft^2). Then, (D_f/B) = 2.5/9.25 = .27. From figure 10.3a again, N_c = $(5.6)(1.25)$ = 7. So, the allowable pressure is

$$p_a = \frac{(.84)(7)}{6} = .98 \text{ tons/ft}^2$$

This is not substantially different (.98 versus .97), so use the 9'3" square footing. The loading on the footing is 79.6/85.6 = .93 tons/ft^2.

The concrete density is approximately 150 pcf. Therefore, the loading from the footing (2') and the slab (6") is

$$(2 + .5)(\frac{150}{2000}) = .19 \text{ tons/ft}^2$$

So, the total load carried by the soil is (.93 + .19) = 1.12 tons/ft^2. However, equation 10.4 gives the allowable load in excess of the soil surcharge. This soil surcharge is

$$(2 + .5)(\frac{115}{2000}) = .14 \text{ tons/ft}^2$$

The net load is (1.12 - .14) = .98 tons/ft^2, the same as the allowable pressure, p_a.

    C. Footings on Sand: For any given sand settlement, the soil pressure will be greatest in intermediate width footings. This is illustrated in figure 10.3b. Since the soil pressure and settlement are related, it is desireable to use the nearly-horizontal part of the curve to avoid large variations in settlement for small variations in soil pressure. A total settling of 1 inch in sand is usually considered the maximum tolerable.

Figure 10.3b
Footings on Sand

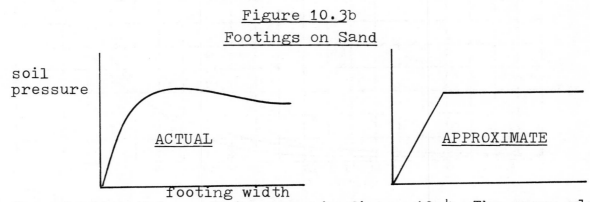

A footing of width B in sand is shown in figure 10.4. The gross ultimate bearing capacity is given by equation 10.5.

$$p_g = \tfrac{1}{2}B\rho N_\gamma + \rho D_f N_q \qquad 10.5$$

N_γ and N_q (see figure 10.5) depend primarily on ϕ. ϕ may be derived from figure 10.5 if N is known. N is the number of blows necessary to drive a standard penetrator one foot into the sand (ASTM procedure D1586). The net ultimate bearing capacity is

$$p_{net} = p_g - \rho D_f = \tfrac{1}{2}B\rho N_\gamma + \rho D_f(N_q-1) \qquad 10.6$$

Figure 10.4
Footings in Sand

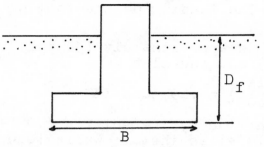

Figure 10.5
Meyerhoff Values of N_γ and N_q for Equation 10.5

(The relationship between N and ϕ is approximate.)

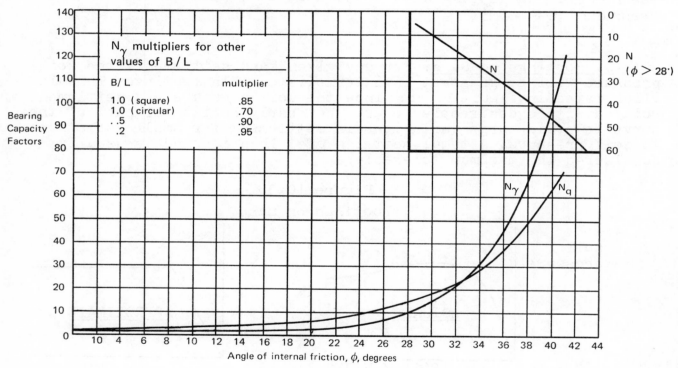

If the water table level is above the footing face (submerged condition), p_{net} should be reduced by 50%.

The allowable sand loading is based on a factor of safety, which is typically taken as 2.

$$p_a = p_{net}/F = \tfrac{1}{2}p_{net} \qquad\qquad 10.7$$

Equations 10.6 and 10.7 can be rewritten as equation 10.8. Since the quantity in brackets is constant for any (D_f/B) ratio, and since N_γ and N_q both depend on ϕ (which depends on N), figure 10.6 may be derived. Figure 10.6 assumes F = 2, ρ = 100 lb/ft^3, and $D_f < B$.

$$p_a = \frac{B}{F}\left[\tfrac{1}{2}\rho N_\gamma + \rho(N_q - 1)\frac{D_f}{B}\right] \qquad 10.8$$

For large values of B, equation 10.8 reduces to

$$p_a \approx (.11)C_n N \quad (tons/ft^2) \qquad 10.9$$

Figure 10.6 (4:309)

Shallow Sand Footing Design Charts

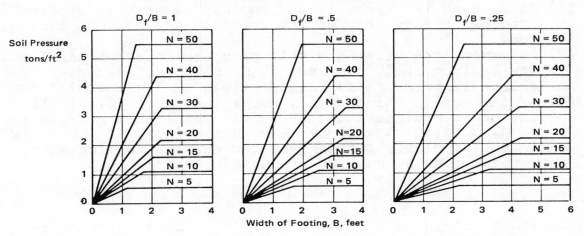

The following corrections may be necessary to use figure 10.6.

(a) No corrections for densities different from 100 lb/ft^3 are usually made.

(b) N is based on the assumption that the original overburden load is approximately 1 ton/ft^2. This means that the N values were derived from data corresponding to depths of 10 to 15 feet below the original surface (not the basement surface). If the footing is to be installed close to the original surface, then a correction factor is required. This factor may be found from table 10.2.

Table 10.2

Overburden Corrections

Overburden $(D_f \rho/2000)$ (ton/ft^2)	C_n
0	2
.25	1.45
.5	1.21
1.0	1.00
1.5	.87
2.0	.77
2.5	.70
3.0	.63
3.5	.58
4.0	.54
4.5	.50
5.0	.46

To use figure 10.6, multiply the field value of N by C_n.

(c) The chart values should be multiplied by .5 if the water level rises to the ground surface. If the water level is D_w above the surcharge level, the multiplicative correction factor can be found from equation 10.10.

$$C_w = .5 + .5(\frac{D_w}{D_f+B})$$

10.10

The following general considerations should be given to footings on sand:

(1) Since sand is permeable and rapidly adjusts to changes in loading, design the footing based on the maximum instantaneous load.

(2) Determine the allowable soil pressure based on the footing with the maximum load, smallest N, deepest surface water, etc. Use this soil pressure for all footings in the building foundation.

(3) If the design point on any curve in figure 10.6 falls on the the horizontal part of the curve, settlement governs and the total settlement will be less than 1 inch. If the design point is on the slope, bearing governs and the footing should be resized for the smaller pressure.

Example 10.3

A foundation constructed on sand (114 lb/ft^3, deep water table) uses two types of footings, each buried to the depth of its thickness.

        column footing: total of live and dead load = 251 tons
                     thickness = $2\frac{1}{2}$ feet
                     N at footing level = 25 (corrected)

        wall footing: total live and dead load = 3.6 tons/ft
                     thickness: 1 foot
                     N at footing level = 24 (corrected)

What size footings should be used?

First, size the column footing because it carries the largest load. From figure 10.6 for N = 25, the allowable pressure is approximately 2.7 tons per square foot. (This may also be derived from equation 10.9.) The footing area required is

$$A = 251/2.7 = 93 \text{ ft}^2.$$

So, use a 9'9" square footing (95.06 ft^2). Then $D_f/B = 2.5/9.75 = .25$. From figure 10.6, the design point is off the scale on the right hand side. This puts it well into the settlement-governs category.

The loading due to the footing weight and any basement slab should be added; and the loading due to the soil and any slab should be subtracted (as was done in example 10.2). However, this will not affect the results much, and the actual footing loading will be approximately (251/95.06) or 2.64 tons/ft^2.

Next, size the wall footing using the soil pressure experienced by the column footing. The footing width is

$$B_{wall} = (3.6/2.64) = 1.36 \text{ feet (say 1.5 feet)}$$

From figure 10.5 for N = 24, N_q = 31 and N_γ = 35. Using equation 10.8 with F = 2,

$$p_a = \frac{1.5}{2}\left[(\tfrac{1}{2})(114)(35) + (114)(30)(\tfrac{1}{1.5})\right]$$

$$= 3206 \text{ psf} = 1.6 \text{ tons/ft}^2$$

The corrected wall footing width is (3.6/1.6) = 2.25 ft. If equation 10.8 is used again with this new value of B, p_a will be 2 tsf showing that B = 2.25 ft is satisfactory.

 D. Footings on Rock: If the bedrock can be reached by excavation, the allowable load is liable to be set by local codes. UBC recommendations specify a safety factor of 5 based on the unconfined compressive strength, although local ordinances may supersede this value. For most rock beds, the design will be based on settlement characteristics, not strength.

4. Rafts

A raft (or 'mat') is a combined footing-slab that covers the entire area beneath a building and supports all walls and columns. A raft foundation should be used (at least for economic reasons) any time the individual footings would constitute half or more of the area beneath a building.

 A. Rafts on Clay: The net ultimate bearing pressure for rafts on clay can be found in the same manner as for footings (equation 10.3). Since the size of the raft is essentially fixed by the building size (plus or minus a few feet), the only method available to increase the loading is to lower the elevation (increase D_f) of the raft.

The factor of safety produced by a raft construction is given by equation 10.11, which may also be solved to give the required D_f if the factor of safety is known. The factor of safety should be at least 3 for normal loadings, but may be reduced to 2 during temporary extreme loading.

$$F = \frac{cN_c}{\dfrac{\text{total load}}{\text{raft area}} - \rho D_f} = \frac{p_a}{p_{actual} - \rho D_f} \qquad 10.11$$

If the denominator in equation 10.11 is small, the factor of safety is very large. If the denominator is zero, the raft is said to be a 'fully compensated foundation'. For D_f less than the fully-compensated depth, the raft is said to be 'partially compensated'.

Example 10.4

A raft foundation is to be designed for a 120' x 200' building with a total dead and live loading of 5.66 EE7 pounds. The clay density is 115 pounds/ft^3 and the clay has an average unconfined compressive strength of .3 tons/ft^2. (a) What should be the raft depth (D_f) for full compensation? (b) What should be raft depth for a factor of safety of 3? (c) What will be the factor of safety if the foundation depth is 10.0 feet?

The loading pressure is $(\frac{5.66\ EE7}{(120)(200)}) = 2.36\ EE3$ lb/ft^2

(a) For full compensation, $2.36\ EE3 = (D_f)(\rho)$

$$D_f = \frac{2.36\ EE3}{115} = 20.5 \text{ ft}$$

(b) Assume D_f = 10 feet. Then $D_f/B = 10/120 \approx .1$. From figure 10.3, N_c = 6.3. From equation 10.11,

$$3 = \frac{(\frac{1}{2})(.3)(2000)(6.3)}{2.36\ EE3 - (115)(D_f)} \quad \text{or } D_f = 15.0 \text{ feet}$$

Although D_f was assumed to be 10 feet, N_c does not significantly change.

(c) $F = \frac{(\frac{1}{2})(.3)(2000)(6.3)}{(2360)-(115)(10.0)} = 1.56$

◆

   B. Rafts on Sand: Rafts on sand are always well protected against bearing capacity failure. Therefore, settlement will govern the design. Since differential settlement will be much smaller for various locations on the raft (due to the raft's rigidity), the allowable soil pressure may be doubled. This increases the allowable settling to approximately 2 inches and reduces the maximum differential settling to approximately 3/4".

Figure 10.6 may be used if the allowable pressure scale is doubled. Alternatively, the allowable pressure may be found from equation 10.12. N should always be at least 5 after correcting for overburden. Otherwise, the sand should be compacted or a pier/pile foundation used.

$$p_a = .22N \quad (5 \leq N < 50) \quad \text{tons/ft}^2 \qquad 10.12$$

The correction for water table levels (equation 10.10) may be used with rafts on sand.

The net soil pressure is

$$p_{net} = \frac{\text{total load}}{\text{raft area}} - \rho D_f \qquad 10.13$$

It is the net soil pressure that should be compared with the allowable pressure.

5. Piers

A pier is a large underground member with a length (depth) greater than its width (diameter). It differs from a pile in its diameter, load carrying capacity, and installation method. A pier is usually constructed after its hole has been dug.

   A. Piers in Clay: If $D_f/B > 4$, then N_c is constant (see equation 10.6). The design of a pier foundation is similar in other respects to a footing design. A factor of safety of 3 is usually used.

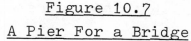

Figure 10.7

A Pier For a Bridge

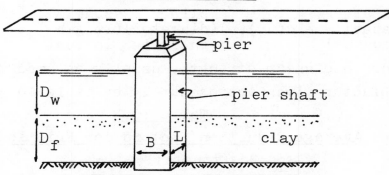

$$p_a = cN_c/F = N_c S_{nc}/2F \quad (S_{nc} \text{ is undrained}) \qquad 10.14$$

$$p_{net} = cN_c \qquad 10.15$$

$$p_{gross} = P_{net} + \rho_{clay}D_f + \rho_w D_w \qquad 10.16$$

Piers derive additional supporting strength from skin friction. This skin friction strength, S_{sf}, is between .3 and .5 (average of .45) times the undrained shear strength of the clay averaged over the pier length. (S_{sf} should not be greater than 1 ton/ft^2. If it is, use 1 ton/ft^2.)

The additional load that the pier can support is the skin friction times the surface area of the pier shaft. If the pier is belled, only the straight part of the pier is used to calculate the skin friction capacity.

If the skin friction is used to support any of the applied load, D_f should be taken as zero in finding N_c from figure 10.3.

B. Piers in Sand: A conservative estimate of safe bearing capacity can be found from the following equation:

$$p_{net} = \tfrac{1}{2}B\rho N_\gamma + \rho D_f(N_q - 1) \qquad 10.17$$

The allowable pressure is given by equation 10.18.

$$p_a = (.11)(C_w)N \qquad (\text{tons/ft}^2) \qquad 10.18$$

C_w is given by equation 10.10.

6. Piles

Piles are small-area members that are usually hammered or vibrated into place. They provide strength to soils that are too weak or compressible to otherwise support a foundation. Piles are often grouped together to provide the required strength to support a column or wall.

Two major pile classifications exist: friction and point-bearing piles. Friction piles derive their load-bearing ability from the friction between the soil and pile. Point-bearing piles derive their strength from the support of the soil near the point.

A. Friction Piles in Clay: The ultimate capacity of a friction pile is given by equation 10.19.

$$\begin{matrix} \text{single} \\ \text{capacity} \\ \text{(pounds)} \end{matrix} = \alpha(\,c\,)(\pi)\begin{matrix} \text{pile} \\ (\text{diameter}) \\ \text{in feet} \end{matrix}\begin{matrix} \text{sunk} \\ (\text{length}) \\ \text{in feet} \end{matrix} \qquad 10.19$$

c is the undrained shear strength, usually assumed to be $\frac{1}{2}S_{nc}$. α is a reduction factor given approximately by table 10.3

<u>Table 10.3</u>

<u>Average Reduction Factors for Friction Piles</u>

$S_{nc}=2c$ (tons/ft^2)	α
.5	.95
1.0	.83
1.5	.69
2.0	.57
2.5	.48
3.0	.44

The capacity of a pile group may be less than the product of the number of piles times the single pile capacity. Total strength of a pile group will be the sum of the total soil shearing resistance and the base capacity, if any. Since a pile group must fail as a group, the shearing capacity must be based on the circumference of the pile group and the pile length. That is,

$$\begin{matrix} \text{group} \\ \text{capacity} \end{matrix} = (c)(\text{perimeter})(\text{sunk length}) \qquad 10.20$$

<u>Figure 10.8</u>

<u>A Pile Group</u>

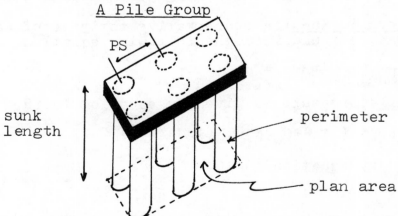

The base capacity can be found from equation 10.3 ($D_f/B = 0$) using the plan area and the soil characteristics at the tips of the piles.

$$\begin{matrix} \text{compressive} \\ \text{group} \\ \text{capacity} \end{matrix} = \begin{pmatrix} \text{plan} \\ \text{area} \end{pmatrix}(p_{net}) \qquad 10.21$$

The pile spacing (PS) should be chosen by trial and error to maximize the sum of these two group capacities. A good starting spacing is (PS) equal to three times the pile diameter.

7. Settlement of Foundations on Clay

The purpose of pre-construction settlement calculations is to determine the magnitude of expected settlement due to an increase in surface loading. However, this method may also be used to find the settlement due to change in any variable, such as a drop in the water table. The following steps constitute the procedure.

step 1: Find the original effective pressure, p_o, at the mid-height of the clay layer. The average effective pressure is the sum of the following items:

* For layers above the clay and above the water table: $p = \left(\begin{smallmatrix}\text{layer}\\\text{thickness}\end{smallmatrix}\right)\left(\begin{smallmatrix}\text{layer}\\\text{density}\end{smallmatrix}\right)$

* For layers above the clay but below the water table: $p = \left(\begin{smallmatrix}\text{layer}\\\text{thickness}\end{smallmatrix}\right)\left(\begin{smallmatrix}\text{layer}\\\text{density}\end{smallmatrix} - 62.4\right)$

    The layer density used for this component is the saturated density.

* For the clay above the water table: $p = \frac{1}{2}\left(\begin{smallmatrix}\text{clay}\\\text{thickness}\end{smallmatrix}\right)\left(\begin{smallmatrix}\text{clay}\\\text{density}\end{smallmatrix}\right)$

* For the clay below the water table: $p = \frac{1}{2}\left(\begin{smallmatrix}\text{clay}\\\text{thickness}\end{smallmatrix}\right)\left(\begin{smallmatrix}\text{clay}\\\text{density}\end{smallmatrix} - 62.4\right)$

    The layer density used for this component is the saturated density.

step 2: Find the increase in pressure, p_v, directly below the foundation and at the midpoint of the clay layer due to the building load. This can be done using Boussinesq's equation, equation 10.22, if the footing width is small compared to the distance z.

$$p_v = \frac{3(\text{load})}{2\pi(z^2)}\left[\frac{1}{1 + (r/z)^2}\right]^{5/2} \qquad 10.22$$

Figure 10.9
Application of Boussinesq's Equation

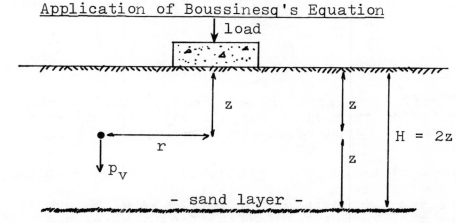

- sand layer -

If the footing is large compared to z, the vertical pressure should be found by use on an influence chart (figure 10.10). This chart is used in the following manner:

step a: Let the distance A-B on the chart correspond to the depth at which the pressure is wanted. Using this scale, draw a plan view of the footing on a piece of tracing paper.

step b: Place the tracing paper over the influence chart. Using the same scale, locate the footing tracing a distance r from the center of the chart. The footing width should be perpendicular to and bisected by one of the radial lines.

step c: Count the number of squares seen under the footing drawing. count partial squares as fractions. Count the pie-shaped areas in the center circle as squares.

step d: Find the pressure from equation 10.23. The units of the applied pressure will be the units of the pressure at point v.

$$\text{pressure} = (\# \text{ squares})(.005)(\text{applied pressure}) \qquad 10.23$$

step 3: Estimate $e_o \approx w(SG)$ $\qquad 10.24$

step 4: Find the compression index of the clay. This can be found from soils tests and equation 10.25. (Other books may use C_c as the symbol for compression index.)

$$I_c = .009(w_1 - 10) \qquad 10.25$$

step 5: Calculate the settlement. For any clay (normally loaded or pre-loaded) for which the original void ratio and change in void ratio is known, equation 10.26 should be used. H is the thickness of the clay layer, irregardless of any overlying layers or surcharges.

$$S* = H\left(\frac{\Delta e}{1+e_o}\right) \qquad 10.26$$

For normally loaded clays only, equation 10.27 can be used, although equation 10.26 also applies. p_v is the change in vertical pressure due to applied loads. Therefore, $p_o + p_v$ is the total pressure after load application or removal.

$$S* = \left(\frac{I_c}{1+e_o}\right)H \log_{10}\left(\frac{p_o+p_v}{p_o}\right) \qquad 10.27$$

Example 10.5

A 40'x 60' raft is constructed as shown in the illustration below. The building rests on sand which has already settled. What long-term settlement can be expected in the clay (a) at the center of the raft? (b) at a corner of the raft?

step 1: The silt pressure: $\quad$ (5)(90) $\qquad = 450 \text{ lb/ft}^2$
$\qquad$ The 'dry' sand layer: (14)(120) $\qquad = 1680$
$\qquad$ The 'wet' sand layer: (22)(120-62.4) $= 1270$
$\qquad$ The clay layer: $\quad \frac{1}{2}$(14)(110-62.4) $= \underline{\quad 330}$
$\qquad\qquad\qquad\qquad\qquad\qquad\qquad\qquad p_o = 3730 \text{ lb/ft}^2$

Figure 10.10
Influence Chart

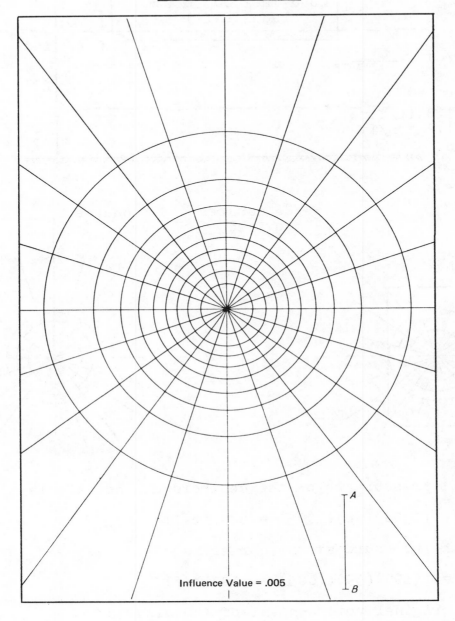

Influence Value = .005

step 2: Use the influence chart (figure 10.10) to calculate the pressure increase due to the building. The distance from the bottom of the raft to the mid-point of the clay layer is 36-3+7=40 feet. Using a scale of 1 inch = 40 feet, the raft is 1"x 1½'. The application of figure 10.10 is known on the next page.

The net load at the base of the raft is the applied live and dead loads minus the 'buoyant' force due to the excavated sand and silt.

$$p_{net} = 2400 - 5(90) - 3(120) = 1590 \text{ lb/ft}^2$$

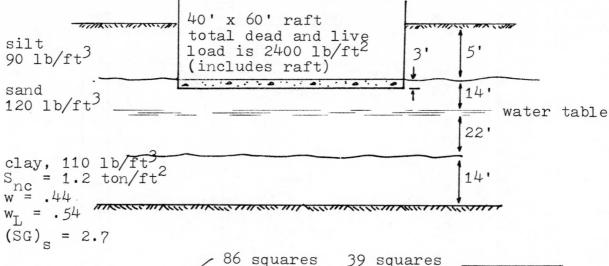

silt
90 lb/ft^3

40' x 60' raft
total dead and live
load is 2400 lb/ft^2
(includes raft)

sand
120 lb/ft^3

water table

clay, 110 lb/ft^3
S_{nc} = 1.2 ton/ft^2
w = .44
w_L = .54
$(SG)_s$ = 2.7

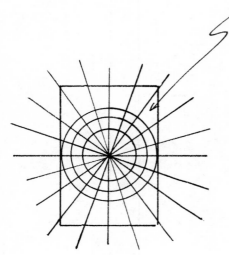

86 squares
covered

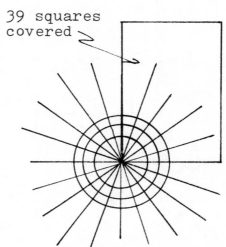

39 squares
covered

The mid-point pressure below the centroid of the raft is

$$p = (1590)(86)(.005) = 680 \text{ lb/ft}^2$$

Similarly, the pressure at the corner is

$$p = (1590)(39)(.005) = 310 \text{ lb/ft}^2$$

step 3: The original void content of the clay was

$$e_o = (.44)(2.7) = 1.188 \text{ (using equation 10.24)}$$

step 4: The compression index is

$$I_c = .009(54-10) = .396$$

step 5: The settlement is

$$S^*_{center} = \frac{.396}{1+1.188}(14) \log(\frac{3730+680}{3730}) = .184 \text{ ft (2.2 inches)}$$

$$S^*_{corner} = \frac{.396}{1+1.188}(14) \log(\frac{3730+310}{3730}) = .087 \text{ ft (1.05 inches)}$$

◆

Settling in clay is a continuous process. The time to reach a specific settlement is given by equation 10.28. z is the layer's half-thickness if drainage is through the top and bottom surfaces. If drainage is from one surface only, z is the layer's full thickness.

$$t = \frac{T_v z^2}{C_v} \qquad \text{Units of t will} \atop \text{depend on } C_v. \qquad 10.28$$

The coefficient of consolidation is assumed to remain constant over small variations in the void ratio, e.

$$C_v = \frac{k(1+e)}{62.4(a_v)} \qquad 10.29$$

a_v can be found from the void ratio and effective stress for any two different loadings.

$$a_v = \frac{e_1 - e_2}{p_2 - p_1} \qquad 10.30$$

T_v is a dimensionless number known as the 'time factor'. T_v depends on the degree of consolidation, U_z. U_z is the percent of the total consolidation (settlement) expected, S*. For U_z less than .60 (60%), T_v is given by equation 10.31.

$$T_v = \tfrac{1}{4}\pi U_z^2 \qquad 10.31$$

Table 10.4 should be used to find T_v for larger values of U_z.

Table 10.4
Approximate Time Factors

U_z	T_v
.65	.34
.70	.40
.75	.48
.80	.55
.85	.70
.90	.85
.95	1.3
1.0	∞

8. Excavation

A. Excavation in Clay: The stability of cuts in clay will depend on the clay shearing strength, depth of cut, and height above a firm stratum. If the clay is of soft-medium consistency, saturated, and is protected from drying out, the following analysis may be used to determine the factor of safety against shear failure.

step 1: Failure will occur along a circular shear plane, so draw a circle of arbitrary length and center, as shown in figure 10.11. If the slope β exceeds 53°, the circle should intersect the toe of the cut. If the slope is less than 53° and the excavation depth is less than one-third of the distance to the firm stratum, the circle is tangent to the firm stratum below and the center is above the midpoint of the slope.

Figure 10.11
Toe and Base Circle Failure Planes

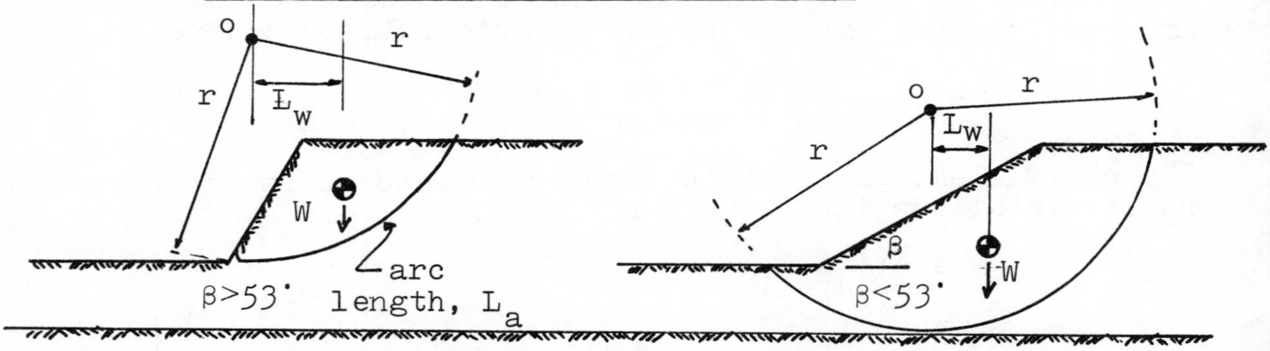

step 2: Determine the weight of clay within the circular area. This may be done graphically or analytically.

step 3: Locate the centroid of the circular area and the length of the moment arm, L_w.

step 4: Calculate the moment about the center of the circle, $M=L_w W$.

step 5: Determine the arc length, L_a.

step 6: The factor of safety is

$$F = \frac{crL_a}{M}$$

10.32

step 7: Repeat from step 1 to find the minimum F over all arbitrarily-chosen centers.

B. Excavations in Sand: If the sand is above the water table, the slopes will be stable if they are kept to 1:1.5 (vertical:horizontal). The water table must be lowered if excavation is required below its original depth. The water table level must be kept below the excavation floor at all times.

9. Retaining Walls

Retaining walls must be safe against settlement. In this regard, their design is similar to footings. They must also have sufficient resistance against overturning and sliding. Of course, the retaining wall must possess adequate structural strength. The method of meeting these requirements is one of trial and error.

The following design characteristics are rules of thumb. They should be used as a starting place for more analytical methods.

base: $.40 < (B/H) < .65$

$\frac{1}{12} < (d/H) < \frac{1}{8}$

For toes projecting (B/3), the toe extension should be one half of the heel extension. That is, $B_1 = \frac{1}{2}B_2$.

stem: Thickness decreases $\frac{1}{4}$" to 3/4" per foot. 10" minimum thickness at the top. Thickness at the bottom equals d.

Figure 10.12
A Cantilever Retaining Wall

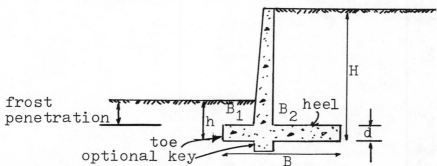

The analysis of a retaining wall's stability requires knowledge of at least six different distributed force systems. These six forces are illustrated in figure 10.13. They are: the forward ('active' or 'tensioned') earth reaction, R_A; the backward ('passive' or 'compressed') earth reaction, R_P; the soil force, R_V; the shear resistance, R_S; the weights of the earth masses, R_T and R_H; and the weight of the wall itself.

Figure 10.13
Pressure Distributions on a Retaining Wall

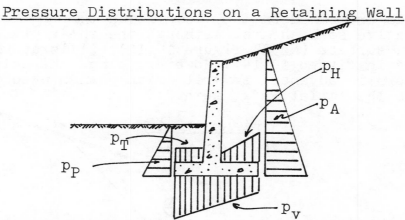

The following steps may be used to design a retaining wall.

step 1: Tentatively design the retaining wall using the guidelines given for the base and stem.

step 2: Determine the active reaction. If it is assumed that the back-fill soil is dry, cohesionless sand having a shearing resistance equal to $\tau = \sigma(\tan\phi)$, then the 'Rankine theory' may be used. At any depth, H, the vertical pressure is

$$p_{vertical} = \rho H \qquad 10.33$$

The horizontal pressure depends on the 'coefficient of earth pressure at rest', k_o, which varies from .4 to .5 for untamped sand.

$$p_{horizontal} = k_o \rho H \qquad 10.34$$

Equations 10.33 and 10.34 apply only to a sand deposit of infinite depth and extent. For sand that is compressed or tensioned (as in around a retaining wall) the pressures are given by equations 10.35 through 10.38.

$$R_A = \tfrac{1}{2}k_A\rho H^2 \qquad\qquad 10.35$$

$$k_A = \frac{1 - \sin\phi}{1 + \sin\phi} \qquad\qquad 10.36$$

$$R_P = \tfrac{1}{2}k_P\rho H^2 \qquad\qquad 10.37$$

$$k_P = 1/k_A \qquad\qquad 10.38$$

k_A and k_P are the coefficient of active and passive earth pressures respectively. R_A is horizontal if the soil above the heel is horizontal.

The Rankine theory is based on infinite, cohesionless soil. It also requires that the soil above the heel be level (not sloping as is shown in figure 10.13.) Modifications may be made to lift these restrictions, as well as to allow a water table above the foundation base. Such modifications are known as 'wedge theories'. 'Coulomb's earth-pressure theory' is one such wedge theory.

The wedge methods are based on the observation that retaining walls fail when the active soil shears. Although the shear plane is actually a slightly curved surface (as in figure 10.11), it is assumed to be linear (line *-* in figure 10.14). However, since the actual shear plane is not known in advance, several trial planes need to be taken. This is known as the 'trial wedge method'.

Figure 10.14

Failure Wedge

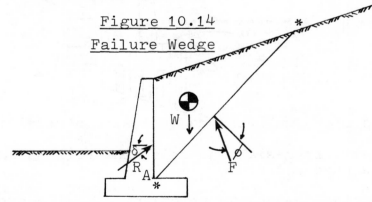

step a: Assume that the resultant active earth pressure acts at an angle δ (see figure 10.14) from the normal to the active retaining wall face. For cantilever retaining walls, choose a direction parallel to the backfill surface. For heavy concrete (gravity) retaining walls without bases (where the failure will be by separation of the wall and fill) choose δ as the angle of friction between the wall and backfill. For concrete walls, this may be approximated by δ= .67ϕ, or table 10.4 may be used.

Table 10.4 (3:7-10-7)
Friction Angles

INTERFACE MATERIALS*	FRICTION ANGLE, δ, DEGREES
Concrete or masonry on the following foundation materials:	
Clean, sound rock	35
Clean gravel, gravel-sand mixtures, and coarse sand	29 - 31
Clean fine to medium sand, silty medium to coarse sand, and silty or clayey gravel	24 - 29
Clean fine sand, and silty or clayey fine to medium sand	19 - 24
Fine sandy silt, and non-plastic silt	17 - 19
Very stiff clay, and hard residual or preconsolidated clay	22 - 26
Medium stiff clay, stiff clay, and silty clay	17 - 19
Steel sheet piles against the following soils:	
Clean gravel, gravel-sand mixtures, and well-graded rock fill with spalls	22
Clean sand, silty sand-gravel mixtures, and single-size hard rock fill	17
Silty sand, gravel or sand mixed with silt or clay	14
Fine sandy silt, and non-plastic silt	11
Formed concrete or concrete sheet piling against the following soils:	
Clean gravel, gravel-sand mixtures, and well-graded rock fill with spalls	22 - 26
Clean sand, silty sand-gravel mixtures, and single-size hard rock fill	17 - 22
Silty sand, and gravel or sand mixed with silt or clay	17
Fine sandy silt, and non-plastic silt	14
Miscellaneous combinations of structural materials:	
Masonry on masonry, igneous and metamorphic rocks:	
Dressed solt rock on dressed soft rock	35
Dressed hard rock on dressed soft rock	33
Dressed hard rock on dressed hard rock	29
Masonry on wood (cross grain)	26
Steel on steel at sheet-steel interlocks	17

*NOTE: Angles given are ultimate values. Sufficient movement is required before failure will occur.

step b: Draw trial wedges, such as *-1, *-2, *-3, etc. behind the wall.

step c: Determine the weight of the soil within all wedges.

Figure 10.15
Trial Wedges

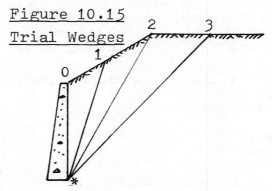

step d: For each wedge, draw a force polygon. The soil weight acts vertically downward; the active force resultant acts at the chosen angle δ; and the resultant of the frictional and normal forces acts at an angle ϕ (angle of internal friction) from the normal to the failure plane.

step e: Combine all wedges and draw the 'earth pressure locus'.

Figure 10.16
Earth Pressure Locus

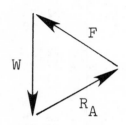

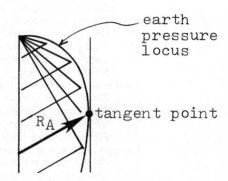

earth pressure locus

tangent point

step f: Draw a vertical tangent to the earth pressure locus. The actual earth pressure is the line passing through the tangent point, parallel to the R_A lines. This also defines the plane of rupture.

step g: If the backfill is plane (horizontal or sloping) and carries no surcharge, the resultant R_A will act H/3 up from the bottom of the base.

If there is a uniform surcharge of q lb/ft^2, two acceptable methods yield slightly different results. The surcharge can be assumed to contribute $R_q = k_A q H$ acting at $\frac{1}{2}H$ above the base (in addition to the regular active load of $R_A = \frac{1}{2}k_A \rho H^2$ acting at H/3.) Alternatively, the surcharge can be converted to an equivalent height of soil, where $H_{eq} = q/\rho$. Then, the total horizontal force on the wall would be $R_A = \frac{1}{2}k_A \rho (H+H_{eq})^2$ acting at $(H+H_{eq})/3$.

For irregularly shaped backfills, draw a line parallel to the failure plane and passing through the centroid of the earth triangle. The intersection of the line and the stem is the application point for R_A.

The wedge theory has been modified to include effects of concentrated and line loads, and to take into account a cohesive soil behaving as $\tau = c + \sigma \tan\phi$. However, since the backfill material is rarely known in detail prior to the design of the retaining wall, charts are ordinarily used to calculate the earth pressure. Charts are applicable as long as H is less than 20 feet. Typical design charts are shown in figures 10.26 and 10.27 which gives the horizontal and vertical parts of R_A.

step 3: Find the passive force, R_p. The passive force can also be found by wedge methods or design charts. However, the passive pressure is usually disregarded on the assumption that the backfill will be in place prior to the front fill, or that the front fill will be removed at some future date for repairs. If necessary, the passive reaction may be found from equation 10.37 or 10.39.

$$R_p = \frac{1}{2}k_p \rho h^2 \qquad \text{(see figure 10.12)} \qquad 10.39$$

step 4: Find the vertical forces against the base. These forces are the weights, W_i, of the areas shown in figure 10.17. Also, find the centroid of each area and the moment arm, r_i, from the centroid of the ith area to point G.

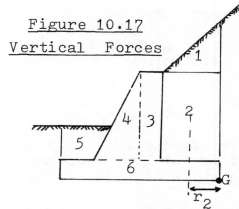

Figure 10.17

Vertical Forces

step 5: Find the moment about point G of all the vertical forces and the active pressure.

$$M_G = \Sigma W_i r_i + R_{Ah} r_A \qquad\qquad 10.40$$

step 6: Find the application point and eccentricity of the vertical force resultant. The eccentricity is the distance from the center of the base. It should be less than (B/6) for the entire base to be in compression.

$$r = \frac{M_G}{\Sigma W_i + R_{Av}} \qquad\qquad 10.41$$

$$e^* = r - \tfrac{1}{2}B \qquad\qquad 10.42$$

step 7: Find the maximum (at the toe) and minimum (at the heel) foundation pressure on the base.

$$p_{v,\text{max \& min}} = \frac{\Sigma W_i + R_{Av}}{B}\left(1 \pm \frac{6e^*}{B}\right) \qquad\qquad 10.43$$

The maximum pressure should not exceed the allowable soil pressure.

step 8: Check the resistance against sliding. Disregarding the passive pressure, the active pressure must be resisted by the shearing strength of the soil or the friction between the base and the soil. Equation 10.44(a) is for use when the wall has a key, and then only for the soil to the left of the key. Equation 10.44(b) is for use with the soil to the right of a key, and for flat-bottomed walls.

$$R_s = (\Sigma W_i + R_{Av})\tan\phi \qquad\qquad 10.44(a)$$
$$ = (\Sigma W_i + R_{Av})\tan\delta \qquad\qquad 10.44(b)$$

In the absense of friction coefficient data, the resistance may be found from equation 10.45 and table 10.5.

$$R_s = k_s(\Sigma W_i + R_{Av}) \qquad\qquad 10.45$$

<u>Table 10.5</u>

<u>Values of k_s</u>

coarse grained soil	
without silt	.55
coarse grained soil	
with silt	.45
silt	.35

<u>step 9</u>: Calculate the factor of safety against sliding. It should be at least 1.5. If it is too low, increase the base length, B, or add a downward projecting spike ('key').

$$F_s = R_s/R_{Ah} \qquad\qquad 10.46$$

Neglecting R_p will give a minimum F_s. If R_p is known or calculated, it can be used to find the maximum F_s.

<u>Example 10.6</u>

A retaining wall is being designed for a soil with $\phi = 30°$, $\delta = 17°$, and a maximum allowable pressure of 3000 psf. The backfill is coarse-grained sand with silt, with a density of 125 lb/ft³. Check the tentative design shown below for stability against sliding.

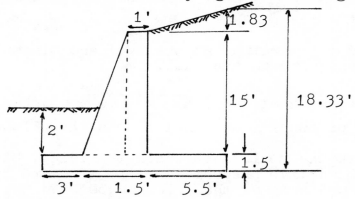

<u>step 1</u>: The retaining wall has be designed.

<u>step 2</u>: Since the backfill is sloped, the Rankine method cannot be used. Use the design chart method with figure 10.26 and assume a type-2 fill.

$$\beta = \arctan(\frac{1.83}{5.5}) = 18.4° \qquad (3:1)$$

From figure 10.26, $k_v \approx 10$ and $k_h \approx 40$. Notice that both k_v and k_h consider the density and geometry of the fill.

$$R_{Av} = \tfrac{1}{2}(10)(18.33)^2 = 1680 \text{ pounds}$$

$$R_{Ah} = \tfrac{1}{2}(40)(18.33)^2 = 6720 \text{ pounds}$$

step 3: Assume $R_P = 0$.

step 4:

i	Area	ρ	W_i(lb)	r_i(ft)	Moment
1	$\tfrac{1}{2}(5.5)(1.83)=5.03$	125	629	1.83	1151
2	$5.5(15)=82.5$	125	10313	2.75	28361
3	$1(15)=15$	150	2250	6.0	13500
4	$\tfrac{1}{2}(.5)(15)=3.75$	150	563	6.67	3755
6	$1.5(10)=15$	150	2250	5.0	11250
5	$2(3)=6$	125	750	8.5	6375
			16755		64393 Totals
R_{Av}			1680	0	0
R_{Ah}			6720	6.11	41060

step 5: $M_G = 64392 + 41060 = 105,450$

step 6: $r = \dfrac{105,450}{16755 + 1680} = 5.72$

$$e^* = 5.72 - \tfrac{1}{2}(10) = .72$$

Since .72 is less than (10/6), the base is in compression everywhere.

step 7: The maximum pressure at the base is

$$p_{max} = \frac{(16755+1680)}{10}(1 + \frac{6(.72)}{10}) = 2640$$

This is less than 3000 lb/ft^2.

step 8: The resisting force against sliding is

$$R_s = (16755 + 1680)\tan17° = 5636$$

$$F_{s,min} = \frac{5636}{6720} = .84 \text{ (no good - use a key)}$$

◆

10. Flexible Bulkheads

A. Anchored Bulkheads: Anchored bulkheads are supported at their bases by virtue of having been sunk into the ground. They are anchored further up with rods projecting back into the soil. These rods may terminate at deadmen, piles, walls, or beams. Backfill is limited to cohesionless soil (usually sand) to minimize the horizontal pressures. Soil below the mud line may be either sand or clay, however.

Figure 10.18

Anchored Bulkheads

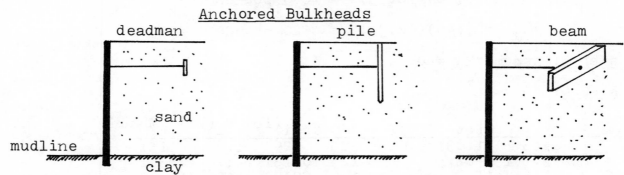

Anchored bulkheads may fail in the following manners:

1. The base clay may fail due to inadequate bearing capacity. The clay will shear along a circular arc passing under the bulkhead. The method of slices may be used to investigate this.

2. The anchorage may fail. There are several ways for this to occur, including rod tension failure and deadman movement.

3. The toe embedment at the mud line may fail.

4. The sheeting may fail (rare).

Figure 10.19

Loads on Sheeting

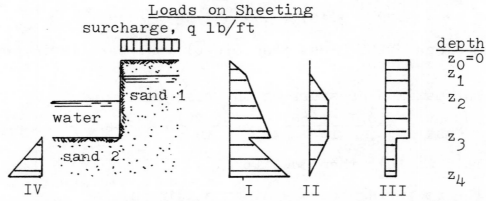

Figure 10.19 shows a bulkhead with sand above and below the mudline. These sands do not need to be the same, and they may have different strength characteristics. Force distribution I is the active earth pressure as calculated from equation 10.36 and equation 10.47.

$$p_A = k_A \rho z \qquad\qquad 10.47$$

$$p_{z0} = 0 \qquad\qquad 10.48$$

$$p_{z1} = k_{A1} \rho_1 z_1 \qquad\qquad 10.49$$

$$p_{z3\,(max)} = p_{z1} + k_{A1} \rho_1' (z_3 - z_1) \qquad\qquad 10.50$$

$$p_{z3\,(min)} = p_{z1} + k_{A2} \rho_1' (z_3 - z_1) \qquad\qquad 10.51$$

$$p_{z4} = p_{z3(min)} + k_{A2} \rho_2' (z_4 - z_3) \qquad\qquad 10.52$$

Notice that $\rho' < \rho$ due to the buoyant effect.

$$\rho' = \rho - 62.4 \qquad\qquad 10.53$$

The pressure distribution II is due to the unbalanced water pressure from the two water levels at z_1 and z_2. Since the sheet ends at level z_4, the pressure difference must be zero at that point if the sand is permeable.

$$p_{z1} = 0 \qquad\qquad 10.54$$

$$p_{z2} = (62.4)(z_2 - z_1) \qquad\qquad 10.55$$

$$p_{z3} = p_{z2} \qquad\qquad 10.56$$

$$p_{z4} = 0 \qquad\qquad 10.57$$

Pressure distribution III is due to any uniformly distributed surcharge of q pounds/foot that might exist on the earth-side surface.

$$p_{z0} = k_{A1} q \qquad\qquad 10.58$$

$$p_{z4} = k_{A2} q \qquad\qquad 10.59$$

Pressure distribution IV is the passive distribution. Its magnitude is reduced by a factor of safety, F.

$$p_{z3} = 0 \qquad\qquad 10.60$$

$$p_{z4} = k_{P2} \rho_2' (z_4 - z_3)/F \qquad\qquad 10.61$$

Figure 10.20

Loads on Sheeting

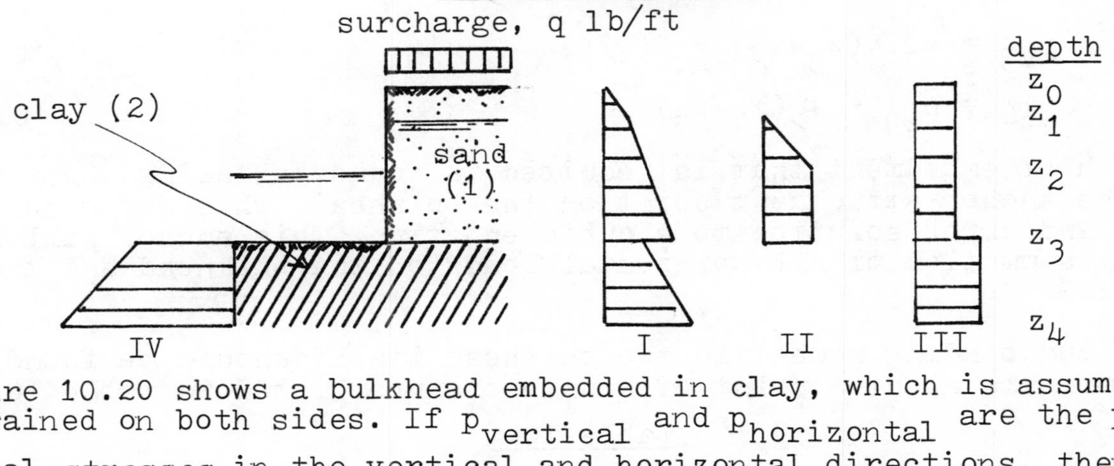

Figure 10.20 shows a bulkhead embedded in clay, which is assumed to be undrained on both sides. If $p_{vertical}$ and $p_{horizontal}$ are the principal normal stresses in the vertical and horizontal directions, then the shear stress for the saturated ($\phi = 0$) case is given by equation 10.62.

$$\tau = c = \tfrac{1}{2}(p_{vertical} - p_{horizontal}) \qquad\qquad 10.62$$

Upon rearranging, the horizontal stress is given by equation 10.63.

$$p_{horizontal} = p_{vertical} - 2c \qquad 10.63$$

For distribution I:

$$p_{z0} = 0 \qquad 10.64$$

$$p_{z1} = k_{A1} \rho_1 z_1 \qquad 10.65$$

$$p_{z3,sand} = p_{z1} + k_{A1} \rho_1'(z_3 - z_1) \qquad 10.66$$

$$p_{z3,clay} = \rho_1 z_1 + \rho_1'(z_3 - z_1) + 62.4(z_2 - z_1) - 2c \qquad 10.67$$

$$p_{z4} = p_{z3,clay} + \rho_2(z_4 - z_3) \qquad 10.68$$

If depth z_4 is unknown, substitute $D = (z_4 - z_3)$ in these equations and carry it along.

For distribution II:

$$p_{z1} = 0 \qquad 10.69$$

$$p_{z2} = 62.4(z_2 - z_1) \qquad 10.70$$

For distribution III:

$$p_{z0} = k_{A1} q \qquad 10.71$$

$$p_{z3,sand} = k_{A1} q \qquad 10.72$$

$$p_{z3,clay} = q \qquad 10.73$$

$$p_{z4} = q \qquad 10.74$$

For distribution IV:

$$p_{z3} = 62.4(z_3 - z_2) + 2(c/F) \qquad 10.75$$

$$p_{z4} = p_{z3} + \rho_2(z_4 - z_3) \qquad 10.76$$

The depth of embedment that is required is found by taking moments about the anchor attachment point on the bulkhead. This may require a trial and error solution to a cubic equation. This anchor pull is found by summation of all horizontal loads on the bulkhead and then adding 20%.

The maximum bending moment in the bulkhead itself should be found by taking moments about a point of counterflexure listed in table 10.6.

Table 10.6

Points of Counterflexure

Embedment material	point
firm and dense	mud line
loose and weak	1 or 2 feet below mud line
soft over hard layer	at hard layer depth

B. Braced Cuts: A braced cut is an excavation in which the load from one bulkhead is used to support the opposite bulkhead's load. Failure in dry soils above the water line generally occurs by wale crippling following by strut buckling. Planned excavations below the water line should be dewatered prior to cutting. The analysis of braced cuts is very heuristic due to extensive bending of the sheeting.

Figure 10.21

A Braced Cut

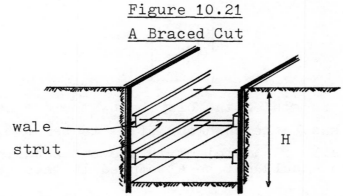

step 1: Tentatively decide on the number of struts and their locations.

step 2: Determine the apparent pressure envelope for the cut material. For dry, moist, and drained sand, the pressure distribution looks like figure 10.22.

$$p_{max} = (.65)\rho(H)\tan^2(45° - \tfrac{1}{2}\phi) \qquad\qquad 10.77$$

Figure 10.22

For undrained clay (typical of cuts made rapidly in comparison to drainage times) where $\phi = 0$, the pressure envelope depends on the undrained shearing strength of the clay in the cut zone. If $(\rho H/c)$ is less than or equal to 4, the pressure envelope is shown by figure 10.23.

Figure 10.23

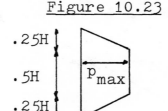

$$p_{max} \approx .3\rho H \qquad\qquad 10.78$$

If $4 < (\rho H/c) < 12$, equation 10.79 and figure 10.24 should be used.

$$p_{max} = H - 4c \qquad\qquad 10.79$$

Figure 10.24

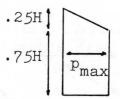

If the shearing strength of the clay below the cut is known (S_b), the quantity ($\rho H/S_b$) should be checked. If it is below 6, the bearing capacity of the soil is sufficient to prevent shearing and upward heave. Simple braced cuts should not be attempted if this quantity exceeds 8.

step 3: Bisect the vertical distance (and pressure distribution) between struts. Determine the area of sheeting supported by each strut based on the horizontal strut spacing and the distance between vertical bisections. Strut loads are calculated by multiplying the tributary strut areas by the pressure from the divided pressure distribution. Referring to figure 10.25, the loads carried by the top and bottom struts are given by equations 10.80 and 10.81 respectively.

$$R_{top} = A_{top} p_{max} H_{top} \qquad\qquad 10.80$$

$$R_{bottom} = A_{bottom} p_{max} H_{bottom} \qquad\qquad 10.81$$

Equations 10.80 and 10.81 must be modified slightly if the pressure distribution is non-uniform (as in figures 10.23 and 10.24).

Figure 10.25
Tributary Strut Areas

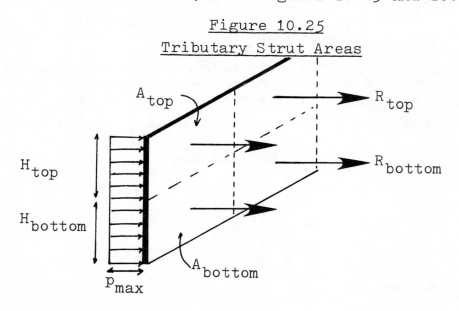

11. Provisions of the Uniform Building Code

The commentary below is abridged from the 1979 Uniform Building Code. This commentary is not complete, but includes only the most important concepts.

2308(b)   Retaining walls for drained earth should be designed for a minimum density of 30 lb/ft^3 of fill.

2903(a)   Permanent excavations shall have slopes less than 2 horizontal to 1 vertical unless approved otherwise.

2903(b)   For excavations less than 12 feet deep, soil from properties adjoining the excavation site should be protected against cave-in and settling. For excavations more than 12 feet deep, foundations on adjoining properties should be extended below the excavation level.

2904(a)   Expansion tendency in a soil shall be classified according to the following table (UBC table 29-C).

Expansion Index	Potential Expansion
0-20	Very Low
21-50	Low
51-90	Medium
91-130	High
above 130	Very High

2904(b)   If the expansion index is 20 or more, special design methods must be used.

2906   Unless higher pressures are substantiated, the following earth pressures are the maximum allowed (UBC table 29-B)

Type of Soil	Allowable Pressure (lb/ft^2)
Massive Crystalline Bedrock	4000
Sedimentary and Foliated Rock	2000
Sandy Gravel and/or Gravel (GW and GP)	2000
Sand, Silty Sand, Clayey Sand, Silty Gravel, and Clayey Gravel (SW, SP, SM, SC, GM, GC)	1500
Clay, Sandy Clay, Silty Clay, and Clayey Silt (CL, ML, MH, and CH)	1000

2907 (a)   Footings shall be constructed of masonry, concrete, or treated wood. All footings shall extend below the frost line. Footings supporting wood shall extend at least 6" above the finished grade. Minimum dimensions (some exceptions) are as given in

UBC table 29-A.

Number of Stories	Width of Footing	Thickness of Footing	Depth of Footing Below Finished Grade
1	12" min.	6" min.	12" min.
2	15	7	18
3	18	8	24

2907 (c) If the ground slopes more than 1 foot in 10 feet, the footings cannot be placed parallel to the ground, but must be placed truly horizontal.

2907 (d) Footing design shall be based on differential settlement.

Bibliography

1. International Conference of Building Officials, Uniform Building Code, 1979 edition, Whittier, CA 90601

2. Merritt, Frederick S., Standard Handbook for Civil Engineers, 2nd ed., McGraw Hill Book Company, Inc., New York, NY 1976

3. Naval Facilities Engineering Command, Design Manual NAVFAC DM-7, Soil Mechanics, Foundations, and Earth Structures, U.S. Government Printing Office, Washington, D.C. 1971

4. Peck, Ralph B., Hanson, Walter E., and Thornburn, Thomas N., Foundation Engineering, 2nd ed., John Wiley and Sons, Inc., New York, NY 1974

Figure 10.26 (3:7-10-14)

Active Components for Low Retaining Walls (Straight Slope Backfill)

$$R_A = \sqrt{R_h^2 + R_v^2}$$

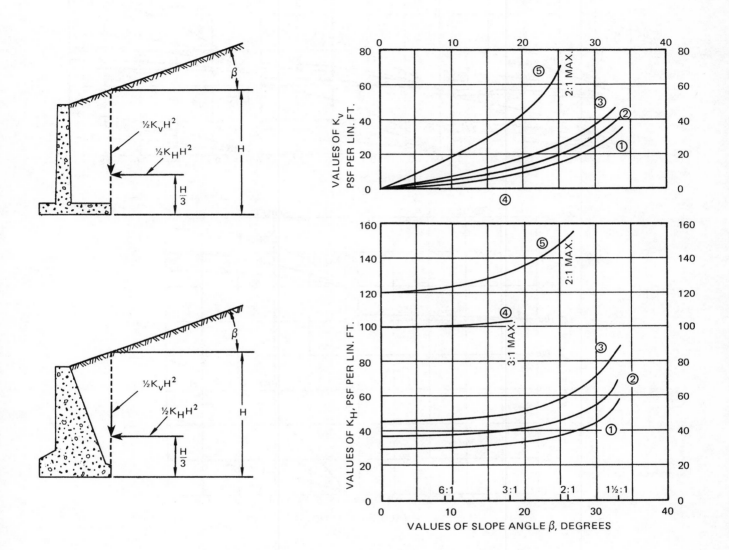

Circled numbers indicate the following soil types. (See bottom of page 9-11.)
  1. Clean sand and gravel: GW, GP, SW, SP.
  2. Dirty sand and gravel of restricted permeability: GM, GM-GP, SM, SM-SP.
  3. Stiff residual silts and clays, silty fine sands, and clayey sands and gravels:
        CL, ML, CH, MH, SM, SC, GC.
  4. Very soft to soft clay, silty clay, organic silt and clay: CL, ML, OL, CH, MH, OH.
  5. Medium to stiff clay deposited in chunks and protected from infiltration: CL, CH.

For type 5 material, H is reduced by 4 feet. The resultant is assumed to act (H-4)/3
above the bottom of the base.

Figure 10.27 (3:7-10-15)

Active Components for Low Retaining Walls (Broken Slope Backfill)

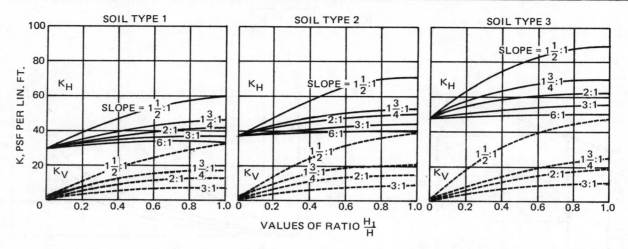

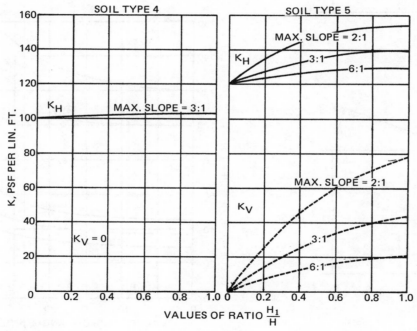

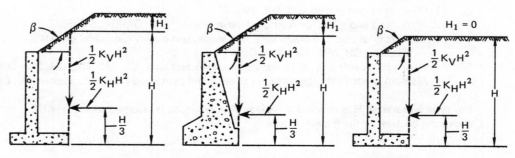

FOR TYPE 5 MATERIAL H IS REDUCED BY 4 FT., RESULTANT ACTS

AT A HEIGHT OF $\frac{(H-4)}{3}$ ABOVE BASE.

Practice Problems: FOUNDATIONS AND RETAINING WALLS

Required:

1. A large manufacturing plant with a floor load of 500 psf is supported on a large mat foundation. A 38 foot thick layer of silty soil is underlain by sand and gravel; and, the plant was constructed on the silty soil 10 feet above the water table. After a number of years (after all settlement of the building had stopped) a series of wells were drilled which dropped the water table from 10 feet below the surface to 18 feet below the surface. The unit weight of the soil above the water table is 100 pcf; it is 120 pcf below the water table. The silt's compression index is .02. The void ratio before the water table was lowered was .6. How much can the building be expected to settle?

2. 360 kips are to be supported by a square footing. The footing is to rest directly on sand which has a density of 121 pcf and an angle of internal friction of 38°. (a) Size the footing. (b) Size the footing if it is to be placed 4 feet below the surface.

3. A reinforced concrete retaining wall is to be 14 feet high. The top surface is horizontal but supports a surcharge of 500 psf. The soil has the following characteristics:

> density: 130 pcf (sandy soil)
> angle of internal friction: 35°
> coefficient of friction with concrete: .5
> Allowable soil pressure: 4500 psf
> frost line: 4 feet below grade

Use a factor of safety of 1.5 against sliding and overturning to design the dimensions of the retaining wall. Neglect structural details.

Optional

4. A 26 foot high wall holds back sand with a 96 pcf dry density. The water table is permanently 10 feet below the top of the wall. The saturated density of the sand is 121 pcf. The angle of internal friction is estimated to be 36°. (a) Disregarding capillary rise, calculate the active earth pressure. (b) Where is the resultant active earth pressure located? (c) Assuming the water table level could be dropped 16 feet to the bottom of the wall, what would be the reduction in overturning moment?

5. The compression index of a soil is .31. When the effective stress on the soil is 2600 psf, the void ratio is 1.04 and the permeability is 4 EE-7 mm/sec. The stress is increased gradually to 3900 psf. (a) What is the change in the void ratio? (b) Compute the settlement of a 16 foot layer. (c) How much time is required for the settlement to be 75% complete?

6. A 30 foot deep excavation is being planned for sand (ϕ = 40°, density of 121 pcf) with a drained water table during construction. The excavation will be 40 feet square. The bracing is to consist of horizontal lagging supported by 8-inch H-beam soldier piles. (a) What is the pressure diagram? (b) If the soldier piles are 8 feet apart,

what bending moment is placed on the lagging?

7. A 2:1 (horizontal:vertical) slope is cut in homogeneous, saturated clay which has a density of 112 pcf and a shear strength of 1.1 EE3 psf. The cut is 43 feet deep, and the clay extends 15 feet below the cut bottom to a rock layer. Compute the safety of this slope.

8. What is the increase in stress at a depth of 10 feet and 8 feet from the center of a 10 foot square footing that exerts a pressure of 3000 psf on the soil?

9. A concentrated vertical load of 6000 pounds is applied at the ground surface. What is the vertical pressure caused by this load at a point 3.5 feet below the surface and 4 feet from the action line of the force?

10. A 10.75" O.D. steel pile is driven 65 feet into a stiff, insensitive clay with a shear strength of 1.3 EE3 psf. The pile has a flat-plate, closed end. The soil density is 115 pcf, and the water table is at the ground surface. What is the ultimate bearing capacity?

NOTES

NOTES

STATICS: DETERMINATE

Nomenclature

a	horizontal distance from point of maximum sag	ft
A	area	ft²
c	parameter in catenary equations	ft
d	distance	ft
F	vertical force	pounds
H	horizontal component of tension	pounds
I	moment of inertia	ft⁴
J	polar moment of inertia	ft⁴
k	radius of gyration	ft
L	cable length from point of maximum sag	ft
M	moment	ft-lb
p	pressure	lb/ft²
R	reaction	pounds
S	maximum sag of cable	ft
T	tension	pounds
w	load per unit weight	lb/ft

Subscripts

c	centroidal
i	the ith component
R	resultant

1. Concentrated Forces and Moments

Forces are vector quantities having magnitude, direction, and location in 3-dimensional space. The direction of a force F is given by its direction cosines, which are cosines of the true angles made by the force vector with the x, y, and z axes. The components of the force are given by equations 11.1, 11.2, and 11.3.

$$F_x = F(\cos\theta_x) \qquad\qquad 11.1$$

$$F_y = F(\cos\theta_y) \qquad\qquad 11.2$$

$$F_z = F(\cos\theta_z) \qquad\qquad 11.3$$

Figure 11.1

Components of a Force F

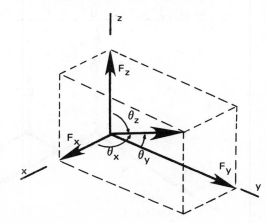

Moments can also be treated as vector quantities, and they are shown
as double-headed arrows for this purpose. Using the right-hand rule as
shown below, the direction cosines are again used to give the x, y, and
z components of a moment vector.

$$M_x = M(\cos\theta_x) \qquad\qquad 11.4$$

$$M_y = M(\cos\theta_y) \qquad\qquad 11.5$$

$$M_z = M(\cos\theta_z) \qquad\qquad 11.6$$

$$M = \sqrt{M_x^2 + M_y^2 + M_z^2} \qquad\qquad 11.7$$

<u>Figure 11.2</u>

<u>Components of a Moment M</u>

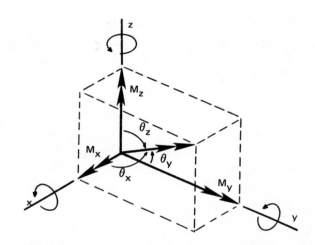

Moment vectors have the properties of magnitude and direction, but not
location (point of application). Moment vectors may be moved from one
location to another without affecting the equilibrium of solid bodies.

If a force is not parallel to an axis, it produces a moment around that
axis. The moment is evaluated by finding the components of the force
and their respective distances to the axis. In figure 11.3, a force
acts through point A (but not through the origin) and produces
moments given by equations 11.8, 11.9, and 11.10.

<u>Figure 11.3</u>

<u>Coordinates of a Point A</u>

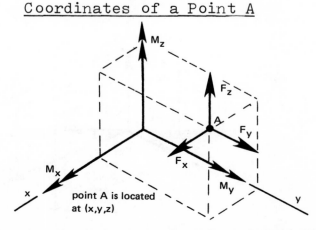

$$M_x = yF_z - zF_y \qquad \text{11.8}$$
$$M_y = zF_x - xF_z \qquad \text{11.9}$$
$$M_z = xF_y - yF_x \qquad \text{11.10}$$

Any two equal, opposite, and parallel forces constitute a couple. A couple is statically equivalent to a single moment vector. In figure 11.4, the two forces F_1 and F_2 of equal magnitude produce a moment vector M_z of magnitude Fy. The two forces can be replaced by this moment vector, which can then be moved to any location on the object.

Figure 11.4

A Couple

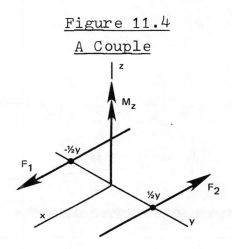

2. Distributed Loads

If a structural member is loaded by its own weight or by another type of continuous loading, it is said to be subjected to a distributed load. Provided that the load, w, is acting in the same direction everywhere, the statically equivalent concentrated load can be found from equation 11.11 by integrating over the line of application.

$$F_R = \int w\ dx \qquad \text{11.11}$$

The location of the resultant is given by equation 11.12.

$$\bar{x} = \frac{\int (wx)\,dx}{F_R} \qquad \text{11.12}$$

Figure 11.5

A Distributed Load and Resultant

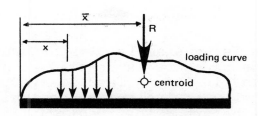

In the case of a straight beam under transverse loading, the magnitude of F equals the area under the loading curve. The location of the resultant F coincides with the centroid of that area. If the distributed load is uniform so that w is constant along the beam, then

$$F = wL \qquad\qquad 11.13$$

$$\overline{x} = \tfrac{1}{2}L \qquad\qquad 11.14$$

If the distribution is triangular and increases as x increases, then

$$F = \tfrac{1}{2}wL \qquad\qquad 11.15$$

$$\overline{x} = \tfrac{2}{3}L \qquad\qquad 11.16$$

Example 11.1

Find the magnitude and location of the resultant of the distributed loads on each span of the beam shown below.

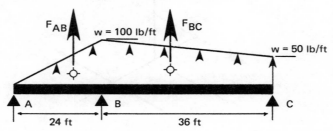

For span A-B: The area under the loading curve is $\tfrac{1}{2}(100)(24)$ = 1200. The centroid of the loading triangle is $(2/3)(24)$ = 16 feet from point A. Therefore, the triangular load on the span A-B can be replaced (for the purposes of statics) with a concentrated load of 1200 pounds located 16 feet from the left end.

For span B-C: The area under the loading curve is

$$(50)(36) + \tfrac{1}{2}(50)(36) = 2700 \text{ pounds}$$

The centroid of the trapezoid is

$$\frac{36[(2)(100) + 50]}{3(100+50)} = 20 \text{ feet from point C}$$

Therefore, the distributed load on span B-C can be replaced (for the purposes of statics) with a concentrated load of 2700 pounds located 20 feet to the left of point C.

3. Pressure Loads

Hydrostatic pressure is an example of a pressure load that is distributed over an area. The pressure is denoted as p pounds per unit area of surface, and is normal to the surface at every point. If the surface is plane, the statically equivalent concentrated load can be found by integrating over the area. The resultant is numerically equal to the pressure times the area. The point of application will be the centroid of the area over which the integration was performed.

4. Resolution of Forces and Moments

Any system (collection) of forces and moments is statically equivalent
to a single resultant force vector plus a single resultant moment
vector in 3-dimensional space. Either or both of these resultants may
be zero.

The x-component of the resultant force is the sum of all of the x-components of the individual forces, and similarly for the y- and z-components of the resultant force.

$$F_{Rx} = \Sigma F_i (\cos\theta_{x,i}) \qquad \text{11.17}$$

$$F_{Ry} = \Sigma F_i (\cos\theta_{y,i}) \qquad \text{11.18}$$

$$F_{Rz} = \Sigma F_i (\cos\theta_{z,i}) \qquad \text{11.19}$$

The determination of the resultant moment vector is more complex. The
resultant moment vector includes the moments of all system forces
around the reference axes plus the components of all system moments.

$$M_{Rx} = \Sigma(y_i F_{z,i} - z_i F_{y,i}) + \Sigma M_i (\cos\theta_{x,i}) \qquad \text{11.20}$$

$$M_{Ry} = \Sigma(z_i F_{x,i} - x_i F_{z,i}) + \Sigma M_i (\cos\theta_{y,i}) \qquad \text{11.21}$$

$$M_{Rz} = \Sigma(x_i F_{y,i} - y_i F_{x,i}) + \Sigma M_i (\cos\theta_{z,i}) \qquad \text{11.22}$$

5. Conditions of Equilibrium

For an object to be in equilibrium, it is necessary that the resultant
force vector and the resultant moment vector both be equal to zero.

$$F_R = \sqrt{F_{Rx}^2 + F_{Ry}^2 + F_{Rz}^2} = 0 \qquad \text{11.23}$$

$$M_R = \sqrt{M_{Rx}^2 + M_{Ry}^2 + M_{Rz}^2} = 0 \qquad \text{11.24}$$

Since the square of any quantity cannot be negative, equations
11.25 through 11.30 follow directly from equations 11.23 and 11.24.

$$F_{Rx} = 0 \qquad \text{11.25}$$

$$F_{Ry} = 0 \qquad \text{11.26}$$

$$F_{Rz} = 0 \qquad \text{11.27}$$

$$M_{Rx} = 0 \qquad \text{11.28}$$

$$M_{Ry} = 0 \qquad \text{11.29}$$

$$M_{Rz} = 0 \qquad \text{11.30}$$

6. Free-Body Diagrams

A free-body diagram is a representation of an object in equilibrium,
showing all external forces, moments, and support reactions. Since
the object is in equilibrium, the resultant of all forces and moments
on the free-body is zero.

If any part of the object is removed and replaced by the forces and
moments which are exerted on the cut surface, a free-body of the

remaining structure is obtained, and the conditions of equilibrium
will be satisfied by the new free-body also.

Figure 11.6
Original and Cut Free-bodies

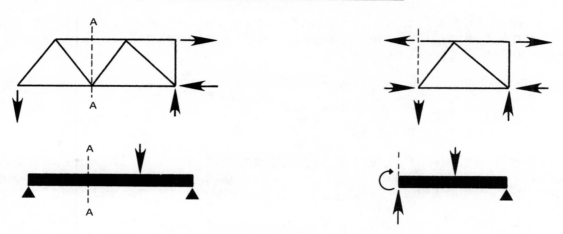

By dividing the object into a sufficient number of free-bodies, the
internal forces and moments can be found at all points of interest,
providing that the conditions of equilibrium are sufficient to give
a static solution.

7. Reactions

A typical first step in solving statics problems is to determine the
supporting reaction forces. The manner in which the structure is
supported will determine the type, location, and direction of the
reactions. Conventional symbols may be used to define the type of
reactions which occur at each point of support. Some examples are
shown in table 11.1.

8. Influence Lines

An influence line ('influence graph') is an x-y plot of the magnitude
of a reaction (any reaction on the object) as it would vary as the load
is placed at different points on the object. The x-axis corresponds to
the location along the object (as along the length of a beam); the
y-axis corresponds to the magnitude of the reaction. For uniformity,
the load is taken as 1 unit. Therefore, for any actual load of P units,
the actual reaction would be given by equation 11.31.

$$\begin{matrix} \text{actual} \\ \text{reaction} \end{matrix} = P \begin{pmatrix} \text{influence graph} \\ \text{ordinate} \end{pmatrix} \qquad 11.31$$

Example 11.2

Draw the influence graphs for the left and right reactions for the beam
shown below.

Type of Support	Symbol	Characteristics
	Table 11.1	
Built-in		Moments and forces in any direction
Simple		Load in any direction; no moment
Roller		Load normal to surface only; no moment
Cable		Load in cable direction; no moment
Guide		No load or moment in guide direction
Hinge		Load in any direction; no moment

If a unit load was at the left end, reaction R_A would be equal to 1.
If the unit load was at the right end, it would be supported entirely by R_B, so R_A would be zero. The influence line for R_A is, then

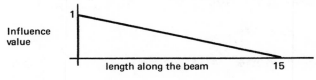

The influence line for R_B is found similarly.

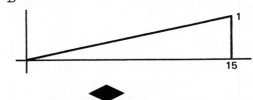

Influence lines may also be drawn for moments, shears, and deflections.

9. Axial Members

A member which is in equilibrium when acted upon by forces at each end and by no other forces or moments is an axial member. For equilibrium to exist, the resultant forces at the ends must be equal, opposite, and co-linear. In an actual truss, this type of loading can be approached through the use of frictionless bearings or pins at the ends of the axial members. In simple truss analysis, the members are assumed to be axial members regardless of the end conditions.

A typical inclined axial member is illustrated in figure 11.7. For that member to be in equilibirum, the following equations must hold:

$$F_{Rx} = F_{Bx} - F_{Ax} = 0 \qquad \qquad 11.32$$

$$F_{Ry} = F_{By} - F_{Ay} = 0 \qquad \qquad 11.33$$

Figure 11.7

An Axial Member

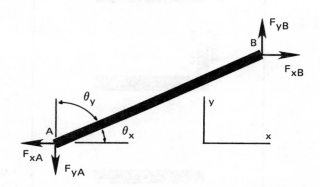

The resultant force, F_R, may be derived from the components by trigonometry and direction cosines.

$$F_{Rx} = F_R \cos\theta_x \qquad \qquad 11.34$$

$$F_{Ry} = F_R \cos\theta_y = F_R \sin\theta_x \qquad \qquad 11.35$$

If, however, the geometry of the axial member is known, similar triangles may be used to find the resultant and/or the components. This is illustrated in the following example.

Example 11.3

A 12-foot long axial member carrying an internal load of 180 pounds is inclined as shown below. What are the x- and y-components of the load?

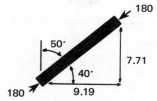

Method 1: Direction Cosines

$$F_x = 180(\cos 40°) = 137.9$$
$$F_y = 180(\cos 50°) = 115.7$$

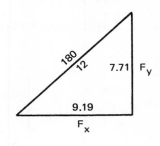

Method 2: Similar Triangles

$$F_x = (\frac{9.19}{12})(180) = 137.9$$

$$F_y = (\frac{7.71}{12})(180) = 115.7$$

10. Trusses

This discussion is directed towards 2-dimensional trusses. The loads in truss members are represented by arrows pulling away from the joints for tension, and by arrows pushing towards the joints for compression.

Figure 11.8
Truss Notation

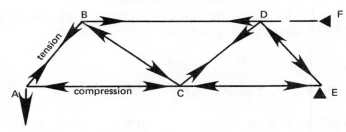

The equations of equilibrium can be used to find the external reactions on a truss. To find the internal resultants in each axial member, three methods may be used. These methods are: method of joints, cut-and-sum, and method of sections.

All joints in a truss will be determinate and all member loads can be found if equation 11.36 holds.

$$\# \text{ truss members} = 2(\# \text{ joints}) - 3 \qquad 11.36$$

If the left-hand side is greater than the right-hand side, methods from chapter 13 must be used to solve the truss. If the left-hand side is less than the right-hand side, the truss is not rigid and will collapse under certain types of loading.

A. Method of Joints

The method of joints is a direct application of equations 11.25 and 11.26. The sums of forces in the x- and y-directions are taken at consecutive joints in the truss. At each joint, there may be up to two unknown axial forces, each of which may have two components. Since there are two equations of equilibrium, a joint with two unknown forces will be determinate.

Example 11.4

Find the force in member BD in the truss shown below.

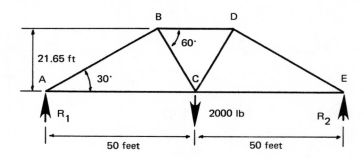

step 1: Find the reactions R_1 and R_2. From Equation 11.29, the sum of moments must be zero. Taking moments (counterclockwise as positive) about point A gives R_2.

$$\Sigma M_A = 100(R_2) - 2000(50) = 0$$

$$R_2 = 1000$$

From equation 11.26, the sum of the forces (vertical positive) in the y direction must be zero.

$$\Sigma F_y = R_1 - 2000 + 1000 = 0$$

$$R_1 = 1000$$

step 2: Although we want the force in member BD, there are three unknowns at joints B and D. Therefore, start with joint A where there are only two unknowns (AB and AC). The free-body of joint A is shown below. The direction of R_1 is known. However, the directions of the member forces are usually not known and need to be found by inspection or assumption. If an incorrect direction is assumed, the force will show up with a negative sign in later calculations.

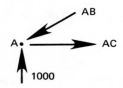

step 3: Resolve all inclined forces on joint A into horizontal and vertical components using trigonometry or similar triangles. R_1 and AC are already parallel to the y and x axes, respectively. Only AB needs to be resolved into components. By observation, it is clear that AB_y = 1000. If this wasn't true, equation 11.26 would not hold.

$$AB_y = AB(\sin 30^\circ)$$
$$1000 = AB(.5) \quad \text{or} \quad AB = 2000$$
$$AB_x = AB(\cos 30^\circ) = 1732$$

<u>step 4</u>: Draw the free-body diagram of joint B. Notice that the direction of force AB is towards the joint, just as it was for joint A. The direction of load BC is chosen to counteract the vertical component of load AB. The direction of load BD is chosen to counteract the horizontal components of loads AB and BC.

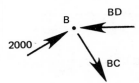

<u>step 5</u>: Resolve all inclined forces into horizontal and vertical components.

$$AB_x = 1732$$

$$AB_y = 1000$$

$$BC_x = BC(\sin 30°) = .5BC$$

$$BC_y = BC(\cos 30°) = .866BC$$

<u>step 6</u>: Write the equations of equilibrium for joint B.

$$\Sigma F_x = 1732 + .5BC - BD = 0$$

$$\Sigma F_y = 1000 - .866BC = 0$$

BC from the second equation is found to be 1155. Substituting 1155 into the first equilibrium condition equation gives

$$1732 + .5(1155) - BD = 0$$

$$BD = 2310$$

Since BD turned out to be positive, its direction was chosen correctly. The direction of the arrow indicates that the member is compressing the pin joint. Consequently, the pin is compressing the member; and member BD is in compression.

◆

B. Cut-and-Sum Method

The cut-and-sum method may be used if a load in an inclined member in the middle of a truss is wanted. The method is strictly an application of the equilibrium condition requiring the sum of forces in the vertical direction to be zero. The method is illustrated in the following example.

<u>Example 11.5</u>

Find the force in member BC for the truss shown in example 11.4.

<u>step 1</u>: Find the external reactions. This is the same step as in example 11.4. $R_1 = R_2 = 1000$.

step 2: Cut the truss through, making sure that the cut goes through only one member with a vertical component. In this case, that member is BC.

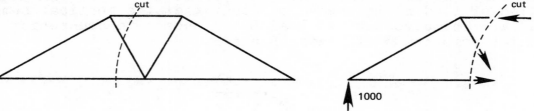

step 3: Draw the freebody of either part of the remaining truss.

step 4: Resolve the inclined load (BC) into vertical and horizontal components.

$$BC_x = .5(BC)$$

$$BC_y = .866(BC)$$

step 5: Sum forces in the y direction for the entire freebody.

$$\Sigma F_y = 1000 - .866(BC) = 0$$

$$BC = 1155$$

◆

C. Method of Sections

The cut-and-sum method will not usually work if the load in a horizontal member is wanted because it may not be possible to cut the truss without going through two members with horizontal components.

The method of sections is a direct approach for finding member loads at any point in a truss. In this method, the truss is cut at an appropriate section, and the conditions of equilibrium are applied to the resulting free-body. This is illustrated in the following example.

Example 11.6

For the truss shown below, find the load in members CE and CD.

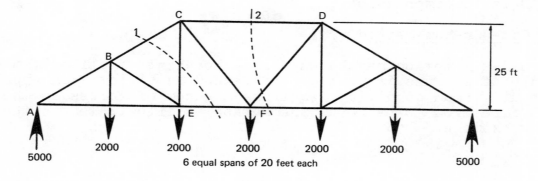

6 equal spans of 20 feet each

For member force CE, the truss is cut at section 1.

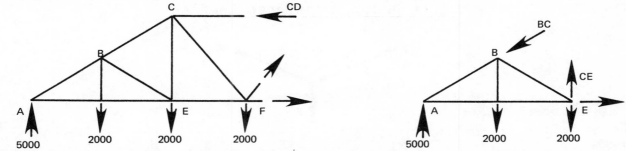

Taking moments about A will eliminate all unknowns except force CE.

$$\Sigma M_A = CE(40) - 2000(20) - 2000(40) = 0$$

$$CE = 3000$$

For member CD, the truss is cut at section 2. Taking moments about point F will eliminate all unknowns except CD.

$$\Sigma M_F = CD(25) + 2000(20) + 2000(40) - 5000(60) = 0$$

$$CD = 7200$$

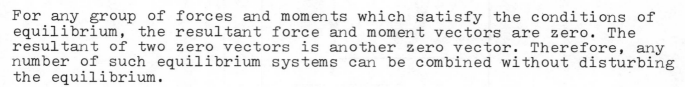

11. Superposition of Loadings

For any group of forces and moments which satisfy the conditions of equilibrium, the resultant force and moment vectors are zero. The resultant of two zero vectors is another zero vector. Therefore, any number of such equilibrium systems can be combined without disturbing the equilibrium.

Superposition methods must be used with discretion in working with actual structures since some structures change shape significantly under load. If the actual structure were to deflect so that the points of application of loads were quite different than in the undeflected structure, then superposition would not be applicable.

In simple truss analysis, the change of shape under load is neglected when finding the member loads. Superposition, therefore, can be assumed to apply.

12. Cables

A. Cables Under Concentrated Loads

An ideal cable is assumed to be completely flexible. It therefore acts as an axial member in tension between any two points of concentrated load application.

The method of joints and sections used in truss analysis apply equally well to cables under concentrated loads. However, no compression members will be found. As in truss analysis, if the cable loads are unknown, some information concerning the geometry of the cable must be known in order to solve for the axial tension in the segments.

<u>Figure 11.9</u>
<u>Cable Under Transverse Loading</u>

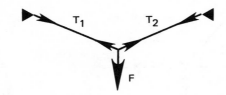

Example 11.7

Find the tension T_2 between points B and C.

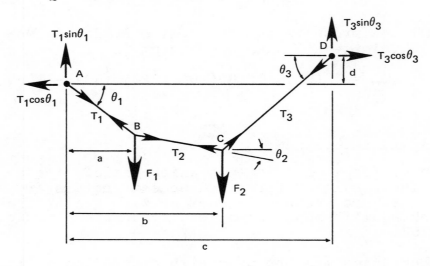

<u>step 1</u>: Take moments about point A to find T_3.

$$\Sigma M_A = aF_1 + bF_2 + dT_3\cos\theta_3 - cT_3\sin\theta_3 = 0$$

<u>step 2</u>: Sum forces in the x direction at point A to find T_1.

$$\Sigma F_x = T_1\cos\theta_1 - T_3\cos\theta_3 = 0$$

<u>step 3</u>: Sum forces in the x direction at point B to find T_2.

$$\Sigma F_x = T_2\cos\theta_2 - T_1\cos\theta_1 = 0$$

B. Cables Under Distributed Loads

An idealized tension cable under a distributed load is similar to a linkage made up of a very large number of axial members. The cable is an axial member in the sense that the internal tension acts in a direction which is along the centerline of the cable everywhere.

Figure 11.10 illustrates a cable under a unidirectional distributed load. A free-body diagram of segment B-C of the cable is also shown. F is the vertical resultant of the distributed load on the segment.

Figure 11.10
Cable Under Distributed Load

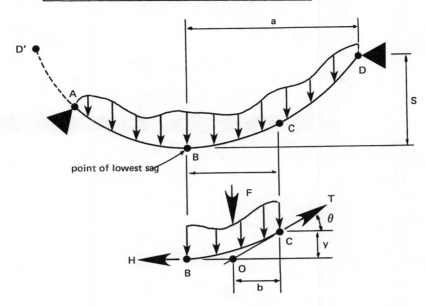

point of lowest sag

T is the cable tension at point C, and H is the cable tension at the point of lowest sag. From the conditions of equilibrium for free-body B-C, it is apparent that the three forces H, F, and T must be concurrent at point O. Taking moments about point C, the following equations are obtained.

$$\Sigma M_C = Fb - Hy = 0 \qquad\qquad 11.37$$

$$H = \frac{Fb}{y} \qquad\qquad 11.38$$

But $\tan\theta = \left(\frac{y}{b}\right)$. So,

$$H = \frac{F}{\tan\theta} \qquad\qquad 11.39$$

From the summation of forces in the vertical and horizontal directions,

$$T\cos\theta = H \qquad\qquad 11.40$$

$$T\sin\theta = F \qquad\qquad 11.41$$

$$T = \sqrt{H^2 + F^2} \qquad\qquad 11.42$$

The shape of the cable is function of the relative amount of sag at point B and the relative distribution (not the absolute magnitude) of the applied running load.

C. Parabolic Cables

If the distributed load per unit length, w, is constant with respect to a horizontal line (as is the load from a bridge floor) the cable will be parabolic in shape. This is illustrated in figure 11.11.

The horizontal component of tension can be found from equation 11.38 using F = wa, b = ½a, and y = S.

Figure 11.11
Parabolic Cable

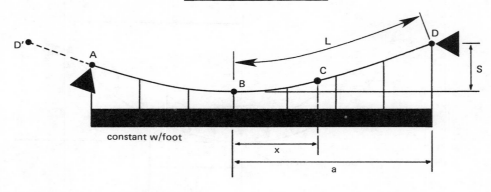

constant w/foot

$$H = \frac{Fb}{y} = \frac{wa^2}{2S} \qquad \text{11.43}$$

$$T = \sqrt{H^2 + F^2} = \sqrt{(wa^2/2S)^2 + (wx)^2} \qquad \text{11.44}$$

$$= w\sqrt{x^2 + (a^2/2S)^2} \qquad \text{11.45}$$

The shape of the cable is given by equation 11.46.

$$y = \frac{wx^2}{2H} \qquad \text{11.46}$$

The approximate length of the cable from the lowest point to the support is given by equation 11.47.

$$L \approx \left[1 + \frac{2}{3}\left(\frac{S}{a}\right)^2 - \frac{2}{5}(S/a)^4\right](a) \qquad \text{11.47}$$

Example 11.8

A pedestrian bridge has two suspension cables and a flexible floor. The floor weighs 28 pounds per foot. The span of the bridge is 100 feet between the two end supports. When the bridge is empty, the tension at point A is 1500 pounds. What is the cable sag, S, at the center? What is the approximate cable length?

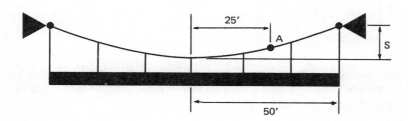

The floor weight per cable is 28/2 = 14 lb/ft. From equation 11.45,

$$1500 = 14\sqrt{(25)^2 + ((50)^2/2S)^2}$$

$$\text{or } S = 12 \text{ feet}$$

From equation 11.47,

$$L = 50\left[1 + (2/3)(12/50)^2 - (2/5)(12/50)^4\right] = 51.9 \text{ ft}$$

D. The Catenary

If the distributed load, w, is constant along the length of the cable (as in the case of a cable loaded by its own weight) the cable will have the shape of a catenary. This is illustrated in figure 11.12.

Figure 11.12

The Catenary

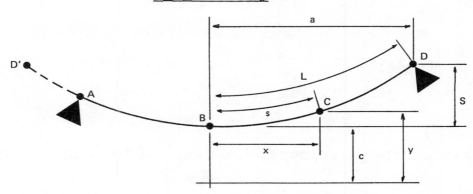

As shown in figure 11.12, y is measured from a reference plane located a distance c below the lowest point of the cable, point B. The location of this reference plane is a parameter of the cable which must be determined before equations 11.48 through 11.53 are used. The value of c does not correspond to any physical distance, nor does the reference plane correspond to the ground.

The equations of the catenary are presented below. Some judgment is usually necessary to determine which equations should be used and in which order they should be used. In order to define the cable shape, it is necessary to have some initial information which can be entered into the equations. For example, if a and S are given, equation 11.51 can be solved by trial and error to obtain c. Once c is known, the cable geometry and forces are defined by the remaining equations.

$$y = c\left(\cosh\left(\frac{x}{c}\right)\right) \qquad\qquad 11.48$$

$$s = c\left(\sinh\left(\frac{x}{c}\right)\right) \qquad\qquad 11.49$$

$$y = \sqrt{s^2 + c^2} \qquad\qquad 11.50$$

$$S = c\left(\cosh\left(\frac{a}{c}\right) - 1\right) \qquad\qquad 11.51$$

$$\tan\theta = \frac{S}{c} \qquad\qquad 11.52$$

$$H = wc \qquad\qquad 11.53$$

$$F = ws \qquad\qquad 11.54$$

$$T = wy \qquad\qquad 11.55$$

$$\tan\theta = \frac{ws}{H} \qquad\qquad 11.56$$

$$\cos\theta = \frac{H}{T} \qquad\qquad 11.57$$

Example 11.9

A cable 100 feet long is loaded by its own weight. The sag is 25 feet and the supports are on the same level. What is the distance between the supports?

From equation 11.50 at point D:

$$c + 25 = \sqrt{(50)^2 + c^2}$$

$$c = 37.5$$

From equation 11.49,

$$50 = 37.5(\sinh(\frac{a}{37.5}))$$

$$a = 41.2 \text{ feet}$$

The distance between supports is: $(2a) = (2)(41.2) = 82.4$ feet

Providing that the lowest point B is known or can be found, the location of the cable supports at different levels does not significantly affect the analysis of cables. The same procedure is used in proceeding from point B to either support. In fact, once the theoretical shape of the cable has been determined, the supports can be relocated anywhere along the cable line without affecting the equilibrium of the supported segment.

Figure 11.13
Non-Symmetrical Segment of Symmetrical Cable

13. 3-Dimensional Structures

The static analysis of 3-dimensional structures usually requires the following steps:

   step 1: Determine the components of all loads and reactions. This is usually accomplished by finding the x, y, and z coordinates of all points and then using direction cosines.

   step 2: Draw three free-bodies of the structure - one each for the x-, y-, and z-components of loads and reactions.

   step 3: Solve for unknowns using $\Sigma F = 0$ and $\Sigma M = 0$.

Example 11.10

Beam ABC is supported by the two cables as shown below. Find the cable tensions T_1 and T_2.

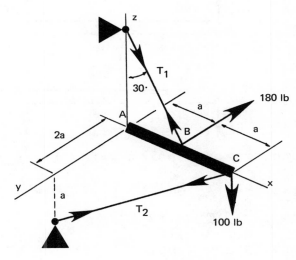

step 1: For the 180 pound load:

$$F_x = 0 \qquad F_y = -180 \qquad F_z = 0$$

For the 100 pound load:

$$F_x = 0 \qquad F_y = 0 \qquad F_z = -100$$

For cable 1:

$$\cos\theta_x = \cos 120° = -.5$$
$$\cos\theta_y = 0$$
$$\cos\theta_z = \cos 30° = .866$$

$$T_{1x} = -.5T_1 \qquad T_{1y} = 0 \qquad T_{1z} = .866T_1$$

For cable 2:

The length of the cable is

$$L = \sqrt{(2a)^2 + (2a)^2 + (-a)^2} = 3a$$

$$\cos\theta_x = \frac{-2a}{3a} = -.667$$

$$\cos\theta_y = \frac{2a}{3a} = .667$$

$$\cos\theta_z = \frac{-a}{3a} = -.333$$

$$T_{2x} = -.667T_2 \qquad T_{2y} = .667T_2 \qquad T_{2z} = -.333T_2$$

step 2:

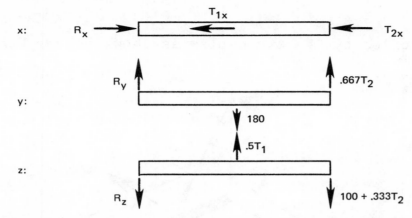

step 3: Summing moments about point A for the y case gives T_2.

$$\Sigma M_{Ay} = .667T_2(2a) - 180(a) = 0$$

$$T_2 = 135$$

Summing moments about point A for the z case gives T_1.

$$\Sigma M_{Az} = .866T_1(a) - .333(135)(2a) - 100(2a) = 0$$

$$T_1 = 335$$

Example 11.11:

In the space framework shown below, points A, B, C, and E are in the x-y plane. Point D is on the z axis. Coordinates of the points are also given. Find the loads in members BA, BC, BD, and BE.

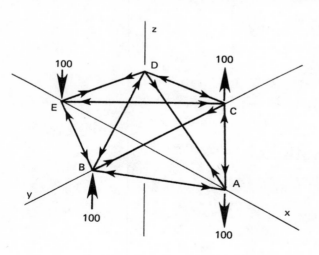

point	coordinates
A	(10.4,0,0)
B	(0,6,0)
C	(0,-6,0)
D	(0,0,6)
E	(-10.4,0,0)

step 1: The direction cosines for all loads on joint B must be found.

BA: length = $\sqrt{(6)^2 + (10.4)^2} = 12$

$\cos\theta_x = (0-10.4)/12 = -.866$

$\cos\theta_y = (6-0)/12 = .5$

$\cos\theta_z = (0-0)/12 = 0$

BC: length = 12

$$\cos\theta_x = (0-0)/12 = 0$$
$$\cos\theta_y = (-6-6)/12 = -1$$
$$\cos\theta_z = (0-0)/12 = 0$$

BD: length = 8.49

$$\cos\theta_x = (0-0)/8.49 = 0$$
$$\cos\theta_y = (6-0)/8.49 = .707$$
$$\cos\theta_z = (0-6)/8.49 = -.707$$

BE: length = 12

$$\cos\theta_x = (0-(-10.4))/12 = .866$$
$$\cos\theta_y = (6-0)/12 = .5$$
$$\cos\theta_z = (0-0)/12 = 0$$

step 2: Draw the freebodies of point B.

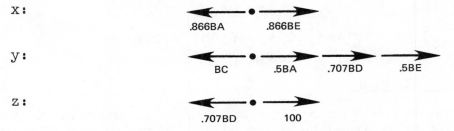

x: .866BA   .866BE

y: BC   .5BA   .707BD   .5BE

z: .707BD   100

step 3: From the equilibrium requirement for the z direction, it is apparent that BD = 141.4 (compression). Since the remaining freebodies are not sufficient to solve for the remaining forces, another point must be evaluated. Find the direction cosines for the loads on joint A.

AB: $\cos\theta_x = (10.4-0)/12 = .866$
$$\cos\theta_y = (0-6)/12 = -.5$$
$$\cos\theta_z = (0-0)/12 = 0$$

AC: length = 12
$$\cos\theta_x = (10.4-0)/12 = .866$$
$$\cos\theta_y = (0-(-6))/12 = .5$$
$$\cos\theta_z = (0-0)/12 = 0$$

AD: $\cos\theta_x = (10.4-0)/12 = .866$
$$\cos\theta_y = (0-0)/12 = 0$$
$$\cos\theta_z = (0-6)/12 = -.5$$

Drawing the freebodies of joint A:

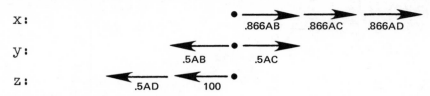

x: .866AB   .866AC   .866AD

y: .5AB   .5AC

z: .5AD   100

From the requirements for equilibrium of joint A in the z direction, AD = -200. Because of the negative sign, the direction of AD was chosen incorrectly when computing the direction cosines. (The direction is shown correctly in the framework illustration.) Therefore, the corrected picture of joint A in the x direction is

x: 173   .866(AB+AC)

However, AB = AC (from the equilibrium requirements of joint A in the y direction).Therefore, AB = AC = 100.

This information can be used to find forces BE and BC:

BE = BA = 100

BC = .5(100) + .707(141.4) + .5(100) = 200

14. General Tripod Solution

The procedure given in the preceding section will work quite well with a tripod consisting of 3 axial members with a load in any direction applied at the apex. However, the tripod problem occurs frequently enough to develop a specialized procedure for solution.

Figure 11.14

A General Tripod

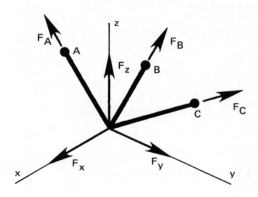

step 1: Use the direction cosines of the force F to find its components.

$$F_x = F(cos\theta_x) \hspace{3cm} 11.58$$

$$F_y = F(cos\theta_y) \hspace{3cm} 11.59$$

$$F_z = F(cos\theta_z) \hspace{3cm} 11.60$$

<u>step 2</u>: Using the x-, y-, and z-coordinates of points A, B, and C (taking the origin at the apex) find the direction cosines for the legs. Repeat the following four equations for each member, observing algebraic signs of x, y, and z.

$$L^2 = x^2 + y^2 + z^2 \qquad\qquad 11.61$$

$$\cos\theta_x = x/L \qquad\qquad 11.62$$

$$\cos\theta_y = y/L \qquad\qquad 11.63$$

$$\cos\theta_z = z/L \qquad\qquad 11.64$$

<u>step 3</u>: Write the equations of equilibrium for joint O. The following simultaneous equations assume tension in all three members. A minus sign in the solution for any member indicates compression instead of tension.

$$F_A\cos\theta_{xA} + F_B\cos\theta_{xB} + F_C\cos\theta_{xC} + F_x = 0 \qquad 11.65$$

$$F_A\cos\theta_{yA} + F_B\cos\theta_{yB} + F_C\cos\theta_{yC} + F_y = 0 \qquad 11.66$$

$$F_A\cos\theta_{zA} + F_B\cos\theta_{zB} + F_C\cos\theta_{zC} + F_z = 0 \qquad 11.67$$

<u>Example 11.12</u>

Find the load on each leg of the tripod shown below. F is parallel to the x-axis.

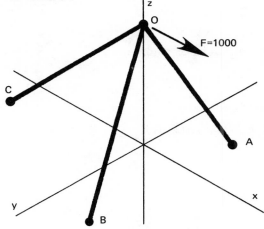

Point	x	y	z
O	0	0	6
A	2	-2	0
B	0	3	-4
C	-3	3	2

Since the origin is not at point (0,0,0), it is necessary to transfer the origin to the apex. This is done by the following equations. Only the z values are actually affected.

$$x' = x - x_0$$
$$y' = y - y_0$$
$$z' = z - z_0$$

The new coordinates with the origin at the apex are:

Point	x	y	z
0	0	0	0
A	2	-2	-6
B	0	3	-10
C	-3	3	-4

The components of the applied force are:

$$F_x = F(\cos\theta_x) = F(\cos(0°)) = 1000$$
$$F_y = 0$$
$$F_z = 0$$

The direction cosines of the legs are found from the following table.

Member	x^2	y^2	z^2	L^2	L	$\cos\theta_x$	$\cos\theta_y$	$\cos\theta_z$
0-A	4	4	36	44	6.63	.3015	-.3015	-.9046
0-B	0	9	100	109	10.44	0	.2874	-.9579
0-C	9	9	16	34	5.83	-.5146	.5146	-.6861

From equations 11.65, 11.66, and 11.67, the equilibrium equations are:

$$.3015F_A + \qquad\qquad -.5146F_C + 1000 = 0$$
$$-.3015F_A + .2874F_B + .5146F_C + \qquad 0 = 0$$
$$-.9046F_A - .9579F_B - .6861F_C + \qquad 0 = 0$$

The solution to this set of simultaneous equations is:

$$F_A = +1531 \text{ (tension)}$$
$$F_B = -3480 \text{ (compression)}$$
$$F_C = +2841 \text{ (tension)}$$

◆

15. Properties of Areas

A. Centroids

The location of the centroid of a 2-dimensional area which is defined mathematically as $y = f(x)$ can be found from equations 11.68 and 11.69. This is illustrated in example 11.13.

$$\bar{x} = \frac{\int x\, dA}{A} \qquad\qquad 11.68$$

$$\bar{y} = \frac{\int y\, dA}{A} \qquad\qquad 11.69$$

$$A = \int f(x)\, dx \qquad\qquad 11.70$$

$$dA = f(x)\, dx = f(y)\, dy \qquad\qquad 11.71$$

Example 11.13

Find the x-component of the area bounded by the x- and y-axes,
x=2, and y = exp(2x).

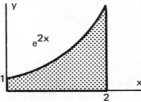

step 1: Find the area. $A = \int_0^2 e^{2x}dx = [\frac{1}{2}e^{2x}]_0^2 = 27.3 - .5 = 26.8$

step 2: Put dA in terms of dx. $dA = f(x)dx = e^{2x}dx$

step 3: Use equation 11.68 to find $\bar{x}$.

$$\bar{x} = \frac{1}{26.8}\int_0^2 xe^{2x}dx = \frac{1}{26.8}[\frac{1}{2}xe^{2x} - \frac{1}{4}e^{2x}]_0^2 = 1.54$$

With very few exceptions, most areas for which the centroidal location
is needed will be either rectangular or triangular. The locations of
the centroids for these and other common shapes are given in figure 11.15.

The centroid of a complex 2-dimensional area which can be divided into
the simple shapes shown in figure 11.15 can be found from equations
11.72 and 11.73.

$$\bar{x}_{composite} = \frac{\Sigma(A_i\bar{x}_i)}{\Sigma A_i} \qquad\qquad 11.72$$

$$\bar{y}_{composite} = \frac{\Sigma(A_i\bar{y}_i)}{\Sigma A_i} \qquad\qquad 11.73$$

Example 11.14

Find the y-coordinate of the centroid for the object shown below.

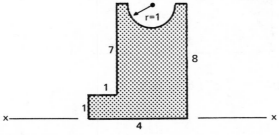

The object is divided into three parts: a 1 x 4 rectangle, a 3 x 7
rectangle, and a half-circle of radius 1. Then, the areas and distances
from the x-x axis to the individual centroids are found.

$$A_1 = (1)(4) = 4 \qquad\qquad \bar{y}_1 = \frac{1}{2}$$
$$A_2 = (3)(7) = 21 \qquad\qquad \bar{y}_2 = 4\frac{1}{2}$$
$$A_3 = (-\frac{1}{2})\pi(1)^2 = -1.57 \qquad \bar{y}_3 = 8 - .424 = 7.576$$

Using equation 11.73,

$$\overline{y} = \frac{(4)(\frac{1}{2}) + (21)(4\frac{1}{2}) - (1.57)(7.576)}{4 + 21 - 1.57} = 3.61$$

◆

B. Moment of Inertia

The moment of inertia, I, of a 2-dimensional area is a parameter which is often needed in mechanics of materials problems. It has no simple geometric interpretation, and its units (length to the fourth power) add to the mystery of this quantity. However, it is convenient to think of the moment of inertia as a resistance to bending.

Thinking of the moment of inertia as a resistance to bending, it is apparent that this quantity must always be positive. Since bending of an object (e.g., a beam) may be in any direction, the resistance to bending must depend on the direction of bending. Therefore, a reference axis or direction must be included when specifying the moment of inertia.

In this chapter, I_x is used to represent a moment of inertia with respect to the x axis. Similarly, I_y is with respect to the y axis. I_x and I_y are not components of the 'resultant' moment of inertia. The moment of inertia taken with respect to a line passing through the area's centroid is known as the centroidal moment of inertia, I_c.

The moments of inertia of a shape which can be expressed mathematically as $y = f(x)$ are given by equations 11.74 and 11.75.

$$I_x = \int y^2 \, dA \qquad\qquad 11.74$$

$$I_y = \int x^2 \, dA \qquad\qquad 11.75$$

Example 11.15

Find I_y for the area bounded by the y axis, $y=8$, and $y^2 = 8x$.

$$I_y = \int_0^8 x^2 \, dA = \int_0^8 x^2(8-y)dx$$

But $y = \sqrt{8x}$

$$I_y = \int_0^8 (8x^2 - \sqrt{8}x^{5/2})dx = 195.04 \text{ inches}^4$$

◆

In general, however, moments of inertia will be found from figure 11.15.

Example 11.16

What is the centroidal moment of inertia of the area shown?

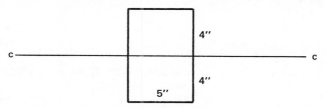

From figure 11.15, $I_c = \dfrac{(5)(8)^3}{12} = 213.3$ inches^4

◆

The polar moment of inertia of a 2-dimensional area can be thought of as a measure of the area's resistance to torsion (twisting). Although the polar moment of inertia can be evaluated mathematically by equation 11.76, it is more expedient to use equation 11.77 if I_x and I_y are known.

$$J_z = \int (x^2 + y^2)dA \qquad\qquad 11.76$$

$$J_z = I_x + I_y \qquad\qquad 11.77$$

The radius of gyration, k, is a distance at which the entire area can be assumed to exist. This distance is measured from the axis about which the moment of inertia was taken.

$$I = k^2 A \qquad\qquad 11.78$$

$$k = \sqrt{I/A} \qquad\qquad 11.79$$

Example 11.17

What is the radius of gyration for the section shown in example 11.16? What is the significance of this value?

$$A = (5)(4+4) = 40$$

From equation 11.79, $k = \sqrt{213.3/40} = 2.31''$

2.31" is the distance from the axis c-c that an infinitely long strip (with area of 40 square inches) would have to be located to have a moment of inertia of 213.3 inches^4.

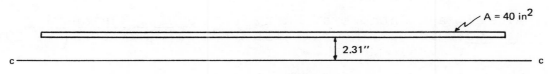

◆

The parallel axis theorem is usually needed to evaluate the moment of inertia of a composite object made up of several simple 2-dimensional shapes. The parallel axis theorem relates the moment of inertia of an area taken with respect to any axis to the centroidal moment of inertia. In equation 11.80, A is the 2-dimensional object's area, and d is the distance between the centroidal and new axes.

$$I_{\text{any parallel axis}} = I_c + Ad^2 \qquad\qquad 11.80$$

Example 11.18

Find the moment of inertia about the x-axis for the 2-dimensional object shown below.

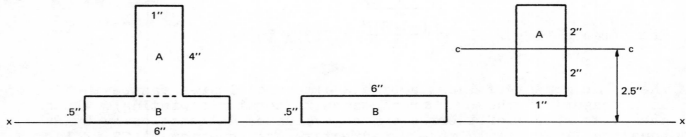

The T-section is divided into two parts: A and B. The moment of inertia of section B can be readily evaluated by using figure 11.15.

$$I_{x-x} = \frac{(6)(.5)^3}{3} = .25$$

The moment of inertia of the stem about its own centroidal axis is

$$I_{c-c} = \frac{(1)(4)^3}{12} = 5.33$$

Using equation 11.80, the moment of inertia of the stem about the x-x axis is

$$I_{x-x} = 5.33 + (4)(2.5)^2 = 30.33$$

The total moment of inertia of the T-section is .25 + 30.33 = 30.58 in^4.

Bibliography

1. Beer, Ferdinand P., and Johnston, E. Russell., _Vector Mechanics for Engineers: Statics and Dynamics_, McGraw-Hill Book Company, New York, NY 1962

2. Housner, George W., and Hudson, Donald E., _Applied Mechanics: Statics_, 2nd ed., D. Van Nostrand Co., Inc., Princeton, NJ 1961

3. Jensen, Alfred, and Chenoweth, Harry H., _Applied Engineering Mechanics_, 3rd ed., McGraw-Hill Book Co., New York, NY 1972.

Figure 11.15
Area Moments of Inertia

SHAPE	DIMENSIONS	CENTROID (x_c, y_c)	AREA MOMENT OF INERTIA
Rectangle	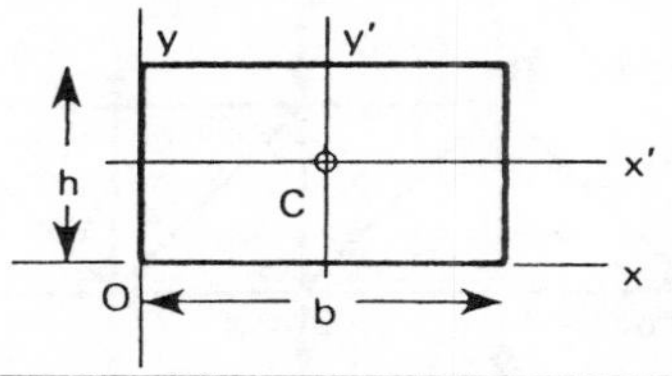	$(\frac{1}{2}b, \frac{1}{2}h)$	$I_{x'} = (1/12)bh^3$ $I_{y'} = (1/12)hb^3$ $I_x = (1/3)bh^3$ $I_y = (1/3)hb^3$ $J_C = (1/12)bh(b^2 + h^2)$
Triangle	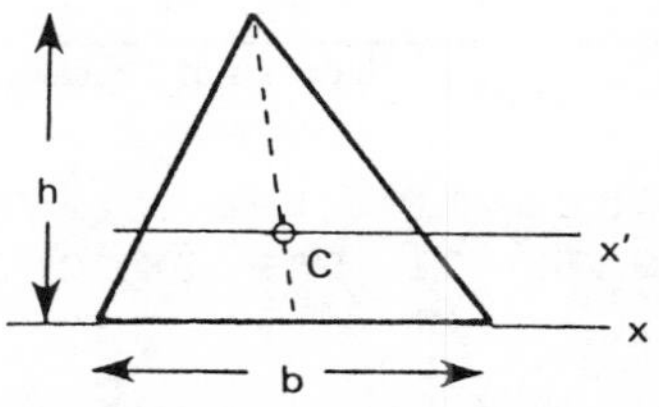	$y_c = (h/3)$	$I_{x'} = (1/36)bh^3$ $I_x = (1/12)bh^3$
Trapezoid	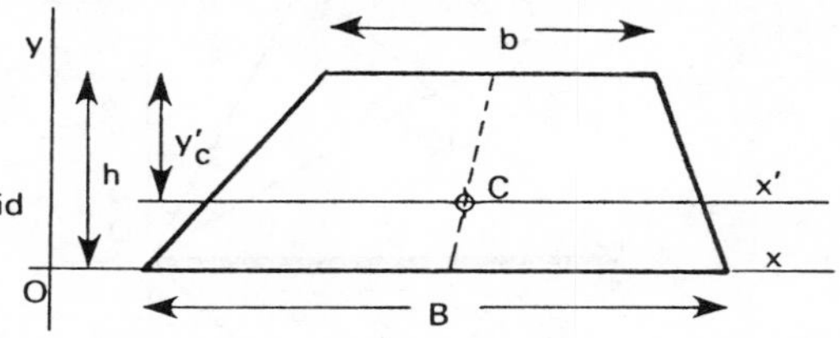	$y'_c = \dfrac{h(2B + b)}{3(B + b)}$ Note that this is measured from the top surface.	$I'_x = \dfrac{h^3(B^2+4Bb+b^2)}{36(B+b)}$ $I_x = \dfrac{h^3(B+3b)}{12}$
Quarter-Circle, of radius r	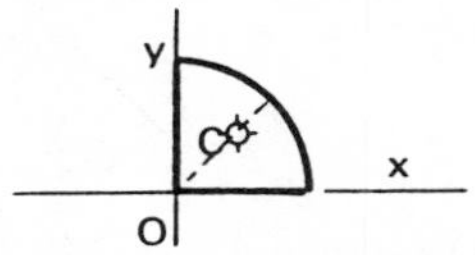	$((4r/3\pi), (4r/3\pi))$	$I_x = I_y = (1/16)\pi r^4$ $J_O = (1/8)\pi r^4$
Half Circle, of radius r	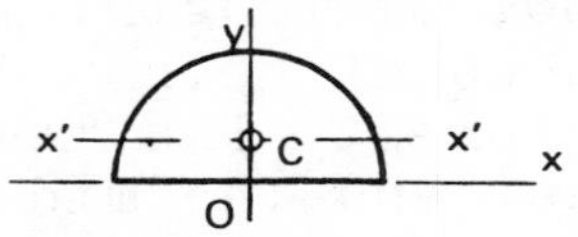	$(0, (4r/3\pi))$	$I_x = I_y = (1/8)\pi r^4$ $J_O = \frac{1}{4}\pi r^4 \quad I_{x'} = .11r^4$
Circle, of radius r	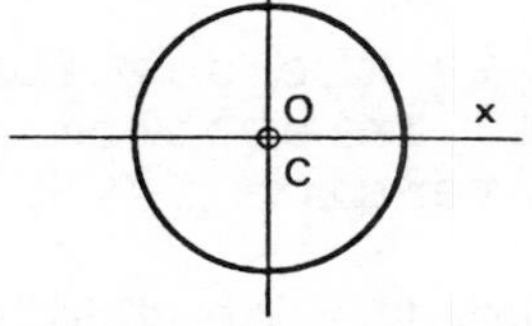	$(0,0)$	$I_x = I_y = \frac{1}{4}\pi r^4$ $J_O = \frac{1}{2}\pi r^4$
Parabolic Area	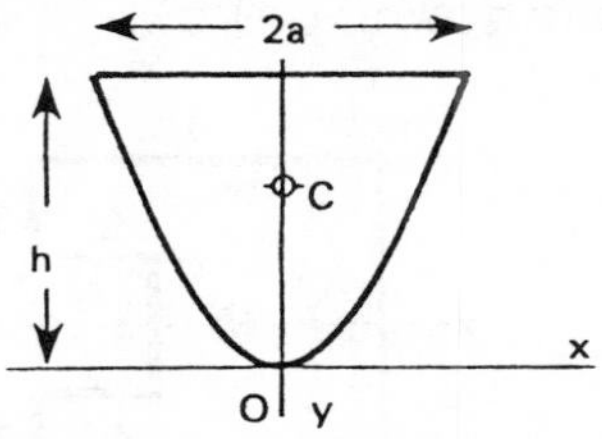	$(0, (3h/5))$	$I_x = 4h^3a/7$ $I_y = 4ha^3/15$
Parabolic Spandrel	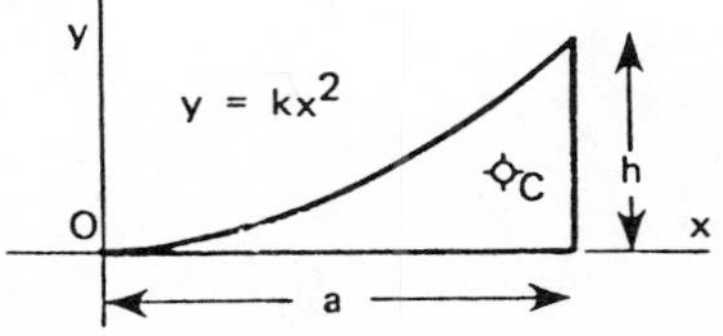	$((3a/4),(3h/10))$	$I_x = ah^3/21$ $I_y = 3ha^3/15$

Practice Problems: STATICS - DETERMINATE

<u>Optional</u>

Required

1. Find the forces in all members of the truss shown.

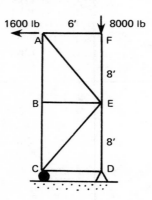

2. Find the forces in each of the legs.

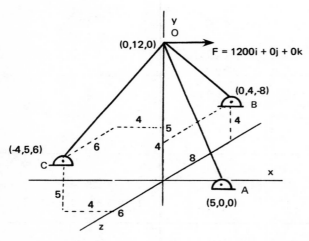

3. Find the centroidal moment of inertia about an axis parallel to the x axis.

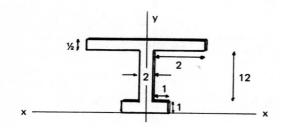

4. Find the forces in members DE and HJ.

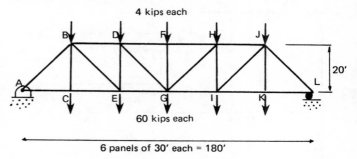

5. What are the x, y, and z components of the forces at A, B, and C?

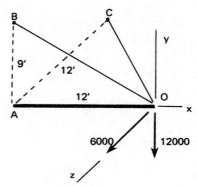

6. A power line weighs 2 pounds per foot of length. It is supported by two equal height towers over a level forest. The tower spacing is 100 feet and the mid-point sag is 10 feet. What are the maximum and minimum tensions?

7. What is the sag for the cable described in problem #6 if the maximum tension is 500 pounds?

8. Locate the centroid of the object shown below.

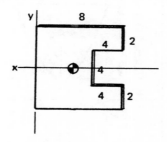

9. Replace the distributed load with three concentrated loads, and indicate the points of application.

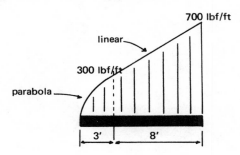

10. Find the forces in all members of the truss shown below.

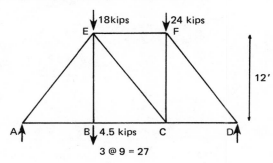

NOTES

Mechanics of Materials

Nomenclature

A	area	in²
c	distance to extreme fiber	in
C	end restraint coefficient	-
D	diameter	in
e	eccentricity	in
E	modulus of elasticity	psi
F	force or load	lb
FS	factor of safety	-
G	shear modulus	psi
I	moment of inertia	in⁴
k	radius of gyration	in
L	length	in
M	moment	in-lb, or ft-lb
n	ratio	-
N	number of cycles	-
r	radius	in
S	strength, or axial load	psi, or lb
T	temperature, or torque	°F, or in-lb
u	virtual truss load	lb
U	energy	ft-lb
V	shear, or volume	lb, or in³
w	load per unit length	lb/in
W	work	ft-lb
x	distance	in, or ft
y	distance from neutral axis to extreme fiber, or deflection	in
Z	section modulus	in³

Symbols

δ	elongation or displacement	in
θ	angle	degrees
σ	normal stress	psi
α	coefficient of linear thermal expansion	1/°F
β	coefficient of volumetric thermal expansion	1/°F
γ	coefficient of area thermal expansion	1/°F
τ	shear stress	psi
ε	strain	-
μ	poisson's ratio	-

Subscripts

a	allowable		o	original
b	bending		t	transformed
c	centroidal, or compressive		th	thermal
e	endurance, or euler		u	ultimate
ext	external		y	yield

1. Properties of Structural Materials

A. The Tensile Test

Many of the required material properties can be derived from the results
of a standard tensile test. In a tensile test, a material sample is
loaded axially in tension, and the elongation is measured as the load
is increased. A graphical representation of typical test data for steel
is shown in figure 12.1 in which the elongation, δ, is plotted against
the applied load, F.

Figure 12.1
Typical Tensile Test Results for Steel

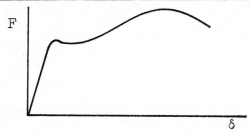

Since the elongation data are only applicable to an object with the same
length and area of the test sample, the data are converted to stresses
and strains by use of equations 12.1 and 12.2. σ is known as the normal
stress, and ε is known as the strain. Strain is the percentage elongation of the sample.

$$\sigma = \frac{F}{A} \qquad\qquad 12.1$$

$$\epsilon = \frac{\delta}{L} \qquad\qquad 12.2$$

The stress-strain data can also be graphed, and the shape of the resulting curve will be the same as figure 12.1 with the scales changed.

Figure 12.2
A Typical Stress-Strain Curve for Steel

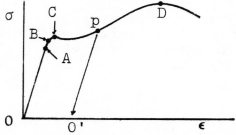

The line O-A in figure 12.2 is a straight line. The relationship
between the stress and the strain is given by Hooke's law, equation
12.3. E is the modulus of elasticity ('Young's modulus') and is the
slope of the line segment O-A. The stress at point A is known as the
proportionality limit.

$$\sigma = E\epsilon \qquad\qquad 12.3$$

Slightly above the proportionality limit is the elastic limit (point B). As long as the stress is kept below the elastic limit, there will be no permanent strain when the applied stress is removed. The strain is said to be 'elastic' and the stress is said to be in the 'elastic region'.

If the elastic limit stress is exceeded before the load is removed, recovery will be along a line parallel to the straight line portion of the curve, as shown in the line segment p-O'. The strain that results (line O-O') is permanent and is known as 'plastic strain'.

The yield point (point C) is very close to the elastic limit. For all practical purposes, the yield stress, S_y, can be taken as the stress which accompanies the beginning of plastic strain. Since permanent deformation is to be avoided, the yield stress is used in calculating safe stresses in ductile materials such as steel.

$$\sigma_a = S_y/(FS) \qquad\qquad 12.4a$$

Some materials, such as aluminum, do not have a well-defined yield point. This is illustrated in figure 12.3. In such cases, the yield point is taken as the stress which will cause a .2% parallel offset.

<u>Figure 12.3</u>

A Typical Stress-Strain Curve for Aluminum

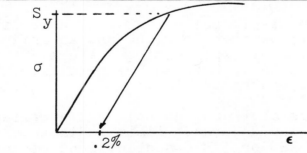

The ultimate tensile strength, point D in figure 12.2, is the maximum load carrying ability of the material. However, since stresses near the ultimate strength are accompanied by large plastic strains, this parameter should not be used for the design of ductile materials such as steel and aluminum.

As the sample is elongated during a tensile test, it will also be decreasing in thickness (width, or diameter). The ratio of the lateral strain to the axial strain is known as poisson's ratio, u. u is typically taken as .3 for steel and .33 for aluminum.

$$u = \frac{\epsilon_{lateral}}{\epsilon_{axial}} = \frac{\Delta D}{D_o} \Big/ \frac{\Delta L}{L_o} \qquad\qquad 12.4b$$

B. Fatigue Tests

A part may fail after repeated stress loadings even if the stress never exceeds the ultimate fracture strength of the material. This type of failure is known as fatigue failure.

The behavior of a material under repeated loadings can be evaluated in a fatigue test. A sample is loaded repeatedly to a known stress, and the number of applications of that stress is counted until the sample fails. This procedure is repeated for different stress levels. The results of many of these tests can be graphed, as is done in figure 12.4.

Figure 12.4
Results of Many Fatigue Tests for Steel

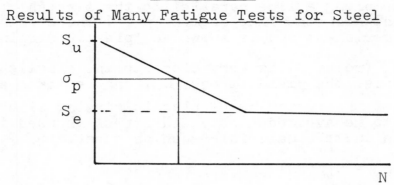

For any given stress level, say σ_p in figure 12.4, the corresponding number of applications of the stress which will cause failure is known as the fatigue life. That is, the fatigue life is just the number of cycles of stress required to cause failure. If the material is to fail after only one application of stress, then the required stress must equal or exceed the ultimate strength of the material.

Below a certain stress level, called the endurance limit or endurance strength, the part will be able to withstand an infinite number of stress applications without experiencing failure. Therefore, if a dynamically loaded part is to have an infinite life, the applied stress must be kept below the endurance limit.

Some materials, such as aluminum, do not have a well-defined endurance limit. In such cases, the endurance limit is taken as the stress that will cause failure at EE8 or 5 EE8 applications of the stress.

Figure 12.5
Fatigue Test Results For Aluminum

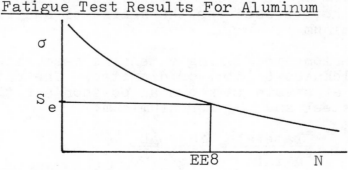

C. Estimates of Material Properties

Although the properties of a material will depend on its classification (ASTM, AISC, etc) average values are given in table 12.1. Yield and ultimate strengths are listed in tables 12.5 and 12.6.

Table 12.1
Average Material Properties

material	E (psi)	G (psi)	u	ρ(pcf)	α (1/°F)
steel (hard)	30 EE6	11.5 EE6	.30	489	6.5 EE-6
steel (soft)	29 EE6	11.5 EE6	.30	489	6.5 EE-6
aluminum alloy	10 EE6	3.9 EE6	.33	173	12.8 EE-6
magnesium alloy	6.5 EE6	2.4 EE6	.35	112	14.5 EE-6
titanium alloy	15.4 EE6	6.0 EE6	.34	282	4.9 EE-6
cast iron	20 EE6	8 EE6	.27	442	5.6 EE-6

2. Deformation Under Loading

Equation 12.2 can be rearranged to give the elongation of an axially loaded member in compression or tension.

$$\delta = L\epsilon = \frac{L\sigma}{E} = \frac{LF}{AE} \qquad 12.5$$

A tension load is taken as positive and a compressive load is taken as negative. The actual length of a member under loading is, then, given by equation 12.6 where the algebraic sign of the deformation must be observed.

$$L_{actual} = L_o + \delta \qquad 12.6$$

The energy stored in a loaded member is equal to the work required to deform it. Within the proportional elastic limit, this energy is given by equation 12.7.

$$U = \tfrac{1}{2}F\delta = \tfrac{1}{2}\left(\frac{F^2 L}{AE}\right) \qquad 12.7$$

3. Thermal Deformation

If the temperature of an object is changed, the object will experience length, area, and volume changes. These changes may be predicted by equations 12.8, 12.9, and 12.10.

$$\Delta L = \alpha L_o (T_2 - T_1) \qquad 12.8$$

$$\Delta A = \gamma A_o (T_2 - T_1) \approx 2\alpha A_o (T_2 - T_1) \qquad 12.9$$

$$\Delta V = \beta V_o (T_2 - T_1) \approx 3\alpha V_o (T_2 - T_1) \qquad 12.10$$

If equation 12.8 is rearranged, an expression for the thermal strain is obtained. Thermal strain is handled in the same manner as strain due to an applied load.

$$\epsilon_{th} = \frac{\Delta L}{L_o} = \alpha (T_2 - T_1) \qquad 12.11$$

For example, if a bar is heated but is not allowed to expand, the stress will be given by equation 12.12.

$$\sigma_{th} = E \, \epsilon_{th}$$ 12.12

4. Shear and Moment Diagrams

It was illustrated in chapter 11 that the sums of forces and moments is equal to zero everywhere for an object in equilibrium. For example, the sum of moments about point A for the beam shown in figure 12.6 is zero.

$$\Sigma M_A = 4(2) + 12(4) - 8(7) = 0$$

Figure 12.6

A Beam in Equilibrium

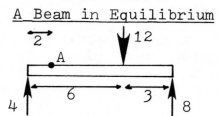

Nevertheless, the beam shown in figure 12.6 will bend under the influence of the forces. This bending (deformation) is evidence of the stress experienced by the beam. The stress and bending will be shown later in this chapter to be proportional to a product of the forces acting on the beam and their moment arms.

Since the sum of moments about any point in equilibrium is zero, the moment used to find stresses and deflection is taken from the point in question to one end of the beam only. This called the 'one-way moment'. The absolute value of the moment will not depend on the end used. This can be illustrated by the beam shown in figure 12.6.

$$\Sigma M_A \text{ (to right end)} = -8(7) + 4(12) = -8$$

$$\Sigma M_A \text{ (to left end)} = 4(2) = +8$$

The moment obtained will depend on the location chosen. A graphical representation of the one-way moment at every point along a beam is known as a moment diagram. The following guidelines should be observed in constructing moment diagrams.

1. Moments should be taken from the left end to the point in question. If the beam is cantilever, place the built-in end at the right.
2. Clockwise moments are positive. The left-hand rule should be used to determine positive moments.
3. Concentrated loads produce linearly increasing lines on the moment diagram.
4. Uniformly distributed loads produce parabolic lines on the moment diagram.
5. The maximum moment will occur when the shear (V) is zero.
6. The moment at any point is equal to the area under the shear diagram up to that point. That is,

$$M = \int V \, dx$$ 12.13

5. The moment is zero at a free end or a hinge.

Similarly, the sum of forces in the y direction on a beam in equilibrium is zero. However, the shearing stress at a point along the beam will depend on the sum of forces and reaction from the point in question to one end only.

A shear diagram is drawn to graphically represent the shear at any point along a beam. The following guidelines should be observed in constructing a shear diagram.

1. Loads and reactions acting up are positive.
2. The shear at any point is equal to the sum of the loads and reactions from the left end to the point in question
3. Concentrated loads produce straight (horizontal) lines on the shear diagram.
4. Uniformly distributed loads produce straight sloping lines on the shear diagram.
5. The magnitude of the shear at any point is equal to the slope of the moment diagram at that point.

$$V = \frac{dM}{dx}$$ 12.14

Example 12.1

Draw the shear and moment diagrams for the following beam.

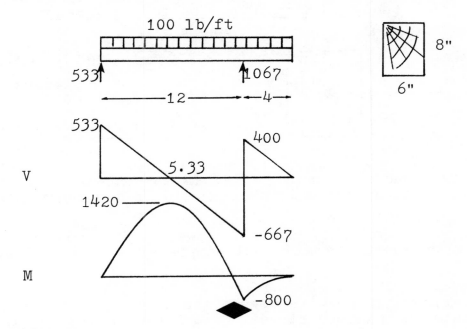

5. Stresses in Beams

A. Normal Stress

Normal stress is the type of stress experienced by a member which is axially loaded. The normal stress is the load divided by the area.

$$\sigma = F/A$$ 12.15

Normal stress also occurs when a beam bends, as shown in figure 12.7.

The lower part of the beam experiences normal tensile stress (which causes lengthening), and the upper part of the beam experiences a normal compressive stress (which causes shortening). There is no stress along a horizontal plane passing through the centroid of the cross section. This plane is known as the neutral plane or neutral axis.

Figure 12.7
Normal Stress Due to Bending

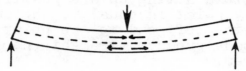

Although it is a normal stress, the stress produced by the bending is usually called 'bending stress' or 'flexure stress'. Bending stress varies with position within the beam. It is zero at the neutral axis, but increases linearly with distance from the neutral axis.

$$\sigma_b = \frac{-My}{I_c} \qquad\qquad 12.16$$

Figure 12.8
Bending Stress Distribution in a Beam

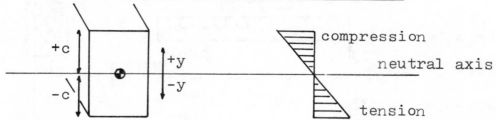

The moment, M, used in equation 12.16 is the one-way moment previously discussed in section 4 of this chapter. I_c is the centroidal moment of inertia of the beam's cross sectional area. The negative sign in equation 12.16 is typically omitted. However, it is required to be consistent with the convention that compression is negative.

Since the maximum stress will govern the design, y may be set equal to c to obtain the maximum stress. c is known as the 'distance from the neutral axis to the extreme fiber.'

$$\sigma_{b,max} = \frac{Mc}{I_c} \qquad\qquad 12.17$$

For any given structural shape, c and I_c are fixed. Therefore, these two terms may be combined into the section modulus, Z.

$$\sigma_{b,max} = \frac{M}{Z} \qquad\qquad 12.18$$

$$Z = I_c/c \qquad\qquad 12.19$$

If an axial member is loaded eccentrically, it will experience axial stress (equation 12.15) as well as bending stress (equation 12.16). This is illustrated by figure 12.9 in which a load is not applied to the centroid of a column's cross sectional area.

Figure 12.9
Eccentric Loading of an Axial Member

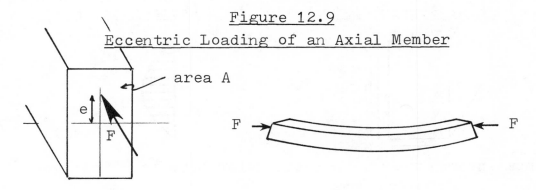

Because the beam bends and supports a compressive load, the stress produced is a sum of bending and normal stress.

$$\sigma_{max,min} = \frac{F}{A} \pm \frac{Mc}{I_c} = \frac{F}{A} \pm \frac{Fec}{I_c} \qquad\qquad 12.20$$

The strain energy stored in a beam experiencing a moment (bending) is

$$U = \tfrac{1}{2}\int \frac{M^2}{EI}dx \qquad\qquad 12.21$$

B. Shear Stress

Normal stress is produced when a load is absorbed by an area normal to it. On the other hand, shear stress is produced by a load being carried by an area parallel to the load. This is illustrated in figure 12.10.

Figure 12.10
Normal and Shear Stresses

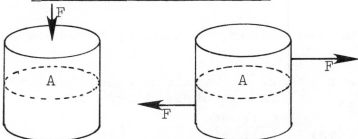

The average shear stress experienced by a pin, bolt, or rivet in single shear (as illustrated in figure 12.10) is given by equation 12.22. Because it gives an average value over the cross section of the shear member, this equation should only be used when the loading is low or when there is multiple redundancy in the shear group.

$$\tau = F/A \qquad\qquad 12.22$$

The actual shear stress in a beam is dependent on the location within the beam, just as was the bending stress. Shear stress is zero at the outer edges of a beam and maximum at the neutral axis. This is illustrated in figure 12.11.

Figure 12.11
Shear Stress Distribution in a Rectangular Beam

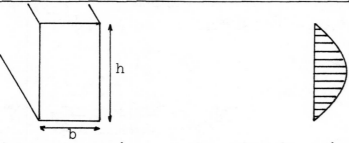

The maximum shear stress in a rectangular beam is

$$\tau_{max} = \frac{3V}{2A} = \frac{3V}{2bh}$$ 12.23

For a round beam of radius r and area A, the maximum shear stress is

$$\tau_{max} = \frac{4V}{3A} = \frac{4V}{3\pi r^2}$$ 12.24

The shear, V, used in equations 12.23 and 12.24 is the one-way shear described in section 4 of this chapter.

Shear stress also occurs when a object is placed in torsion. The shear stress at the outer surface of a bar of radius r which is torsionally loaded by a torque T is

$$\tau = \frac{Tr}{J}$$ 12.25

The total strain energy due to torsion is

$$U = \frac{T^2 L}{2GJ}$$ 12.26

Example 12.2

What are the maximum shear and bending stresses for the beam shown in example 12.1?

From the shear diagram, the maximum shear is -667 pounds. From equation 12.23, the maximum shear stress is

$$\tau_{max} = \frac{(3)(-667)}{(2)(6)(8)} = -20.8 \text{ psi}$$

From the moment diagram, the maximum moment is +1420 ft-lbs. The centroidal moment of inertia is

$$I_c = \frac{(6)(8)^3}{12} = 256 \text{ in}^4$$

Therefore, from equation 12.17, the maximum bending stress is calculated on the next page.

$$\sigma_{b,max} = \frac{(1420)(12)(4)}{256} = 266.3 \text{ psi}$$

◆

6. Stresses in Composite Structures

A composite structure is one in which two or more different materials are used, both of which carry a part of the loading. Unless the various materials used all have the same strength (i.e., modulus of elasticity) the stress analysis will be dependent on the assumptions made.

Some simple composite structures can be analyzed using the assumption of consistent deformations (or 'consistent strain'). This is illustrated in examples 12.3 and 12.4. The technique used to analyze structures for which the strains are consistent is known as the transformation method.

step 1: Determine the modulus of elasticity for each of the materials used in the structure.

step 2: For each of the materials used, calculate the ratio

$$n = E/E_{weakest} \qquad 12.27$$

$E_{weakest}$ is the lowest modulus of elasticity of any of the materials used in the composite structure.

step 3: For all of the materials except the weakest, multiply the actual material stress area by its n ratio. Consider this expanded ('transformed') area to be the same composition as the material with the lowest modulus of elasticity.

step 4: If the structure is a tension or compression member, the distribution or placement of the transformed areas is not important. Just assume that the transformed areas carry the the axial load. For beams in bending, the transformed area can add to the width of the beam, but it cannot change the depth of the beam or the thickness of the reinforcing.

step 5: For compression or tension members, calculate the stresses in the weakest and stronger materials.

$$\sigma_{weakest} = F/A_t \qquad 12.28$$

$$\sigma_{stronger} = \frac{nF}{A_t} \qquad 12.29$$

step 6: For beams in bending, proceed through step 9. Find the centroid of the transformed beam.

step 7: Find the centroidal moment of inertia of the transformed beam, I_{ct}.

step 8: Find V_{max} and M_{max}.

step 9: Calculate the stresses in the weakest and stronger materials.

$$\sigma_{weakest} = \frac{Mc_{weakest}}{I_{ct}}$$ 12.30

$$\sigma_{stronger} = \frac{nMc_{stronger}}{I_{ct}}$$ 12.31

Example 12.3

Find the stress in the steel inner cylinder and the copper tube which surrounds it if a uniform compressive load is applied axially. The copper and steel are well bonded. Use E_{steel} = 3 EE7 psi and E_{copper}= 1.75 EE7 psi.

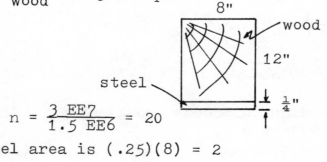
↓100 kips

10"

5"

$$n = \frac{3 \ EE7}{1.75 \ EE7} = 1.714$$

The actual steel area is $\frac{1}{4}\pi(5)^2$ = 19.63 in^2

The actual copper area is $\frac{1}{4}\pi[(10)^2-(5)^2]$ = 58.9 in^2

The transformed area is A_t = 58.9 + 1.714(19.63) = 92.55 in^2

$$\sigma_{copper} = \frac{100,000}{92.55} = 1080.5 \ psi$$

$$\sigma_{steel} = (1.714)(1080.5) = 1852.0 \ psi$$

◆

Example 12.4

Find the maximum bending stress in the steel-reinforced wood beam shown below at a point where the moment is 40,000 ft-lb. Use E_{steel} = 3 EE7 psi and E_{wood} = 1.5 EE6 psi.

8"

wood

12"

steel

$\frac{1}{4}$"

$$n = \frac{3 \ EE7}{1.5 \ EE6} = 20$$

The actual steel area is (.25)(8) = 2

The area of the steel is expanded to 20(2) = 40. Since the depth of beam and reinforcement cannot be increased, the width must increase. The transformed beam is shown below. The 160" dimension is arrived at by dividing the area of 40 square inches by the thickness of $\frac{1}{4}$".

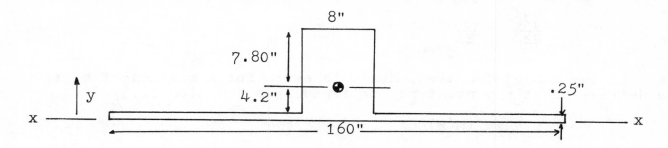

From the method presented in chapter 11, the centroid is located at
$\bar{y}$ = 4.45 inches from the x-x axis. The centroidal moment of inertia
of the transformed section is I_c = 2211.5 in^4. Then, from equations
12.30 and 12.31,

$$\sigma_{max,wood} = \frac{(40,000)(12)(7.8)}{(2211.5)} = 1692 \text{ psi}$$

$$\sigma_{max,steel} = \frac{(20)(40,000)(12)(4.45)}{2211.5} = 19320 \text{ psi}$$

7. Allowable Stresses

Once the actual stresses are known, they must be compared to allowable
stresses. If the allowable stress is calculated, it should be based on
the yield stress and a reasonable factor of safety. This is known as
the 'allowable stress design method'.

$$\sigma_a = S_y/(FS) \qquad\qquad 12.32$$

For steel, the factor of safety typically ranges from 1.5 to 2.5
depending on the type of steel and application.

However, failure of a structural member may occur at a stress far
below the yield strength, as in the case of a slender column or
fatigue-loaded beam. Therefore, the allowable stress method is being
supplemented in structural work by the 'load factor design method',
also known as the 'ultimate strength method' and 'plastic design
method'. In this method, the applied loads are multiplied by a load
factor. The product must be less than the structural member's strength,
as usually determined from a table.

8. Beam Deflections

A. Double Integration Method

The deflection and slope of a loaded beam are related to the applied
moment and shear by equations 12.33 through 12.36.

$$y = \text{deflection} \qquad\qquad 12.33$$

$$\frac{dy}{dx} = \text{slope} \qquad\qquad 12.34$$

$$\frac{d^2y}{dx^2} = \frac{M}{EI} \qquad\qquad 12.35$$

$$\frac{d^3y}{dx^3} = \frac{V}{EI} \qquad\qquad 12.36$$

Thus, if the moment function, $M(x)$, is known for a section of beam, the deflection at any point may be found from equation 12.37:

$$y = \frac{1}{EI}\int\int M(x)\ dx \qquad\qquad 12.37$$

In order to find the deflection, constants need to be introduced during the integration process. These constants may be found by use of table 12.2. If nothing is listed in table 12.2, then no generalization is possible about that particular case.

Table 12.2

Beam Boundary Conditions

end condition	y	y'	y"	V	M
simple support	0				
built-in support	0	0			
free end			0	0	0
hinge					0

Example 12.5

Find the tip deflection of the beam shown below. EI is 5 EE10 lb-in^2 everywhere on the beam.

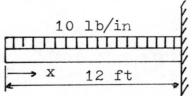

The moment at any point x from the left end of the beam is

$$M(x) = (-10)(x)(\tfrac{1}{2}x) = -5x^2 \quad (\text{in-lb})$$

This is negative by the left-hand rule convention. From equation 12.35,

$$y" = \frac{M}{EI}$$

So, $\qquad EIy" = -5x^2$

$$EIy' = \int -5x^2\ dx = -\tfrac{5}{3}x^3 + C_1$$

Since $y' = 0$ at a built-in support (table 12.2) and $x = 144$ inches at the built in support,

$$0 = -\tfrac{5}{3}(144)^3 + C_1 \qquad \text{or } C_1 = 4.98\ EE6$$

$$EIy = \int(-\tfrac{5}{3}x^3 + 4.98\ EE6)dx = \frac{-5}{12}x^4 + (4.98\ EE6)x + C_2$$

Again, $y = 0$ at $x = 144$, so $C_2 = -5.38\ EE8$.

Therefore, the deflection as a function of x is

$$y = (\frac{1}{EI})[(-\frac{5}{12})x^4 + (4.98 \text{ EE6})x - 5.38 \text{ EE8}]$$

At the tip, x = 0, so the deflection is

$$y_{tip} = \frac{-5.38 \text{ EE8}}{5 \text{ EE 10}} = -.0108 \text{ inches}$$

◆

B. Moment Area Method

The moment area method is a semi-graphical technique which is applicable whenever slopes of deflection beams are not too great. This method is based on the following two theorems.

Theorem I: The angle between tangents at any two points on the elastic line of a beam is equal to the area of the moment diagram between the two points divided by EI. That is,

$$\theta = \int \frac{M(x)dx}{EI} \qquad 12.38$$

Theorem II: One point's deflection away from the tangent of another point is equal to the statical moment of the bending moment between those two points divided by EI. That is,

$$y = \int \frac{xM(x)}{EI}dx \qquad 12.39$$

The application of these two theorems is aided by the following two comments.

* If EI is constant, the statical moment $\int xM(x)dx$ may be calculated as the product of the total moment diagram area times the distance from the point whose deflection is wanted to the centroid of the moment diagram.

* If the moment diagram has positive and negative parts (areas above and below the zero line) the statical moment should be taken as the sum of two products, one for each part of the moment diagram.

Example 12.6

Find the deflection, y, and the angle, θ, at the free end of the cantilever beam shown below.

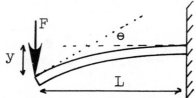

The deflection angle, θ, is the angle between the tangents at the free and built-in ends (Theorem I). The moment diagram is shown below.

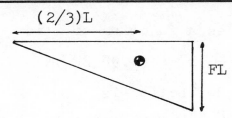

The area of the moment diagram is $\frac{1}{2}(FL)(L) = \frac{1}{2}FL^2$

From Theorem I, $\theta = \dfrac{FL^2}{2EI}$

From Theorem II, $y = \dfrac{FL^2}{2EI}(\frac{2}{3}L) = \dfrac{FL^3}{3EI}$

Example 12.7

Find the deflection of the free end of the cantilever beam shown below.

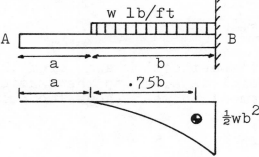

The distance from point A (where the deflection is wanted) to the centroid is (a + .75b). The area of the moment diagram is $(wb^3/6)$. So, from Theorem II,

$$y = \frac{wb^3}{6EI}(a + .75b)$$

C. Strain Energy Method

The deflection at a point of load application may be found by the strain energy method. This method equates the external work to the total internal strain energy as given by equations 12.7, 12.21, and 12.26. Since work is a force moving through a distance (which in this case is the deflection) we can write equation 12.40.

$$\frac{1}{2}Fy = \Sigma U \qquad\qquad 12.40$$

Example 12.8

Find the deflection at the point of application of the load on the stepped beam shown below.

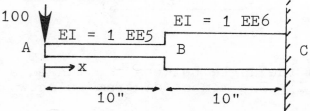

Underline section A-B: $M = 100x$ (in-lb)

From equation 12.21, $U = \frac{1}{2}\int_0^{10} \frac{(100x)^2}{1\ EE5} dx = 16.67$ in-lb

Underline section B-C: $M = 100x$

$U = \frac{1}{2}\int_{10}^{20} \frac{(100x)^2}{1\ EE6} dx = 11.67$ in-lb

Equating the internal work (U) and the external work,

$16.67 + 11.67 = \frac{1}{2}(100)y$

$y = .567$ inches

◆

D. Conjugate Beam Method

The conjugate beam method transforms a deflection problem into one of drawing moment diagrams. The method has the added advantage of being able to handle beams of varying cross sections and materials. It has the disadvantage of not being able to handle beams with two built-in ends. The following steps constitute the conjugate beam method.

step 1: Draw the moment diagram for the beam as it is actually loaded.

step 2: Construct the (M/EI) diagram by dividing the value of M at every point along the beam by the product of (EI) at that point. If the beam is of constant cross section, (EI) will be constant and the (M/EI) diagram will have the same shape as the moment diagram. However, if the beam cross section varies with x, then I will change. In this case, the (M/EI) diagram will not look the same as the moment diagram.

step 3: Draw a conjugate beam of the same length as the original beam. The material and cross sectional area of this conjugate beam are not relevant.

(a) If the actual beam is simply supported at its ends, the conjugate beam will be simply supported at its ends.
(b) If the actual beam is simply supported away from its ends, the conjugate beam has hinges at the support points.
(c) If the actual beam has free ends, the conjugate beam has built-in ends.
(d) If the actual beam has built-in ends, the conjugate beam has free ends. (Thus, if the actual beam has two built-in ends, the conjugate beam will have two free ends!)

step 4: Load the conjugate beam with the (M/EI) diagram. Find the conjugate reactions by methods of statics. Use the superscript * to indicate conjugate parameters.

step 5: Find the conjugate moment at the point where the deflection is wanted. The deflection is numerically equal to the moment as calculated from the conjugate beam forces.

Example 12.9

Find the deflections at the two load points. EI has a constant value of 2.356 EE7 lb-in^2.

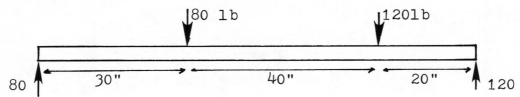

step 1: The moment diagram for the actual beam is:

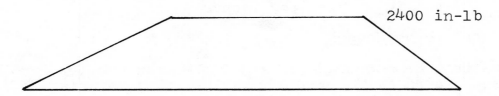

2400 in-lb

steps 2, 3, and 4: Since the cross section is constant, the conjugate load has the same shape as the original moment diagram. The peak load on the conjugate beam is

$$\frac{2400 \text{ in-lb}}{2.356 \text{ EE7 lb-in}^2} = 1.019 \text{ EE-4 (1/in)}$$

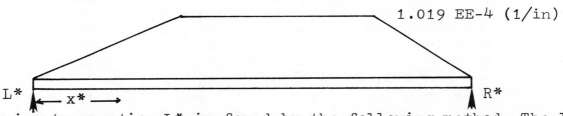

1.019 EE-4 (1/in)

The conjugate reaction L* is found by the following method. The loading diagram is assumed to be made up of a rectangular load and two 'negative' triangular loads. The area of the rectangular load (which has a centroid at x* = 45)is (90)(1.019 EE-4) = 9.171 EE-3

Similarly, the area of the left triangle (which has a centroid at x* = 10) is ½(30)(1.019 EE-4) = 1.529 EE-3. The area of the right triangle (which has a centroid at x* = 83.33) is ½(20)(1.019 EE-4) = 1.019 EE-3.

$$\Sigma M^*_{(L^*)} = 90R^* + (1.019 \text{ EE-3})(83.3) + (1.529 \text{ EE-3})(10)$$
$$- (9.171 \text{ EE-3})(45) = 0$$

So   R* = 3.472 EE-3 (1/in)

Then, L* = (9.171 - 1.019 - 1.529 - 3.472) EE-3

= 3.151 EE-3 (1/in)

step 5: The conjugate moment at x* = 30 is

M* = (3.151 EE-3)(30) + (1.529 EE-3)(30-10)
        - (9.171 EE-3)(30/90)(15)

= 6.266 EE-2 inches

The conjugate moment at the right-most load is

M* = (3.472 EE-3)(20) + (1.019 EE-3)(13.3)
        - (9.171 EE-3)(20/90)(10)

= 6.266 EE-2 inches

E. Table Look-Up Method

Table 12.4 is an extensive listing of the most commonly needed beam formulas. The use of these formulas is recommended whenever they may be applied singly or as part of a superposition solution.

F. Method of Superposition

If the deflection at a point is due to the combined action of two or more loads, the deflections at that point due to the individual loads can be added to find the total deflection.

9. Truss Deflections

A. Strain Energy Method

The deflection of a truss at the point of a single load application may be found by the strain energy method if all member forces are known.

Example 12.10

Find the vertical deflection of point A under the external load of 707 pounds. AE = 10 EE5 pounds for all members. The internal forces have been determined and are listed in the table below.

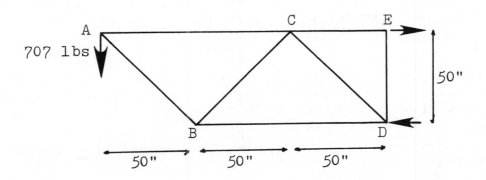

The length of member AB is $\sqrt{(50)^2 + (50)^2} = 70.7$ inches. From equation 12.7, the internal strain energy in member AB is

$$U = \frac{(-1000)^2(70.7)}{2(10\ EE5)} = 35.4 \text{ in-lb}$$

Similarly, the energy in all members can be determined, as listed in the following table.

Member	L	F	U
AB	70.7	-1000	+35.4
BC	70.7	+1000	+35.4
AC	100	+707	+25.0
BD	100	-1414	+100.0
CD	70.7	-1000	+35.4
CE	50	+2121	+112.5
DE	50	+707	+12.5
			356.2

The external work is $W_{ext} = \frac{1}{2}(707)y$, so

$$(\tfrac{1}{2})(707)y = 356.2$$

$$y = 1 \text{ inch}$$

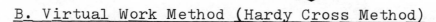

B. Virtual Work Method (Hardy Cross Method)

An extension of the strain-energy method results in an easy procedure for computing the deflection of any point on a truss.

step 1: Draw the truss twice.

step 2: On the first truss, place all of the actual loads.

step 3: Find the forces, S, in all of the members due to the actual applied loads.

step 4: On the second truss, place a dummy 1 pound load in the direction of the desired displacement.

step 5: Find the forces, u, in all members due to the one pound dummy load.

step 6: Find the desired displacement from equation 12.41.

$$\delta = \Sigma\frac{SuL}{AE} \qquad\qquad 12.41$$

In equation 12.41, the summation is over all truss members which have non-zero forces in both trusses.

Example 12.11

What is the horizontal deflection of point F on the truss shown on the next page? Use E = 3 EE7 psi. Joint A is restrained horizontally. (Member areas are given in the table for step 5.)

steps 1 and 2: Use the truss as drawn.

step 3: The forces in all of the truss members are summarized below.

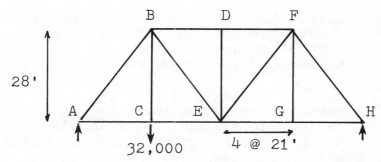

<u>step 4</u>: Draw the truss and load with a unit horizontal force at point F.

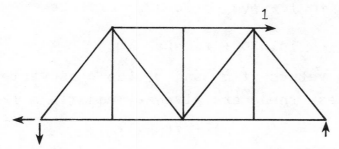

<u>step 5</u>: Find the forces, u, in all members of the second truss. These are summarized in the following table. Notice the sign convention: + for tension and - for compression.

member	S(lbs)	u	L(ft)	A(in^2)	$\frac{SuL}{AE}$ (ft)
AB	-30,000	5/12	35	17.5	-8.33 EE-4
CB	32,000	0	28	14	0
EB	-10,000	-5/12	35	17.5	2.78 EE-4
ED	0	0	28	14	0
EF	10,000	5/12	35	17.5	2.78 EE-4
GF	0	0	28	14	0
HF	-10,000	-5/12	35	17.5	2.78 EE-4
BD	-12,000	1/2	21	10.5	-4.00 EE-4
DF	-12,000	1/2	21	10.5	-4.00 EE-4
AC	18,000	3/4	21	10.5	9.00 EE-4
CE	18,000	3/4	21	10.5	9.00 EE-4
EG	6,000	1/4	21	10.5	1.00 EE-4
GH	6,000	1/4	21	10.5	1.00 EE-4
					12.01 EE-4 (ft)

Since 12.01 EE-4 is positive, the deflection is in the direction of the dummy unit load. In this case, the deflection is to the right.

10. Combined Stresses

Most practical cases of combined stresses have normal stresses on two perpendicular planes and a known shear stress acting parallel to these two planes. Based on knowledge of these stresses, the shear and normal stresses on all other planes can be found from conditions of equilibrium.

Under any condition of stress at a point, a plane can be found where the shear stress is zero. The normal stresses on this plane are known as the principal stresses. The principal stresses are the maximum and minimum stresses at the point in question.

The normal and shear stresses on a plane inclined an angle θ from the horizontal are given by equations 12.42 and 12.43.

$$\sigma_\theta = \tfrac{1}{2}(\sigma_x+\sigma_y) + \tfrac{1}{2}(\sigma_x-\sigma_y)\cos 2\theta + \tau\sin 2\theta \qquad 12.42$$

$$\tau_\theta = -\tfrac{1}{2}(\sigma_x-\sigma_y)\sin 2\theta + \tau\cos 2\theta \qquad 12.43$$

The maximum and minimum values of σ_θ and τ_θ (as θ is varied) are known as the principal stresses. These are given by equations 12.44 and 12.45.

$$\sigma(\max,\min) = \tfrac{1}{2}(\sigma_x+\sigma_y) \pm \tau(\max) \qquad 12.44$$

$$\tau(\max,\min) = \pm\, \tfrac{1}{2}\sqrt{(\sigma_x-\sigma_y)^2 + (2\tau)^2} \qquad 12.45$$

The angle of the planes on which the normal stresses are minimum and maximum are given by equation 12.46. θ is measured from the x axis, clockwise if negative, and counter-clockwise if positive. Equation 12.46 will yield two angles. These angles must be used in equation 12.42 to determine which angle corresponds to the minimum normal stress and which angle corresponds to the maximum normal stress.

$$\theta = \tfrac{1}{2}\arctan\!\left(\frac{2\tau}{\sigma_x-\sigma_y}\right) \qquad 12.46$$

The angles of the planes on which the shear stress is minimum and maximum are given by equation 12.47. The same angle sign convention used for equation 12.46 applies to equation 12.47.

$$\theta = \tfrac{1}{2}\arctan\!\left(\frac{\sigma_x-\sigma_y}{-2\tau}\right) \qquad 12.47$$

Proper sign convention must be adhered to when using equations 12.42 through 12.47. Normal tensile stresses are positive; normal compressive stresses are negative. Shear stresses are positive as shown in figure 12.12.

Figure 12.12

Plane of
Principal Stress

Sign Convention

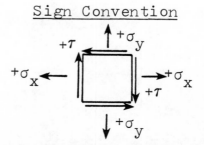

Example 12.12

For the part of a structural member shown below, find the maximum shear stress and the maximum normal stress.

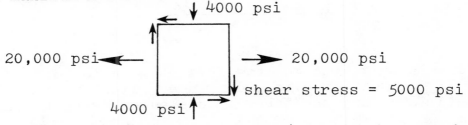

By the sign convention of figure 12.12, the 4000 psi is negative. From equation 12.45, the maximum shear stress is

$$\tau_{max} = \tfrac{1}{2}\sqrt{[20,000 - (-4000)]^2 + [(2)(5000)]^2} = 13,000 \text{ psi (tension)}$$

From equation 12.44, the maximum normal stress is

$$\sigma_{max} = \tfrac{1}{2}(20,000 + (-4000)) + 13,000 = 21,000 \text{ psi}$$

◆

11. Dynamic Loading

If a load is applied suddenly to a structure, the transient response may create stresses greater than would normally be calculated from the concepts of statics and mechanics of materials alone. Although a dynamic analysis of the structure is appropriate, the procedure is extremely lengthy and complicated. Therefore, arbitrary dynamic factors are applied to the static stress. For example, if the load is applied quickly compared to the natural period of the structure, a dynamic factor of 2 may be used. This assumes the load is applied as a ramp function.

12. Influence Diagrams

Shear, moment, and reaction influence diagrams (influence lines) can be drawn for any point on a truss. This is a necessary step in the evaluation of stresses induced by moving loads. It is important to realize, however, that the influence line applies to only one point on the truss.

To begin, it is necessary to know if the loads are transmitted to the truss members at the lower chords (a 'through truss') or at the upper chords (a 'deck truss'). If the truss is a through truss, the moving load is assumed to move along the lower chords.

Example 12.13

Draw the influence diagram for vertical shear in panel DF of the through truss shown on the next page.

First, allow a unit load to move from joint L to joint G along the lower chords. If the unit vertical load is at a distance x from point L, the right reaction will be +(1 - (x/120)). The unit load itself has a value of (-1), so the shear at distance x is just (-x/120).

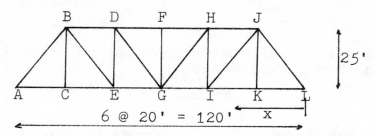

6 @ 20' = 120'

Next, allow a unit load to move from joint A to joint E along the lower chords. If the unit load is a distance x from point L, the left reaction will be (x/120), and the shear at distance x will be ((x/120) - 1).

These two lines can be graphed, as is done below.

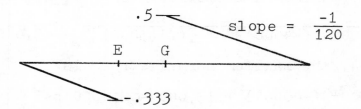

The influence line is completed by connecting the two lines as shown below. Therefore, the maximum shear in panel DF will occur when a load is at point G on the truss.

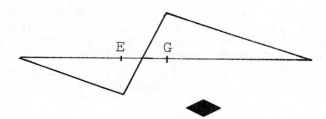

Example 12.14

Draw the moment influence diagram for panel DF on the truss shown in example 12.13

The left reaction is (x/120) where x is the distance from the unit load to the right end. If the unit load is to the right of point G, the moment can be found by summing moments from point G to the left. The moment is (x/120)(60) = .5x.

If the unit load is to the left of point E, the moment will be

$$(\frac{x}{120})(60) - (1)(x-60) = 60 - .5x$$

These two lines can be graphed. The moment for a unit load between points E and G is obtained by connecting the two end points of the lines derived above.

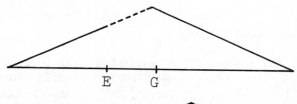

13. Moving Loads on Beams

If a beam supports a single moving load, the maximum bending and shearing stresses at a point can be found by drawing the moment and shear influence diagrams for that point. Once the positions of maximum moment and maximum shear are known, the stresses at the point in question can be found from equations 12.17 and 12.23.

If a simply-supported beam carries a set of moving loads (which remain equidistant as they travel across the beam) the following procedure may be used to find the dominant load. The dominant load is the one which occurs directly over the point of maximum moment.

> step 1: Calculate and locate the resultant of the load group.

> step 2: Assume one of the loads is dominant. Place the group on the beam such that the distance from one support to the assumed dominant load is equal to the distance from the other support to the resultant of the load group.

> step 3: Check to see that all loads are on the span and that the shear changes sign under the assumed dominant load. If the shear does not change sign under the assumed dominant load, the maximum moment may occur when only some of the load group is on the beam. If it does change sign, calculate the bending moment under the assumed dominant load.

> step 4: Repeat steps 2 and 3 assuming the other loads are dominant.

> step 5: Find the maximum shear by placing the load group such that the resultant is a minimum distance from a support.

14. Columns

The Euler load is the theoretical maximum load that an initially straight column can support without buckling. For columns with pinned ends, this load is given by equation 12.48.

$$F_e = \frac{\pi^2 EI}{L^2} = \frac{(k\pi)^2 EA}{L^2} \qquad\qquad 12.48$$

The corresponding column stress is

$$\sigma_e = \frac{F_e}{A} = \frac{\pi^2 E}{(L/k)^2} \qquad\qquad 12.49$$

Equations 12.48 and 12.49 assume that the column is long so that the Euler stress is reached before the yield stress is reached. If the column is short, then the yield stress of the material may be less than the Euler stress. In that case, short-column curves based on test data are used to predict the allowable column stress.

The value of L/k at the point of intersection of the short column and Euler curves is known as the critical (or transitional) slenderness ratio. The critical slenderness ratio becomes lower as the compressive yield stress increases. The region in which the short column formulas apply is determined by tests for each particular type of column and material. Typical critical slenderness ratios range from 80 to 120.

In general, the Euler allowable stress formula may be used if the the stress obtained from equation 12.49 does not exceed one-half of the compressive yield stress.

Example 12.15

An S-type, 4 x 9.5 steel I-beam 8.5 feet long is used as a column. What is the working stress for a safety factor of 3?

From table 12.5 , the yield stress for A36 steel is 36,000 psi. Also, use E = 2.9 EE7 psi. The required properties of the I beam are:

$$A = 2.79 \text{ in}^2$$
$$I = .903 \text{ in}^4$$
$$k = .569 \text{ in}$$

From equation 12.49, the Euler stress is

$$\sigma_e = \frac{\pi^2(2.9 \text{ EE7})}{((8.5)(12)/.569)^2} = 8907 \text{ psi}$$

Since 8907 is less than one-half of 36,000, the Euler formulas are valid. The allowable working stress is

$$\sigma_a = \frac{8907}{3} = 2969 \text{ psi} \blacklozenge$$

An ultimate load for any column can be found using the 'secant formula'. The secant formula is particularly suited for use when the column is short or intermediate in length.

$$\sigma = \frac{F}{A} = \frac{S_y}{(1 + \frac{ec}{k^2})\sec\theta} \tag{12.50}$$

$$\theta = \frac{1}{2}(\frac{L}{k})\sqrt{F/AE} \quad \text{(radians)} \tag{12.51}$$

The formula is solved by trial and error for (F/A) with the given eccentricity, e. If the value of e is not known, the eccentricity ratio is taken as .25. Substituting this value and E = 2.9 EE7 for steel and 1.0 EE7 for aluminum respectively, the following formulas result which converge quickly to the known (L/k) ratio when assumed values of (F/A) are substituted.

$$\theta = \arccos(\frac{1.25}{S_y}(\frac{F}{A})) \quad \text{(radians)} \tag{12.52}$$

$$\frac{L}{k} = 2\theta\sqrt{EA/F} \tag{12.53}$$

$$(\frac{L}{k})_{steel} = \frac{10,770(\theta)}{\sqrt{F/A}} \tag{12.54}$$

$$(\frac{L}{k})_{aluminum} = \frac{6325(\theta)}{\sqrt{F/A}} \tag{12.55}$$

Example 12.16

A steel column has a compressive yield strength of 90,000 psi and an (L/k) ratio of 75. Find the working stress assuming a factor of safety of 2.5.

From equation 12.49, the Euler stress is

$$\sigma_e = \frac{\pi^2(2.9\ EE7)}{(75)^2} = 50,880 \text{ psi}$$

Since this exceeds one-half of the compressive yield strength, a short column formula should be used. Therefore, use the secant formula. Assume that the maximum stress (F/A) is 30,000. Substituting into equations 12.52 and 12.54,

$$\theta = \arccos(\frac{1.25(30,000)}{90,000}) = 1.141 \text{ radians}$$

$$\frac{L}{k} = \frac{(10,770)(1.141)}{\sqrt{30,000}} = 70.9$$

Since (F/A) will be lower when (L/k) is higher, try (F/A) = 28,000. If this is done, θ = 1.1714 radian and (L/K) = 75.4 (close enough). Therefore, 28,000 is the ultimate strength of the column. Applying a factor of safety,

$$\sigma_a = \frac{28,000}{2.5} = 11,200 \text{ psi}$$

◆

All of the preceding column formulas are for columns with frictionless round or pinned ends. For other end conditions, the effective length L' should be used in place of L.

$$L' = CL \hspace{4cm} 12.56$$

C is the end restraint coefficient which varies from .5 to 2. For practical columns, C smaller than .7 should not be used since infinite stiffness of the support structure is not normally achievable.

Table 12.3

End-Restraint Coefficients

end conditions	C
both ends pinned	1
both ends built in	.5
one end pinned, one end built in	.7
one end built in, one end free	2

Table 12.4

Beam Formulas

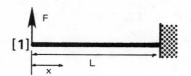

[1]

[4]

[7]

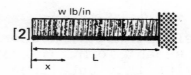

[2]

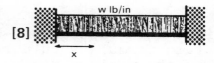

[5]

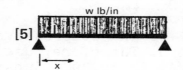

[8]

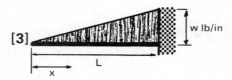

[3]

[6]

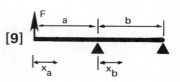

[9]

CASE	MOMENT	DEFLECTION
1	$M = Fx$ $M_{max} = FL$	$y = (F/6EI)(2L^3 - 3L^2x + x^3)$ $y_{max} = FL^3/3EI$
2	$M = \tfrac{1}{2}wx^2$ $M_{max} = \tfrac{1}{2}wL^2$	$y = (w/24EI)(3L^4 - 4L^3x + x^4)$ $y_{max} = wL^4/8EI$
3	$M = wx^3/6L$ $M_{max} = wL^2/6$	$y = (w/120EIL)(4L^5 - 5L^4x + x^5)$ $y_{max} = wL^4/30EI$
4	$M = -\tfrac{1}{2}Fx$ $M_{max} = -\tfrac{1}{4}FL$	$y = (Fx/48EI)(3L^2 - 4x^2)$ $y_{max} = FL^3/48EI$
5	$M = (\tfrac{1}{2}wx)(x - L)$ $M_{max} = -wL^2/8$	$y = (wx/24EI)(L^3 - 2Lx^2 + x^3)$ $y_{max} = 5wL^4/384EI$
6	$M = (-wx/6L)(L^2 - x^2)$ $M_{max} = -.064wL^2$ at $x = .5774L$	$y = (wx/360EIL)(7L^4 - 10L^2x^2 + 3x^4)$ $y_{max} = .00652wL^4/EI$ at $x = .5193L$
7	$M = \tfrac{1}{2}F[(\tfrac{1}{4}L) - x]$ $M_{max} = FL/8$ at $x = 0$ $M_{max} = -FL/8$ at $x = \tfrac{1}{2}L$	$y = (Fx^2/48EI)(3L - 4x)$ $y_{max} = FL^3/192EI$
8	$M = (\tfrac{1}{2}wL^2)[(1/6) - (x/L) + (x/L)^2]$ $M_{max} = wL^2/12$ at $x = 0$ and $x = L$ $M = -wL^2/24$ at $x = \tfrac{1}{2}L$	$y = (wx^2/24EI)(L - x)^2$ $y_{max} = wL^4/384EI$
9	$M_a = Fx_a$ $M_b = (Fa/b)(b - x_b)$ $M_{max} = Fa$ at $x_a = a$	$y_a = (F/3EI)[(a^2+ab)(a-x_a) + (x_a/2)(x_a^2 - a^2)]$ $y_b = (Fax_b/6EI)[3x_b - (x_b^2/b) - 2b]$ $y_{tip} = (Fa^2/3EI)(a + b)$ (max up) $y_{max} = (0.06415)Fab^2/EI$ at $x_b = .4226b$ (max down)

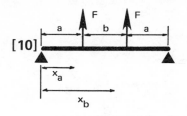

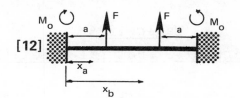

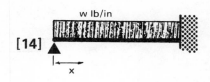

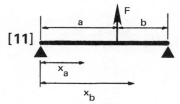

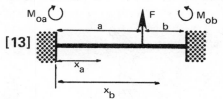

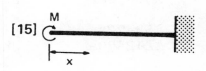

CASE	MOMENT	DEFLECTION
10	$M_a = -Fx_a$ $M_b = -Fa$ $M_{max} = -Fa$ (everywhere between loads)	$y_a = (Fx_a/6EI)[(3a)(L-a) - x_a^2]$ $y_b = (Fa/6EI)[3x_b(L - x_b) - a^2]$ $y_{max} = (Fa/24EI)(3L^2 - 4a^2)$
11	$M_a = -Fbx_a/L$ $M_b = -Fa(L - x_b)/L$ $M_{max} = -Fab/L$ at $x_a = a$	$y_a = (Fbx_a/6EIL)(L^2 - b^2 - x_a^2)$ $y_b = (Fb/6EIL)[(L/b)(x_b-a)^3 + (L^2 - b^2)x_b - x_b^3]$ $y = Fa^2b^2/3EIL$ at $x_a = a$ $y_{max} = (.06415Fb/EIL)(L^2 - b^2)^{3/2}$ at $x = \sqrt{a(L+b)/3}$
12	$M_a = (Fa/L)(L - a) - Fx_a$ $M_b = Fa^2/L$ $M_o = (Fa/L)(L - a)$	$y_a = (Fx_a^2/2EI)[a(1 - (a/L)) - (x_a/3)]$ $y_b = (Fa^2/2EI)[x_b - (x_b^2/L) - (a/3)]$ $y_{max} = (Fa^2/24EI)(3L - 4a)$ at $x = \frac{1}{2}L$
13	$M_a = (Fb^2/L^3)[aL - x_a(L + 2a)]$ $M_b = (Fa^2/L^3)[bL - (L - x_b)(L + 2b)]$ $M_{oa} = Fab^2/L^2$ (max when $a<b$) $M_{ob} = Fa^2b/L^2$ (max when $a>b$) $M = -2Fa^2b^2/L^3$ at $x_a = a$	$y_a = (Fx_a^2b^2/6EIL^3)[3aL - x_a(3a + b)]$ $y_b = (F(L-x_b)^2a^2/6EIL^3)[3bL - (L-x_b)(3b + a)]$ $y = Fa^3b^3/3EIL^3$ at $x_a = a$ $y_{max} = 2Fa^3b^2/[3EI(L+2a)^2]$ at $x = 2aL/(L+2a)$
14	$M = (3wLx/8) - \frac{1}{2}wx^2$ $M_{max} = wL^2/8$ at $x = L$	$y = (wx/48EI)[L^3 - 3Lx^2 + 2x^3]$ $y_{max} = wL^4/185EI$ at $x = .4215L$
15	$M = M$ everywhere	$y = Mx^2/2EI$ $y_{max} = ML^2/2EI$ at free end

Table 12.5
Properties of Structural Steel (ksi)

Designation	Application	S_u	S_y	Approximate S_e
A36-70a	shapes	58-80	36	29-40
	plates	58-80	36	29-40
A53-72a	pipe	60	35	30
A242-70a	shapes	70	50	35
	plates to 3/4"	70	50	35
A440-70a	shapes	70	50	35
	plates to 3/4"	70	50	35
A441-70a	shapes	70	50	35
	plates to 3/4"	70	50	35
A500-72	tubes	45	33	22
A501-71a	tubes	58	36	29
A514-70	plates to 3/4"	115-135	100	55
A529-72	shapes	60-85	42	30-42
	plates to ½"	60-85	42	30-42
A570-72	sheet/strip	55	40	27
A572-72	shapes	60	42	30
	plates	60	42	30
A588-71	shapes	70	50	35
	plates to 4"	70	50	35
A606-71	hot rolled sheet	70	50	35
	cold rolled sheet	65	45	32
A607-70	sheet	60	45	30
A618-71	shapes	70	50	35
	tubes	70	50	35

Table 12.6
Properties of Structural Aluminum (ksi)

Designation	Application	S_u	S_y	Approximate S_e (EE8 cyc.)
2014-T6	shapes/bars	63	55	19
6061-T6	all	42	35	14.5

Table 12.7
Properties of Structural Magnesium (ksi)

Designation	Application	S_u	S_y	Approximate S_e (EE7 cyc.)
AZ31	shapes	38	29	19
AZ61	shapes	45	33	19
AZ80	shapes	55	40	

Bibliography

1. Borg, Sidney F., and Gennaro, Joseph J., <u>Advanced Structural Analysis</u>, D. Van Nostrand Company, Inc., Princeton, NJ 1960

2. Jensen, Alfred, <u>Applied Strength of Materials</u>, McGraw-Hill Book Company, New York, NY 1957

3. Kurtz, Max, <u>Comprehensive Structural Design Guide</u>, McGraw-Hill Book Company, New York, NY 1969

4. Sutherland, Hale, and Bowman, Harry L., <u>Structural Theory</u>, 2nd. ed., John Wiley and Sons, New York, NY 1935

5. Timoshenko, S., <u>Strength of Materials</u>, part 1, 2nd ed., D. Van Nostrand Company, Inc., New York, NY 1940

6. Timoshenko, S., and Young, D.H., <u>Theory of Structures</u>, 1st ed., McGraw-Hill Book Company, Inc., New York, NY 1945

Practice Problems: MECHANICS OF MATERIALS

Required

1. A 14 foot long simple beam is uniformly loaded with 200 pounds per foot over its entire length. If the beam is 3.625" wide and 7.625" deep, what is the maximum bending stress? What is the maximum shear stress?

2. A reinforced concrete beam is illustrated below. It is subjected to a maximum moment of 8125 ft-lb. The total cross sectional area of steel is 1 square inch. Take the modulus of elasticity for concrete to be 2 EE6 psi. Disregard the area of concrete in tension and use the transformation method to calculate the maximum stress in the steel and concrete.

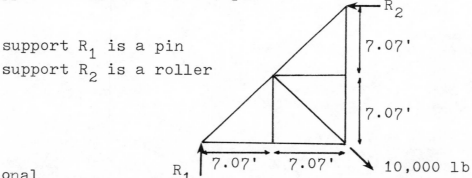

3. A steel truss with pinned joints in constructed as shown below. The length/area ratio for each member is 50 using consistent units. What is the vertical deflection at the point where the 10 kip load is applied? Use E = 2.9 EE7 psi.

support R_1 is a pin

support R_2 is a roller

Optional

4. A beam 25 feet long is simply supported at the left end and 5 feet from the right end. A uniform load of 2 kips/ft extends over a 10 foot length starting from the left end. There is also a concentrated 10 kip load at the right end. Draw the shear and moment diagrams.

5. A 40 foot long steel beam with moment of inertia I is reinforced along the middle 20 feet, leaving the two 10 foot ends unreinforced. The moment of inertia of the reinforced section is 2I. A 20,000 pound load is applied mid-span along the beam, 20 feet from each end. What is the deflection in terms of EI?

6. The truss shown on the next page carries a moving uniform live load of 2 kips per foot and a moving concentrated live load of 15 kips. What are the maximum forces in members Bb and BC?

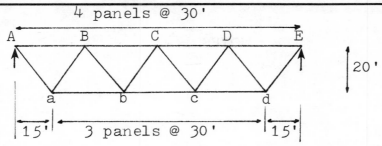

7. A 24 foot long, 4" x 4" white oak timber (E = 1.5 EE6 psi) is used to support a sign. One end is fixed in a deep concrete base, the other end supports a sign 24 feet above the ground. Neglect wind effects and find the critical buckling sign weight.

8. What is the mid-span deflection for the steel beam shown below? The cross sectional moment of inertia is 200 inches^4.

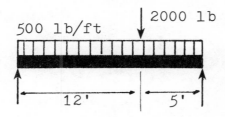

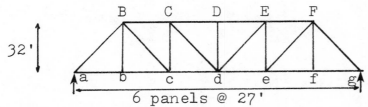

9. The truss shown below carries a group of live loads along its bottom chord. What is the maximum force in member CD?

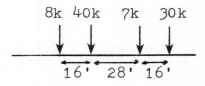

10. The allowable stress in a steel beam is 20,000 psi. If the maximum applied moment is 1.5 EE5 ft-lb, what is the required section modulus?

NOTES

INDETERMINATE STRUCTURES

Nomenclature

A	area	ft^2
C	carry-over factor	-
COM	carry-over moment	ft-lb
D	distribution factor	-
E	modulus of elasticity	lb/ft^2
F	fixity factor	-
I	moment of inertia	in^4
k	spring constant	ft-lb/radian
K	stiffness factor	ft-lb
L	length	ft
M	moment	ft-lb
P	load	lb
R	relative stiffness, or reaction	ft-lb, or lb
S	force in truss member	lb
T	temperature	°F
u	dummy force in truss member	lb
w	continuous load	lb/ft
x*	distance to centroid of moment diagram	ft

Symbols

δ	deflection	ft
σ	stress	lb/ft^2
α	linear coefficient of thermal expansion	1/°F
θ	angle	radians

Subscripts

BM	bending moment	
c	concrete	
M	due to moment M	
P	due to force P	
s	steel	

1. Introduction

A structure that is statically indeterminate (redundant) is one for which the equations of statics are not sufficient to determine all reactions, moments, and internal stress distributions. Additional formulas involving deflection relationships are required to completely determine these unknowns. Although there are a large number of problem types which are statically indeterminate, this chapter is primarily concerned with the following cases:

* Beams on more than two supports (continuous beams)
* Trusses with more than 2(# joints) -3 members
* Rigid frames

The degree of redundancy is equal to the number of reactions or members that would have to be removed in order to make the structure statically determinate. For example, a 2-span beam on three simple supports is redundant to the first degree.

2. Simple Structures: Consistent Deformation

The method of consistent deformation may be used to evaluate simple structures consisting of two or three members in tension or compression. Although this method cannot be proceduralized, it is very simple to learn and to apply. The method makes use of geometry to develop relationships between the deflections (deformations) between different members or locations on the structure.

Example 13.1

A pile is constructed of concrete with a steel jacket. What is the stress in the steel and concrete if a load P is applied? Assume the end caps are rigid and the steel-concrete bond is perfect.

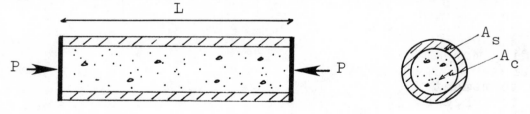

Let P_c and P_s be the loads carried by the concrete and steel respectively. Then,

$$P_c + P_s = P \qquad\qquad I$$

The deformation of the steel is

$$\delta_s = \frac{P_s L}{A_s E_s}$$

Similarly, the deflection of the concrete is

$$\delta_c = \frac{P_c L}{A_c E_c}$$

But $\delta_c = \delta_s$ since the bonding is perfect. Therefore,

$$\frac{P_c}{A_c E_c} - \frac{P_s}{A_s E_s} = 0 \qquad\qquad\qquad \text{II}$$

Equations I and II are solved simultaneously to determine P_c and P_s. The respective stresses are:

$$\sigma_s = P_s / A_s$$

$$\sigma_c = P_c / A_c$$

Example 13.2

A uniform bar is clamped at both ends and the axial load applied near one of the supports. What are the reactions?

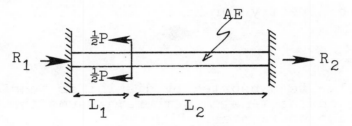

Clearly the first required equation is

$$R_1 + R_2 = P \qquad\qquad\qquad \text{III}$$

The shortening of section 1 due to the reaction R_1 is

$$\delta_1 = \frac{-R_1 L_1}{AE}$$

The elongation of section 2 due to the reaction R_2 is

$$\delta_2 = \frac{R_2 L_2}{AE}$$

However, the bar is continuous, so $\delta_1 = -\delta_2$. Therefore,

$$R_1 L_1 = R_2 L_2 \qquad\qquad\qquad \text{IV}$$

Equations III and IV are solved simultaneously to find R_1 and R_2.

Example 13.3

The non-uniform bar shown below is clamped at both ends. What are the reactions at both ends if a temperature change of ΔT is experienced?

The thermal deformations of sections 1 and 2 may be calculated directly.

$$\delta_1 = \alpha_1 L_1 \Delta T \qquad\qquad V$$

$$\delta_2 = \alpha_2 L_2 \Delta T \qquad\qquad VI$$

The total deformation is $\delta = \delta_1 + \delta_2$. However, the deformation may be calculated from the principles of mechanics of material.

$$\delta = \frac{RL_1}{A_1 E_1} + \frac{RL_2}{A_2 E_2} \qquad\qquad VII$$

Combining equations V, VI, and VII gives

$$(\alpha_1 L_1 + \alpha_2 L_2)\Delta T = (\frac{L_1}{A_1 E_1} + \frac{L_2}{A_2 E_2})R$$

This can be solved directly for R.

◆

Example 13.4

The beam shown below is supported by dissimilar tension members. What are the reactions in the tension members? Assume the horizontal bar is rigid and remains horizontal.

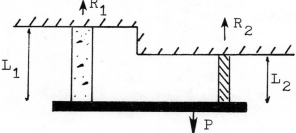

The required equilibirum condition is

$$R_1 + R_2 = P \qquad\qquad VIII$$

The elongations of the two tension members are

$$\delta_1 = \frac{R_1 L_1}{A_1 E_1}$$

$$\delta_2 = \frac{R_2 L_2}{A_2 E_2}$$

If the horizontal bar remains horizontal, then $\delta_1 = \delta_2$. Therefore,

$$\frac{R_1 L_1}{A_1 E_1} = \frac{R_2 L_2}{A_2 E_2} \qquad\qquad IX$$

Equations VIII and IX are solved simultaneously to find R_1 and R_2.

◆

Example 13.5

Find the forces in the three tension members.

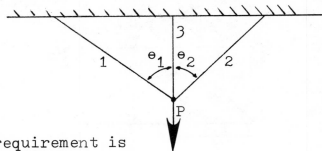

The equilibrium requirement is

$$P_{1y} + P_3 + P_{2y} = P$$

or
$$P_1 \cos\theta_1 + P_3 + P_2 \cos\theta_2 = P \qquad\qquad X$$

The vertical elongations of all three tension members are the same at the junction.

$$\frac{P_1 L_1}{A_1 E_1}(\cos\theta_1) = \frac{P_3 L_3}{A_3 E_3} = \frac{P_2 L_2}{A_2 E_2}(\cos\theta_2) \qquad\qquad XI$$

Equations X and XI may be solved simultaneously to find P_1, P_2, and P_3. It may be necessary to work with the x-components of the deflections in order to find a third equation.

3. Continuous Beams: Moment Distribution Method

The moment distribution method is extremely powerful. It can be used on beams of almost any complexity. This includes beams with built in ends and beams with yielding supports. It converges rapidly to a solution, despite the fact that the method is essentially an iterative process. Furthermore, the moment distribution method is easily learned, even if the underlying theory is not understood.

Consider the simply-supported beam shown in figure 13.1. The type of loading is not important, as only general relationships are being developed. The deflection angles θ_A and θ_B may be found from the moment-area method discussed in chapter 12. A_{BM} is the area of the bending moment diagram and x^* is the distance from the left end to the centroid of the bending moment diagram. By convention, a clockwise rotation is positive, hence the negative sign in equation 13.2.

$$\theta_A = \frac{A_{BM}(L-x^*)}{EIL} \qquad\qquad 13.1$$

$$\theta_B = \frac{-A_{BM}(x^*)}{EIL} \qquad\qquad 13.2$$

Figure 13.1

Simply Supported Beam with Concentrated Loads

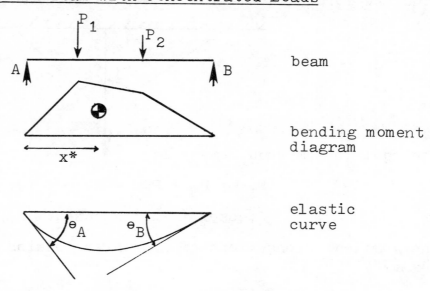

beam

bending moment
diagram

elastic
curve

Similarly, the deflection angles can be calculated for a simply
supported beam with a uniformly distributed load of w pounds per unit
length over its entire length.

$$\theta_A = \frac{wL^3}{24EI}$$ 13.3

$$\theta_B = \frac{-wL^3}{24EI}$$ 13.4

Finally, if a simply-supported beam is acted upon by a clockwise
couple (a moment) at one end as shown in figure 13.2, the angles of
rotation will be given by equations 13.5 and 13.6.

$$\theta_A = \frac{M_A L}{3EI}$$ 13.5

$$\theta_B = \frac{-M_A L}{6EI}$$ 13.6

The convention followed in this chapter is that any moment or couple
which would rotate an end of a beam clockwise is positive. Similarly,
any moment which would rotate an interior joint clockwise is positive.

Figure 13.2

Simply-Supported Beam with a Couple

Figure 13.2 is particularly important because the equilibrium of the forces and moment is not dependent on whether the moment is applied and the forces react, or if the forces are applied and the moment reacts. If the forces are applied to the beam in figure 13.2, the moment develops to keep the beam end horizontal. A moment that is required to keep a beam end horizontal is known as a <u>fixed-end moment</u>.

<u>Example 13.6</u>

What fixed-end moment at the left end is required to keep the left end of the beam shown below horizontal?

From equation 13.1, if the left end of the beam were free to rotate, the angle of rotation would be

$$\theta_{A,P} = \frac{A_{BM}(L-x^*)}{EIL}$$

To keep the left end horizontal will require a counter-clockwise moment at the left end. The angle produced at end A by a moment at end A is given by equation 13.5.

$$\theta_{A,M} = \frac{M_A L}{3EI}$$

Since the total angle θ_A is zero,

$$\frac{A_{BM}(L-x^*)}{EIL} + \frac{M_A L}{3EI} = 0$$

$$M_A = \frac{-3A_{BM}(L-x^*)}{L^2}$$

M_A is negative which is consistent with the convention that counter-clockwise moments at the end of a beam are negative. The bending moment of the beam is given below.

The area of the bending moment diagram is $\frac{1}{2}(L)\frac{1}{4}PL = PL^2/8$. Also, $x^* = \frac{1}{2}L$. Therefore, the moment at A required to keep the left end horizontal is

$$M_A = \frac{-3(\frac{PL^2}{8})(L-\frac{1}{2}L)}{L^2} = \frac{-3PL}{16}$$

Being able to find the fixed end moment for a loaded beam is absolutely necessary to the success of the moment distribution method. Luckily, extensive tables are available which contain the fixed end moments for almost every conceivable loading. Once such collection is shown in table 13.1.

<u>Table 13.1</u>

<u>Fixed End Moments</u>

(This table assumes that both ends remain horizontal.)
(w in lbf/ft)

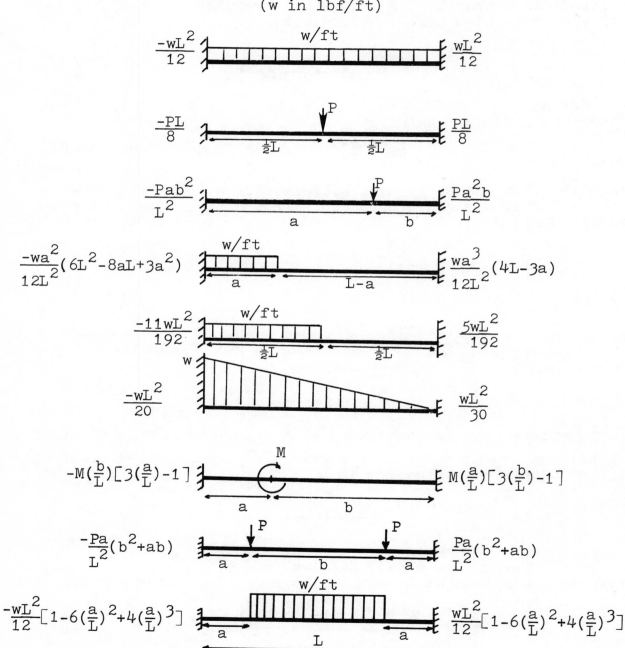

Example 13.7

Find the fixed-end moment for the left end of the simply-supported beam shown below. The beam is acted upon by a moment of magnitude M_B at the right end.

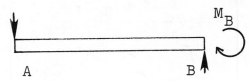

From equation 13.6, the angle of rotation at the left end due to a moment at the right end is

$$\theta_{A,M_B} = \frac{-M_B L}{6EI}$$

The angle at the left end due to the fixed end moment at the left end is given by equation 13.5.

$$\theta_{A,M_A} = \frac{M_A L}{3EI}$$

However, $\theta_A = 0$, so

$$\frac{M_B L}{6EI} = \frac{M_A L}{3EI} \qquad \text{or } M_A = \tfrac{1}{2}M_B$$

◆

From the above example, it is seen that a moment at one end of a simply-supported beam is partially transmitted to a fixed end. This transmitted moment is known as a <u>carry-over moment</u>, and the ratio of the carry-over moment to the original moment is known as the <u>carry-over factor</u>, C. In the above example, C was equal to $\tfrac{1}{2}$.

The carry-over factor can be determined for various degrees of fixity. The fixity is specified by a fixity factor, F, which varies from 0 (for a pin end) to 1 (for a built-in end).

$$C = \frac{2F}{3+F} \qquad\qquad 13.7$$

If the responding end is partially constrained by a spring or a yielding support, its fixity factor may be calculated from equation 13.8 and the spring constant, k, in units of in-lb/radian.

$$F = \frac{3k}{\frac{12EI}{L} + 3k} \qquad\qquad 13.8$$

The product EI is known as the <u>stiffness</u> of the beam. The <u>relative stiffness</u> is

$$R = \frac{EI}{L} \qquad\qquad 13.9$$

The <u>stiffness factor</u> is defined as

$$K = (3+F)R \qquad 13.10$$

At this point, it is possible to generalize about the moment required to produce a given angle of rotation. Equation 13.5 gives the moment for a simply supported beam.

$$M_A = \frac{3EI\theta}{L} \qquad 13.11$$

Equation 13.11 could have been derived by using equation 13.10. Since F = 0 for a pinned end or simply-supported end,

$$M_A = K\theta = 3R\theta = \frac{3EI\theta}{L} \qquad 13.12$$

If the beam is built-in at an opposite end, the moment required to produce an angle θ can be derived using equation 13.10 with F = 1.

$$M_A = \frac{4EI\theta}{L} \qquad 13.13$$

Therefore, we conclude that beams with built-in ends are (4/3) as stiff as beams with pinned ends.

Now, consider the joint B illustrated in figure 13.2(a). The joint rigidly connects a complex beam that is simply supported at ends A and D, but is fixed at end C. The relative stiffnesses are R_1, R_2, and R_3 respectively for beams BA, BC, and BD. Let a moment of magnitude U be applied to the joint B. The joint will rotate as shown in figure 13.2(b).

<u>Figure 13.2</u>

<u>A General Joint</u>

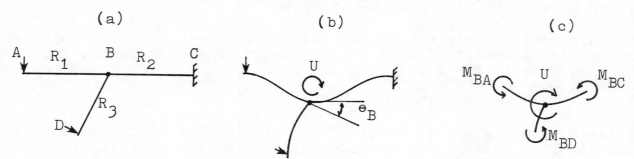

A free-body diagram of joint B is shown in figure 13.2(c). It is clear that the rotation will continue until the moment U is resisted by the stiffnesses of the beams:

$$U = M_{BA} + M_{BC} + M_{BD} \qquad 13.14$$

However, knowing θ_B gives the values of M_{BA}, M_{BC}, and M_{BD} from equations 13.12 and 13.13.

$$U = 3R_1\theta_B + 4R_2\theta_B + 3R_3\theta_B \qquad 13.15$$

The individual resisting moments, M_{BA}, M_{BC}, and M_{BD} can be found from the following analysis. Take M_{BA} for example.

$$\frac{M_{BA}}{U} = \frac{3R_1\Theta_B}{3R_1\Theta_B + 4R_2\Theta_B + 3R_3\Theta_B} \tag{13.16}$$

or

$$M_{BA} = U\left(\frac{K_1}{K_1 + K_2 + K_3}\right) \tag{13.17}$$

Similarly,

$$M_{BC} = U\left(\frac{K_2}{K_1 + K_2 + K_3}\right) \tag{13.18}$$

$$M_{BD} = U\left(\frac{K_3}{K_1 + K_2 + K_3}\right) \tag{13.19}$$

From equations 13.17, 13.18, and 13.19, it is apparent that the couple U is distributed to the three spans in proportion to the ratios of the stiffness factors. The quantities in parentheses are known as distribution factors since they distribute the applied moment, U, over all resisting beams.

At this time, it is possible to summarize the procedure for the moment distribution method.

step 1: Divide the beam into independent spans, with all ends assumed to be built in. If the beam is symmetrical, split it in half and work only with one half. Assume the beam is built-in where the cut in half is made.

step 2: Calculate the relative stiffness of each member. Over-hanging spans have R=0 by definition. Both E and I can be given a value of 1.0 if they are constant along the entire span.

$$R = \frac{EI}{L} \tag{13.20}$$

step 3: Calculate the stiffness factors, K, and the distribution factors, D, for each span meeting at a joint. F is associated with the actual end constraint (not the assumed built-in) for the end which looks into the joint. F=0 for simply-supported ends. F=1 for built-in ends and at continuous supports. (Wherever F=0, the moment will also be zero.)

$$K = (3+F)R \tag{13.21}$$

$$D = \frac{K}{\Sigma K} \tag{13.22}$$

step 4: Calculate the carry-over factor for each end of all members leading out of a joint. In general, C is $\frac{1}{2}$ except at joints which look out to a simply-supported or free end, in which case C=0. C is also $\frac{1}{2}$ for a simply supported end looking into an interior support on a continuous beam. (As in step 3, F is associated with the end which looks into the joint.)

$$C = \frac{2F}{3+F} \tag{13.23}$$

step 5: Calculate the fixed-end moment (FEM) using table 13.1 and the actual transverse loading on each span.

step 6: At each joint, balance any unbalanced moments by distributing a counter moment among the connecting members according to their distribution factors. At beam ends, add the following counter moments for balancing:

at a fixed end:	zero
at a pinned or free end:	-(unbalance)
at a partially restrained end:	-(1-F)(unbalance)

step 7: These distributed balancing moments produce carry-over moments at the opposite ends of members equal to the distributed moment times the carry-over factor, with the same sign as the distributed balancing moment. The carry-over moments appear at the opposite end from the location where the balancing moment was applied.

step 8: Repeat steps 6 and 7 until sufficient accuracy is obtained. Unless there is just one distribution (as in a 2-span beam), end the distribution process with a distribution, not a carry-over. The final moment at the end of any member is equal to the algebraic sum of the original fixed-end moments and all distributed balancing and carry-over moments. The final moment is the moment on the beam end, not the moment on the support due to the loads.

Example 13.8

Find the moments at points A, B, and C.

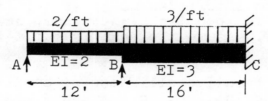

step 1: Divide the beam into spans A-B and B-C.

step 2: $R_{BA} = 2/12 = 1/6$

$R_{BC} = 3/16$

step 3: Since end C of the actual beam is fixed, $F_C = 1$, and from equation 13.21,

$$K_{BC} = (3+1)(\frac{3}{16}) = \frac{12}{16} = \frac{3}{4}$$

End A is simply supported, so $F_A = 0$

$$K_{BA} = (3+0)(\frac{1}{6}) = \frac{1}{2}$$

The only joint in the structure is at point B. From equation 13.22,

$$D_{BC} = \frac{\frac{3}{4}}{\frac{3}{4} + \frac{1}{2}} = 3/5 = .6$$

$$D_{BA} = \frac{\frac{1}{2}}{\frac{3}{4} + \frac{1}{2}} = 2/5 = .4$$

step 4: $C_{AB} = \frac{1}{2}$ $C_{BA} = 0$ $C_{BC} = \frac{1}{2}$ $C_{CB} = \frac{1}{2}$

step 5: Work with span BC, assuming both ends are fixed. From table 13.1, the fixed end moment at joint B is

$$M_{BC} = \frac{-wL^2}{12} = \frac{-(3)(16)^2}{12} = -64$$

Notice the sign convention. The moment at B is caused by the applied load of 3/ft. The applied moment is clockwise, and therefore it is positive. However, fixed-end moments are the resisting moments, and the resisting moment at B is counterclockwise.

$$M_C = +64$$

Now, work with span BA, still assuming both ends are fixed.

$$M_A = \frac{-(2)(12)^2}{12} = -24$$

$$M_{BA} = +24$$

At this time, it is necessary to start a table to keep track of the results of the above and all subsequent calculations.

	M_A	M_{BA}	M_{BC}	M_C
FEM	-24	24	-64	64

step 6: At joint A, there is a -24 moment which is not balanced by anything. Since joint A is a pinned end (which cannot carry any moment) the moment there must be removed. This is done by adding +24 to that joint. Since joint A has only one 'side', the entire +24 is added to M_A.

	M_A	M_{BA}	M_{BC}	M_C
FEM	-24	24	-64	64
Balance	+24			

At joint B, there is an unbalance of (24-64) = -40. So, this must be balanced by adding +40 to joint B. However, joint B goes in two directions (towards A and C), so the +40 must be distributed to spans BA and BC in proportion to their stiffnesses. The distribution factors were previously found to be .4 and .6 for M_{BA} and M_{BC} respectively.

$$\text{Balance } M_{BA} = .4(40) = 16$$
$$\text{Balance } M_{BC} = .6(40) = 24$$

	M_A	M_{BA}	M_{BC}	M_C
FEM	-24	24	-64	64
Balance	24	16	24	

The support at joint C was assumed fixed in step 1. Since it actually is fixed, no correction is required.

	M_A	M_{BA}	M_{BC}	M_C
FEM	-24	24	-64	64
Balance	24	16	24	0

step 7: The balancing moments produce carry-over moments. If we add +24 to M_A, we must also add $\frac{1}{2}(24) = 12$ to M_{BA} since the carry-over factor $C_{AB} = \frac{1}{2}$. Similarly, the rest of the following table is prepared.

	M_A	M_{BA}	M_{BC}	M_C
FEM	-24	24	-64	64
Balance	24	16	24	0
COM	0	12	0	12

step 8: However, now the addition of the carry-over moments has unbalanced the beam again. So, the process must be repeated from step 6.

step 6, revisited: At joint A: No carry-over moment was applied, so no balancing is needed.

At joint B: The total carry-over moment applied was (12+0) = 12. So, a balancing moment of -12 is needed. The -12 is distributed to M_{BA} and M_{BC} according to

$$M_{BA} = .4(-12) = -4.8$$
$$M_{BC} = .6(-12) = -7.2$$

At joint C: Joint C is actually fixed, so the carry-over moment of 12 is balanced by the beam's support. No balancing moment is needed.

step 7, revisited: Using the carry-over factors, we add 0 to M_A and -3.6 to M_C.

	M_A	M_{BA}	M_{BC}	M_C
FEM	-24	24	-64	64
1st Balance	24	16	24	0
1st COM	0	12	0	12
2nd Balance	0	-4.8	-7.2	0
2nd COM	0	0	0	-3.6

step 8, revisited: Since the -3.6 is balanced by the external support at joint C, and since all other terms are zero, no further work is required.

$$\Sigma M_A = -24 + 24 = 0$$
$$\Sigma M_B = 24 + 16 + 12 - 4.8 = \pm 47.2$$
$$\Sigma M_C = 64 + 12 - 3.6 = 72.4$$

Example 13.9

Draw the moment diagram for the beam shown below.

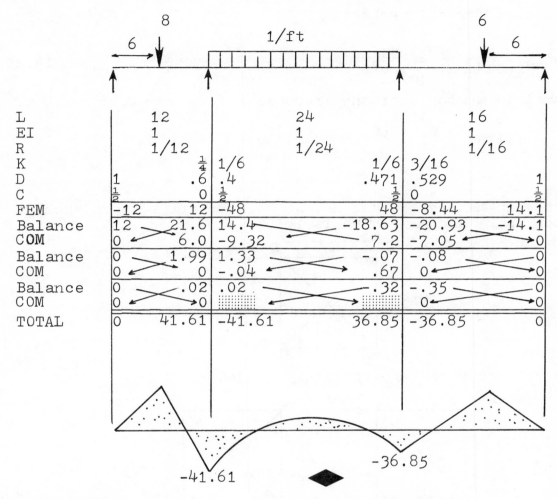

4. Indeterminate Trusses: Dummy Unit Load Method

Due to the time required for solutions, it is unlikely that an indeterminate truss will have more than one redundant member. Therefore, the following method is written specifically for trusses which are indeterminate to the first degree.

> step 1: Draw the truss twice. Omit the redundant member on both trusses.

> step 2: Load the first truss (which is now determinate) with the actual loads.

> step 3: Calculate the forces, S, in all of the members.

> step 4: Load the second truss with two unit forces acting colinearly towards each other along the line of the redundant member.

> step 5: Calculate the force, u, in each of the members.

> step 6: Calculate the force in the redundant member from equation 13.24.

$$S_{redundant} = \frac{-\Sigma(\frac{SuL}{AE})}{\Sigma(\frac{u^2L}{AE})}$$ 13.24

If AE is the same for all members,

$$S_{redundant} = \frac{-\Sigma SuL}{\Sigma u^2 L}$$ 13.25

The true force in member J of the truss is

$$P_{j,true} = S_j + (S_{redundant})u_j$$ 13.26

Example 13.10

Find the force in members BC and BD. AE = 1 for all members except for CB (which is 2) and AD (which is 1.5).

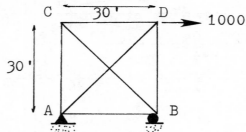

The two trusses are shown appropriately loaded.

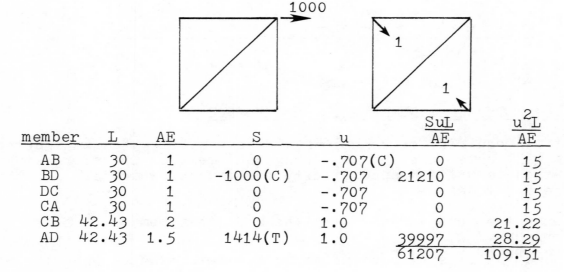

member	L	AE	S	u	$\frac{SuL}{AE}$	$\frac{u^2L}{AE}$
AB	30	1	0	-.707(C)	0	15
BD	30	1	-1000(C)	-.707	21210	15
DC	30	1	0	-.707	0	15
CA	30	1	0	-.707	0	15
CB	42.43	2	0	1.0	0	21.22
AD	42.43	1.5	1414(T)	1.0	39997	28.29
					61207	109.51

From equation 13.24,

$$S_{BC} = \frac{-61207}{109.51} = -558.9 \; (C)$$

From equation 13.26,

$$P_{BD,true} = -1000 + (-558.9)(-.707) = -604.9 \; (C)$$

◆

5. Continuous Frameworks Not in a Straight Line

The moment distribution method is capable of handling all continuous frameworks. It is not necessary for the beams to be in straight lines, nor is it necessary for the members to meet at joints with any particular angle (such as 90°, 180°, etc.). It is only necessary that the moment at the joint be known.

The following example showns a rigid frame that is not loaded transversely on the beam, but rather carries a moment applied directly to a joint.

Example 13.11

Find the bending moments at the ends of members OA and OB, both of which are steel. Joint 0 is rigid. The vertical load of 6000 is applied on a lever arm of length 3.

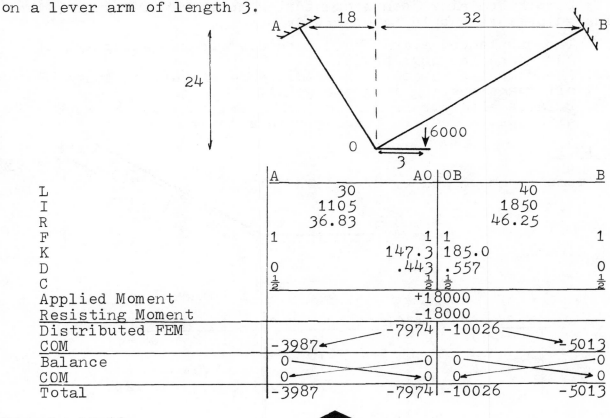

	A	AO	OB	B
L		30		40
I		1105		1850
R		36.83		46.25
F	1	1	1	1
K		147.3	185.0	
D	0	.443	.557	0
C	½	½	½	½
Applied Moment		+18000		
Resisting Moment		−18000		
Distributed FEM		−7974	−10026	
COM	−3987			−5013
Balance	0	0	0	0
COM	0	0	0	0
Total	−3987	−7974	−10026	−5013

Example 13.12

Repeat example 13.11 for end A being pinned.

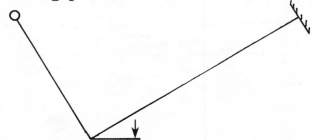

	A	A0	0B	A
L		30	40	
I		1105	1850	
R		36.83	46.25	
F	0	1	1	1
K		110.5	185	
D		.374	.626	
C	½	0	½	½
Applied Moment		18,000		
Resisting Moment		-18,000		
Distributed FEM		-6732	-11208	
COM	0			-5604
Total	0	-6732	-11208	-5606

◆

If the framework contains joints connecting more than two spans, the following guidelines should be observed:

    1. The unbalanced moment should be distributed among all of the spans leaving the joint.

    2. The moment on a joint is produced by all of the loads on all spans leading into the joint.

    3. Balance all members during each iteration. Do not work with just 2 members leading into a joint.

Example 13.13

Find the moments at points A, B, and C.

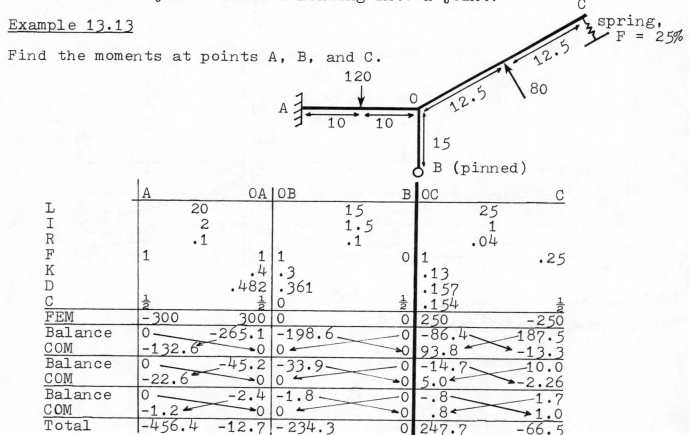

	A	0A	0B	B	0C	C
L		20	15		25	
I		2	1.5		1	
R		.1	.1		.04	
F	1	1	1	0	1	.25
K		.4	.3		.13	
D		.482	.361		.157	
C	½	½	0	½	.154	½
FEM	-300	300	0	0	250	-250
Balance	0	-265.1	-198.6	0	-86.4	187.5
COM	-132.6	0	0	0	93.8	-13.3
Balance	0	-45.2	-33.9	0	-14.7	10.0
COM	-22.6	0	0	0	5.0	-2.26
Balance	0	-2.4	-1.8	0	-.8	1.7
COM	-1.2	0	0	0	.8	1.0
Total	-456.4	-12.7	-234.3	0	247.7	-66.5

◆

6. Rigid Portal Frames

The rigid portal frame or 'bent' is a fundamental structural unit. Therefore, an ability to solve portal frames is essential.

Figure 13.3

A Rigid Frame

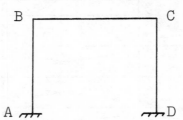

Because of the rigid joints at B and C in figure 13.3, it follows that all members meeting at a joint rotate through the same angle at that joint. Furthermore, any moment generated on a member will be transmitted through the joint to a connecting member.

A. No Sidesway

If a portal frame is symmetrical with repect to its leg lengths and loading, it will not sway or tend to move horizontally. If the frame or loading is unsymmetrical, it will sway unless prevented from doing so. Frames which are prevented from swaying are easily solved with the moment distribution method.

Figure 13.4

A Rigid Frame with Sidesway

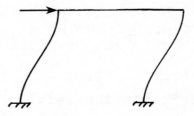

Example 13.14

Find the horizontal reactions at points A and D. The frame is kept from moving laterally by a passive force at C. That is, the force R at C resists lateral movement, but does not add any load or moment to the structure.

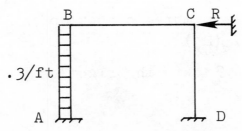

	A	AB	BC	BC	CD	D
L		25		70		25
I		1000		10,500		1000
R		40		150		40
F	1	1	1	1	1	1
K	160	160	600	600	160	160
D	0	.21	.79	.79	.21	0
C	½	½	½	½	½	½
FEM	-15.63	15.63	0	0	0	0
Balance	0	-3.28	-12.35	0	0	0
COM	-1.64	0	0	-6.18	0	0
Balance	0	0	0	4.88	1.30	0
COM	0	0	2.44	0	0	.65
Balance	0	-.51	-1.93	0	0	0
COM	-.25	0	0	-.97	0	0
Balance	0	0	0	.77	.20	0
COM	0	0	.38	0	0	.10
Balance	0	-.08	-.30	0	0	0
COM	-.04	0	0		0	0
Total	-17.57	11.76	-11.76	-1.50	1.50	.75

The free-bodies of the sections are given below.

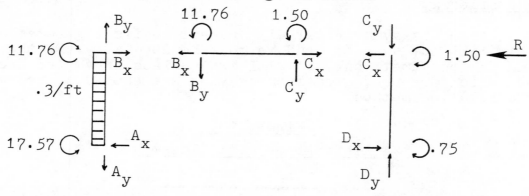

Work with member AB. To find B_x, sum moments about point A, choosing clockwise as positive.

$$\Sigma M_A = \tfrac{1}{2}(25)^2(.3) + 11.76 - 17.57 + 25(B_x) = 0$$
$$B_x = -3.52 \text{ (to the left)}$$

Similarly, to find A_x:

$$\Sigma M_B = -\tfrac{1}{2}(25)^2(.3) + 11.76 - 17.57 + 25(A_x) = 0$$
$$A_x = 3.98$$

Work with member DC. Clearly, $C_x = -B_x = 3.52$ (to the right). To find D_x, sum moments about point C.

$$\Sigma M_C = 1,50 + .75 - 25(D_x) = 0$$
$$D_x = .09 \quad \text{(to the right)}$$

B. Frames with Sidesway

If an unsymmetrically-loaded frame is not braced against sidesway, lateral movement will occur until the generated shear force in the vertical columns just equals the horizontal force needed to prevent sidesway.

Example 13.15

For the frame used in example 13.14, determine the shear required to prohibit further sidesway.

Summing moments about point D,

$$\Sigma M_D = .75 + 1.50 - 25(R) + 25(3.52) = 0$$

$$R = 3.61 \text{ (to the left)}$$

◆

The following procedure may be used to calculate the fixed-end moments for a frame for which sidesway is allowed.

step 1: Solve the frame assuming sidesway is prohibited. This was illustrated in example 13.14. Notice that the reaction R was not used in the moment distribution method.

step 2: Calculate the reaction required to prevent sidesway. This was illustrated in example 13.15.

step 3: Remove the actual loads and replace them with an arbitrary total fixed end moment such as 100 or 1000. Divide this fixed end moment between the vertical members in proportion to (EI/L^2). If both vertical members are the same, then each carries half. The signs of the moment at both ends of both vertical members are the same.

step 4: Solve the frame using the moment distribution method.

step 5: Find the reaction required to prevent sidesway.

step 6: Calculate corrections for each end and joint.

$$\Delta M = (\frac{\text{reaction from step 2}}{\text{reaction from step 4}})(\text{moment from step 4}) \quad 13.27$$

step 7: Calculate the true moment at all joints.

$$M = (\text{moment from step 1}) - \Delta M \qquad 13.28$$

Example 13.16

Find the moment at end D for the frame shown in example 13.14. Sidesway is permitted.

step 1: See example 13.14.

step 2: See example 13.15.

step 3: Load the frame with a total moment of 100. (Continued next page.)

For member DC, $(EI/L^2) = 1.6$. For member AB, $(EI/L^2) = 1.6$. So, the proportion of the 100 moment taken by member AB is

$$\frac{1.6}{1.6 + 1.6} = .5$$

Similarly, member DC is loaded with 50.

step 4:

	A	AB	BC	BC	CD	D
L		25		70		25	
I		1000		10,500		1000	
R		40		150		40	
F	1	1	1	1	1	1
K	160	160	600	600	160	160
D	0	.21	.79	.79	.21	0
C	½	½	½	½	½	½
I/L^2		1.6	not loaded		1.6	
FEM	-50	-50	0	0	-50	-50
Balance	0	10.5	39.5	39.5	10.5	0
COM	5.25	0	19.75	19.75	0	5.25
Balance	0	-4.15	-15.60	-15.60	-4.15	0
COM	-2.07	0	-7.8	-7.8	0	-2.07
Balance	0	1.64	6.16	6.16	1.64	0
COM	.82	0	3.08	3.08	0	.82
Balance	0	-.65	-2.43	-2.43	-.65	0
COM	-.32	0		-2.43	0	-.32
Total	-46.32	-42.66	42.66	42.66	-42.66	-46.32

step 5: Take member AB as a freebody.

$B_x \longrightarrow$ ↻ 42.66

↻ 46.32

$$\Sigma M_A = -42.66 - 46.32 + B_x(25) = 0$$

$$B_x = 3.56 \text{ (to the right)}$$

Take member DC as a freebody. $C_x = -B_x = -3.56$ (to the left)

$C_x \longleftarrow$ | $\longrightarrow R$ ↻ 42.66

↻ 46.32

$$\Sigma M_D = -42.66 - 46.32 - 3.56(25) + R(25)$$

$$R = 7.12 \text{ (to the right)}$$

step 6: $\Delta M_D = (\frac{-3.61}{7.12})(-46.32) = (-.51)(-46.32) = 23.62$

> The negative sign was used in front of the 3.61 because of the difference in directions of the two reactions, 3.61 and 7.12.

step 7: $M_D = .75 - 23.62 = -22.87$

C. Frame Complications

1. Frames with Inclined Members

If a frame has an inclined member, figure 13.5 may be used to find the fixed-end moments. As long as the joints remain rigid, the fixed-end moments depend only on the horizontal moment arm and the vertical component of force. (Use the actual length, L, however in calculating the relative stiffness.)

Figure 13.5

Frame with Inclined Member

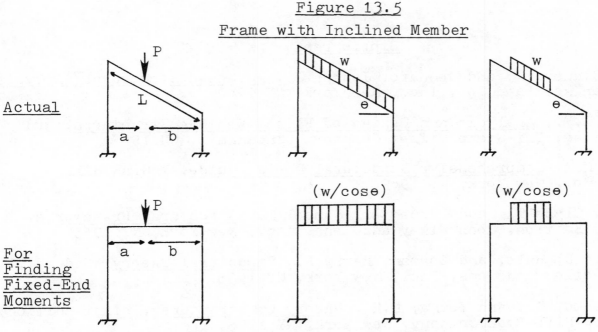

2. Frames with Concentrated Joint Loads

A frame which carries a concentrated load at a joint, as in figure 13.6, does not develop any fixed-end moments due to transverse loading since there are no transverse loads. However, the load does cause the frame to sway, and hence, moments are generated at the joints.

Figure 13.6

(a)   (b)

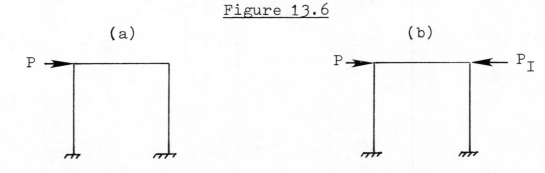

The solution procedure is similar to that used to correct moments on a frame which has sidesway. Place an imaginary load, P_I, at a joint which will resist the sidesway. Load both columns with an assumed moment divided in proportion to the column (EI/L^2) values. The sign of the assumed moments at both ends of the two columns should be the same. Calculate the force, R, required to counteract the assumed moment load. The actual moments are, then,

$$M = -(P_I/R)\binom{\text{moment from assumed}}{\text{moment loading}} \qquad 13.29$$

Watch the signs carefully. Do they make sense?

Bibliography

1. Borg, Sidney F., and Gennaro, Joseph J., <u>Advanced Structural Analysis</u>, D. Van Nostrand Co., Inc., New York, NY 1960

2. Bruhn, E.F., <u>Analysis and Design of Flight Vehicle Structures</u>, 3rd printing, Tri-State Offset Company, Cincinnati, OH 1965

3. Kurtz, Max, <u>Comprehensive Structural Design Guide</u>, McGraw Hill Book Co., New York, NY 1969

4. Morris, Clyde T., and Carpenter, Samuel T., <u>Structural Frameworks</u>, First Edition, John Wiley and Sons, Inc., New York, NY 1955

5. Sutherland, Hale, and Bowman, Harry L., <u>Structural Theory</u>, 2nd ed., John Wiley and Sons, Inc., New York, NY 1935

6. Timoshenko, S., and Young, D.H., <u>Theory of Structures</u>, First edition, McGraw Hill Book Company, New York, NY 1945

MOMENT DISTRIBUTION WORKSHEET

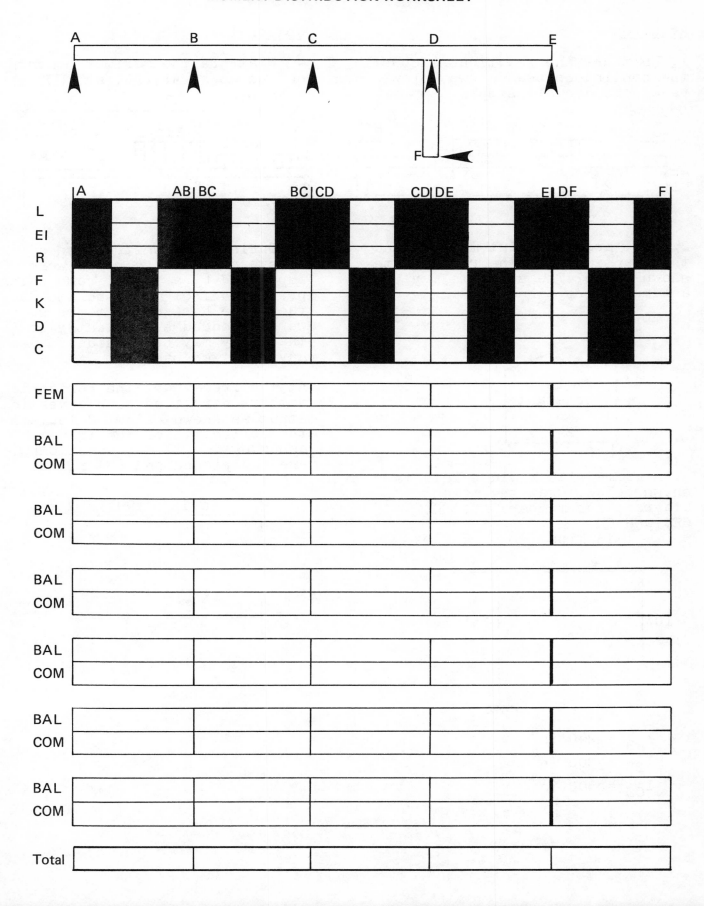

Practice Problems: INDETERMINATE STRUCTURES

<u>Required</u>

1. What are the reactions supporting the continuous beam shown below? The reactions are all simple.

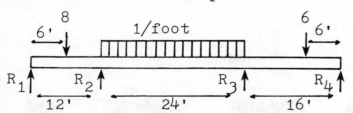

2. What are the joint moments and the reactions for the rigid frame shown below? The supports may be assumed to be pinned.

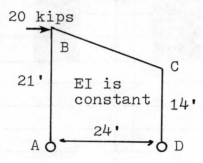

3. A frame with rigid joints is shown below. Draw the moment diagrams for members AB, BC, CE, and CD.

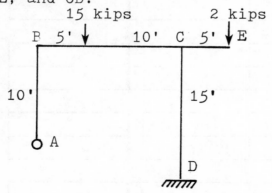

$I_{AB} = 240 \text{ in}^4$

$I_{BE} = 300 \text{ in}^4$

$I_{CD} = 300 \text{ in}^4$

<u>Optional</u>

4. What are the maximum moments on the spans AB, BC, and CD?

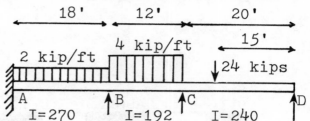

5. During the construction of Shasta dam, access ramps were supported by a steel frame which projected from the concrete dam face. The horizontal members were embedded sufficiently so that they may be considered as having fixed bases.

At the free ends, the horizontal members were tied with a vertical strut as shown below. Calculate the deflection of the frame assuming the tied connections are (a) pinned and (b) rigid.

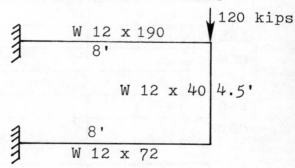

NOTES

NOTES

CONCRETE DESIGN

Nomenclature

a	short side length	in
A	area	in^2
b	long side length, or beam width	in
B	footing width	in
C	total compressive load	lb
d	top cover on reinforcement	in
D	diameter	in
DL	dead load	lb
E	modulus of elasticity	psi
f'_c	ultimate strength of concrete	psi
f_s	allowable stress in steel	psi
f_y	yield stress for steel	psi
H	soil height	in
j	a constant or ratio	-
k	a constant or ratio	-
K	a factor	psi
l_d	development length	in
L	length	ft
LL	live load	lb
M	moment	in-lb
n	ratio of (E_s/E_c)	-
p	ratio of steel to concrete area	-
p_g	ratio of steel area to gross area	-
p_{net}	net factored soil pressure	psf
P_{max}	maximum allowable load	lb
q	a ratio	-
R	reduction factor	-
R_h	horizontal soil reaction	lb
R_n	coefficient of resistance	psi
T	total tensile load	lb
U	ultimate load	lb
v	actual shear stress	psi
v_c	maximum allowable shear stress in concrete	psi
w	density, or load per unit length*	pcf, or lb/in
WL	wind load	lb

Symbols

ϕ	capacity reduction factor	-
β_c	ratio of long side to short side	-
θ	angle from the horizontal	degrees

---

*The ACI code uses 'w' to represent density. Although ρ is used in other chapters, w is used here to be consistent with the ACI code.

Subscripts

b	reinforcing bar
c	concrete, or column
f	footing
g	gross column
i	spiral bar
o	ideal at zero eccentricity
s	steel
u	actual ultimate
v	for stirrup steel

1. Introduction

This chapter is written for the civil engineer who does not work in the field of concrete design but who nevertheless wants to be able to solve some of the more standard concrete design and analysis problems.

Due to the wide variety of problems, variations, and complications, however, this chapter should be considered adequate only for the types of problems narrowly defined in the following pages. Of course, there is more than enough material for a 'non-concrete' engineer to study in a reasonable amount of time. This chapter is based on ACI 318-77.

There is no substitute for the ACI code. Civil engineers who daily work in concrete design should add the ACI code or the Uniform Building Code to their list of required references.

2. Definitions

Absorption: The process by which a liquid is drawn into and tends to fill permeable pores in a porous body. Also, the increase in weight of a porous solid body resulting from the penetration of liquid into its permeable pores.

Admixture: A material other than water, aggregates, and portland cement that is used as an ingredient of concrete and is added to the batch immediately before or during its mixing.

Aggregate: Inert material which is mixed with portland cement and water to produce concrete.

Aggregate, coarse: Aggregate which is retained on a #4 sieve.

Aggregate, fine: Aggregate passing the #4 sieve and retained on the #200 sieve.

Aggregate, lightweight: Aggregate having a dry, loose weight of 70 pounds per cubic foot or less.

Bleeding: The autogenous flow of mixing water within, or its emergence from, concrete freshly placed.

Cement factor: The number of bags or cubic feet of cement per cubic yard of concrete.

Column: An upright compression member the length of which exceeds three times its least lateral dimension.

Combination column: A column in which a structural steel member, designed to carry the principal part of the load, is encased in concrete which carries the remainder of the load.

Composite column: A column in which a steel or cast-iron structural member is completely encased in concrete containing spiral and longitudinal reinforcement.

Composite concrete flexural construction: A precast concrete member and cast-in-place reinforced concrete so interconnected that the component elements act together as a flexural member.

Concrete: A mixture of portland cement, fine aggregate, coarse aggregate, and water.

Concrete, normal weight: Concrete having a hardened density of approximately 150 pcf which is made from aggregate of approximately the same density.

Concrete, plain: Concrete that is not reinforced with steel.

Concrete, precast: A plain or reinforced concrete element cast in other than its final position in the structure.

Concrete, prestressed: Reinforced concrete in which there have been introduced internal stresses of such magnitude and distribution that the stresses resulting from service loads are counteracted to a desired degree.

Concrete, reinforced: Concrete containing steel reinforcement.

Concrete, structural lightweight: A concrete containing lightweight aggregate.

Crushed gravel: The product resulting from artificial crushing of gravel with substantially all fragments having at least one fracture face.

Crushed stone: The product resulting from the artificial crushing of rocks, boulders, or large cobblestones, substantially all faces of which have resulted from the crushing operation.

Deformed bar: A reinforcing bar with ridges to increase bonding with the concrete.

Double reinforcement: A concrete beam with steel on both sides of the neutral axis to resist tension and compression.

Effective area of concrete: The area of a section which lies between the centroid of the tension reinforcement and the compression face of the flexural member.

Effective area of reinforcement: The area obtained by multiplying the the cross sectional area of the reinforcement by the cosine of the angle between its direction and the direction for which the effectiveness is to be determined.

Fineness modulus: An empirical factor obtained by adding the total percentages of a sample of the aggregate retained on each of a specified series of sieves, and dividing the sum by 100.

Flat slab: A 2-way slab supported at its corners only.

Formwork: The wood molds used to hold concrete during the curing and pouring processes.

Gravel: Granular material retained on a #4 sieve which is the result of natural disintegration of rock.

Heat of hydration: The exothermic heat given off by concrete as it cures.

Hydration: The chemical reaction which occurs when the cement ions attach themselves to the water molecules of crystallization.

Monolithic construction: Constructed as one piece.

One-way: Constructed with reinforcing steel running in one direction only.

Pedestal: An upright compression member whose height does not exceed three times its average least lateral dimension.

Plain bar: Reinforcement that does not conform to the definition of a deformed bar. That is, reinforcing bars without raised ridges.

Rebar: Steel reinforcing bar.

Sand: Granular material passing through a #4 sieve but predominantly retained on a #200 sieve.

Saturated Surface Dry: A condition of an aggregate which holds as much water as it can without having any water within the pores between the aggregate particles.

Slab: A cast concrete floor.

Slump: The decrease in height of wet concrete when a supporting mold is removed.

Spiral column: A column with a continuous spiral of wire around the longitudinal steel.

SSD: See 'Saturated Surface Dry'

Surface water: Water carried by an aggregate except that held by absorption within the aggregate particles themselves. Water in addition to SSD water.

Tie: A stirrup.

Tied column: A column which has individual loops of wire around the longitudinal steel.

Two-way: Construction with steel reinforcing running in two perpendicular directions

Water-cement ratio: The number of gallons of water per 94 pound sack of cement.

3. Concrete Mixing

A. Types of Concrete

Concrete is a mixture of mineral aggregates locked ('cemented') into a solid structure by a binding material. The concrete is produced by adding water to the aggregate and binder, and then casting the mixture in place. The semi-fluid mixture hardens by chemical action to concrete.

The binding material is known as cement. There are two types of cement: bituminous and non-bituminous. Asphalt and tar are the most common bituminous cements. These cements harden upon cooling; they do not require water.

The most common non-bituminous cements are alumina cement and portland cement. Since portland cement is the most widely used, the terms 'cement' and 'concrete' are frequently used interchangeably. It is assumed in this chapter that the binding agent is portland cement.

Portland cement is manufactured from lime, silica, and alumina in the appropriate proportions. After being ground and blended, the mixture is kilned at approximately 2700°F. The resulting clinkers are ground and mixed with a small amount of plaster of paris.

There are five types of portland cement in common use, although special cements are available upon special order. The type of cement used is dependent on the application intended for it.

> Type 1: Normal portland cement: This is a general-purpose cement used whenever sulfate hazards are absent and when the heat of hydration will not produce objectionable rises in temperature. Typical uses are for sidewalks, pavement, beams, columns, and culverts.

> Type 2: Modified portland cement: This cement has a moderate sulfate resistance, but is generally used in hot weather

in the construction of large concrete structures. Its heat rate and total heat generation are lower than for normal portland cement.

Type 3: High-early-strength portland cement: This type of cement develops its strength quickly. It is suitable for use when the structure must be put into early use or when long-term protection against cold temperatures is not feasible. Its shrinkage rate, however, is higher than for types 1 and 2; and, extensive cracking may result.

Type 4: Low-heat portland cement: For extensive concrete structures, such as gravity dams, low-heat cement is required to minimize the curing heat. The ultimate strength also develops more slowly than for the other types.

Type 5: Sulfate-resistant portland cement: This type of cement is applicable when exposure to severe sulfate concentration is expected. This typically occurs in some western states having highly alkaline soils.

Table 14.1
Relative Strengths of Concrete Types

	Compressive strength, % of normal strength concrete		
	3 days	28 days	3 months
Type 1	100	100	100
Type 2	80	85	100
Type 3	190	130	115
Type 4	50	65	90
Type 5	65	65	85

B. Aggregates

Cement itself is very fine. More than 90% of it will pass through a #325 sieve. However, the bulk of cast-in-place concrete consists of sand and rock particles that have been added to the cement-water mixture to increase the weight and bulk. These sand and rock particles are known as aggregate.

Sand and other particles that will pass through a #4 sieve (less than .25") are known as fine aggregate. Any particles that are larger than this are known as coarse aggregate. Coarse aggregate is produced from crushed stone and blast-furnace slag.

Regardless of its size, a good aggregate should be clean, hard, strong, and weather resistant. The cost of aggregate is much less than the cost of cement, so increasing the amount of aggregate will decrease the concrete cost. However, there are strength and workability limits on the proportions of cement, aggregate, and water used.

Aggregate having a density of 70 pcf or less is known as lightweight aggregate. This type is used in the production of lightweight concrete.

(Foamed concrete does not fall into this category.) Lightweight concrete in which only the coarse aggregate is lightweight is known as 'sand-lightweight concrete.' If both the coarse and fine aggregates are lightweight, it is known as 'all-lightweight concrete.' Unless noted otherwise, concrete in this chapter is assumed to be normal weight concrete.

As was done with soils, aggregate can be successively screened to determine a particle size distribution. The standard screens used as numbers 4, 8, 16, 30, 50, and 100 for fine aggregate, and 6", 3", 1½", 3/4", 3/8", and number 4 for coarse aggregate. The fineness modulus is used to describe the distribution. It is computed by adding the cumulative percentages retained on the six standard screens and then dividing by 100. Typical values range from 2.5 to 3.0.

C. Proportioning Concrete

The quality, cost, workability, and strength of the finished concrete will depend on the proportions used to produce the concrete. Since there are many variables involved, and since some of the desired properties are conflicting, samples of concrete produced with different proportions should be tested prior to the start of a job.

The proportions of a concrete mixture are usually designated as a ratio of cement, fine aggregate, and coarse aggregate, in that order. For example, 1:2:3 means that one part of cement, two parts of fine aggregate, and three parts of coarse aggregate are to be combined. The ratio may be either in terms of weight or volume. Weight ratios are more common. However, the method of comparison (weight or volume) should be specified.

The amount of water used is called out in terms of gallons of water per 94 pound sack of cement. This is known as the water-cement ratio.

D. Admixtures

Anything else added to the concrete to improve its workability, hardening, or strength characteristics is known as an admixture. Hydrated lime, diatomaceous silica, fly ash, and bentonite are added to concrete which has too little fine aggregate. These admixtures separate the coarse aggregate and reduce the friction of the mixture. Calcium chloride may be added as a curing accelerator. It is also an antifreeze, but its use as such is not recommended. (The ACI code does not allow artificially-introduced chloride ions in pre-stressed concrete or concrete with aluminum embedments. Natural chloride ions are limited to about 500 ppm.)

Sulfonated soaps and oils, as well as natural resins, increase air entrainment. This increases the durability of the concrete while decreasing the strength and weight only slightly.

Styrofoam may be added to increase the insulating values of the concrete.

E. The Trial Mix Method

Since aggregates vary considerably in size, grading, particle shape, absorption, and surface smoothness, it is best to develop the required proportions of the four mix ingredients (cement, fine aggregate, coarse aggregate, and water) by the trial mix method.

To use this method, start with dry aggregates and follow these steps:

step 1: Decide on a desired slump for the application. (See table 14.2)

step 2: Consider three cement:water ratios such as 1.0, 1.5, and 2.0 by weight.

step 3: If two test cylinders are to be used, take 7 pounds of water and 7 pounds of cement, and make the cement/water paste. If three test samples are to be used, take 10 pounds each.

step 4: Add sand and coarse aggregate until the desired slump and consistency is obtained.

step 5: Make two (or three) standard test cylinders for ultimate strength testing.

step 6: Repeat steps 3, 4, and 5 twice with the same cement/water ratio, but varying the ratio of sand to coarse aggregate from that used in the first mix. Keep track of the improvement or degradation in the workability.

step 7: Repeat steps 3 through 6 for the remaining cement/water ratios.

step 8: Cure the cylinders and perform compressive tests.

step 9: Plot the average strength versus cement/water ratio for the mixes with the best workabilities. The strength will not vary much with the aggregate ratios.

stpe 10: Deduce the desired mix from the workability and strength data based on the desired final ultimate compressive strength.

F. Ordering Out Concrete

The amount of concrete that can be made from mixing known proportions of ingredients can be found from the absolute or 'solid' volume method. This method assumes that there will be no voids in the placed concrete. Therefore, the amount of concrete made will be the sum of the solid volumes of the cement, sand, coarse aggregate, and water.

To use the absolute volume method, it is necessary to know the solid densities of the constituents. In the absence of any other information, the following data may be used for solid densities: cement, 195 pcf; fine aggregate, 165 pcf; coarse aggregate, 165 pcf; water, 62.4 pcf.

Also, it is necessary to recall that a sack of cement weighs 94 pounds and that 7.48 gallons of water make a cubic foot.

If the mix proportions are volumetric, it will be necessary to multiply the ratio values by the bulk densities to get the weights of the constituents. Then, weight ratios may be calculated and the absolute volume method applied directly.

Example 14.1

A mix was designed as 1:1.9:2.8 by weight. The water-cement ratio was chosen as 7 gallons of water per sack of cement. (a) What is the concrete yield in cubic feet per sack of cement? (b) How much of each constituent is needed to make 45 cubic yards of concrete?

(a) The solution can be tabulated as follows:

material	ratio	weight per sack cement	solid density		absolute volume $(ft^3/sack)$
cement	1.0	94	195	94/195 =	.48
sand	1.9	179	165	179/165 =	1.08
coarse	2.8	263	165	263/165 =	1.60
water				7/7.48 =	.94
					4.10

Therefore, the yield is 4.1 cubic feet of concrete per sack cement.

(b) The number of one-sack batches is

$$\frac{(45) \text{ yd}^3 \ (27) \text{ ft}^3/\text{yd}^3}{(4.1) \text{ ft}^3/\text{sack}} = 296.3 \text{ sacks (say 297)}$$

So, order 297 sacks of cement.

$$\frac{(297)(1.9)(94)}{2000} = 26.5 \text{ tons of sand}$$

$$\frac{(297)(2.8)(94)}{2000} = 39.1 \text{ tons of coarse aggregate}$$

$$(297)(7) = 2079 \text{ gallons of water}$$

The problem may be complicated by air entrainment requirements and/or the water content of the aggregate. The following guidelines may be used:

1. The absolute volume yield is increased by the addition of air. This can be accounted for by dividing the solid yield by (1 - %air).

2. Any water content in the aggregate above the 'saturated surface dry' (SSD) water content must be subtracted from the water requirements.

3. Any porosity (water affinity) below the SSD water content must be added to the water requirement.

4. There are 239.7 gallons per ton of water.

5. The densities used in the calculation of the yield should be the SSD densities.

Example 14.2

50 cubic feet of $1:2\frac{1}{2}:4$ concrete are to be produced. The constituents have the following properties:

constituent	SSD density (pcf)	moisture (from SSD)
fresh cement	197	-
sand	164	5% excess
coarse aggregate	168	2% dry

What are the required order quantities if 5.5 gallons of water are to be used per sack and the mixture is to have 6% entrained air?

constituent	ratio	weight per sack cement	SSD density	absolute volume
cement	1.0	94	197	.477
sand	2.5	235	164	1.433
coarse	4.0	376	168	2.238
water		5.5/7.48		= .735

$$4.883 \text{ ft}^3/\text{lbm}$$

The yield with 6% air is $(\frac{4.883}{1-.06}) = 5.19 \text{ ft}^3/\text{sack}$.

The number of one sack batches is $(50/5.19) = 9.63$.

The required sand weight (ordered as is, not SSD) is

$$\frac{(9.63)(1.05)(94)(2.5)}{2000} = 1.19 \text{ tons}$$

The required coarse aggregate weight (ordered as is, not SSD) is

$$\frac{(9.63)(.98)(94)(4)}{2000} = 1.77 \text{ tons}$$

The excess water contained in the sand is

$$(1.19)(\frac{.05}{1.05})(239.7) \text{ gal/ton} = 13.58 \text{ gallons}$$

The water needed to bring the coarse aggregate to SSD conditions is

$$(1.77)(\frac{.02}{.98})(239.7) \text{gal/ton} = 8.66 \text{ gallons}$$

The total water needed is $(5.5)(9.63) + 8.66 - 13.58 = 48.0 \text{ gallons}$

◆

4. Properties of Concrete

A. The Slump Test (ASTM C-143)

The slump test is a measure of consistency of the plastic concrete mass prior to hardening. This test consist of filling a slump cone in three layers of about one-third of the mold volume. Each layer is rodded with 25 strokes of 5/8" round rod. The mold is then removed by raising it carefully in the vertical direction. The slump is the difference in the mold height and the resulting concrete pile height.

Concrete mixtures that do not slump appreciably are known as stiff mixtures. Stiff mixtures are inexpensive because of the large amount of coarse aggregate. However, placing time and workability are impaired. Mixtures with large slumps are known as 'wet' or 'watery' mixtures. Such mixtures are needed for thin casting and structures with extensive reinforcing.

Recommended slumps for concrete that is hand vibrated are given in table 14.2. If high-frequency vibrators are used, the values given should be reduced by about one-third.

Table 14.2

Recommended Slumps

| Application | Slumps, inch | |
	Maximum	Minimum
Reinforced foundations and footings	3	1
Plain footings and substructure walls	3	1
Slabs, beams, and reinforced walls	4	1
Columns, reinforced	4	1
Pavement & slabs	3	1
Heavy mass construction	2	1

B. Compressive Strength Test (ASTM C-39)

Two or three 6" x 12" cylinders of the concrete are cast as specified by the procedure for each mixture ratio to be tested. The casting procedure is similar to that used for the slump test. After curing, the cylinders are placed in a compressive tester. A load is applied at a constant rate, using .05 inches/minutes or 20 - 50 psi/second. The compressive strength is found by dividing the maximum load carried by the sample by the area.

Typical ultimate strengths, f_c', vary from 2000 psi to 8000 psi at 28 days, although 6000 psi is a common upper limit.

Since the ultimate strength of concrete increases with time, all values of f_c' should be referenced to the age. If no age is given a standard 28 day age is assumed.

An estimate of the 28-day strength can be obtained from the 7-day strength by using equation 14.1.

$$f'_{c,28} = f'_{c,7} + 30\sqrt{f'_{c,7}}$$

14.1

Figure 14.1

Compressive Strength vs. Age

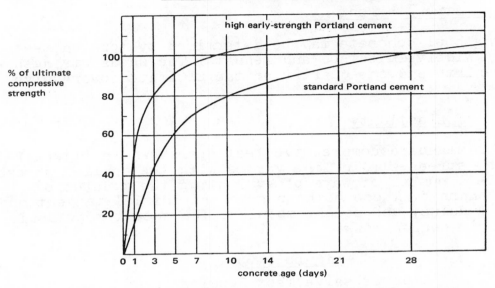

The ultimate compressive strength is also highly dependent on the water/cement ratio, and it is fairly independent of the proportion of the mixes. Evidence towards this regard was developed by Dr. Duff Abrams in 1918, after whom Abram's Strength Law is named: The compressive strength of concrete varies directly with the cement/water ratio, provided that the mix is of a workable consistency. This law is illustrated in figure 14.2.

Figure 14.2

Compressive Strengths vs. Water Content

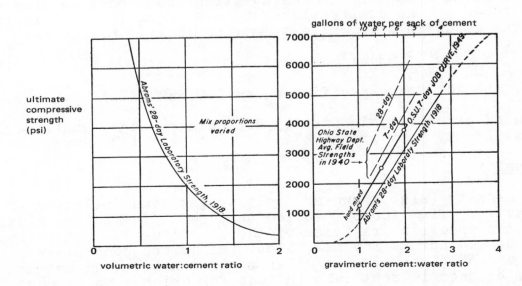

C. Tensile Load Test

Because concrete is not used to resist tension, tensile tests are seldom performed on normal weight concrete. However, the ACI code does specify a splitting tensile test for structural lightweight concrete. The ultimate tensile strength usually varies from between 7% to 10% of the ultimate compressive strength.

D. Shearing Strength Test

The shear strength of concrete may be determined by torsion tests. Such tests vary widely in method and results. The shear strength will be between one-sixth and one-quarter of the ultimate compressive strength.

E. Modulus of Elasticity

The results of a standard compressive test are shown in figure 14.3. The slope of the stress-strain line varies with the applied stress, and hence, there are a number of ways of evaluating the modulus of elasticity of concrete. These methods give the initial tangent modulus, actual tangent modulus, and secant modulus. The secant modulus is most frequently used in design work.

Figure 14.3
Compressive Test Results

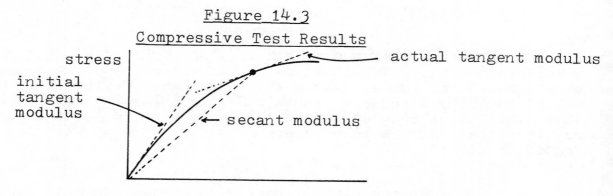

The ACI code gives a formula for calculating the secant modulus of elasticity based on the compressive strength. Equation 14.2 is for normal weight concrete only, with a density of w=90 to 155 pcf.

$$E_c = w^{1.5}(33)\sqrt{f'_c}$$

14.2

(For steel, the ACI code specifies E_s as 2.9 EE7 psi.)

The modulus of elasticity is greatly dependent on age, quality, and proportions. It may vary from 1 EE6 to 5 EE6 psi at 28 days.

F. Density

The weight density (also known as 'unit weight' and 'weight') of concrete may vary from 100 pcf to 160 pcf, depending on the mixture ratios and the specific gravities of the constituents. Generally, the range will be 140 pcf to 160 pcf. Steel reinforced concrete will be between 3% and 5% higher than similar plain concrete. An average of 150 pcf may be used in most calculations for which a value must be

assumed. These values are bulk densities, which include entrained air.
Solid densities exclude entrained air and are 5% to 10% higher.

F. Shrinkage Coefficients

Volume changes occur during water and temperature content changes.
The thermal coefficient of linear expansion is usually assumed to be
6 EE-6 (1/°F). The percentage change in volume during curing may
vary from .02% to .5%.

5. Allowable Stresses

Where the working stress method is employed, the ACI code specifies
allowable stresses for concrete and steel. For concrete, the allowable
stresses depend on the application and the ultimate compressive strength.
See tables 14.3 and 14.4.

Table 14.3

Allowable Stresses for Concrete
(not for use with lightweight concrete)

Application	f_c
flexure	
compressive stress	$.45f'_c$
shear	
beams (concrete carries all shear)	$1.1\sqrt{f'_c}$
1-way footings and 1-way slabs	$1.1\sqrt{f'_c}$
2-way footings and 2-way slabs	$2.0\sqrt{f'_c}$
2-way footings and 2-way slabs minimum of	$\begin{cases} 2.0\sqrt{f'_c} \\ (1+\frac{2}{\beta})\sqrt{f'_c} \end{cases}$
bearing	$.30\sqrt{f'_c}$

Table 14.4

Allowable Stress for Steel

Application	f_s
beams	
grades 40, 50	20,000 psi
grade 60 and steels with f_y above 60,000 psi	24,000 psi
slabs, one-way, less than 12 foot span	
#3 bars or smaller main reinforcement, use the smaller of	$\frac{1}{2}f_y$ or 30,000 psi

6. Steel Reinforcing

Since concrete is essentially incapable of resisting tension, steel reinforcing bars are used. Figure 14.4 shows three typical types of reinforcing in concrete beams.

Figure 14.4
Beam Reinforcement

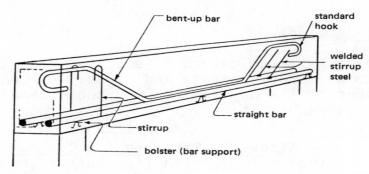

Straight and bent-up bars resist flexural tension in the central part of the beam. Since the bending moment is smaller near the ends of the beam, less reinforcing is necessary there. Some of the bar is bent up in order to resist diagonal shear near the beam ends. In continuous beams, the horizontal upper parts of the bent-up bars are continued on to the next span.

Since the bent-up bars cannot resist all of the diagonal tension, stirrups are used. These pass underneath the bottom steel for anchoring or are welded to the bottom steel. The stirrups may be placed at any convenient angle, as shown.

The horizontal steel is supported on bolsters ('chairs') of which there are a variety of designs and heights.

Columns are reinforced with longitudinal steel. This steel runs the full length of the column and is usually placed around the periphery or at the corners of a rectangular columns. The longitudinal steel is wrapped with individual circumferential bars (tied columns) or with a continuous spiral bar (spiral columns).

Reinforcing steel bars, known as 'rebars', are available in a large number of sizes, as well as in the form of wire for spiral wrapping and wire mesh for shrinkage and thermal expansion reinforcement. The steel comes in several grades, depending on its yield strength. In older codes, there were three grades: structural, intermediate, and hard.

Figure 14.5
Old Code Steel Grades

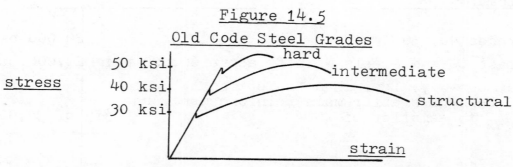

In the current code, the steel is specified with respect to its yield strength. Grade 60 is most common, although grade 40 was widely used in the past, and grades 50, 75, and 80 are also available. However, not all sizes of steel bar may be available in every grade. (There is essentially no difference between ACI and ASTM grading.)

Table 14.5
Standard Reinforcing Bars

Size	weight lb/ft	diameter inches	area sq. inches	perimeter inches
ASTM A-615 (40 and 60 ksi)				
#2	.167	.250	.05	.786
#3	.376	.375	.11	1.178
#4	.668	.500	.20	1.571
#5	1.043	.625	.31	1.963
#6	1.502	.750	.44	2.356
#7	2.044	.875	.60	2.749
#8	2.670	1.000	.79	3.142
#9	3.400	1.128	1.00	3.544
#10	4.303	1.270	1.27	3.990
#11	5.313	1.410	1.56	4.430
ASTM A-615 (60 ksi only)				
#14S	7.65	1.693	2.25	5.32
#18S	13.60	2.257	4.00	7.09

Although rebar steel may have a yield strength of 40,000 psi or above, its working stress is limited by the ACI code to less than this. Generally, a factor of safety of between 1.8 to 2.5 is used to keep the allowable stress in the 20,000 to 24,000 psi range.

7. Ultimate versus Working Strength Design

In the past, design of reinforced concrete members has been based on the working stress design (WSD) method. The WSD procedures calculate the actual stress in a structure under loading. This stress is then compared to the allowable stress, as given in table 14.3. The design is acceptable if the actual stress is less than the allowable stress.

The ultimate strength design (USD) method is now in widestread use. It is known as the "strength design method" in the ACI code, where it has been a part since 1963. In fact, the current ACI codes don't even cover WSD methods to any great extent. In the USD method, the actual loads are increased by multiplicative safety factors. The stresses calculated from these 'factored' loads are then compared to the loads or stresses that would cause failure in the structure. The design is acceptable if the factored stress is less than the ultimate stress.

The ultimate stress is also multiplied by a capacity reduction factor, ϕ, to account for workmanship and understrength in the materials. Values of ϕ are given in table 14.6.

Inasmuch as both methods are acceptable and are in use, both WSD and USD methods are covered in this chapter.

Table 14.6
Strength Reduction Factors

Type of Stress	ϕ
flexure	.90
axial tension	.90
shear	.85
torsion	.85
axial compression with spiral reinforcement	.75
axial compression with tied reinforcement	.70
bearing on concrete	.70

8. Column Footing Design

Note: This section is suitable for designing square and rectangular footings. It is not sufficient to completely design combined or wall footings, or for footings carrying moments.

A. Failure Mechanisms

The design and reinforcement of individual column footings is based, to a large extent, on empirical observations. Improvement in design techniques since the original footing experiments in the early 1900's has been associated with the increased strength of concrete and steel.

Column footings may fail in several modes. A typical mode is known as the 'stage of primary failure'. This occurs at some stress less than the ultimate when the deformations and cracks are so extensive that the footing becomes useless for its intended purpose. This can occur if the tension reinforcement along the bottom is inadequate (allowing the steel to yield or the bars to slip). In this case, ultimate failure occurs due to shear around the loaded zone.

Figure 14.6
Shear Failure of a Footing

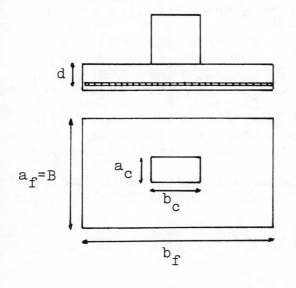

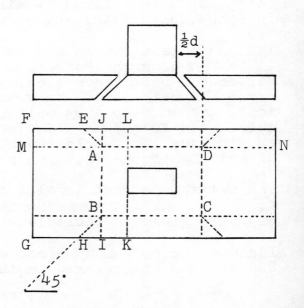

The critical section for two-way shear is assumed to be d/2 from the column or wall face, where d is the depth of the steel reinforcement cover. Therefore, the distance AB (or BC, CD, or DA) is equal to the column width plus 2(d/2) or a_c + d. For square footings, the total shear used to find the unit stress on critical section AB is the sum of the forces on area ABHGFE. This is known as 'two-way' shear. For long rectangular footings in one-way shear (beam bending), the total forces on section IGFJ must also be checked on the critical section, assumed to be at distance d (not d/2) from the column face.

The footing may also fail in flexure. The critical section for bending is at the face of the column (i.e., line KL).

B. Reinforcement Schemes

Because of their multi-mode failure capacity, footings are reinforced in both horizontal directions. Bars in the long direction should be placed uniformly across the width a_f. For bars in the short direction, the ACI code specifies that the fraction of steel concentrated in a center band of width a_f shall be $2/[(b_f/a_f)+1]$. 9" to 12" spacings are typical. Hooks may be used to achieve development length.

Figure 14.7
Footing Reinforcement

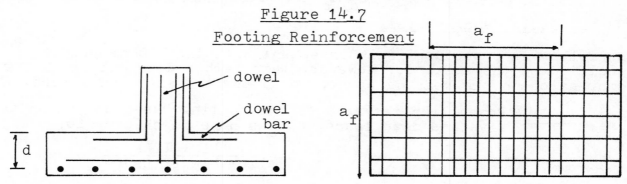

In addition, vertical 'dowels' or 'dowel bars' may be used to provide resistance to horizontal shear and to transfer the column load to the footing by bond instead of bearing.

Rebar sizes used are dependent on the loads that must be carried. However, typical sizes are #5 to #9 for horizontal steel, #6 for column dowels, and #9 for dowel bars.

Stirrups are not usually used, as the allowable shear stress is kept very low.

A minimum of 3" of concrete should be used below the bars. The minimum cover, d, for reinforced concrete on soil is 6". The two steel layers will be in contact, as no clear distance between them is required.

C. Allowable Loads and Stresses

The factored load to be resisted according to the ACI code is given by equations 14.3, 14.4, and 14.5. The largest value is to be used as the design load.

$$U = 1.4(DL) + 1.7(LL) \qquad\qquad 14.3$$

$$U = .75(1.4(DL) + 1.7(LL) + 1.7(WL)) \qquad 14.4$$

$$U = 0.9(DL) + 1.3(WL) \qquad 14.5$$

Usually equation 14.3 will hold. If the ratio of dead and live loads is not known (as in the case of a multi-story building) an average factor of 1.6 can be used.

The maximum shear stress for footings without special shear reinforcement is specified by the ACI code in equations 14.6 and 14.7. Notice that these allowable stresses are multiplied by the capacity reduction factor, ϕ, which has a value of .85 for shear (see table 14.6). Use equation 14.7 for square footings. Both equations must be used when checking rectangular footings.

$$v_c = \frac{V}{bd} = 2\phi\sqrt{f'_c} \qquad \text{(one-way shear)} \qquad 14.6$$

$$v_c = \frac{V}{bd} = (2 + \frac{4}{\beta_c})\phi\sqrt{f'_c} \qquad \text{(two-way shear)} \qquad 14.7$$

$$\beta_c = b_c/a_c \text{ (not less than 2)} \qquad 14.8$$

D. Design and Analysis

The following procedure is in agreement with the USD method as specified by the ACI code. However, this method does not check for compression on the bearing block, nor does it design dowel bars.

step 1: Proportion the area of the footing based on the unfactored live and dead loads, as was done in chapter 10.

$$A_f = B^2 = \frac{DL + LL}{\text{allowable pressure}}$$

step 2: Apply load factors to the column loads using equation 14.3 as appropriate.

step 3: Calculate the maximum allowable shear stress, v_c, in psi, from equation 14.6 or 14.7.

step 4: Calculate the net factored soil pressure (in psf) from the factored load (step 2) and the footing area.

$$p_{net} = \frac{U}{A_f}$$

step 5: Calculate the following ratios.

$$k_1 = a_c/B \qquad \text{(square footing)} \qquad 14.9$$

$$k_2 = \frac{p_{net, \text{ factored, psf}}}{(576)v_{c, \text{ psi}}} \qquad 14.10$$

For rectangular (not square) footings, also check for

$$k'_1 = a_c/a'_f \qquad 14.11$$

$$a_f' = a_f\sqrt{2\beta_f - 1} \qquad\qquad 14.12$$

$$\beta_f = b_f/a_f \qquad\qquad 14.13$$

step 6: Calculate a trial value of d. Remember that $a_f = B$ for square footings. Also, d must be 6" or greater from the ACI code. Add $3" + \frac{1}{2}D_b$ for steel cover on the bottom.

$$d = \frac{a_f\left[\sqrt{4k_2^2 + 4k_2 + k_1^2(1 + 4k_2 + 4k_2^2)} - k_1(1+2k_2)\right]}{2(1+k_2)} \qquad 14.14$$

d is measured from the center of the top row of steel.

step 7: Check the shear stress at the critical section.

$$V_u = p_{net,\ factored}(\text{critical shear area})$$

The concrete below the steel is not considered to resist shear.

$$v_u = \frac{V_u}{(d)(\text{critical section length})} \qquad\qquad 14.15$$

Compare v_u with v_c from step 3. If $v_u < v_c$, then no special shear reinforcing is required.

step 8: Calculate the bending moment on the footing. For square footings,

$$M_u = p_{net,factored}(B)(\tfrac{1}{8})(B-a_c)^2 \qquad\qquad 14.16$$

step 9: Select reinforcement based on the trial value of d.

$$A_s = pBd \qquad\qquad 14.17$$

$$p = \frac{.85f_c'}{f_y}(1 - \sqrt{1 - (2R_u/.85f_c')}) \qquad\qquad 14.18$$

$$\boxed{\begin{array}{c}\text{number of bars =}\\ 1 + \dfrac{(B" - 6)}{\text{spacing"}}\end{array}}$$

$$R_u = M_u/\phi Bd^2 \qquad (\phi = .90 \text{ for bending}) \qquad\qquad 14.19$$

step 10: Check the development length of the steel, l_d. Assuming #11 bars or smaller will be used,

$$l_d = (.04)(A_b)f_y/\sqrt{f_c'} \qquad\qquad 14.20$$

However, l_d must not be less than the minimum of 12" and $(.0004)(\text{bar diameter})(f_y)$. l_d must be less than $(\frac{1}{2}(B-a_c)-3")$ unless hooks are used. If the bars are spaced on 6" centers or greater, the development length may be reduced 20% (i.e., multiplied by 0.8).

<u>Example 14.3</u>

A square (111" x 111") footing carries 82,560 pounds dead load and 75,400 pounds live load transmitted through a 12" x 12" square column. Design the footing for 2-way shear, using 4000 psi concrete and 60,000 psi steel.

<u>step 1</u>: The footing has been proportioned.

<u>step 2</u>: The factored load is found from equation 14.3.

$$F_u = 1.4(82,560) + 1.7(75,400) = 243,770 \text{ lb}$$

<u>step 3</u>: β_c = 12/12 = 1 (must be 2 or greater)

$$v_c = (2 + \frac{4}{2})(.85)\sqrt{4000} = 215 \text{ psi}$$

<u>step 4</u>: $p_{net,factored} = \dfrac{243,770}{(\frac{111}{12})^2} = 2849 \text{ psf}$

<u>step 5</u>: k_1 = 12/111 = .108

k_2 = 2849/(576)(215) = .023

<u>step 6</u>: Using equation 14.14,

$$d = (111)(\frac{\sqrt{.1069} - .113}{2.046}) = 11.61" \text{ (say 12")}$$

The thickness of the footing, including the steel cover on the bottom and an allowance for the bar is 12+3+.5 = 15.5".

<u>step 7</u>: $\frac{1}{2}d$ = 6"

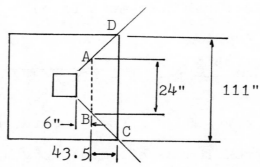

The critical area contributing to 2-way shear failure is ABCD. That area is

$$(24)(43.5) + (2)(\tfrac{1}{2})(43.5)(\frac{111-24}{2}) = 2936 \text{ in}^2$$

$$V_u = (2849)(\frac{2936}{144}) = 58,088 \text{ lb}$$

$$v_u = V_u/A = 58,088/(24)(12) = 201.7 \text{ psi}$$

This is okay since it is less than 215 psi from step 3. Therefore, no shear reinforcing is required.

step 8: The applied bending moment is

$$M_u = (\frac{2849}{144})(111)(49.5)(\frac{49.5}{2}) = 2.69 \text{ EE6 in-lb}$$

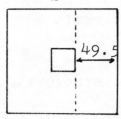

step 9: $R_u = \dfrac{2.69 \text{ EE6}}{(.9)(111)(12)(12)} = 187 \text{ psi}$

$$p = \frac{(.85)(4000)}{60,000}(1 - \sqrt{ 1 - \frac{(2)(187)}{(.85)(4000)} }) = .00321$$

$$A_s = (.00321)(111)(12) = 4.28 \text{ in}^2$$

Using a 10" spacing would require $1 + [111-(2)(3)]/10$
or 12 bars. 4.28/12 = .36 square inches. So choose #6
bars with .44 square inches each. Space the bars evenly
in both directions since the footing is square.

step 10: $l_d = \dfrac{(.04)(.44)(60,000)}{\sqrt{4000}} = 16.7"$

This is less than (49.5 - 3) which is okay.

9. Beams

A. WSD Design: The Elastic Load Theory

The elastic load theory is a design procedure based on the following
assumptions.

    (a) Plane sections remain plane during bending.
    (b) Both steel and concrete remain in the elastic range.
    (c) Concrete carries all compressive loads in an area
        above the neutral axis.
    (d) Steel carries all tension.
    (e) Stress in the concrete is proportional to the distance
        from the neutral axis.

Figure 14.8

A Beam with Tensile Reinforcement Only

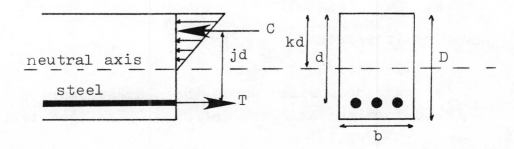

The ratio of moduli of elasticity is

$$n = E_s/E_c \quad \text{(round to the nearest integer)} \qquad 14.21$$

Also, define the reinforcement ratio as

$$p = A_s/bd \qquad 14.22$$

For a given or assumed maximum stress in the concrete, f_c, and steel, f_s, define

$$k = \frac{1}{1 + (f_s/nf_c)} = \sqrt{2np + (np)^2} - np \qquad 14.23$$

k also defines the location of the neutral axis, as shown in figure 14.8.

The total compressive force is

$$C = \tfrac{1}{2}f_c bkd \qquad 14.24$$

The total tensile force is

$$T = A_s f_s = C \qquad 14.25$$

Since the resultant, C, acts (1/3) from the top of its distribution,

$$jd = d - \frac{kd}{3} \qquad 14.26$$

or $\qquad j = 1 - \frac{k}{3} \qquad 14.27$

The moments that the concrete and steel can withstand are

$$M_c = Cjd = \tfrac{1}{2}f_c jkbd^2 \qquad 14.28$$

$$M_s = Tjd = A_s f_s jd = f_s jpbd^2 \qquad 14.29$$

The required steel area is

$$A_s = M_s/f_s jd \qquad 14.30$$

If the design is 'balanced', then $M_c = M_s$. Therefore,

$$\tfrac{1}{2}f_c jkbd^2 = f_s jpbd^2 \qquad 14.31$$

If K is defined as

$$K = \tfrac{1}{2}f_c jk = f_s jp \qquad 14.32$$

then $\qquad M_s = M_c = Kbd^2 \qquad 14.33$

Good design practices are:

- 1.75 < (d/b) < 2.0
- round beam sizes up to the nearest quarter inch
- determine the number of steel bars from table 14.7.

Table 14.7
Minimum Beam Widths

The following table is easily derived from the clear spacing requirements and depth of cover specifications. The table assumes that #3 stirrups are used, and cover is provided for them. If no stirrups are used, deduct 3/4" from the figures shown. For additional bars, increase width of beam by adding dimension in last column. Remember, clear space between bars must be at least 4/3 times the size of the largest aggregate.

Bar size, No.	Number of bars in a single layer of reinforcing							Add for each added bar
	2	3	4	5	6	7	8	
4	5¾	7¼	8¾	10¼	11¾	13¼	14¾	1½
5	6	7¾	9¼	11	12½	14¼	15¾	1 5/8
6	6¼	8	9¾	11½	13¼	15	16¾	1¾
7	6½	8½	10¼	12¼	14	16	17¾	1 7/8
8	6¾	8¾	10¾	12¾	14¾	16¾	18¾	2
9	7¼	9½	11¾	14	16¼	18½	20¾	2¼
10	7¾	10¼	12¾	15¼	17¾	20¼	23	2 5/8
11	8	11	13¾	16½	19½	22¼	25	2 7/8

Example 14.4

Find the stresses in the steel and concrete due to a 700,000 in-lb moment. Use n = 12.

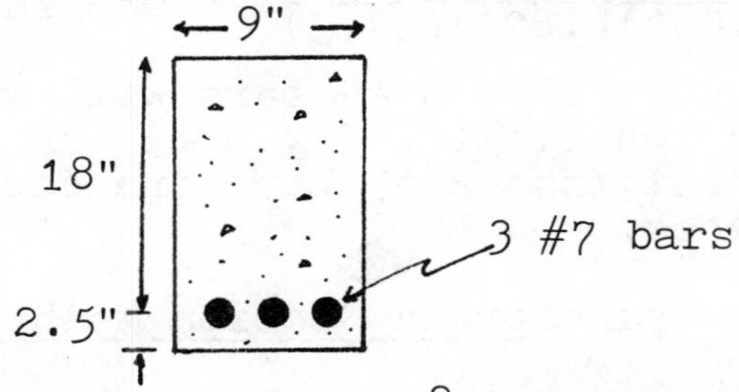

3 #7 bars

The steel area is $A_s = 3(.6) = 1.8$ in^2. From equations 14.22, 14.23, and 14.27,

$$p = \frac{1.8}{(9)(18)} = .0111$$

$$k = \sqrt{[2(12)(.0111)] + [(.0111)(12)]^2} - (12)(.0111)$$
$$= .400$$

$$j = 1 - \frac{.400}{3} = .867$$

From equations 14.28 and 14.29,

$$f_c = \frac{2(700,000)}{(.400)(.867)(9)(18)^2} = 1384.4 \text{ psi}$$

$$f_s = \frac{700,000}{(1.8)(.867)(18)} = 24,919 \text{ psi (excessive)}$$

Example 14.5

Design a balanced beam to withstand a moment of 250,000 in-lb using
the following data:

$$f_{c, \text{ allowable}} = 1000 \text{ psi} \qquad E_c = 2.5 \text{ EE6}$$

$$f_{s, \text{ allowable}} = 20,000 \text{ psi} \qquad E_s = 3.0 \text{ EE7}$$

From equations 14.21, 14.23, and 14.27,

$$n = \frac{3 \text{ EE7}}{2.5 \text{ EE6}} = 12$$

$$k = \frac{1}{1 + \frac{20,000}{12(1000)}} = .375$$

$$j = 1 - \frac{.375}{3} = .875$$

From equation 14.28,

$$bd^2 = \frac{(2)(250,000)}{(1000)(.375)(.875)} = 1523.8$$

If d = 2b, then d = 14.5 and b = 7.25.

Allow 2 inches of cover for the steel, so the beam is (7.25 x 16.5).
The steel area required is

$$A_s = \frac{250,000}{(20,000)(.875)(14.5)} = .985 \text{ inches}^2$$

From table 14.5, it looks like 3 #4 bars will work. The area of steel
is 3(.2) = .6. This is too little. Try 3 #5 bars: 3(.31) = .93, still
too little. It is evident that 2 #7 bars are needed. Total area is
2(.6) = 1.2 square inches.

◆

B. Analysis - The Transformed Section Method

The method of transformed sections was presented in chapter 12 as a
method of evaluating the stresses in composite structures. This method
can be used with steel reinforced concrete beams with only one modification: If the beam contains reinforcement in the compressive zone
(i.e., the beam is doubly reinforced) the area of the compression
steel is to be expanded by (2n) instead of n. This is a requirement
of the ACI code used with the WSD method.

Figure 14.9

Transformation of a Reinforced Beam

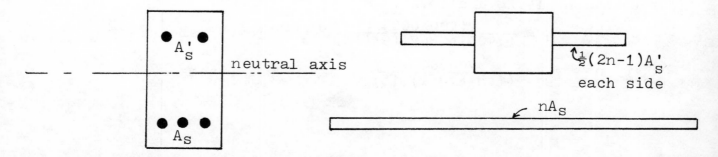

Of course, the moment of inertia and the location of the neutral axis are calculated based on the transformed beam.

For T-beams, any part of the stem below the flange but above the neutral axis may be omitted from the calculation as a simplication.

C. Web Reinforcement (USD)

A diagonal tension equal to the shear stress and inclined at 45° from the horizontal exists at every point along the span of the reinforced beam. This stress must be resisted by vertical stirrups attached to the longitudinal reinforcement and extending into the compression region.

Stirrup reinforcement (under the USD method) is required if the nominal shear stress exceeds v_c. Use factored loads in equation 14.34.

$$v_{nominal} = \frac{V}{bd\phi} \tag{14.34}$$

ϕ = .85 for shear (table 14.6).

$$v_c = 2\sqrt{f'_c} \tag{14.35}$$

Stirrup reinforcement should extend d inches past the point where $v_{nominal}$ decreases below v_c. The practical (not specified by code) minimum spacing is about 4 inches. The code specifies the maximum spacing as

$$(spacing)_{max} = \tfrac{1}{4}d \text{ if } v_{nominal} > 4\sqrt{f'_c} \tag{14.36}$$

$$= \tfrac{1}{2}d \text{ otherwise} \tag{14.37}$$

Spacing is not to exceed 24 inches.

The area of stirrup steel (where the stirrup is inclined at an angle θ from the horizontal) is

$$A_{stirrup} = \frac{(spacing)(v_c - v_{nominal})b}{f_y(\sin\theta + \cos\theta)} \tag{14.38}$$

θ is usually 45° or 90°, but it cannot be less than 45°. If the stirrups are bent in the shape of a U, each bent stirrup contributes twice its area. If a single bar is used, the area of that bar is

$$A_{stirrup} = \frac{(v_c - v_{nominal})bd}{(\sin\theta)f_y} \tag{14.39}$$

If reinforcing is required, then the stirrup area must exceed

$$A_{stirrup} > \frac{(50)(b)(spacing)}{f_y} \tag{14.40}$$

Other special restrictions are:

(a) Any 45° line starting from a point (.5)d from the top of the beam and slanting down at 45° towards

a support must cross at least 1 line of shear
reinforcement.
(b) Only the center (3/4) of an inclined longitudinal
bar may be included for shear reinforcement.
(c) Maximum $(v_c - v_{nominal})$ cannot exceed $8\sqrt{f_c'}$.

D. Bond Stresses (USD)

The adhesion between the concrete and steel must be able to resist
the longitudinal shear. This is accomplished by specifying the
minimum length of deformed bar that is to withstand the tensile
force. This minimum length is known as the 'development length'.

The basic development length for unhooked bars #11 or smaller is

$$l_d = (.04)A_b f_y / \sqrt{f_c'} \qquad\qquad 14.41$$

However, l_d is not to be less than $(.0004)(f_y)(\text{bar diameter})$ 14.42

The basic development length may also be multiplied by .8 if the bar
spacing is 6" or more and at least 3" of cover is provided to the
edge of the member. If more steel reinforcing is used than is required,
the development length may be multiplied by the ratio of actual to
required steel area.

Under no circumstances is the actual development length to be less
than 12".

If standard hooks are used, the development length (not to be less than
$8D_b$ or 6") is

$$l_d = \frac{1200D_b f_y}{6000\sqrt{f_c'}} \qquad\qquad 14.43$$

Development length for deformed bars in compression is calculated as
$(.02)D_b f_y / \sqrt{f_c'}$, not to be less than $(.0003)D_b f_y$ or 8".

E. Ultimate Strength Design (USD)

Note: These procedures are not complete enough to handle T-beams or
doubly-reinforced beams.

Since failure of reinforced beams does not occur when the steel reaches
its yield point, inelastic effects must be considered to predict the
ultimate strength of beams. Since the location of the neutral axis
shifts as the beam is stressed in the plastic region, much of the USD
procedure is empirical.

The ACI code permits the assumption that the compressive stress distribution in concrete is any shape (not just triangular) that yields
results in agreement with compressive tests. One common assumption
(the 'Whitney assumption') is that the distribution is rectangular
with a value of $(.85)f_c'$ at failure.

Figure 14.10
USD Stress Distribution

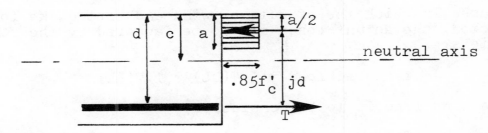

It follows that

$$a = \frac{A_s f_y}{(.85)(f'_c)b} = \frac{p(f_y)d}{(.85)(f'_c)} \qquad 14.44$$

where

$$p = A_s/A_c = A_s/bd \qquad 14.45$$

The ultimate moments are

$$M_s = T j d = C j d = A_s f_y (d - \tfrac{1}{2}a) \qquad 14.46$$

$$M_c = C(d - \tfrac{1}{2}a) = .85 f'_c ab(d - \tfrac{1}{2}a) \qquad 14.47$$

For balanced design, a = .537d. Therefore, for balanced design of rectangular beams with tension reinforcement only,

$$M_c = .333 f'_c bd^2 \qquad 14.48$$

The ACI code is similar to the above analysis, adding ϕ and the restriction that p should not exceed $.75(p_{balanced})$. A minimum value of p is $(200/f_y)$ unless the actual reinforcement used is one-third greater than called for by the calculations.

The ultimate moment that the beam can carry is

$$M_u = bd^2 R_n \phi = bd^2 f'_c q(1 - .59q)\phi = A_s f_y(d - \tfrac{1}{2}a)\phi \qquad 14.49$$

$\phi = .90$ for flexure. R_n is defined in equation 14.59. q is

$$q = p f_y/f'_c = \frac{A_s f_y}{f'_c bd} \qquad 14.50$$

$$a = \frac{A_s f_y}{(.85)f'_c b} \qquad 14.51$$

The balanced reinforcement ratio is

$$p_{balanced} = \frac{(.85)\beta \, f'_c}{f_y}\left[\frac{87,000}{87,000 + f_y}\right] \qquad 14.52$$

$$\beta = .85 - .05(\frac{f'_c - 4000}{1000}) \qquad 4000 < f'_c < 8000 \qquad 14.53$$

$$\beta = .85 \qquad\qquad f'_c < 4000 \qquad 14.54$$

A good assumption in designs is q = .2.

For cases in which the effects of wind and earthquake forces can be neglected, the actual loads are to be factored by the following equations:

$$\text{factored load} = 1.4(DL) + 1.7(LL) \qquad\qquad 14.55$$

If the wind load is to be considered,

$$\text{factored load} = .75(1.4(DL) + 1.7(LL) + 1.7(WL) \qquad 14.56$$
$$= .9(DL) + 1.3(WL) \qquad\qquad 14.57$$

The greater of equations 14.56 and 14.57 is to be used when wind loads are present, except that the factored load is not to be less than that from equation 14.55. If earthquake loads are present, substitute 1.1(EL) for WL in the above equations.

Example 14.6

What is the maximum uniform live load that the beam shown below can carry on a 10-foot span with built-in ends? Assume f'_c = 3500 psi and f_y = 40,000 psi.

total steel area
.531 in^2

12"

6"

2"

Note that this is a mid-length section. Steel is required at the top near the ends.

$$p = \frac{.531}{(12)(6)} = .00738$$

From equation 14.44,

$$a = \frac{(.00738)(40,000)(6)}{(.85)(3500)} = .595 \text{ inch}$$

From equations 14.46 and 14.47,

$$M_s = (.531)(40,000)(6 - \tfrac{1}{2}(.595)) = 121,121 \text{ in-lb}$$

$$M_c = (.85)(3500)(.595)(12)(6 - \tfrac{1}{2}(.595)) = 121,130 \text{ in-lb}$$

Usually these will not be so close together. In this case, steel governs.

The ultimate moment for a uniform load occurs at the end. The fixed-end moment is

$$M = \frac{wL^2}{12}$$

Since L = 10 feet = 120 inches, and ϕ = .90 for flexure,

$$w = \frac{(.90)(12)(121,121)}{(120)^2} = 90.8 \text{ lb/inch} = 1090 \text{ lb/ft}$$

Assuming 150 pcf concrete and ignoring the weight of the steel, the dead load of the beam per foot can be calculated.

$$DL = \frac{(12)(8)(150)}{(144)(1)} = 100 \text{ lb/ft}$$

From equation 14.55, maximum LL = $(1090 - 1.4(100))/1.7 = 559$ lb/ft.

The following procedure may be used to design a rectangular beam with tension reinforcement. This procedure does not check for beam deflection which should be done to check the values of b and d chosen.

    <u>step 1</u>: Select an appropriate tension reinforcement ratio, p, equal to or less than 75% of the balanced p, but greater than $(200/f_y)$. There is nothing wrong with a ratio of 75%. However, cracking will be minimized by keeping the ratio to less than half the maximum (i.e., .375 of the balanced p), particularly when grade 40 steel is used.

    <u>step 2</u>: Calculate the required value of bd^2 from equation 14.49. An alternative form of this equation is

$$bd^2 = \frac{M_u}{\phi R_n} \qquad\qquad 14.58$$

M_u is the factored ultimate moment from equation 14.55 and ϕ = .90 for flexure.

$$R_n = pf_y(1 - \frac{pf_y}{2(.85)f_c'}) \qquad\qquad 14.59$$

    <u>step 3</u>: Size the member so that bd^2 is approximately equal to the required value obtained from equation 14.58. b can be assumed (e.g., 9" or 10" is good when bd^2 is between 1700 and 2000 in^3.) Another good choice is to keep (d/b) between 1.75 and 2.0. Regardless, the choice of b and d must ultimately be checked by calculating the deflection of the beam under loading.

    Size the beam to the nearest $\frac{1}{4}$".

    <u>step 4</u>: Compute a revised p. In this and subsequent calculations, bd^2 is found from the dimensions chosen, not the results of step 2.

$$p_{revised} = \frac{(.85)f_c'}{f_y}\left[1 - \sqrt{1 - \frac{2R_n}{(.85)f_c'}}\right] \qquad\qquad 14.60$$

$$R_n = \frac{M_u}{\phi(bd^2)} \qquad\qquad 14.61$$

    <u>step 5</u>: Calculate the required steel area.

$$A_s = (p_{revised})(bd) \qquad\qquad 14.62$$

    <u>step 6</u>: Select reinforcement to satisfy distribution requirements of the code. Refer to table 14.7.

    <u>step 7</u>: If necessary, check the capacity of the beam as was done in example 14.6.

step 8: Design stirrups. This was covered previously in this chapter.

step 9: Calculate the concrete cover depth below the steel. For a beam with stirrups (figure 14.11), the required cover below the steel assuming exterior exposure is given by equation 14.63.

$$d_c = 1.5 + (\frac{\text{bar diameter}}{2}) + \text{stirrup bar diameter} \quad 14.63$$

For other exposures and for large bar sizes, refer to the detailing specifications at the end of this chapter.

<u>Figure 14.11</u>

<u>Reinforced Beam with Stirrups</u>

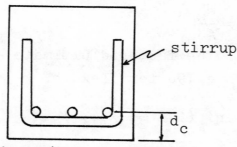

step 10: Check for cracking. (Not done here.)

Example 14.7

Design a tension-reinforced rectangular beam to carry service loads of 34,300 ft-lb (dead) and 30,000 ft-lb (live). Use f_y = 40,000 psi and f_c' = 3500 psi.

step 1: $p_{balanced} = \frac{(.85)(.85)(3500)}{40,000}(\frac{87,000}{87,000 + 40,000}) = .0433$

$p = (0.75)(.0433) = .0324$

step 2: $R_n = (.0324)(40,000)(1 - \frac{(.0324)(40,000)}{(2)(.85)(3500)}) = 1013.7$

$M_u = ((1.4)(34,300) + (1.7)(30,300))12 = 1.194 \text{ EE6 in-lb}$

$bd^2 = \frac{1.194 \text{ EE6}}{(.90)(1013.7)} = 1309$

step 3: Let $(d/b) = 1.8$. Then

$d = ((\frac{d}{b})(bd^2))^{.333} = ((1.8)(1309))^{.333} = 13.3$ (say 13.25)

$b = 13.3/1.8 = 7.39$ (say 7.5)

step 4: $R_n = \frac{1.194 \text{ EE6}}{(.9)(7.5)(13.25)^2} = 1008$

$$p = \frac{(.85)(3500)}{40,000}(1 - \sqrt{1 - \frac{(2)(1008)}{(.85)(3500)}}) = .0321$$

.0321 is less than $.75(p_{balanced})$.

<u>step 5</u>: $A_s = (.0321)(7.5)(13.25) = 3.19$ in^2

<u>step 6</u>: Choose 3 #10 bars or 2 #9 and 1 #10 in combination.

<u>step 7</u>: Not done, but a good idea.

<u>step 8</u>: Assume #4 stirrups for the sake of this example.

<u>step 9</u>: $d_c = 1.5 + \frac{1}{2}(1.27) + .5 = 2.635"$ (say 2.75")

Total beam dimensions will be (7.5" x 16")

◆

10. Slabs

Concrete floor slabs are usually poured monolithically with the beams. If the floor slab is reinforced in both horizontal directions (at right angles) and is supported on its edges, it is known as a 'two-way slab'. If it is reinforced in two directions, but is supported at its corners only, it is known as a 'flat slab'. If the slab is reinforced in one direction only (perpendicular to the supporting beams) it is known as a 'one-way slab'.

Simply-supported one-way slabs can be designed and evaluated as beams, as illustrated in figure 14.12. Minimum slab thicknesses apply - see section 13.

Figure 14.12

Simple One-Way Slab

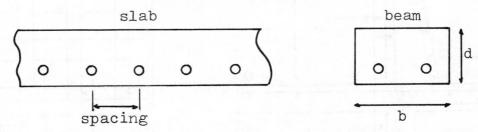

In the analysis, b is taken as 12 inches so that the ultimate moment can be computed per foot of slab width.

The area of steel in the beam is taken as

$$A_s = (\frac{12}{spacing})(\text{area of one bar}) \qquad 14.64$$

One-way slabs with monolithic beams (figure 14.13) can be treated as T-beams. A WSD analysis of T-beams can be easily performed with the transformed-area method. However, neither this nor the ACI code provisions for USD are covered in this chapter. (For those who care to try, the ACI code specifies that the effective overhanging flange width is to be less than or equal to the smaller of (a) one-half of the clear distance between adjacent beams, or (b) eight times the slab thickness, t.)

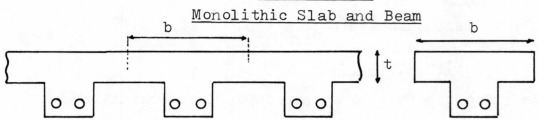

Figure 14.13
Monolithic Slab and Beam

11. Retaining Walls

A. Failure Mechanism

A cantilever retaining wall is most likely to fail structurally at its base due to the applied moment. In this regard, it is similar to a cantilever beam with a non-uniform distributed load.

B. Reinforcing Schemes

Typical reinforcement of a retaining wall is shown in figure 14.14

Figure 14.14
Retaining Wall Reinforcement

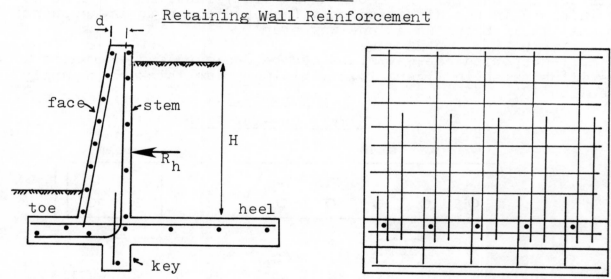

C. WSD Procedure

A WSD procedure for designing a cantilever retaining wall as a cantilever beam is given below.

step 1: Calculate the maximum moment on the stem, which occurs at the base. Use the horizontal component of active pressure. Neglect the passive pressure.

$$M_{stem} = (\tfrac{H}{3})R_h \qquad\qquad 14.65$$

step 2: Assume a balanced design. Calculate K.

$$K = \tfrac{1}{2}f_c jk \quad (f_c \text{ is from table } 14.3) \qquad 14.66$$

j and k can be found from equations 14.23 and 14.27.

step 3: Assuming moment governs, calculate d for a 12" width of retaining wall. That is, let b = 12".

$$d^2 = \frac{M_{stem}}{Kb} \qquad\qquad 14.67$$

Calculate d assuming shear governs. It usually doesn't for the stem.

$$d = \frac{R_h}{(b)(j)(v_{allowable})} \qquad\qquad 14.68$$

$v_{allowable}$ can be taken as 90 psi.

Choose the larger d and round up to the nearest quarter inch.

step 4: Add 3" to d for cover because of the cast-in-place concrete. This is the total stem thickness at the base.

step 5: Check the shear stress, $v_{nominal}$, against the code maximum, v'. b = 12" again. (ϕ = .85 for shear.)

$$v_{nominal} = R_h/bd \qquad\qquad 14.69$$

$$v_c = 2\sqrt{f_c'} \qquad\qquad 14.70$$

If needed, vertical shear reinforcement shall not be less than .0012 (#5 bars or smaller with $f_y \geq 60,000$ psi), and .0015 otherwise.

step 6: Calculate the steel area.

$$A_s = M_{stem}/f_s jd \qquad\qquad 14.71$$

step 7: Select bar spacing based on table 14.7.

step 8: Check bond strength and development length.

step 9: d can be varied along the stem length. As a general rule,

$$d_{top} = \tfrac{1}{2}(d_{bottom}) \qquad\qquad 14.72$$

This does not include the 3" + $\tfrac{1}{2}D_b$ cover, nor does it check for the minimum cover on the passive side.

step 10: Not all bars need to be extended the full height of the stem. Analytical methods similar to steps 1 through 8 can (should) be used to find the steel area at any height above the base. As a conservative rule, extend (1/3) of the steel all of the way up, (2/3) of the steel (2/3) of the way up, and all of the steel (1/3) of the way up.

The long vertical bars may be spliced into dowels. The dowels may be key dowels or they may be bent-up steel from the toe. In any case, the dowel length extending into the stem should considerably exceed the length for tensile splices.

step 11: Design the heel as a cantilever beam with the vertical component of active pressure, any surcharge, and the heel dead weight uniformly distributed on top of the heel. However, some of the downward pressure is cancelled by the overturning moment. If this moment is great enough, a situation might exist where the upward heel pressure becomes zero.

Some rational attempt should be made to account for this reduction. One such short-cut method is to neglect heel pressure and to multiply the moment and shear from active load, surcharge, and dead weight by (2/3). Of course, actual pressure distributions can also be used.

Steps 1 through 8 are used. Shear should be checked in step 3, as it may govern d.

Since the heel and toe are cast against the earth, it is a good idea to allow 1"-2" extra above the required cover to compensate for uneven excavation. The heel thickness is usually constant along its length.

step 12: Design the toe as a cantilever beam acting under the influence of the passive earth (downward) load, dead weight, and toe soil pressure (upward). The toe depth, d, or thickness, t, should be the same as for the heel. For ease of construction, however, d and t may taper to the tip.

step 13: Design the horizontal steel in the stem, toe, and heel based on the ACI code requirements for shrinkage, creep, and temperature changes. The total horizontal steel in the stem is not be be less than

$$.0025 \left(\begin{matrix} \text{stem} \\ \text{vertical height} \end{matrix} \right) \left(\begin{matrix} \text{average} \\ \text{stem thickness} \end{matrix} \right) \qquad 14.73$$

Equation 14.73 assumes that the horizontal steel is #6 or larger or $f_y < 60,000$ psi. For #5 bars or smaller and $f_y \geq 60,000$ psi, the coefficient is .0020. Between $\frac{1}{2}$ and 2/3 of the horizontal steel should be placed on the passive (outer) side of the stem. The remaining $\frac{1}{2}$ or 1/3 should face the active soil.

D. USD of Retaining Walls

As was illustrated in the WSD section, retaining walls are designed as cantilever beams. Therefore, beam design procedures should be followed if USD is specifically called for. Many of the comments in the WSD section, however, are still applicable.

12. Columns

A. General Code Provisions

Most column design procedures are based on experimental results which are based on ultimate strengths. Thus, stress relationships are only valid when the column is loaded to nearly its ultimate strength. Since a column should not be loaded past its elastic limit, no attempt is made to calculate the stress in the column, and the amount of reinforcing for ties and spirals is based on empirical data.

Some of the important ACI code requirements are given here.

(1) Spacing of bars: Longitudinal bars shall have a clear distance of at least $1\frac{1}{2}$ times the bar diameter and not less than $1\frac{1}{2}$".

(2) The minimum spiral wire diameter is (3/8)".

(3) The minimum tie wire for tied columns is #3 (#4 if longitudinal steel is #11 or larger).

(4) The clear distance between spirals should not exceed 3", but it should not be less than 1".

(5) The concrete covering shall be at least $1\frac{1}{2}$".

(6) At least 4 bars are to be used for square columns. At least 6 bars are to be used for round columns.

(7) The ratio of longitudinal steel area to the gross column area must be between .01 and .08.

(8) Center-to-center spacing of the ties must not exceed the smaller of 16 longitudinal bar diameters, 48 tie bar diameters, or the least gross (outside) column dimension.

(9) Every corner and alternating longitudinal bar shall be supported by a tie corner in tied columns.

(10) In tied columns, no bar shall be more than 6" away from a tie-corner supported bar.

Although the ACI code no longer requires it, the longitudinal steel size should be #5 or larger. Maximum practical tie size is #5. Maximum practical spiral wire size is (5/8)".

The ACI code no longer specifies a minimum eccentricity that must be considered when designing short columns. (A short column is defined by ACI 10.11.4 as one having a slenderness ratio less than 22.) Therefore, the following formulas can be used for small eccentricities. Although the ACI code does not define a small eccentricity, use of the following formulas should be limited to spiral columns with $e < (.05)$(width) and tied columns with $e < (.10)$(width). Interaction diagrams should be used in other cases.

The ACI code also allows use of WSD methods. However, the allowable loads must be reduced by 60% (i.e., multiplied by .4) if WSD is used.

B. Axially Loaded Spiral Columns: USD

The actual loads carried by the column should be factored by using equation 14.74. (Equation 14.74 ignores wind and earthquake loads.)

$$P_{factored} = 1.4(DL) + 1.7(LL) \qquad 14.74$$

The factored load is then compared to the ultimate column load, P_{max}. The design is acceptable if P_{max} exceeds $P_{factored}$.

$$P_o = A_g[.85(1-p_g)f'_c + f_y p_g] \qquad 14.75$$

$$P_{max} = .85\phi P_o \qquad 14.76$$

From table 14.6, $\phi = .75$ for spirally-reinforced columns.

A_g is the gross area of the column. p_g is the ratio of the area of longitudinal steel to the column gross area. The ratio p_g is limited to the range of .01 to .08 by the code. A value of .02 is typical.

The code also specifies that the reinforcement ratio for spiral columns not be less than

$$p_s = .45(\frac{A_g}{A_c} - 1)\frac{f'_c}{f_y} \qquad 14.77$$

A_c is the area of the circle made by the longitudinal steel core.

If the spiral bar diameter is known or assumed, the spiral pitch can be found from equation 14.78.

$$\text{spiral pitch} = \frac{4A_i}{p_s(\text{core diameter})} \qquad 14.78$$

A_i is the cross sectional area of the spiral wire. The clear distance between spirals will be (spiral pitch - spiral wire diameter).

Example 14.8

Calculate the maximum factored load that the short spiral column shown below can carry. Assume $f_y = 40,000$ psi and $f'_c = 3500$ psi.

$$A_g = \tfrac{1}{4}\pi(22)^2 = 380.1 \text{ in}^2$$
$$A_{\#8 \text{ bar}} = .79 \text{ in}^2, \text{ so } A_{steel} = 8(.79) = 6.32 \text{ in}^2$$

$$p_g = \frac{6.32}{380.1} = .0166$$

From equations 14.75 and 14.76,

$$P_o = 380.1[.85(1-.0166)3500 + 40,000(.0166)] = 1.364 \text{ EE6}$$

$$P_{max} = (.85)(.75)P_o = 869,800 \text{ lb}$$

Example 14.9

Design a spiral column to carry an axial load of 375,000 pounds using 3000 psi concrete and 40,000 psi steel. The load is factored.

Assume $p_g = .02$, which is in the allowable range. Larger values of p_g will decrease the column gross area. (Since slenderness effects are neglected in this analysis, it is desired to keep the column as fat as possible.)

From equations 14.75 and 14.76,

$$375,000 = A_g(.85)(.75)[.85(1-.02)3000 + 40,000(.02)]$$

$$A_g = 178.3 \text{ in}^2$$

$$D_g = 15.07 \text{ inches (say 15.25 inches)}$$

With a $1\frac{1}{2}$" cover, the core diameter will be 15.25-3.0 = 12.25".

The required steel is $(.02)(178.3) = 3.57 \text{ in}^2$.

The code requires at least 6 bars be used. Some possibilities are: 6 #7 bars, 9 #6 bars, and 12 #5 bars. Choose the #6 bars after checking the clear spacing.

From equation 14.77, substituting D^2 for A, the ratio of spiral reinforcement is

$$p_s = .45[(\frac{15.25}{12.25})^2 - 1]\frac{3,000}{40,000} = .0186$$

Since .02 is greater than .0186, this is okay. Assume 3/8" (#3) spiral wire. Then, $A_i = .11$. The spiral pitch is given by equation 14.78.

$$\text{spiral pitch} = \frac{(4)(.11)}{(.02)(12.25)} = 1.80$$

This pitch should be checked to see if it meets the 1" and clear spacing requirements. Since 1"<(1.8-3/8")<3", it does.

C. Axially Loaded Tied Columns: USD

The maximum load for axially loaded tied columns is

$$P_{max} = (.80)\phi P_o \qquad 14.79$$

As given in table 14.6, ϕ = .70 for tied columns.

13. Miscellaneous ACI Detailing Requirements

A. Cover on Non-prestressed Steel (minimums, cast-in-place only)

- Cast against and permanently exposed to earth ....3"
- Exposed to earth or weather
    #5 bars or smaller.................................$1\frac{1}{2}$"
    #6 bars or larger.................................2"
- Not exposed to weather or ground contact
    Slabs, walls, or joists
        #11 bars or smaller..........................3/4"
        #14 and #18 bars............................$1\frac{1}{2}$"
    Beams, girders, or columns......................$1\frac{1}{2}$"

B. Minimum Horizontal Clear Distance Between Bars in a Layer

- Beams: The maximum of one bar diameter or.........1"

- Walls: The maximum of one bar diameter or........1"

- Columns: Refer to page 14-35

C. Minimum Vertical Clear Distance Between Bars in a Layer

- Beams:...1"

D. Maximum Rebar Spacing (clear distance)

- Walls and Slabs: The minimum of the three times
    the wall or slab thickness, or................18"

E. Splices (Ratio of lap length of development length, l_d)

Tension splices
    Class A:.......................................1.0
    Class B:.......................................1.3
    Class C:.......................................1.7

The minimum tension lap length is 12"

Compression splices (f'_c exceeds 3000)
    The minimum of 12", (.0005)(f_y)(bar diameter), and
    one development length

Many other provisions apply to both tension and compression lap splices. Welded splices are butt-welds developing an ultimate tensile strength of 125% of the yield strength of the bar.

F. Minimum Bend Radii (In bar diameters)

    #3 to #8 bars....................................6
    #9, #10, and #11 bars............................8
    #14 and #18 bars................................10

G. Development Length

See page 14-26

H. Maximum Deflections

The ACI code does not really set limits to deflection, other than saying that the deflection cannot exceed an amount adversely affecting strength and serviceability. However, it does specify minimum moments of inertia. Where the deflection is not checked by the moment of inertia method, the values given in the following table are applicable.

element	when	is element attached to non-structural items likely to be damaged by large deflections?	limit
flat roof	immediate	no	L/180
floor	immediate	no	L/360
roof/floor	sustained	no	L/240
roof/floor	sustained	yes	L/480

These limits are fractions of the distance between supports.

I. Minimum Beam and Slab Thickness (D in figure 14.8) if deflections are not computed. These thicknesses apply to normal weight concrete (density greater than 120 pcf) and grade 60 steel. For other grade steels, multiply the thickness by

$$.4 + \frac{f_y}{100,000}$$

These fractions are the fraction of span length between supports.

	Simply Supported	One End Continuous	Both Ends Continuous	Cantilever
Solid, one-way slabs	1/20	1/24	1/28	1/10
Beams or ribbed one-way slabs	1/16	1/18.5	1/21	1/8

J. Maximum Aggregate Size

(1/5) of the narrowest dimension between sides of the forms, or (1/3) the slab depth, or (3/4) of the clear spacing between bars, whichever is greatest.

Bibliography

1. American Society for Testing and Materials, <u>Selected ASTM Standards for Civil Engineering Students</u>, ASTM, Philadelphia, PA 1968

2. Cernica, John N., <u>Fundamentals of Reinforced Concrete</u>, Addison-Wesley Publishing Co., Inc., Reading, MA 1964

3. International Conference of Building Officials, <u>Uniform Building Code</u>, 1979 Edition, Whittier, CA 1979

4. Kurtz, Max, <u>Comprehensive Structural Design Guide</u>, McGraw-Hill Book Company, New York, NY 1969

5. Large, George E., <u>Basic Reinforced Concrete Design</u>, Ronald Press, Co., New York, NY 1950

6. Peck, Ralph; Hanson, Walter E., and Thornburn, Thomas H., <u>Foundation Engineering</u>, 2nd ed., John Wiley and Sons, Inc., New York, NY 1974

7. Portland Cement Association, <u>Notes on ACI 318-77, Building Code Requirements for Reinforced Concrete with Design Applications</u>, Skokie, IL 1978

8. Rice, Paul F., and Hoffman, Edward S., <u>Structural Design Guide to the ACI Building Code</u>, Van Nostrand Reinhold Company, New York, NY 1979

Practice Problems: CONCRETE DESIGN

Required:

1. 1.50 cubic yards of portland cement concrete are needed. Using the specifications below, determine the number of pounds of water, cement, fine aggregate, and coarse aggregate needed.

> cement content: 6.5 bags per cubic yard
> water cement ratio: 5.75 gallons per bag
> cement specific gravity: 3.10
> fine aggregate specific gravity: 2.65
> coarse aggregate specific gravity: 2.00
> aggregate grading: 30% fine, 70% coarse by volume
> free moisture: 1.5% in fine aggregate
> coarse aggregate will absorb 3%
> entrained air: 5%

2. Design a square 2-way footing to carry 240,000 pounds which are transmitted through a 16" square column. The allowable soil pressure is 4000 psf. Use 3000 psi concrete and intermediate grade steel.

3. Design a rectangular beam with tension reinforcement only using the ultimate strength method. The dead load moment is 50,000 ft-lb and the live load moment is 200,000 ft-lb. Use 3000 psi concrete and 50,000 psi steel.

Optional

4. Design a balanced beam meeting the following specifications. Use the working strength design method.

> maximum moment: 50.4 ft-kips
> maximum steel stress: 20,000 psi
> maximum concrete stress: 1350 psi
> E_s/E_c: 10

5. Design a spiral column to carry a dead load of 175,000 pounds and a live load of 300,000 pounds. The loads are axial. Use 3000 psi concrete and 40,000 psi steel.

6. Design a short tied column to carry a 100,000 pound dead load and a 125,000 pound live load. Include specifications for the ties. Use 3500 psi concrete and 40,000 psi steel.

7. An 18" square column supports 200,000 pounds and 145,000 pounds dead and live loads respectively. The allowable soil pressure is 4000 psf. Use the ultimate strength design method with 3000 psi concrete and 40,000 psi steel to design the footing. The column steel consists of 10 #9 bars.

8. A beam is being designed to withstand an ultimate factored moment of 400,000 ft-lb. 4000 psi concrete and 40,000 psi steel are available. Specify values of b, d, and the required steel area. How many layers of steel are needed?

9. 3000 psi concrete is used in the floor slab shown below. The slab must carry 20,000 ft-lbf per foot of width. Use the transformed area method to evaluate the stresses in the concrete and steel.

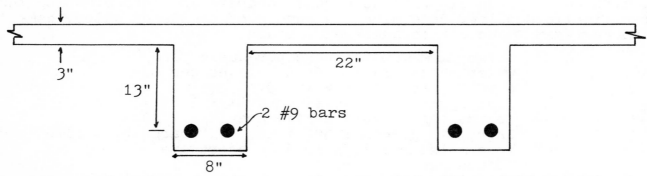

10. 100 pounds of aggregate were sieve graded. The weights retained on each sieve are shown below. What is the fineness modulus of the aggregate?

sieve	weight retained
4	4
8	11
16	21
30	22
50	24
100	17
dust	1

NOTES

NOTES

STEEL DESIGN

Nomenclature

A	area	in^2
b	depth	in
B	eyebar diameter	in
c	distance from neutral axis to extreme fibers	in
C_c	critical slenderness ratio	-
C_m	a coefficient	-
C_t	a reduction factor	-
d	depth, or diameter	in
DL	dead load	lb
E	modulus of elasticity	psi
EL	earthquake load	lb
f	actual stress	psi
F	strength or allowable stress	psi
g	gage spacing	in
G	shear modulus	psi
I	area moment of inertia	in^4
J	polar moment of inertia	in^4
K	end restraint coefficient	-
l	length between supports	in
L	length	in
LL	live load	lb
M	moment	in-lb
P	load	lb
Q	statical moment	in^3
r	radius of gyration, radius, or distance	in
R	radius	in
s	pitch spacing	in
S	section modulus	in^3
t	thickness	in
T	torque	in-lb
T_e	effective weld thickness	in
u	poisson's ratio	-
V	total shear	lb
w	weld size	in
WL	wind load	lb
Z	plastic modulus	in^3

Symbols

α	coefficient of thermal expansion	$1/°F$
ϕ	shape factor	-

Subscripts

a	axial
b	bracing, or bending
c	maximum bracing for $(F_b/F_y) = .66$
e	effective, or endurance
f	flange

g   gross
n   net
p   bearing
t   tension
u   ultimate, or maximum bracing for $(F_b/F_y) = .60$
v   shear
w   web
y   yield

1. Introduction

This chapter presents the subject of steel design as a blend of
theoretical mechanics and code requirements. This chapter cannot serve
as a complete substitute for the AISC Manual of Steel Construction
(hereinafter referred to as 'AISCM') or the AISC Specifications
(hereinafter referred to as 'AISCS') or the Uniform Building Code.
Nevertheless, this chapter will enable you to work most of the common
steel analysis and design problems, as well as some of the more
obscure problem types.

In a noticeable departure from this book's policy of providing all
necessary data to work problems, this chapter does not contain listings
of structural shapes. Therefore, it is absolutely necessary that you
obtain listings of such shapes prior to working steel problems.
However, due to recent consolidations of rolled shapes, the listing
you obtain should have been made after 1977.

2. Conversions

The following abbreviations are used in this conversion table:

kPa - kiloPascals
m   - meter
N   - Newton
Pa - Pascal

Multiply	By	To obtain
ft-lb	1.356	N-m
in-lb	1.13 EE-1	N-m
kPa	1000	Pa
ksi	6.895 EE6	Pa
N-m	.7376	ft-lb
N-m	8.851	in-lb
N/m^2	1.0	Pa
Pa	1.0	N/m^2
Pa	.001	kPa
Pa	1.45 EE-7	ksi
Pa	1.45 EE-4	psi
Pa	2.089 EE-2	psf
psf	4.788 EE1	Pa
psi	6.895 EE3	Pa

3. Definitions

Beam-column: A structural member that carries both axial and lateral loads.

Compact section: A structural shape used as a beam or column that meets the tests presented in section 15.9

Effective area: The area which is assumed to carry all of the load. Generally, the effective area is calculated as the gross area minus the area of any voids or breaks in the cross section.

Girder: A structural shape usually used horizontally to support a floor or roof.

Joist: A horizontal member extending from one wall to the opposite wall and used to support a floor.

Plate girder: A large girder which is constructed from flat plates welded or riveted together and used when the strength or length of standard shapes are insufficient.

Ponding: Collection of rain water on a roof.

Shoring: Temporary bracing used to support steel joists while a wet concrete floor slab hardens. Until the concrete hardens, the steel joists and shores must support the weight of the concrete. After hardening, the concrete is capable of carrying its own weight if properly reinforced.

Upset ends: An increase in diameter or thickness due to an application of an impact load.

4. Types of Steels

ASTM A36 is the designation given to the all-purpose, carbon steel used for most bridge and building projects. ASTM A36 replaces the A7 and A373 steels. Most A36 shapes are hot rolled.

Other steels are available that have higher strengths. Their use results in lower dead weights but higher material costs. The commonly available steels and their applications are listed in table 15.1. The only 50 ksi steel available as rolled (not requiring heat treatment) is A588.

Table 15.1

Structural Steels

DESIGNATION	TYPE	USE
A36	carbon steel	shapes, plates, and bars
A53	carbon steel	pipe
A242	corrosion resistant, high strength, low alloy steel	shapes, plates, and bars
A375	high strength, low alloy steel	sheet and strip
A440	high strength steel	shapes, plates, and bars
A441	high strength, low alloy steel	shapes, plates, and bars
A500	carbon steel	cold formed tubing
A501	carbon steel	hot formed tubing
A514	quenched and tempered alloy steel	plates
A529	carbon steel	thin plates, shapes
A570	carbon steel	sheet and strip
A572	high strength, low alloy steel	shapes, plates, and bars
A588	corrosion resistant, high strength, low alloy steel	shapes, plates, and bars
A606, A607	corrosion resistant, high strength, low alloy steel	sheet and strip
A618	high strength, low alloy steel	seamless tubing

All of the steel grades listed in table 15.1 are weldable except A440.

5. Steel Properties

Some properties of steel (such as the modulus of elasticity and density) are essentially independent of the type of steel. Other properties (such as the ultimate and yield strengths) depend not only of the type of steel but also upon the size or thickness of the piece. The properties below apply to A36 steel. Refer to the AISCM and AISCS for other steels. (See also table 12.5.)

Table 15.2
Properties of A36 Steel

E, modulus of elasticity: 2.9 EE7 psi (up to 100°F)

G, shear modulus: 11.5 EE6 psi

α, coefficient of thermal expansion: 6.5 EE-6 1/°F

u, poisson's ratio: .30

density: 490 pcf

F_y, yield strength: 36,000 psi (to 8" thickness inclusive)
$\qquad\qquad\qquad\qquad$ 32,000 psi (over 8" thickness)

F_u, ultimate strength: 58,000 psi minimum

F_e, endurance limit: approximately 30,000 psi

weldability: excellent

6. Allowable Stresses

AISCS limits stresses to the following:

TENSION:
eye bars with pin-holes	.45 F_y on net area
all other cases (except bending)	.60 F_y on gross area
	and .50 F_u on net area

SHEAR:
all cases except plate girder webs	.40 F_y on gross area
bolts and rivets	see table 15.6

COMPRESSION:
axially loaded members:	Refer to section 15.10
on gross area of plate girder stiffeners	.60 F_y
on web of rolled shapes at the fillet toe	.75 F_y (no web stiffeners)

BENDING:

symmetrical compact sections loaded in the
plane of their minor axes and meeting the
requirements of section 15.9 $.66\ F_y$

doubly-symmetrical W and S shapes, solid
round and square bars, and solid rectangular bars bent about the weak axis
and meeting the first 2 requirements in
in section 15.9 $.75\ F_y$

all other cases (e.g., beams with
$$L_u \geq L_b \geq L_c)$$ $.60\ F_y$

BEARING: (Use the smallest F_y or F_u of all pieces.)

on milled surfaces, including stiffeners
and pins in holes $.90\ F_y$

on projected area of bolts and rivets in
shear connections $1.50\ F_u$

If the load is to be applied and removed less than 20,000 times (as
would be the case in a conventional building), no provision for repeated
loading is necessary. (20,000 times is roughly equivalent to two times
a day for 25 years.) However, some designs, such as for crane runway
girders and supports, must consider the effects of fatigue. AISCS
gives the allowable stress range and maximum stress in such cases.

Stress concentration factors are not normally applied to connections
with multiple redundancy.

Allowable stresses may be rounded up or down to the nearest ksi.

The effects of impact are included by increasing the actual live load
(but not the dead load) by the percentages contained in table 15.3.

Table 15.3

Impact Loading Factors

Supports for	Live load increase
elevators	100%
cab operated travel cranes	25%
travel operated travel cranes	10%
shaft or motor driven machinery	20% minimum
reciprocating machinery	50% minimum
floors and balconies	33%

Most allowable stresses may be increased (1/3) for transitory wind and
earthquake loading.

7. Structural Shapes

Many different structural shapes are available. These are shown in
figure 15.1 along with their AISC dimensional nomenclature. The
identifying dimensions and weight must be appended to the shape
designations to uniquely identify the shape. For example, W 30 x 132
means a W-shape with an overall depth of approximately 30 inches
which weighs 132 pounds per foot.

Figure 15.1
Structural Shapes

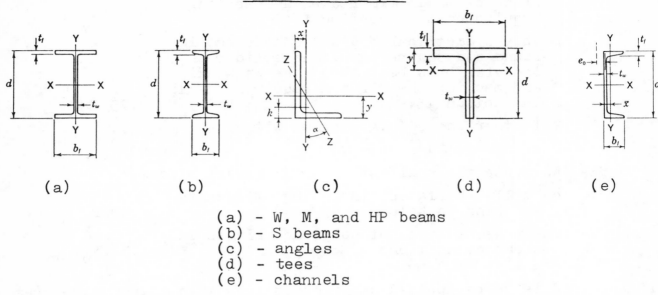

(a)   (b)   (c)   (d)   (e)

(a) - W, M, and HP beams
(b) - S beams
(c) - angles
(d) - tees
(e) - channels

Table 15.4 gives the old and new structural shape designations.

Table 15.4
Old and New Structural Shape Designations

Shape	Old designation	New designation
wide flange beams	WF	W
light beams	B	W
standard flanged beams	I	S
misc. flanged beams	M, JR	M
American std. channels	[	C
bearing piles	BP	HP
angles (all)	∠	L
tees	ST	XT (cut from X)
plate	PL	PL
bar	bar	bar
pipe	pipe	pipe
structural tubing	tube	TS

8. Design of Tension Members

Typical tension members are wire cables, rods with upset ends, eye-bars (pin-connected plates), and structural shapes. Axially loaded members are designed so that the nominal stress is less than the allowable stress. The nominal stress is just the average stress, calculated by dividing the area into the design load.

$$f = \frac{P}{A}$$

15.1

For pin-connected members, A is the actual net area at the pin-hole. This is shown in figure 15.2. (The gross area is not checked.) For riveted or bolted connections, the area is taken as the gross area for checking against the yield strength and it is taken as the effective net area for checking against the ultimate strength.

Figure 15.2
Areas for Tensile Calculations

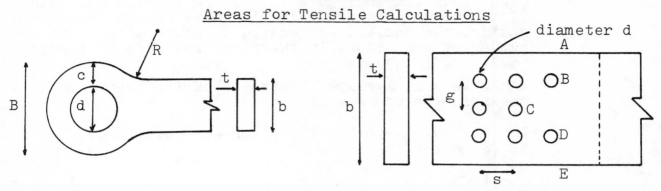

For multiply-redundant connections, such as the riveted connection shown in figure 15.2, the gross section is

$$A_g = bt \qquad\qquad 15.2$$

The effective area, which for splice and gusset plates is not to exceed 85% of the gross area, depends on the connection design. For a chain of holes extending across an element,

$$A_e = C_t A_n \qquad\qquad 15.3$$

A_n is the actual net area, calculated as the net width of the part times the thickness.

$$A_n = t b_n \qquad\qquad 15.4$$

$$b_n = b - \Sigma\binom{\text{hole diameters}}{\text{in the chain}} + \Sigma\frac{s^2}{4g} \qquad\qquad 15.5$$

The last summation term in equation 15.5 is taken over each gage space (2 in figure 15.2) in a diagonal path. s is the pitch (the longitudinal spacing) and g is the gage (transverse spacing).

The chain of holes to be used is the one which will give the minimum length. This minimum length chain sometimes must be found by trial and error. On figure 15.2, chains ABDE and ABCDE must both be checked.

C_t is 1.0 if all connectors are in the same plane. For structural shapes, C_t is .90, although the AISCS has some restrictions. For connections to built-up sections, C_t is .85 if there are 3 or more fasteners per line, and C_t is .75 if there are only 2 fasteners per line.

The following special rules and observations also apply:

* The nominal hole diameters should be increased by 1/16" each to account for manufacturing tolerances.

* The nominal hole diameter (clearance hole) is 1/16" larger than the nominal fastener diameter.

* A_e may not exceed $(.85)A_g$ in splice and gusset plates
* If the first row of connectors has fewer than subsequent rows (as in figure 15.2), the tensile strength of the second row net effective area should be checked against the total load reduced in proportion to the number of connectors in the first row (2/8 in figure 15.2).

* A complete analysis will also check for bearing and shear in the connectors.

* The following eye bar standards are specified by AISCS. Refer to figure 15.2.

  - $(b/t) \leq 8$ 15.6
  - $t \geq \frac{1}{2}$"   15.7
  - $1.33 < \dfrac{(B-d)t}{(b)(t)} < 1.5$ (constant thickness eyebars)   15.8
  - $d_{pin} \geq (7/8)b$ 15.9
  - $R \geq B$ 15.10
  - $(d - d_{pin}) \leq (1/32)$"   15.11
  - $d \leq 5t$ if F_y exceeds 70,000 psi   15.12

* Where structural shapes are used in tension, the following non-mandatory slenderness ratios are specified by AISCS as 'preferable'.

  - $(\frac{KL}{r})$ less than 240 for main members

       less than 300 for bracing and secondary members

Example 15.1

Design a plate eye bar to carry a static tensile load of 500,000 pounds. Use A36 steel.

The allowable stress on the net area of the eyebar is

$$F_{te} = .45(36,000) = 16,200 \text{ psi}$$

The allowable stress on the gross section of the bar is

$$F_{tg} = .60(36,000) = 21,600 \text{ psi}$$

This is less than $.5(F_u)$ so the 21,600 psi governs.

The required bar area is

$$A = \frac{500,000}{21,600} = 23.15 \text{ in}^2$$

Assume $1\frac{1}{2}$" plate is used. Use PL $1\frac{1}{2}$ x 15.5 for an actual bar area of 23.25 square inches. Check the (b/t) ratio.

$$\frac{b}{t} = \frac{15.5}{1.5} = 10.33$$

This exceeds 8. So try $1\frac{3}{4}$ plate, $13\frac{1}{4}$ inches wide with an actual area of 23.19 in^2.

$$\frac{b}{t} = \frac{13.25}{1.75} = 7.57 \text{ (okay)}$$

Now, calculate a trial eye area.

$$A = \frac{500,000}{16,200} = 30.86 \text{ in}^2 \text{ in tension}$$

The minimum eye area is limited by equation 15.8

$$(1.33)(23.19) = 30.84 \text{ in}^2$$

The 30.86 governs, so use it. The eye width, c, is

$$c = \frac{30.86}{(2)(1.75)} = 8.82 \text{ (say 9")}$$

Then, the actual eye area will be

$$(2)(1.75)(9) = 31.5 \text{ in}^2$$

31.5 is less than 1.5(23.19) = 34.8, so it is okay.

The pin diameter must exceed (7/8)(13.25) = 11.59, so use a 11.75" pin. The hole diameter will be 11.75 + (1/32) = 11.78". Therefore,

$$B = 11.78 + 2(9) = 29.78 \text{ (say 29.75")}$$

The minimum transition radius is 29.75 (say 30")

As a final step, the bearing stress should be calculated.

$$F_p = .90(36,000) = 32,400$$

$$f_p = \frac{500,000}{(11.75)(1.75)} = 24,320 \text{ (okay)}$$

◆

Example 15.2

Choose a 25 foot long W shape to carry dead and live loads of 198,000 and 270,000 pounds respectively. The shape will be used as a main member whose ends are free to rotate. Use A36 steel.

For a structural shape without holes, the gross and net areas are the same. The allowable tensile stress is the minimum of

$$(.6)(36,000) = 21,600 \text{ psi (governs)}$$
$$(.5)(58,000) = 29,000 \text{ psi}$$

Using $C_t = .90$, the required area is

$$\frac{198,000 + 270,000}{21,600(.90)} = 24.07 \text{ in}^2$$

Try a W 21 x 82 which has an area of 24.2 and r_{min} = 1.99 inch.

The slenderness ratio is $\frac{(25)(12)}{1.99}$ = 150. This is less than 240, so it meets the suggested standards also.

Example 15.3

What is the allowable tensile force for the PL $\frac{1}{2}$ x 9 plate connection shown below? Disregard the shear strength of the connectors which are $\frac{1}{2}$" bolts. The steel is A36.

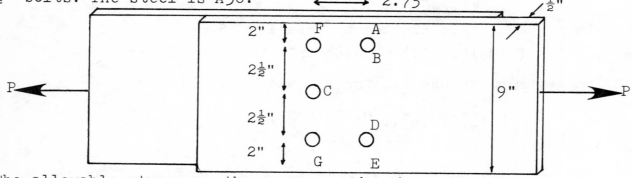

The allowable stress on the gross section is

$$F_t = (.6)(36,000) = 21,600 \text{ psi}$$

The allowable stress on the net effective section is

$$F_t = (.5)(58,000) = 29,000 \text{ psi}$$

The gross area of the plate is ($\frac{1}{2}$ x 9) = 4.5 in^2. The effective hole diameter (1/16" for clearance, 1/16" for manufacturing) is $\frac{1}{2}$ + 2(1/16) = .625 in. The net area of the plate must be evaluated three ways: lines ABDE, ABCDE, and FCG. The net area of ABDE is

$$.5(9 - 2(.625)) = 3.875 \text{ in}^2$$

To check area ABCDE, it is necessary to calculate

$$\frac{s^2}{4g} = \frac{(2.75)^2}{(4)(2.5)} = .756"$$

The net area of ABCDE is

$$.5(9 - 3(.625) + 2(.756)) = 4.319 \text{ in}^2$$

The net area of FCG is

$$.5(9 - 3(.625)) = 3.5625 \text{ in}^2$$

In this case, FCG governs, but it doesn't always. The net effective area is given by equation 15.3. But C_t = 1 because all of the connectors are in the same plane. So, $A_e = A_n = 3.5625$.

The capacity of the connection is

$$(21,600)(4.5) = 97,200 \text{ lb (gross section check)}$$
$$(29,000)(3.5625) = 103,313 \text{ lb (net section check)}$$

The capacity is 97,200 lbs. Check the section ABDE assuming holes F, C, and G carry (3/5) of the load. The remaining load is

$$(\tfrac{2}{5})(97,200) = 38,880 \text{ lb}$$

The tensile stress in section ABDE is $(\frac{38,880}{3.875})$ = 10,034 psi, which is okay. Similarly, if holes BD carry (2/5) of the load, the stress in section FCG will be (3/5)(97,200)/3.5625 = 16,371 psi, also okay.

9. Beam Design

This chapter assumes that a beam is loaded so that bending will occur in one plane only. Adequate lateral bracing is assumed for all shapes which are symmetrical about one axis only. This chapter is also not adequate for designing beams with flange stiffeners or beams which are box or tube shapes.

A. Elastic Design

Simple one-span beams are usually designed by the allowable stress method, analogous to the WSD method for concrete beams. The allowable stress method is also known as the 'elastic design method'. There is no advantage to using plastic design procedures for one-span beams.

Elastic design of one-span beams is customarily done with simple beam bending theory. That is, the flexure stress equation is used to size the beam. For bending in one plane only,

$$f_{b,max} = \frac{M_{max}c}{I} = \frac{M_{max}}{S} \qquad 15.13$$

Unless the beam is extremely short, it should be sized using equation 15.13, but it should be checked for shear according to equation 15.14.

$$f_v = \frac{VQ}{It} \qquad 15.14$$

For rolled W shapes and similar, equation 15.14 simplifies to

$$f_v = \frac{V}{dt_w} \qquad 15.15$$

The 'basic' allowable bending stress in tension or compression is $(.6)F_y$, just as it was for simple tension. The allowable bending stress for compact sections (which have thicker webs) is $.66F_y$. A compact section is one which meets <u>all</u> of the following requirements.

requirement #1: The flanges must be continuously connected to the web. A built-up section or plate girder which is constructed with intermittent welds does not qualify.

requirement #2: For standard rolled shapes without flange stiffeners,

$$\frac{b_f}{2t_f} \le 65.0/\sqrt{(F_y/1000)} \qquad 15.16$$

<u>requirement #3</u>: $\dfrac{d}{t_w} \leq 640/\sqrt{F_y/1000}$ 15.17

<u>requirement #4</u>: Except for boxes and tubes, the compression flange is supported laterally at intervals of L_b (or less).

$$L_b \leq L_c = \frac{(76.0)b_f}{\sqrt{(F_y/1000)}} \qquad \text{inches} \qquad 15.18$$

$$\text{and } L_b \leq L_u = \frac{(20,000)b_f t_f}{d(F_y/1000)} \qquad \text{inches} \qquad 15.19$$

Such lateral support would normally be supplied if the beam supported a continuous roof or floor.

If the shape meets all of the 4 requirements, it is a compact section and the design stress may be taken as $.66F_y$. If it meets all requirements except #2, and if

$$\frac{b_f}{2t_f} \leq \frac{95.0}{\sqrt{(F_y/1000)}} \qquad 15.20$$

then the allowable stress is

$$F_b = F_y \left[.79 - .002 \left(\frac{b_f}{2t_f}\right)\sqrt{(F_y/1000)} \right] \qquad 15.21$$

For doubly-symmetrical W and S sections (except those constructed of A514 steel) meeting the first two requirements for compact sections, and bending about their minor axis, the allowable bending stress is

$$F_b = .75F_y \qquad 15.22$$

If a doubly-symmetrical shape meets the first requirement but not the second, and if equation 15.20 also holds, then the allowable stress is

$$F_b = F_y \left[1.075 - .005\left(\frac{b_f}{2t_f}\right)\sqrt{(F_y/1000)} \right] \qquad 15.23$$

The shear area for rolled and fabricated shapes is calculated as the product of overall depth and web thickness. That is,

$$A_v = (d)(t_w) \qquad 15.24$$

Deflections are limited to $(1/360)$ of the span for beams supporting a plastered ceiling. However, relationships based on AISC <u>Commentary</u> suggest, but do not require, that deflections for uniformly loaded beams not exceed (span/290) for floors and (span/232) for roofs. Special calculations are needed to check for ponding on flat roofs.

Also of value are the following fine points:

* Static loading is assumed as long as the number of loading cycles is less than 20,000.

* If the live load is known but the dead load is not, it will be necessary to assume a beam weight prior to the selection.

Example 15.4

Select a W shape beam (F_y = 36,000 psi) to carry a maximum moment of 140,000 ft-lb if the compression flange is braced every 6 feet.

Assume a compact section so that the allowable bending stress is

$$F_b = (.66)(36,000) = 23,760 \text{ psi}$$

The required section modulus is

$$S = \frac{M}{F_b} = \frac{(140,000)(12)}{23,760} = 70.7 \text{ in}^3$$

Searching the W shape lists (or the allowable stress design selection table) for the lightest beam yields

W 21 x 44: d = 20.66
b_f = 6.500
t_f = .451
t_w = .348

Now, check to see if the beam meets all of the requirements for a compact section.

requirement #2: $\frac{6.500}{2(.451)} < 65.0/\sqrt{36}$

$\quad\quad\quad 7.21 < 10.8 \text{ (okay)}$

requirement #3: $\frac{20.66}{.348} < 640/\sqrt{36}$

$\quad\quad\quad 59.4 < 106.7 \text{ (okay)}$

requirement #4:

equation 15.18: $(76.0)(6.500)/\sqrt{36} = 82.3$ inches

equation 15.19: $\frac{(20,000)(.451)(6.500)}{(20.66)(36)} = 78.8$ inches

72 inches (6 feet) is less than both of these.

If the total shear, V, was known, it would be necessary to check for the shear stress also. ◆

B. Plastic Design

Plastic design (also known as 'inelastic design' and 'ultimate strength design') is similar in concept to the USD method used in chapter 14. The design procedure is based on a factored load which is compared to an ultimate load for the beam. The design is accomplished without stress calculations.

Plastic design is ideally suited to continuous beams. It should not be used with non-compact sections or for crane runway sections. Such beams should be designed using the elastic theory. Also, F_y must be less than or equal to 65 ksi to use the plastic design method.

The plastic design method is summarized below:

step 1: Multiply the working loads by a design load factor to obtain the plastic moment capacity, M_p, at failure. The factored load is the maximum of the following equations.

$$\text{factored load} = 1.7(DL) + 1.7(LL) \hspace{2cm} 15.25$$
$$= 1.3(DL) + 1.3(LL) + 1.3(WL) \hspace{1cm} 15.26$$
$$= 1.3(DL) + 1.3(LL) + 1.3(EL) \hspace{1cm} 15.27$$

step 2: Calculate the required plastic modulus.

$$Z = M_p/F_y \hspace{3cm} 15.28$$

An alternate method is to multiply the section modulus, S, by a shape factor (if it is known).

$$Z = S\phi \hspace{4cm} 15.29$$

step 3: Select a beam with the required plastic modulus.

step 4: Check for compactness.

step 5: Check for shear strength.

$$V_u < (.55)F_y t_w d \hspace{3cm} 15.30$$

Example 15.5

Select a W shape (A36 steel) to support dead and live loads totalling 4000 pounds/ft over the beam shown below. All of the supports are simple.

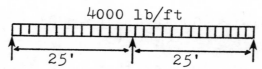

4000 lb/ft

25'   25'

step 1: The factored load is $(1.7)(4000) = 6800$ lb/ft. The maximum moment and maximum shear at ultimate failure in a uniformly loaded, simply-supported end span are $(.086)wL^2$ and $(.586)wL$ respectively.

$$M_p = (.086)(6800)(25)^2 = 3.66 \text{ EE5 ft-lb}$$

$$V_{max} = (.586)(25)(6800) = 99,620 \text{ lb}$$

step 2:

$$Z = \frac{(3.66 \text{ EE5})(12)}{36,000} = 122 \text{ in}^3$$

step 3: Try a W24 x 55: d = 23.57; b_f = 7.005; t_f = .505

$$t_w = .395; \quad A = 16.2; \quad Z = 134$$

<u>step 4</u>: The beam is compact. This can be checked by using equations 15.16 through 15.19.

<u>step 5</u>: The ultimate shear strength is

$$(.55)(36,000)(23.57)(.395) = 184,340 \text{ lb}$$

This exceeds 99,620, so the beam is adequate. Bracing length should also be checked.

◆

10. Column Design with Axial Loading

Ideal columns were previously investigated in chapter 12. The Euler buckling stress was found to be

$$\frac{\text{buckling}}{\text{stress}} = \frac{\text{Euler load}}{\text{Area}} = \frac{\pi^2 E}{(\frac{KL}{r})^2} \qquad 15.31$$

Recommended values of K may be found from table 15.5.

Table 15.5
End Restraint Coefficients

End #1	End #2	Theoretical	AISC Recommended
built-in	built-in	.5	.65
built-in	pinned	.7	.80
built-in	rotation fixed, translation free	1.0	1.2
built-in	free	2.0	2.1
pinned	pinned	1.0	1.0
pinned	rotation fixed, translation free	2.0	2.0

Equation 15.31 should be used to calculate the Euler stress or load if equation 15.32 holds.

$$(\frac{KL}{r}) \geq C_c \qquad 15.32$$

$$C_c = \sqrt{2\pi^2 E/F_y} \qquad 15.33$$

Intermediate columns $(\frac{KL}{r} < C_c)$ should be designed to keep the allowable column stress below F_a from equation 15.34.

$$F_a = F_y(1 - \frac{1}{2C_c^2}(\frac{KL}{r})^2) \qquad 15.34$$

The following suggestions and additional code requirements should be noted:

    (a) For long columns, the Euler stress (equation 15.31) is to be divided by a factor of safety of 1.917. This is a code requirement.

(b) For intermediate columns ($KL/r < C_c$), equation 15.34 is to be divided by the factor of safety calculated from equation 15.35.

$$\text{factor of safety} = \frac{5}{3} + \frac{3(\frac{KL}{r})}{8C_c} - \frac{(\frac{KL}{r})^3}{8(C_c)^3} \qquad 15.35$$

(c) If an axially-loaded member is a secondary member and (L/r) is greater than 120, then the allowable stress may be increased by dividing it by equation 15.36.

$$1.6 - \frac{L}{200r} \quad \text{(K is set to 1.0 to calculate } F_a) \qquad 15.36$$

(d) The allowable stress is to be used with the gross area of the column.

(e) Lacing bars carry no column load. They may be used to maintain column shape and to resist shear.

(f) Cover plates may be used to carry column load.

(g) (KL/r) may not exceed 200.

(h) To prevent local buckling, the following width-thickness ratio must be met:

$$\frac{b_f}{t} < H/\sqrt{F_y/1000} \quad \text{where H is defined below} \qquad 15.37$$

Elements supported on one edge only	H
single angle struts or separated double angle struts	76
stems of tees	127
double angles in contact	95
flanges of beams	95
flanges of tees (only $\frac{1}{2}b_f$ may be used)	95
flanges of I-beams (only $\frac{1}{2}b_f$ may be used)	95

Elements supported on both edges	
square, rectangular, and tubular sections	238
cover plates with access holes	317
all other uniformly compressed elements	253
webs of beams - see item (k) below	

(i) The AISCS has provisions for reducing the allowable column stress when the above width-thickness ratios are not met.

(j) The solution is very trial and error if the AISCM column tables are not used.

(k) For webs of plate girders and other shapes in flexure, special width-thickness ratio minimums are specified by the AISCS.

Example 15.6

Design a 25 foot long A36 W shape main member column to support a 375,000 pound live load. The base is rigidly framed in both directions. The top is rigidly framed in the weak direction, but translation in the strong direction is possible.

Assume the column dead weight will be about 2000 pounds. Then, the actual load will be 375,000 + 2,000 = 377,000 lb.

From the information about framing, the end restrain coefficients are K_y = .65 and K_x = 1.2.

The effective lengths are

$$L_{e,weak} = (.65)(25) = 16.25 \text{ ft}$$

$$L_{e,strong} = (1.2)(25) = 30 \text{ ft}$$

From the AISCM column design tables (part 3), enter with 16 feet and 377 kips. Try W12x79 with the following characteristics:

$$A = 23.2 \qquad b_f/2t = 8.2 \qquad d/t_w = 26.3$$

$$r_x = 5.34 \qquad r_x/r_y = 1.75$$

The equivalent effective length for the strong axis is 30/1.75 = 17.1. Since 17.1>16.25, the strong axis controls, and it is necessary to reenter the table with 17.1 ft and 377 kips. W12x79 is chosen again. The design is complete at this point.

Since the strong axis controls, the maximum slenderness ratio is

$$C_x = \frac{(1.2)(25)(12)}{5.34} = 67.42$$

The actual column stress is

$$f_a = \frac{375,000 + 25(79)}{23.2} = 16249 \text{ psi}$$

From equation 15.33, the critical slenderness ratio is

$$C_c = \sqrt{(2)\pi^2(2.9 \text{ EE7})/36,000} = 126.1$$

Since C_x is less than 126.1, this is an intermediate column. The allowable stress can be found from the allowable stress tables (in appendix A, AISCS) or from equations 15.34 and 15.35. The stress before the factor of safety is applied is

$$36,000(1 - \frac{1}{2(126.1)^2}(67.42)^2) = 30,855$$

The factor of safety is

$$\frac{5}{3} + \frac{3(67.42)}{8(126.1)} - \frac{(67.42)^3}{8(126.1)^3} = 1.848$$

The allowable stress is

$$F_a = \frac{30,855}{1.848} = 16,696$$

Since the actual stress is less than the allowable, the beam is safe as a column.

Now, check the width/thickness ratio. The maximum $(b_f/2t)$ value for the flange (unsupported member) is $95/\sqrt{36} = 15.83$. The actual value is 8.2 (okay).

11. Beam-Columns

An axial member which carries an axial load and a bending moment is known as a beam-column. The bending moment may be due to an eccentric load or to a lateral load. There are three methods of designing and analyzing beam columns.

A. Allowable Stress Design

If the slenderness ratio (KL/r) is less than $(95/\sqrt{(F_y/1000)})$, a beam-column may be evaluated on the basis of the standard combined stress equation. For bending in one plane only, the stress in a short beam-column is given by equation 15.38.

$$f_{max} = \frac{P}{A} + \frac{Mc}{I} \qquad\qquad 15.38$$

For compact sections, f_{max} is $.66F_y$. Otherwise, it is $.60F_y$. To design a beam, a trial shape is chosen which sets A, c, and r. This is a very trial and error procedure.

B. Interaction Formula Method

Equation 15.38 may be rewritten as

$$f_{max} = f_a + f_b \qquad\qquad 15.39$$

If $(f_a/F_a) < .15$, AISCS permits the following analysis equation to be used to determine if a column is safe.

$$\frac{f_a}{F_a} + \frac{f_{bx}}{F_{bx}} \le 1 \qquad\qquad 15.40$$

F_a is the axial stress that would be allowed if the axial force acted alone. This is dependent on the slenderness ratio. Similarly, F_b is the bending stress that would be allowed if the moment existed alone. That is, $F_b=(.60)F_y$, or if $L_b \le L_c$, then $F_b=(.66)F_y$, exclusive of any wind loading allowance.

If $(f_a/F_a) > .15$, then both of the following must hold.

$$\frac{f_a}{F_a} + \frac{C_{mx}f_{bx}}{(1 - \frac{f_a}{F'_{ex}})F_{bx}} \leq 1 \qquad\qquad 15.41$$

$$\frac{f_a}{(.60)F_y} + \frac{f_{bx}}{F_{bx}} \leq 1 \qquad\qquad 15.42$$

where $\qquad F'_{ex} = \dfrac{\pi^2 E}{(1.917)(KL/r_y)^2} \qquad\qquad 15.43$

C_m is a coefficient with the following values:

$\qquad C_m$ = .85 for members subject to sidesway (joint translation).

$\qquad C_m$ = .6 - .4(M_1/M_2) but not less than .4 for members not subject
   to sidesway and not subject to transverse loading between
   their supports in the plane of bending. (M_1/M_2) is the ratio
   of the smaller to larger moments at the ends of that portion
   of the member unbraced in the plane of bending under consideration. (M_1/M_2) is positive when the member is bent in
   reverse curvature and negative when it is bent in single
   curvature.

$\qquad C_m$ = For compression members in frames braced against joint translation in the plane of loading and subjected to transverse
   loading between their supports, the value of C_m may be taken
   as .85 for members with restrained ends and as 1.0 for
   members with unrestrained ends.

C. Equivalent Axial Compression Method

In order that the beam selection process not be a trial and error
procedure, equations 15.40 - 15.42 are modified to allow the use of
the column tables in the AISCM. An initial quick trial selection may be
made on the basis of equation 15.44. P_{eq} is the required tabular load
from the column selection tables. B_x is also tabulated in the tables.

Because this method is able to handle bending in two planes so easily,
the following equations have been expanded to allow for such bending.

$$P_{eq} = P + B_x M_x + B_y M_y \qquad\qquad 15.44$$

Equation 15.44 will always oversize the beam-column, so further analysis
should start with a slightly weaker section.

Example 15.7

Select an initial trial W shape using A36 steel to carry 200,000 pounds axially and a 200,000 ft-lb moment about its strong axis. The member's unsupported length is 20 feet. Assume K = 1.0 for this problem, and assume sidesway is permitted in the direction of bending.

KL = 20 feet. Turn to the column tables in the AISCM. Try W 10 x 60. For this column, B_x = .264, and the allowable 20 foot load is 244,000 pounds. The equivalent load is

$$P_{eq} = 200,000 + .264(200,000)(12)\frac{in}{ft}$$

$$= 833,600$$

Since 833,600 exceeds 244,000, the column is too small. Try W14 x 145. $P_{max,20}$ = 743,000.

$$P_{eq} = 200,000 + .184(200,000)(12) = 641,600$$

This has excess capacity. So, try W14 x 120 with $P_{max,20}$ = 601,000.

$$P_{eq} = 200,000 + .186(200,000)(12) = 646,400$$

This is close enough since this method always oversizes the column.

Equation 15.44 oversizes columns, but it serves as a starting point for a more accurate analysis. When $(f_a/F_a) \leq .15$, equation 15.45 may be used directly with the AISCM column tables in place of equation 15.40.

$$P_{eq} = P + B_x M_x (\frac{F_a}{F_{bx}}) + B_y M_y (\frac{F_a}{F_{by}}) \qquad 15.45$$

If $(f_a/F_a) > .15$, equations 15.46 and 15.47 must be used.

$$P_{eq} = P + \left[B_x M_x C_{mx} (\frac{F_a}{F_{bx}})(\frac{a_x}{a_x - (KL)^2(P/1000)}) \right]$$

$$+ \left[B_y M_y C_{my} (\frac{F_a}{F_{by}})(\frac{a_y}{a_y - (KL)^2(P/1000)}) \right] \qquad 15.46$$

$$P_{eq} = P(\frac{F_a}{.6F_y}) + \left[B_x M_x (\frac{F_a}{F_{bx}}) \right] + \left[B_y M_y (\frac{F_a}{F_{by}}) \right] \qquad 15.47$$

Example 15.8

Check the selection of the W14 x 120 beam in example 15.7.

From the AISCM column table,

$$A = 35.3 \qquad r_y = 3.74 \qquad L_u = 44.1 \qquad B_x = .186 \qquad a_x = 204.8 \text{ EE6}$$

$$L_c = 15.5$$

The allowable bending stress is $.60F_y$. Since 20 is greater than L_c, $.66F_y$ cannot be used as the bending stress.

$$F_b = (.60)(36,000) = 21,600 \text{ psi}$$

$$(KL/r) = \frac{(1)(20)(12)}{3.74} = 64.2 \text{ (say 64)}$$

From equations 15.34 and 15.35, or from the allowable column stress table (AISCS - Appendix A), $F_a = 17,040$ psi.

Now, check to see if equation 15.45 can be used.

$$f_a = \frac{P}{A} = \frac{200,000}{35.3} = 5666 \text{ psi}$$

$$\frac{f_a}{F_a} = \frac{5666}{17040} > .15$$

Therefore, equations 15.46 and 15.47 must be used.

$C_{mx} = .85$ since sidesway is permitted.

$$\frac{P(KL)^2}{1000} = (\frac{200,000}{1000})((1)(20)(12))^2 = 1.152 \text{ EE7} = 11.5 \text{ EE6}$$

Then, from equation 15.46,

$$P_{eq} = 200,000 + \left[(.186)(200,000)(12)(.85)(\frac{17040}{21600})(\frac{204.8}{204.8-11.5})\right]$$

$$= 517,144$$

From equation 15.47,

$$P_{eq} = 200,000(\frac{17,040}{.6(36,000)}) + \left[(.186)(200,000)(12)\frac{17040}{21600}\right]$$

$$= 509,938$$

The larger is 517,144. This is less than 601,000, so the column is safe. However, it is proven to be considerably oversized.

◆

12. Bending with Axial Tension

Equation 15.42 is to be used to size sections that are subject to axial tension and bending.

13. Reinforcement of Mill Shapes

One type of problem that has occasionally appeared is that of reinforcing a standard beam with plates. It is possible to solve this type of problem by use of moments of inertia only. No stress calculations are necessary.

The following guidelines are applicable to the choice of cover plates.

* Plate widths may be larger or smaller than b_f, but they should not be the same as b_f due to welding difficulties.

* Plates are usually available in widths of even inches. Consult with the mill.

* Available plate thicknesses are:
     1/32" increments up to $\frac{1}{2}$"
     1/16" increments from 9/16" to 1"
     1/8" increments from 1(1/8)" to 3"
     $\frac{1}{4}$" increments $3\frac{1}{4}$" and up

* The usual cover plate thicknesses are $\frac{1}{2}$", 1", and $1\frac{1}{2}$".

* If continuous welding is used, the depth/thickness ratios may be used to maximize the allowable stresses.

Example 15.9

A W 30 x 124 beam must be reinforced to the bending strength of a W 30 x 173 by welding plates to both flanges. All size plates are available. Assume continuous fastening to the support, and do not design the welding. Use A36 steel.

For ease of welding, assume the plate thickness will be approximately the same as t_f, which is about one inch. For a W 30 x 124, d = 30.17", so $\frac{1}{2}$d = 15.085".

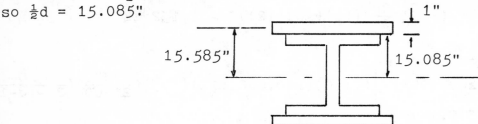

I for a W 30 x 173 is 8200. I for a W 30 x 124 is 5360. The difference in bending resistance is 2840.

The centroidal moment of inertia of the two plates is

$$I_{c,p} = 2\left(\frac{w(1)^3}{12}\right) = \frac{w}{6}$$

By the parallel axis theorem, the moment of inertia of the two plates about axis x-x is

$$I_{x-x,p} = \frac{w}{6} + 2(w)(1)(15.585)^2 = 486.0w$$

Since $I_{x-x,p}$ = 2840, w = $\frac{2840}{486}$ = 5.84 (say 6")

14. Bolted and Riveted Connections

A concentric connection is one for which the applied load passes through the centroid of the bolt or rivet group. At low loading, the distribution of forces among the fasteners is very non-uniform since friction carries some of the load. However, at higher stresses (near yielding) the load is carried equally by all fasteners in the group.

A. Riveted and Bolted Concentric Tension Connections

The number of connectors is determined by checking both the shear in the connectors and the bearing on the plates. The maximum of equations 15.48 and 15.49 should be used.

$$\text{\# connectors} = \frac{\text{total load}}{\left(\substack{\text{allowable shear} \\ \text{stress}}\right)\left(\substack{\text{connector} \\ \text{area}}\right)} \qquad 15.48$$

$$\text{\# connectors} = \frac{\text{total load}}{\left(\substack{\text{allowable bearing} \\ \text{stress}}\right)\left(\substack{\text{bearing} \\ \text{area}}\right)} \qquad 15.49$$

The allowable bearing stress is given in section 15.6. The allowable connector stresses are given in the following table.

Table 15.6

Allowable Connector Stresses (ksi) (note e)

TYPE OF CONNECTOR	F_t (note f)	F_v FRICTION CONNECTION (note d)	F_v BEARING CONNECTION (note d)
A 502, hot driven rivets, grade 1	23.0 (note a)	don't use	17.5 (note b)
grade 2	29.0 (note a)	don't use	22.0 (note b)
A 307, ordinary bolts	20.0 (note a)	don't use	10.0 (notes b, c)
A 325, structural bolts			
- no threads in the shear plane	44.0	17.5	30.0 (note b)
- threads in the shear plane	44.0	17.5	21.0 (note b)
A 490, structural bolts			
- no threads in the shear plane	54.0	22.0	40.0 (note b)
- threads in the shear plane	54.0	22.0	28.0 (note b)
A 449 bolts, and other threaded parts	refer to the AISCS		

Notes: (a) Static loading only. (b) Reduce 20% for connector pattern longer than 50 inches. (c) Threads are permitted in the shear plane. (d) Reductions are required for oversize holes. See AISCS. (e) A (1/3) increase is allowed for transitory seismic and wind loading. (f) Use with stress area calculated from the following formula:

$$\text{stress area} = \tfrac{1}{4}\pi\left[\text{major thread diameter} - \left(\frac{.9743}{\text{threads/in.}}\right)\right]^2$$

Some of the standards that must be adhered to in connector design are given here:

* Minimum distance between centers of fastener holes must not be less than (8/3) times the nominal fastener diameter.

* Along a line of transmitted force, the center-to-center hole spacing must not be less than $(2P/F_u t) + \tfrac{1}{2}d$. P is the load on the fastener, t is the part thickness, and d is the nominal diameter.

* The approximate minimum distance from the hole center to the side of a part is 1.75 times the nominal diameter for sheared edges and 1.25 times the diameter for rolled or gas-cut edges.

* Along a line of transmitted force, the distance from a hole center to an edge must not be less than $(2P/F_u t)$.

* The maximum edge distance is the minimum of 12 plate thicknesses or 6 inches.

Example 15.10

Two PL $\frac{1}{4}$ x 8 A36 plates carrying 35,000 pounds are to be joined in a lap joint using (3/4)" A502 grade 1 rivets. Design the connection.

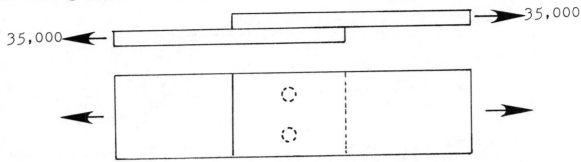

The area of a (3/4)" rivet is $\frac{1}{4}\pi(.75)^2$ = .442 in². The allowable rivet shear stress from table 15.6 is 17,500 psi. So, the number of rivets needed is given by equation 15.48.

$$\text{\# rivets} = \frac{35,000}{(17,500)(.442)} = 4.52 \text{ (governed by shear)}$$

The bearing area is (.75)(.25) = .1875 in² per rivet. The allowable bearing stress is

$$f_b = (1.5)(58,000) = 87,000 \text{ psi}$$

So, the number of rivets required is given by equation 15.49.

$$\text{\# rivets} = \frac{35,000}{(87,000)(.1875)} = 2.15 \text{ (governed by bearing)}$$

Shear governs. 5 rivets are needed. Use 6 for symmetry.

The spacing shown below satisfies both the edge and spacing requirements.

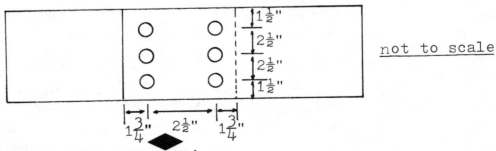

not to scale

B. Riveted and Bolted Beam Connections

Beam-to-column and beam-to-girder connections which do not transmit moments (Type-2 connections) are known as flexible beam connections. Seated beam connections which attach to beam flanges are also included in this category. These connections are designed by using the AISCM beam connection tables. The design method is not covered in this chapter.

C. Riveted and Bolted Moment-Resisting Beam Connections

Rigid (type 1) connections designed to transmit moments from beams to columns are not covered in this chapter. Refer to AISCM.

D. Riveted and Bolted Eccentric Torsion Connections

An eccentric torsion connection is illustrated in figure 15.3(a). This type of connection gets its name from the tendency that the load has to rotate the bracket. This rotation must be resisted by the shear stress in the connectors.

Figure 15.3

(a)   Torsion Resistance   (b)

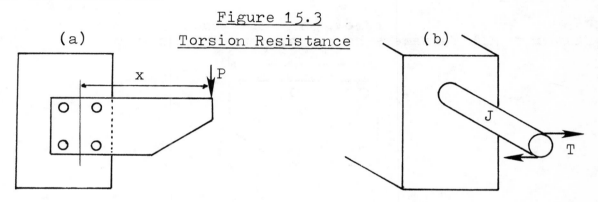

Cases (a) and (b) of figure 15.3 are very similar. The governing stress equation of a shaft loaded in torsion by a torque T (case b) was given in chapter 12 as

$$\text{shear stress} = \frac{Tr}{J} \qquad\qquad 15.50$$

This same equation may be used to help evaluate the eccentric torsion connection. However, the following changes in definition must be made:

* 'Shear stress' is replaced by the 'allowable shear stress' for the rivets or bolts. These are found from table 15.6.

* The torque, T, is replaced by the moment on the bracket. This moment is the product of the eccentric load, P, and the distance from the load to the centroid of the fastener group, x.

* r is taken as the distance from the centroid of the fastener group to the critical fastener. The critical fastener is found by inspection. For loads vertically downward, the critical fastener is usually in the right-most column. (Specifically, the critical fastener is the one for which the vector sum of the vertical and torsional shear stresses is the greatest.)

* J is based on the parallel-axis theorem. As bolts and rivets have little resistance to twisting in their holes, their polar moments of inertia (J_i) are omitted. The J_i should be included, however, for secure connections such as welds and welded studs.

$$J = \Sigma(r_i^2 A_i) \qquad \text{BOLTS AND RIVETS} \qquad 15.51$$

$$J = \Sigma(J_i + r_i^2 A_i) \qquad \text{WELDED STUDS} \qquad 15.52$$

r_i is the distance from the fastener group centroid to the ith fastener, which has an area of A_i.

* The vertical shear stress in the critical fastener must be added in a vector sum to the torsional shear stress. This vertical shear stress is

$$\text{vertical shear stress} = \frac{PA_{critical}}{\Sigma A_i} \qquad 15.53$$

Example 15.11

For the bracket shown below, find the load on the most critical fastener. All fasteners have a nominal ½" diameter.

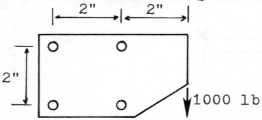

Since the fastener group is symmetrical, the group centroid is centered within the 4 fasteners. This makes the eccentricity of the load equal to 3 inches. Each fastener is located r from the centroid, where

$$r = \sqrt{(1)^2 + (1)^2} = 1.414$$

Also, $\qquad A_i = \frac{1}{4}\pi(.5)^2 = .1963$

Using the parallel axis theorem for polar moments of inertia,

$$J = 4(.1963(1.414)^2) = 1.570 \text{ in}^4$$

The torsional stress on each fastener is

$$f_{v,T} = \frac{(1000)(3)(1.414)}{(1.570)} = 2702 \text{ psi}$$

This torsional shear stress is directed perpendicularly to a line connecting each fastener with the centroid.

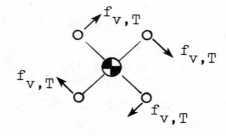

$f_{v,T}$ can be divided into horizontal stresses of $f_{v,T,x}$ and vertical stresses of $f_{v,T,y}$. Both of these components are equal to 1911 psi. In addition, each fastener carries a vertical shear load equal to $(1000/4) = 250$ pounds. The vertical shear stress due to this load is $(250/.1963) = 1274$.

The two right fasteners have vertical downward components of $f_{v,T}$ which add to the vertical downward stress of 1274. Thus, both of the two right fasteners are critical. The total stress in each of these fasteners is

$$f_v = \sqrt{(1911)^2 + (1911+1274)^2} = 3714 \text{ psi}$$

◆

15. Welded Connections

A. Introduction

The most common weld type is the fillet weld shown in figure 15.4. Such welds are commonly used to connect one plate to another. The applied load is assumed to be carried by the effective weld throat which is related to the weld size, w, as shown.

Figure 15.4

Fillet Lap Weld and Symbol

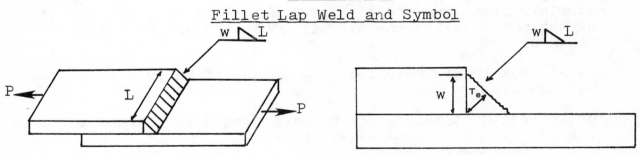

The effective weld throat size for submerged arc processes can be found from equations 15.54 and 15.55.

$$T_e = (.707)w + .11 \qquad (w \geq 7/16) \qquad \text{15.54}$$

$$= w \qquad (w \leq 3/8) \qquad \text{15.55}$$

For manual shielded arc welding processes,

$$T_e = (.707)w \qquad \text{15.56}$$

Weld sizes (w) of (3/16), (1/4), and (5/16) inch are desirable because they can be made in a single pass. However, fillet welds from (3/16)" to $\frac{1}{2}$" in (1/16)" increments are available. The increment is (1/8)" for larger welds.

Allowable stress on the weld is the minimum of equations 15.57 and 15.58 (shear stress on fillet throat only).

$$f_{a,weld} = (.30)F_{ut,welding\ rod} \qquad \text{15.57}$$

$$f_{a,weld} = (.40)F_{y,base\ metal} \qquad \text{15.58}$$

The ultimate strength of the welding rod is part of the rod designation. Thus, F_{ut} for an E70 welding rod is 70,000 psi. The following rods are available: E60, E70, E80, E90, E100, and E110.

Some special restrictions that apply to fillet welds are given here.

* Minimum weld sizes (w) depend on the thickness of the thickest of the two parts joined.

Table 15.7

Minimum Fillet Weld Size

Larger Part Thickness	Minimum w
to $\frac{1}{4}$" inclusive	1/8"
over $\frac{1}{4}$ to $\frac{1}{2}$	3/16
over $\frac{1}{2}$ to 3/4	1/4
over 3/4	5/16

* The maximum weld size is equal to the edge thickness for materials less than $\frac{1}{4}$" thick. For materials thicker than $\frac{1}{4}$", the maximum size is generally 1/16" less than the material thickness.

* The minimum length weld for full strength analysis is 4 times the weld size, w.

* For intermittent welds, the minimum length is 4 times the weld size, w, with a minimum of $1\frac{1}{2}$ inches.

* For lap joints (see figure 15.4), the minimum weld length is 5 times the thinner plate's thickness, and not less than 1 inch.

B. Welded, Concentric Tension Connections

If the weld group centroid is in a line with the applied load, equation 15.59 may be used to design or evaluate the connection.

$$f_v = P/A_{weld} = \frac{P}{(L_{weld})(T_e)} \qquad 15.59$$

Example 15.12

Two A36 PL $\frac{1}{2}$ x 8 plates are lap welded (E70) to carry 50,000 pounds. Size the welds.

The total length of weld will be 2(8) = 16 inches. This meets the 5 thicknesses minimum specification. From table 15.7, the minimum weld size is 3/16" for a $\frac{1}{2}$" plate, so try a 3/16" weld. From equation 15.59,

$$f_v = \frac{50,000}{(16)(.707)(3/16)} = 23,574$$

The allowable stress is the minimum of $(.4)(36,000)$ and $(.3)(70,000)$ or $14,400$. Therefore, a larger weld is necessary. Try 5/16".

$$f_v = \frac{50,000}{(16)(.707)(5/16)} = 14,144 \text{ psi (okay)}$$

◆

C. Welded Beam Connections

Refer to AISCM for both type 1 and type 2 connections.

D. Welded Eccentric Torsion Connections

Welded eccentric torsion connections can be handled in a manner very similar to that used for eccentric rivet connections. That is, the maximum shear stress in the weld group is

$$f_v = \frac{\text{load}}{(\text{total weld area})} + \frac{Mr}{J} \qquad 15.60$$

The following modifications need to be made to use equation 15.60.

* The total weld area is the total weld length times T_e.

* M is calculated as the product of the load times the eccentricity, as measured to the centroid of the weld group.

* r is the distance from the centroid of the weld group to the critical weld point. For vertically downward loads, the critical weld point is on the same side of the weld group (with respect to a vertical line through the centroid) as the load. The critical weld point maximizes r.

* J of the weld group is calculated using the parallel axis theorem and the area moments of inertia.

$$J = I_x + I_y \qquad 15.61$$

Example 15.13

An A36 plate bracket is welded with E70 electrodes to the face of a column as shown below. What size fillet weld is required? Neglect buckling and bending effects of the plate and column.

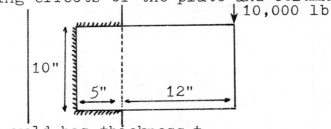

step 1: Assume the weld has thickness t.

step 2: Find the centroidal location of the weld group. By inspection, $\overline{y}_c = 0$. For the three welds,

$$A_1 = 5t \qquad A_2 = 10t \qquad A_3 = 5t$$

$$\overline{x}_1 = 2.5 \qquad \overline{x}_2 = 0 \qquad \overline{x}_3 = 2.5$$

So, $\qquad \overline{x}_c = \dfrac{5t(2.5) + 10t(0) + 5t(2.5)}{5t + 10t + 5t} = 1.25$

step 3: Determine the centroidal moment of inertia of the weld group about the x axis. Use the parallel axis theorem for areas 1 and 3.

$$I_x = \frac{t(10)^3}{12} + 2(\frac{5(t)^3}{12} + 5t(5)^2)$$

$$= 333.33t + .833(t)^3$$

Since t will be small (probably less than .5") the t^3 term can be neglected. So, $I_x = 333.33t$.

step 4: Determine the centroidal moment of inertia of the weld group about the y axis.

$$I_y = \frac{10(t)^3}{12} + (10t)(1.25)^2 + 2(\frac{t(5)^3}{12} + (5t)(1.25)^2)$$

$$= .833t^3 + 52.08t \approx 52.08t$$

step 5: The polar moment of inertia is

$$J = I_x + I_y = 333.33t + 52.08t = 385.4t$$

step 6: By inspection, the maximum shear stress will occur at point a.

$$r = \sqrt{(3.75)^2 + (5)^2} = 6.25$$

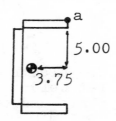

step 7: The applied moment is

$$M = (10,000)(12 + 3.75) = 157,500 \text{ in-lb}$$

step 8: The torsional shear stress is

$$f_{v,T} = \frac{Mr}{J} = \frac{(157,500)(6.25)}{385.4t} = \frac{2554.2}{t} \text{ psi}$$

This shear stress is directed at right angles to the line r. The x and y components of the stress can be determined from geometry.

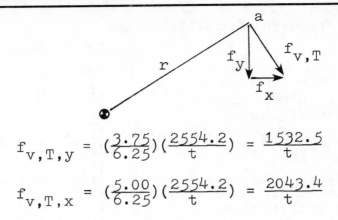

$$f_{v,T,y} = \left(\frac{3.75}{6.25}\right)\left(\frac{2554.2}{t}\right) = \frac{1532.5}{t}$$

$$f_{v,T,x} = \left(\frac{5.00}{6.25}\right)\left(\frac{2554.2}{t}\right) = \frac{2043.4}{t}$$

step 9: In addition, there is a vertical downward shear due to the vertical support required for the 10,000 pounds. The vertical shear stress is

$$\frac{10,000}{5t + 10t + 5t} = \frac{500}{t}$$

step 10: The resultant shear stress at point a is

$$f_v = \sqrt{\left(\frac{2043.4}{t}\right)^2 + \left(\frac{1532.5+500}{t}\right)^2} = \frac{2882.1}{t}$$

step 11: From equations 15.57 and 15.58, the allowable stress is the minimum of $(.30)(70,000) = 21,000$ and $(.40)(36,000) = 14,400$ psi. Therefore, the weld throat is

$$T_e = 2882.1/14,400 = .200 \text{ in.}$$

step 12: The weld size required is $(.200/.707) = .283"$ (say 5/16").

Bibliography

1. American Institute of Steel Construction, Inc., Manual of Steel Construction, 8th ed., Chicago, IL 1980

2. Crawley, Stanley W., and Dillon, Robert M., Steel Buildings, Analysis and Design, 2nd ed., John Wiley & Sons, New York, NW, 1977

3. Johnston, Bruce G., Lin, Fung-Jen, and Galambos, T.V., Basic Steel Design, 2nd ed., Prentice-Hall, Inc., Englewood Cliffs, NJ 1980

Practice Problems: STEEL DESIGN

Required

1. A 25 foot long beam is simply supported at its left end and 7 feet from its right end. A load of 3000 pounds per foot is uniformly distributed over its entire length. Lateral support is provided only at the reactions. Choose an economical W shape for this application. Use A36 steel.

2. Select the lightest W section of A36 steel to serve as a main member 30 feet long and to carry an axial load of 160,000 pounds. The member is pinned at its top and bottom. It is supported at mid-height in its weak direction. The member is vertical.

3. Determine the stresses in bolts A, B, C, and D. All connectors are 3/4" bolts

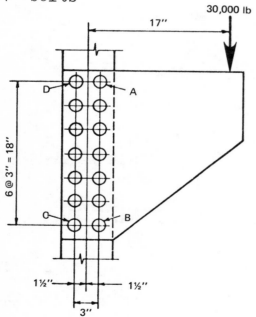

Optional

4. Determine the size of the fillet weld required to connect the plate bracket to the column face. The steel is A36; the electrodes are E70XX.

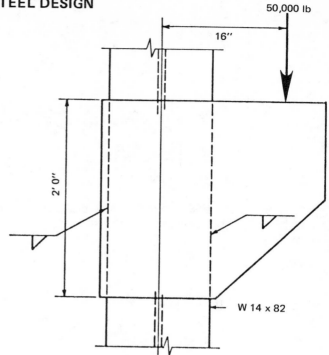

5. Design the interior columns of the frame shown below. Columns are braced at both ends in the plane perpendicular to the frame. Beams are W12 x 96. Use A36 steel.

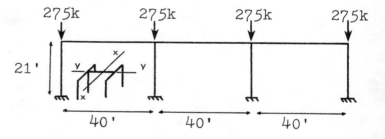

6. Design a plate eyebar to carry a static tensile load of 300,000 pounds. Use A36 steel.

7. Determine the tensile capacity of the connection shown. A325, 7/8" diameter bolts are used. The steel is grade 50. The connection is a friction type with no threads in the shear plane.

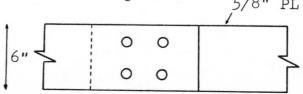

8. Select an A36 W shape with lateral support at $5\frac{1}{2}$ foot intervals to span 20 feet and carry a uniformly distributed load of 1000 lb/ft. The load includes a uniform dead load allowance of 40 lb/ft. Limit the maximum deflection to (L/240).

9. Select an economical A36 W shape (completely laterally braced) to span 24 feet such that the maximum deflection is (L/300). The span carries a uniformly distributed load of 800 lb/ft over its entire length. This load does not include the dead weight of the beam.

10. A 16 foot column is acted upon by an axial gravity load of 17,000 lb and a uniform wind load of w lb/ft. The W 12 x 58 column is made of A 441 steel. The lower end is built in. (a) Find the required spacing of lateral bracing for maximum lateral (uniform) loading. (b) Find the maximum lateral (uniform) loading allowed with the spacing calculated in part a.

NOTES

TRAFFIC ANALYSIS AND HIGHWAY DESIGN

Nomenclature

a	acceleration	ft/sec^2
ADT	current average daily traffic	vpd
B	bus adjustment factor	-
D	directional factor	-
DDHV	directional design hourly volume	vph
DHV	design hourly volume	vph
EAL	equivalent axle load	lb
EWL	equivalent wheel load	lb
f	coefficient of friction	-
F	force	lb
g	acceleration due to gravity (32.2), or grade	ft/sec^2, or decimal
G	grade	decimal
GE	gravel equivalent	ft
HP	horsepower	hp
k	soil stiffness (modulus of subgrade reaction)	psi/in
LSF	load safety factor	-
m	mass	slugs
MPH	vehicle speed	mph
N	number of lanes	
r	curve radius	ft
R	soil resistance value	-
s	distance	ft
S	sight distance	ft
t	time	seconds
T	truck adjustment factor	-
%T	percent of non-car traffic	decimal
TI	traffic index	-
v	vehicle velocity	fps
w	vehicle weight	lb
W	width adjustment factor	-

Symbols

θ	angle	degrees

Subscripts

c	centrifugal
f	frictional
n	net
o	initial
p	reaction and perception
t	tangential
w	at the wheels

1. Introduction

The number of problem types in the field of traffic analysis and high-way design is considerable. Furthermore, many traffic questions are qualitative in nature, and they do not lend themselves well to numer-ical solutions. Nevertheless, there are some important subjects that every civil engineer should be familiar with. Those subjects are covered in this chapter.

This chapter draws heavily on the practices and policies of CALTRANS, the California Department of Transportation. In particular, the proce-dures used to calculate roadway structural sections are used by CALTRANS, and these procedures may vary considerably from those used or specified by AASHTO.

2. Definitions

Abandonment: The reversion of title to the owner of the underlying fee where an easement for highway purposes is no longer needed.
Access control: See 'Control of access.'
Acquisition: The process of obtaining right of way.
Arterial highway: A general term denoting a highway primarily for through traffic usually on a continuous route.
Auxiliary lane: The portion of the roadway adjoining the traveled way for truck climbing, speed change, or for other purposes supple-mentary to through traffic movement.
Base course: The bottom portion of a pavement where the top and bottom portions are not of the same composition.
Base: A layer of selected, processed, or treated aggregate material of planned thickness and quality placed immediately below the pavement and above the subbase or basement soil.
Belt highway: An arterial highway for carrying traffic partially or entirely around an urban area or portion thereof.
CBD: Abbreviation for 'central business district.'
Cement treated base: A base layer constructed with good quality, well-graded aggregate mixed with up to 6% cement.
Channelization: The separation or regulation of conflicting traffic movements into definite paths of travel by use of pavement markings, raised islands, or other means.
Condemnation: The process by which property is acquired for public purposes through legal proceedings under power of eminent domain.
Control of access: The condition where the right of owners or occupants of abutting land or other persons to access in connection with a highway is fully or partially controlled by public authority.
Controlled access highway: In situations where it has been determined advisable, a highway may be designated as a 'controlled access highway' instead of a 'freeway.'
Divided highway: A highway with separated roadbeds for traffic in opposing directions.
Easement: A right to use or control the property of another for designated purposes.
Embankment: A raised structure constructed of natural soil from excavation or borrow sources.
Eminent domain: The power to take private property for public use without the owner's consent upon payment of just compensation.

Encroachment: Use of the highway right-of-way for non-highway structures or other purposes.

Flexible pavement: A pavement having sufficiently low bending resistance to maintain intimate contact with the underlying structure
yet having the required stability furnished by aggregate inter-
lock, internal friction, and cohesion to support traffic.

Freeway: A divided arterial highway with full control of access.

Frontage road: A local street or road auxiliary to and located on the
side of an arterial highway for service to abutting property and
adjacent areas and for control of access.

Gore: The area immediately beyond the divergence of two roadways
bounded by the edges of those roadways.

Inverse condemnation: The legal process which may be initiated by a
property owner to compel the payment of just compensation where
his property has been taken or damaged for a public purpose.

Median: The portion of a divided highway separating the traveled ways
for traffic in opposite directions.

Median lane: A speed change lane within the median to accomodate left-
turning vehicles.

Parkway: An arterial highway for non-commercial traffic, with full or
partial control of access, and usually located within a park or a
ribbon of parklike development.

Penetration treatment: Application of light liquid asphalt to the roadbed material. It is used primarily as a dust reducer on detours,
medians, and parking areas.

Plant mix: An asphalt concrete mixture that is not prepared at the
paving site.

Prime coat: The initial application of a low viscosity liquid asphalt
to an absorbent surface, preparatory to any subsequent treatment,
for the purpose of hardening or toughening the surface and
promoting adhesion between it and the superimposed constructed
layer.

Resurfacing: A supplemental surface or replacement placed on an existing
pavement to restore its riding qualities or increase its strength.

Right of access: The right of an abutting land owner for entrance to
or exit from a public road.

Rigid pavement: A pavement having sufficiently high bending resistance
to distribute loads over a comparatively large area.

Road-mixed asphalt surfacing: A lower-quality surfacing used when plant
mixes are not available or not economically feasible. Liquid
asphalts are normally used. It is used on low-traffic volume
roads where higher quality surfacing is not required for traffic
volume.

Roadbed: That portion of the roadway extending from curb line to curb
line or shoulder line to shoulder line. Divided highways are
considered to have two roadbeds.

Seal coat: A bituminous coating with or without aggregate applied to
the surface of a pavement for the purpose of water-proofing and
preserving the surface, rejuvenating a previous bituminous surface,
altering the surface texture of the pavement, providing delineation
or, providing resistance to traffic abrasion.

Structural section: The planned layers of specified materials, normally
consisting of subbase, base, and pavement placed over the basement
soil.

Subbase: A layer of aggregate of planned thickness and quality placed on
the basement soil as a foundation for the base.

Subgrade: The portion of a roadbed surface, which has been prepared as specified, upon which a subbase, base, base course, or pavement is to be placed.

Tack coat: The initial application of bituminous material to an existing surface to provide bond between the existing surface and the new material.

Traveled way: The portion of the roadway for the movement of vehicles exclusive of shoulders and auxiliary lanes.

3. Translational Dynamics

Newton's second law can be used to relate the net tractive force on a vehicle to its acceleration.

$$F_n = ma \qquad\qquad 16.1$$

where

$$m = w/g \qquad\qquad 16.2$$

The net tractive force is the difference between the applied force at the wheels and the frictional force.

$$F_n = F_w - F_f \qquad\qquad 16.3$$

The net force can be found directly from the velocity of the vehicle and the horsepower expenditure at that velocity.

$$F_n = \frac{(550)(HP)}{v} \qquad\qquad 16.4$$

The frictional force is a combination of dynamic, rolling, turning, and aerodynamic forces which act to oppose motion.

The relationships between position, velocity, and acceleration as functions of time for linear motion are given below.

$$a = \frac{dv}{dt} = \frac{d^2s}{dt^2} \qquad\qquad 16.5$$

$$v = \frac{ds}{dt} = \int a\, dt \qquad\qquad 16.6$$

$$s = \int v\, dt = \int\int a\, dt \qquad\qquad 16.7$$

If the acceleration is uniform, table 16.1 may be used to determine values of unknown variables for various types of problems. Acceleration is negative for vehicles with decreasing velocities.

4. Simple Roadway Banking

If a vehicle travels in a circular path with instantaneous radius r and tangential velocity v_t, it will experience an apparent centrifugal force given by equation 16.8.

$$F_c = mv_t^2/r \qquad\qquad 16.8$$

Table 16.1
Uniform Acceleration Formulas

to find	given these	use this equation
t	a v_o v	$t = \dfrac{v - v_o}{a}$
t	a v_o s	$t = \dfrac{\sqrt{2as + v_o^2} - v_o}{a}$
t	v_o v s	$t = \dfrac{2s}{v_o + v}$
a	t v_o v	$a = \dfrac{v - v_o}{t}$
a	t v_o s	$a = \dfrac{2s - 2v_o t}{t^2}$
a	v_o v s	$a = \dfrac{v^2 - v_o^2}{2s}$
v_o	t a v	$v_o = v - at$
v_o	t a s	$v_o = \dfrac{s}{t} - \tfrac{1}{2}at$
v_o	a v s	$v_o = \sqrt{v^2 - 2as}$
v	t a v_o	$v = v_o + at$
v	a v_o s	$v = \sqrt{v_o^2 + 2as}$
s	t a v_o	$s = v_o t + \tfrac{1}{2}at^2$
s	a v_o v	$s = \dfrac{v^2 - v_o^2}{2a}$
s	t v_o v	$s = \tfrac{1}{2}t(v_o + v)$

This centrifugal force must be resisted by a combination of roadway banking (superelevation) and sideways friction. If it is desireable to bank the roadway so that little or no friction is required to resist the centrifugal force, the angle of superelevation is given by equation 16.9. Generally, it is not desireable to rely on roadway banking alone, since such banking will be applicable only at one speed.

$$\tan \theta = \frac{v^2}{gr} \qquad\qquad 16.9$$

Example 16.1

A 4000 pound car travels at 40 mph around a banked curve with a radius of 500 feet. What should be the banking angle such that tire friction is not needed to prevent the car from sliding?

$$v = (40)(1.467)\frac{fps}{mph} = 58.68 \text{ fps}$$

From equation 16.9,

$$\theta = \arctan\left(\frac{(58.68)^2}{(32.2)(500)}\right) = 12.07°$$

◆

The basic formula used for determining superelevation when friction is relied upon to counteract some of the centrifugal force is equation 16.10.

$$\tan \theta = \frac{v^2}{gr} - f \qquad\qquad 16.10$$

If v is expressed in mph, then equation 16.10 becomes

$$\tan \theta = \frac{(MPH)^2}{15r} - f \qquad\qquad 16.11$$

f is the side friction factor. It is usually assumed to be .16 for speeds of 30 mph and under. For higher speeds,

$$f = .16 - .01\left(\frac{MPH - 30}{10}\right) \qquad\qquad 16.12$$

Since the maximum $\tan\theta$ is usually .08 or .10, equation 16.10 may be used to calculate the minimum curve radius if the speed is known.

5. Design Speeds and Volumes

Most elements of roadway design depend on the design speed. The design speed is the maximum safe maintainable speed on a roadway under the design conditions. Typical minimum design speeds are given in table 16.2.

Table 16.2
Minimum Design Speeds, MPH (4:606)

Type of Facility	Level	Rolling	Mountainous
Freeways			
Rural	70	60	50
Urban	50	50	50
Rural arterial highways			
50<ADT≤750, DHV<200	50	40	30
DHV>200	70	60	40
Urban arterial highways	30-40	30-40	30-40
Suburban arterial highways	40-50	40-50	40-50
Rural roads and streets			
ADT<250	40	30	20
250<ADT<400	50	40	20
ADT>400, DHV>100	50	40	30
Urban roads and streets			
Collectors	30-40	30-40	30-40
Local	20-30	20-30	20-30

The volume of use (design volume) is also needed to design a structural section. There are actually six different volume parameters. Not all of these are needed in every capacity decision.

ADT: The current average daily traffic.

DHV: The design hourly volume in the (possibly future) design year. DHV is usually the 30th highest hourly expected volume in the design year. It is not an average or a maximum.

D: A directional factor. D is the percentage in the dominant flow direction. D may range up to 80% for rural roadways at peak hours and down to 50% for central business district traffic.

DDHV: The directional design hourly volume. It is calculated as the product of the directional factor, D, and DHV. That is,

$$DDHV = (D)(DHV) \qquad 16.13$$

%T: The percentage of truck and bus traffic in the design hour. Values between 7% and 10% are typical with the higher values applicable to highways carrying municipal transit buses and truck freight lines. 3% to 5% is typical for central business districts.

Design Capacity: The maximum volume of traffic which the roadway can handle. Typical values are given in table 16.3. Higher values are possible, but only for short durations with ideal conditions. In level terrain, commercial vehicles may be assumed to be equivalent to 2 passenger cars on multi-lane highways and to $2\frac{1}{2}$ cars on two-lane facilities.

Table 16.3

Maximum Capacities

Highway Type	Capacity (vph)
2 or more lanes in each direction	1500-2000 passenger cars in urban areas; 1000-1500 in rural - per lane
2 lanes in 2 directions	2000 total both directions
3 lanes in 2 directions	4000 total both directions
one lane, from stopped condition	1500

Since volumes near capacity may result in poor safety, maneuverability, and speed, the quality of driving may be specified by a service level. Levels B and C should be selected for design.

The level or service can be found by calculating the volume/capacity (v/c) ratio. The volume is usually known. The capacity is found from equation 16.14.

$$Capacity = (2000)NWTB \quad (vph) \qquad 16.14$$

N is the number of lanes. W is the width adjustment factor from table 16.5. T and B are the truck and bus adjustment factors, both of which require an extensive table to be evaluated. Typical values for rolling and mountainous terrain are given in table 16.6.

Table 16.4
Service Level Parameters for Multi-Lane Highways

Service Level	Normalized Speed (base 70 mph)	Normalized Volume (base 2000 vph)	Maximum Service Volume at Average Highway Speed per lane		
			70 mph	60 mph	50 mph
A	.91	.35	600		
B	.83	.55	1000	400	
C	.75	.75	1500	1000	500
D	.67	.90	1800	1700	1400
E	.33	1.00	2000	2000	2000
F	<.33	----	----	----	----

Table 16.5 (4:328)
Width Adjustment Factor

Distance (ft) from traffic lane to obstruction on right side only	Two Lanes in Each Direction				Three or More Lanes in Each Direction			
	12-ft Lanes	11-ft Lanes	10-ft Lanes	9-ft Lanes	12-ft Lanes	11-ft Lanes	10-ft Lanes	9-ft Lanes
6	1.00	.95	.89	.77	1.00	.95	.89	.77
4	.98	.94	.88	.76	.99	.94	.88	.76
2	.95	.92	.86	.75	.97	.93	.86	.75
0	.88	.85	.80	.70	.94	.90	.83	.72
With additional obstruction on left side								
2	.94	.91	.75		.96	.92	.85	.75
0	.81	.79	.66		.91	.87	.81	.70

Table 16.6 (4:323)
Sample Values of Bus and Truck Factors

Condition	% bus or truck volume			
	3%	5%	10%	20%
Trucks on low grades	varies with grade and length			
Trucks in rolling terrain	.92	.87	.77	.63
Trucks in mountainous terrain	.83	.74	.59	.42
Buses in rolling terrain	.94	.91	.83	.71
Buses in mountainous terrain	.89	.83	.71	.56

Example 16.2

A rural four-lane undivided highway is constructed with 11-foot lanes, no shoulders, and retaining walls at the pavement edge. The average vehicle speed is 60 mph. The terrain is rolling. The actual service volume is 1500 vehicles per hour, with 3% buses and 5% trucks. What is the level of service?

base volume = 2000 per lane (from table 16.4).
N = 2
W = .85 (from table 16.5)
T = .87 (from table 16.6)
B = .94 (from table 16.6)

Capacity volume = (2000)(2)(.85)(.87)(.94) = 2780

$$(v/c) = \frac{1500}{2780} = .54$$

From table 16.4, this (v/c) ratio corresponds to service level B.

6. Sight and Stopping Distances

Sight distance is the length of roadway that the driver can see. It is assumed that the driver's eyes are 3.75 feet above the surface of the roadway. The sight distance should be long enough to allow a driver traveling at the maximum speed to stop before coming upon an observed object. This required distance is known as the stopping sight distance. It is assumed that the object being observed has a height of .5 feet.

Since distance is covered during the driver's reaction period as well as during the deceleration period, the stopping sight distance includes both of these distances. The coefficient of friction is usually evaluated for wet pavement. For straight-line travel on a constant grade, g, equation 16.15 may be used. g is a decimal, and it is negative if the roadway is downhill.

$$S = (1.47)(t_p)(MPH) + \frac{(MPH)^2}{(30)(f + g)} \qquad 16.15$$

If the design speed is used in equation 16.15, S is known as a 'desirable value'. If the speed is less than the design value, S is known as a 'minimum value.' The minimum value speed to be used is

$$(MPH)_{min} = (MPH)_{design} - .2((MPH)_{design} - 20) \qquad 16.16$$

The desireable value should be used in most cases. These are listed in table 16.7 for various design speeds.

The braking reaction-perception time, t_p in equation 16.15, has a median value of approximately .90 seconds for unexpected (not anticipated) events. However, this time varies widely from subject to subject. Individuals with slow reactions may require up to 2.0 seconds.

The passing sight distance is applicable only to 2-lane, 2-way highways. It is the length of roadway ahead necessary to pass without meeting an oncoming vehicle. Minimum passing sight distances are given in table 16.7. The values should be increased 18% for downgrades steeper than 3% and longer than one mile.

If a vehicle locks its brakes and skids to a stop, the deceleration will be $(f)(g_o) = (f)(32.2)$ ft/sec^2. The skidding distance will be

$$\frac{\text{skidding}}{\text{distance}} = \frac{(MPH)^2}{(30)(f + g)} \qquad 16.17$$

Table 16.7
Minimum Sight Distances

Design Speed	Passing Sight Distance	Stopping Sight Distance
30 mph	1,100 ft	200 ft
40	1,500	275
50	1,800	350
60	2,100	525
65	2,300	600
70	2,500	750

Table 16.8 (4:26)
Coefficients of Skidding Friction

BC: bituminous concrete, dry
SA: sand asphalt, dry
RA: rock asphalt, dry
CC: portland cement concrete, dry
wet: AASHTO recommended for all wet pavements

condition	BC	SA	RA	CC	wet
new tires					
11 mph	.74	.75	.78	.76	
20	.76	.75	.76	.73	.40
30	.79	.79	.74	.78	.36
40	.75	.75	.74	.76	.33
50					.31
60					.30
70					.29
badly worn tires					
11 mph	.61	.66	.73	.68	
20	.60	.57	.65	.50	.40
30	.57	.48	.59	.47	.36
40	.48	.39	.50	.33	.33
50					.31
60					.30
70					.29

If a vehicle does not lock its brakes, its deceleration will be dependent on its brakes. The distance traveled during deceleration is given by equation 16.18.

$$\text{stopping distance} = \frac{v^2}{2a} = \frac{(1.08)(MPH)^2}{a} \qquad 16.18$$

7. Roadway Construction

A. Types of Pavement

1. Portland Cement Concrete

Non-reinforced portland cement concrete is used almost exclusively where rigid pavement is called for. Typical applications are traffic lanes, freeway-to-freeway connections, and freeway exit ramps which experience heavy traffic.

Cement concrete pavement has, as its primary advantages, excellent durability and long service life. It provides good contrast with asphalt surfaces. It will also withstand repeated flooding and sub-surface water without deterioration.

It has three primary disadvantages: (1) It may lose its original non-skid surface during use. (2) It must be used with an even subgrade and only where uniform settling is expected. (3) It may rise (fault) at some transverse joints.

2. Asphalt Concrete

Typical uses of asphalt concrete are for traffic lanes, auxiliary lanes, ramps, parking areas, frontage roads, and shoulders.

Asphalt concrete pavement has the advantage of adjusting to limited amounts of differential settlement. It is easily repaired, and additional thicknesses can be placed at any time to withstand increased usage and loading. Its non-skid properties do not deteriorate to any great extent.

However, asphalt concrete loses some of its flexibility and cohesion with time, and it will have to be resurfaced sooner than would cement concrete. It would not normally be chosen over cement concrete where the presence of water could be expected

B. Parameters for Designing Pavement

The pavement thickness will be dependent on the axle loadings that are estimated for the truck traffic predicted for the design period. The effects of passenger cars, pickups, and two-axle trucks with single rear tires are not considered. Both the number of trucks and the axle loading of those trucks must be known.

The usual sources of truck volume data are traffic counts. These may be actual counts on existing roadways needing resurfacing, or they may be redistributed counts from other highways. The trucks are classified according to the number of axles, with trucks having six or more axles being classified as five-axle trucks. In the past, truck traffic counts were reported as two-way traffic. One-way traffic counts are now generally used.

Pavement thickness should be chosen to serve the estimated one-way truck traffic for a period of 20 years. A shorter period, not to be less than 10 years, may be used for temporary construction or for other justifiable reasons.

Logical methods, including straight-line extrapolation, should be used in the estimation of future traffic counts. Expansion factors should be determined for each of the axle classifications. Considerable judgment is needed to develop realistic expansion factors.

It is also necessary to estimate the distribution of truck traffic on the various lanes of a multi-lane facility. Traffic is usually lightest in the inside lane (lane 1, or the 'fast lane'). The following lane distribution factors may be used.

Table 16.9
Lane Distribution Factors for Multi-lane Roads

Number of lanes in one direction	Lane 1	Lane 2	Lane 3	Lane 4
1	1.0			
2	1.0	1.0		
3	.2	.8	.8	
4	.2	.2	.8	.8

The quality of the subgrade and basement soil, as measured by their R-values, is also a required factor. R-values are usually given in the materials report. Since there may be a considerable range in these values, a design R-value must be chosen. If the range of R is small, the lowest R-value should be selected. If there are a few exceptionally low values which come from one area, it may be possible to specify replacing that area's soil with borrow soil. If there are changing geological formations along the route which modify the R-value, it may be necessary to design different pavement sections to match the R-values.

C. Compaction

Bases, subbases, and earthwork should be compacted to 95% relative compactness down to a minimum depth of ½ foot below the grading plane. Attempts to make compensating design modifications should be avoided due to the extreme scarcity of test data on this subject.

In addition, 95% relative compactness should be achieved for a minimum depth of 2½ feet below the finished grade for the width of the roadway plus 3 feet on each side. Some exceptions to this specification may be justifiable.

D. Standard H Truck Loadings

Figure 16.1 illustrates standard H truck loadings, with the variables being taken from table 16.10. The symbol V implies a variable distance between 14 and 30 feet that should be selected to produce the maximum stress in a structure.

Table 16.10

Standard Truck Loadings

Load Designation	F_1	F_2	F_3	d_1	d_2
H20-44	8000	32,000	0	14'	
H15-44	6000	24,000	0	14'	
H10-44	4000	16,000	0	14'	
HS20-44	8000	32,000	32,000	14'	V
HS15-44	6000	24,000	24,000	14'	V

Figure 16.1

Standard Truck Loadings

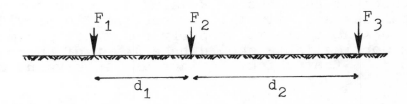

8. Designing Portland Cement Concrete Pavement

A. Introduction

The method for designing structural sections using portland cement concrete pavement uses the following data:

- (a) soil survey data (R-value)
- (b) number of lanes
- (c) predicted truck traffic
- (d) loadometer survey data
- (e) load repetition constants (estimated, or derived from the loadometer data)

On projects with three or more lanes in one direction, separate designs are usually made for the inside and outside lanes. This results in steps at the bottoms of pavement and base. It is cheaper to construct stepped sections than uniform or tapered sections, which result in increased soil removal. However, in order to provide a uniform grading plane, it is permissible to increase the thickness of the subbase under the inside lanes. This total thickness of subbase should be used to design the pavement for the inside lanes.

Figure 16.2

Typical Portland Concrete Cement Section

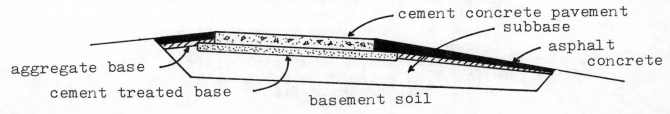

B. Design Procedure

Note: This procedure does not include corrections for frost.

step 1: From the R-value of the basement soil, determine the support reaction k-value from figure 16.3. A reading of k to the nearest 5 units is sufficient. Remember that R can be estimated from CBR values or the soil classification. See chapter 9.

step 2: Choose a subbase thickness. Use figure 16.4 to obtain a new k-value for the basement soil/subbase combination.

Cement treated bases should be supported by a minimum depth of .5 foot of subbase material with a minimum R-value of 40 under the weight of one foot of cover. (If the basement soil itself has an R-value of 40 or more, the subbase may be omitted.)

When the basement soil has an R-value of 40 or more but consists of cohesionless sand, it will be necessary to provide a coarse aggregate working table to properly construct a cement-treated base. This working table should be a layer of untreated aggregate from .25 to .35 feet thick. No increase in k-value should be allowed for this working table.

step 3: Choose a base thickness. Read the total k-value from figure 16.5.

The pavement should rest on a layer of cement-treated base. A minimum thickness of .35 foot should be used. Due to manufacturing tolerances, a safety factor of .10 foot should be added for a total nominal thickness of .45 foot. If no high subgrade is allowed at the top of the subbase, a safety factor of .05 foot should be used for a total nominal thickness of .40 foot.

The base should extend one foot outside the pavement edge on each side.

step 4: Determine the modulus of rupture for the concrete to be used. A minimum of 550 psi for a 28-day strength should be specified.

step 5: Choose a load safety factor (LSF) appropriate to the type of facility. This corrects for impact and provides a margin of safety.

Table 16.11

Load Safety Factors

Type of Facility	LSF
Outside lanes of multi-lane facilities with high truck traffic	1.3
Inside lanes of multi-lane facilities with high truck traffic; and all lanes with moderate truck traffic	1.2
Minor highways, frontage roads, and all streets with low truck traffic	1.1
Residential streets or roads with occasional truck traffic	1.0

Figure 16.3 (1:7-641.4A)

k-Value vs. R-Value

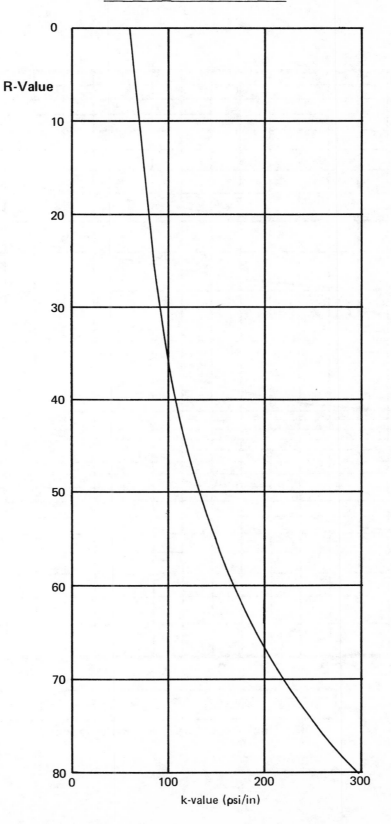

Figure 16.4 (1:7-641.4B)

k-Values for Base and Basement Soil Combination

(Do not use if subbase is omitted)

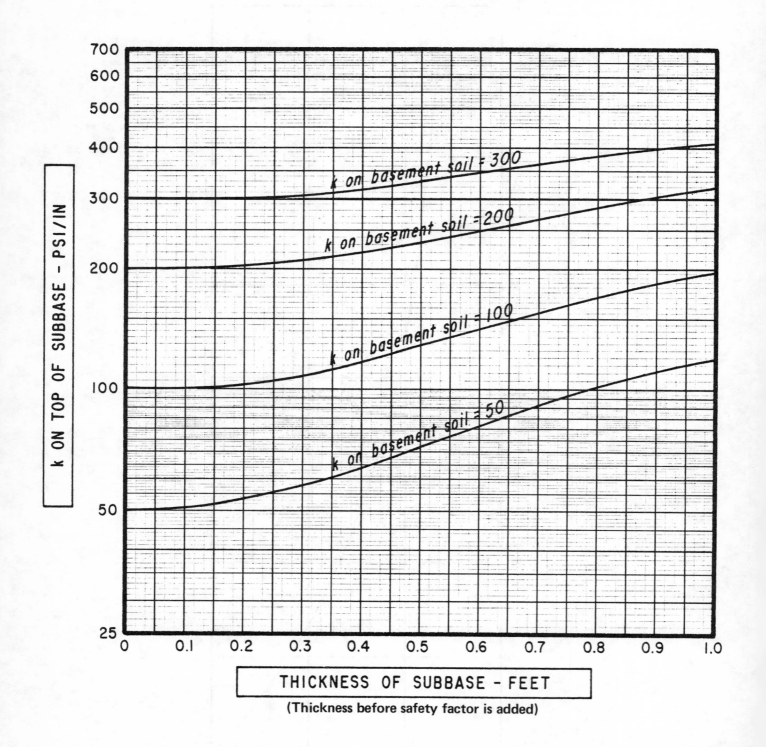

THICKNESS OF SUBBASE - FEET

(Thickness before safety factor is added)

Figure 16.5 (1:7-641.4C)

Total k-Values

(k values are for basement soil if subbase is omitted)

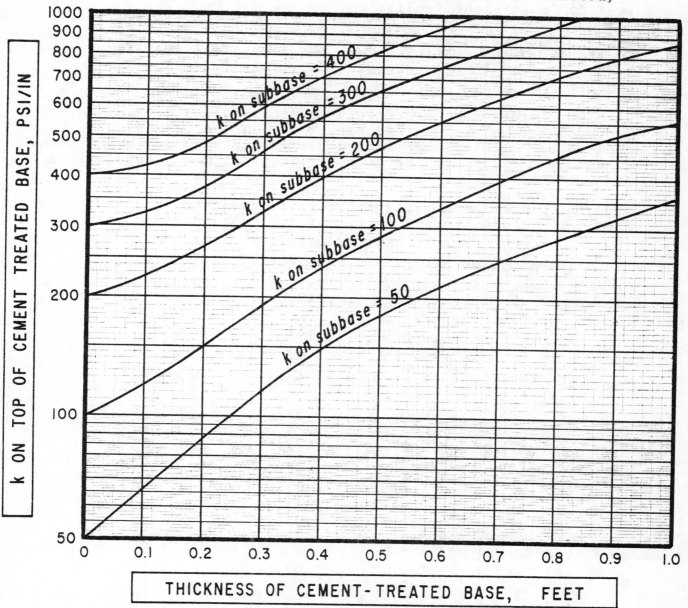

THICKNESS OF CEMENT-TREATED BASE, FEET

(Thickness before safety factor is added)

step 6: For all values of axle loading that are expected (from the loadometer survey), multiply the axle loading times the LSF.

step 7: Estimate the required concrete thickness. Start with .75 feet for multi-lane facilities.

step 8: For each axle loading (multiplied by the LSF), use figure 16.6 or 16.7 to find the stress induced in the concrete. The stress is read to the nearest 5 psi. Some interpolation is usually required.

step 9: Divide each stress value by the modulus of rupture. Record these values to the nearest .01. No values of .5 or less need to be calculated or recorded since this corresponds to the endurance strength of the concrete, and unlimited repetitions are allowed.

step 10: Determine the allowable repetitions of each stress ratio from table 16.12.

Table 16.12 (1:7-641.5)

Allowable Load Repetitions for Various Stress Ratios

Stress Ratio	Allowable Repetitions	Stress Ratio	Allowable Repetitions
.51	400,000	.71	1,500
.52	300,000	.72	1,100
.53	240,000	.73	850
.54	180,000	.74	650
.55	130,000	.75	490
.56	100,000	.76	360
.57	75,000	.77	270
.58	57,000	.78	210
.59	42,000	.79	160
.60	32,000	.80	120
.61	24,000	.81	90
.62	18,000	.82	70
.63	14,000	.83	50
.64	11,000	.84	40
.65	8,000	.85	30
.66	6,000	.86	23
.67	4,500	.87	17
.68	3,500	.88	13
.69	2,500	.89	10
.70	2,000	.90	8

step 11: Estimate the number of axle repetitions for the design life of the project. If the design life is 20 years, figure 16.8 may be used to calculate the repetitions for multi-lane freeways. Figure 16.9 should be used for city streets.

step 12: Divide the estimated repetitions from step 11 by the allowable repetitions from step 10. Add these ratios for all axle loadings. A good thickness will yield a total greater than 1.0 but less than 1.25.

Figure 16.6 (1:7-641.4F)

Stress Chart for Single Axle Loads

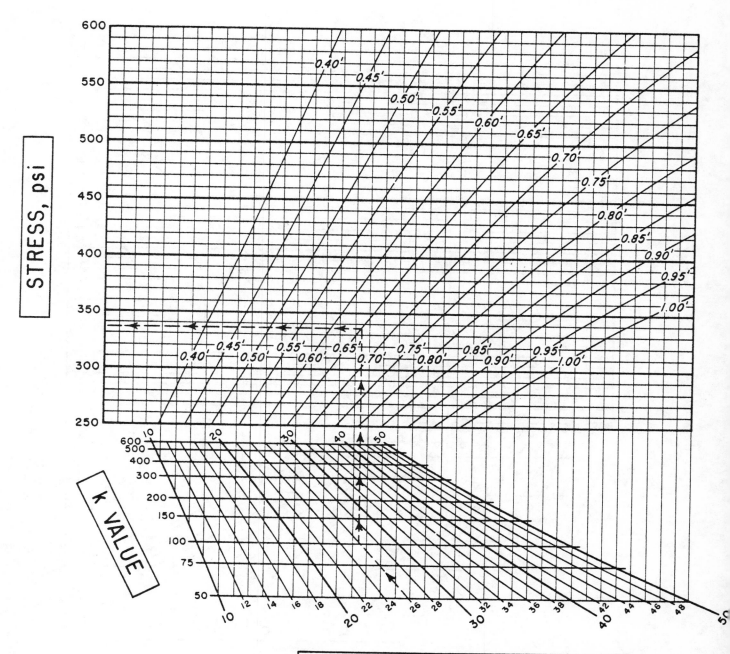

STRESS, psi

K VALUE

SINGLE AXLE LOAD, kips

Figure 16.7 (1:7-641.4G)

Stress Chart for Tandem Axle Loads

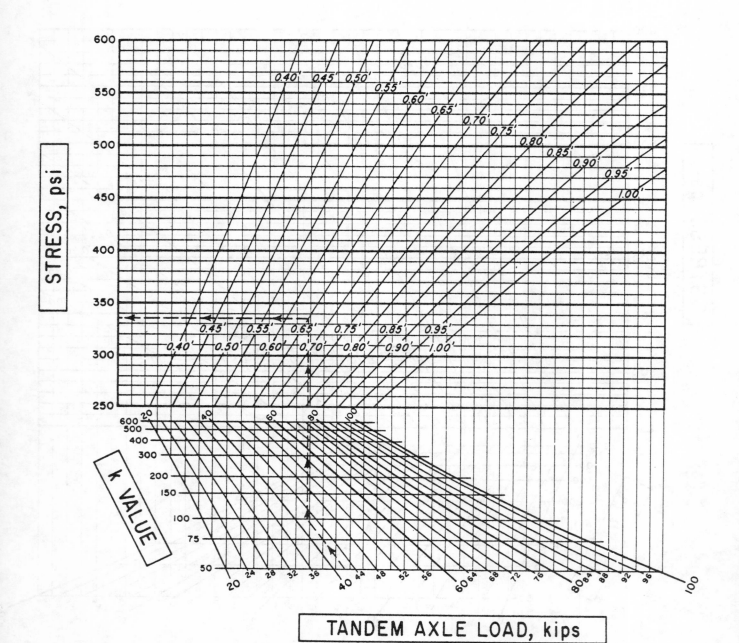

Figure 16.8

Load Repetition Constants
(For use with one-way truck counts)

Axle Loads Kips	2-Axle ADTT= Constant	2-Axle Repetitions	3-Axle ADTT= Constant	3-Axle Repetitions	4-Axle ADTT= Constant	4-Axle Repetitions	5 or more-Axle ADTT= Constant	5 or more-Axle Repetitions	Total 20-year Repetitions
SINGLE AXLE									
35									
34			9.57		26.9		2.63		
32	4.53		9.57		25.3		0.45		
30	4.23				17.9		0.58		
28	4.23				17.9		0.45		
26	4.23		29.9				2.04		
24	8.76		29.9		17.9		5.99		
22	17.5		39.6		155		58.1		
20	219		265		519		1341		
18	298		530		1245		3114		
TANDEM AXLE									
55							3.80		
54							3.80		
52							3.80		
50			8.67				2.49		
48			8.67				2.34		
46			8.67		8.95		6.74		
44			8.67		8.95		6.74		
42			17.4		8.95		19.0		
40			17.4		8.95		29.6		
38			60.7		8.95		69.5		
36			165		17.9		449		
34			130.1		64.1		1435		
32			121.4		8.95		1361		
30			121.4		119.3		525		
28			120.4		119.3		525		
26					119.3		525		

Figure 16.9

Load Repetition Constants for City Streets

SINGLE AXLE

Axle Loads Kips	2 Axle ADT =		3 Axle ADT =		4 Axle ADT =		5 or more Axle ADT =		Total 20 year Repetitions
	Constant	Repetitions	Constant	Repetitions	Constant	Repetitions	Constant	Repetitions	
26									
24	6.72								
22	80.8		2.70						
20	226.		56.0		147.		63.5		
18	372.		348.		526.		242.		
16	372.		542.		562.		1831.		
14	868.		540.		548.		968.		
12	868.		1612.		2098.		963.		
10			1606.		2092.		2032.		
							2030.		

TANDEM AXLE

Axle Loads Kips	2 Axle ADT =		3 Axle ADT =		4 Axle ADT =		5 or more Axle ADT =		Total 20 year Repetitions
	Constant	Repetitions	Constant	Repetitions	Constant	Repetitions	Constant	Repetitions	
40			34.6						
38			30.8						
36			96.2				25.5		
34			164.		147.		226.		
32			123.		147.		413.		
30			133.		147.		999.		
28			133.		204.		652.		
26			133.		204.		648.		
24			154.		204.		646.		
22			154.		428.		208.		
20			154.		428.		205.		
					422.		205.		

Figure 16.10
Cement Concrete Worksheet

Load Safety Factor (L.S.F.) _____   Basement Soil R-Value _____

Subbase Depth _____   Cement treated Base Depth _____

k-Values: Basement _____ Subbase _____ C.T.B. _____

Modulus of Rupture (M.R.) _____   Trial Depth PCC _____

(1)	(2)	(3)	(4)	(5)	(6)	(7)
Axle Load Kips	Axle Load x L.S.F. Kips	Stress psi	Stress Ratio Col. 3 / M.R.	Allowable Repetitions No.	Estimated Repetitions No.	Fatigue Resistance Used col. 6÷col. 5 %

SINGLE AXLE LOADS

35						
34						
32						
30						
28						
26						
24						
22						
20						
18						

TANDEM AXLE LOADS

55						
54						
52						
50						
48						
46						
44						
42						
40						
38						
36						
34						
32						
30						
28						

Total % Fatigue Used

9. Designing Asphalt Concrete Pavements

A. Introduction

The design of asphalt concrete pavements is based on the same parameters as were used in the portland cement concrete design procedure. The major difference is the use of equivalent axle load constants in place of load repetition constants. The constants used in this procedure are equivalent 18,000 pound axle loads (EAL) instead of the 5000 pound equivalent wheel loads (EWL) that were used in the past. However, EWL can be used to calculate the EAL from equation 16.19.

$$EAL = \frac{EWL}{11.8}$$
16.19

Multi-lane facilities using asphalt-concrete may be designed for different thicknesses of pavement in the outer and inner lanes.

Figure 16.11
Asphalt Concrete Pavement Cross Section

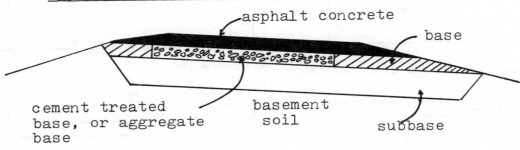

B. Design Procedure

Note: This procedure does not include corrections for frost or impact, nor is it similar to the AASHTO procedure.

step 1: Multiply the expanded average daily truck volumes by the appropriate EAL constants for 20-year operation. For shorter periods, the EAL constants may be reduced in direct proportion to the number of years.

Table 16.13
20-Year Equivalent Axle Load Constants

Vehicle Type	EAL Constant
2 axle trucks	1,380
3	3,680
4	5,880
5	13,780

Add up the products of the daily truck volumes and the EAL constants to give the Equivalent Axle Load (EAL).

step 2: Use equation 16.20 to calculate the traffic index, TI. TI should be calculated to the nearest .5. Greater accuracy is not justified.

$$TI = (9.0)(\frac{EAL}{1,000,000})^{.119}$$

16.20

step 3: Determine the base type for the first trial design. The following types of bases are used:

Aggregate base: This is the most commonly used. A typical R-value is 78.

Cement-treated base: This is a layer constructed with good quality, well-graded aggregate mixed with up to 6% cement. It provides a high slab strength, spreading the load over a large area. Class A CTB is used under asphalt to provide added strength under heavier traveled roads. Class B CTB is used to increase the R-value of the structural section. An R-value of 80 (minimum) is used for class B CTB. No R-values have been assigned to class A CTB.

Other bases: Other bases which are used to treat poor soils are lime stabilized, bituminous, and soil cement bases.

step 4: Calculate the gravel equivalent (GE) from equation 16.21.

$$GE = .0032(TI)(100 - R)$$

16.21

step 5: If TI exceeds 8, or if the subgrade R exceeds 40 and the subbase is omitted, add a safety factor to the GE value. For class A CTB, the increased thickness is applied to the base layer. For all other base materials, the safety factor is applied to the base and pavement layers. The total GE of the three layers should be the same with and without the safety factors.

Table 16.14

GE Safety Factors

base type	GE increase
class A CTB	.24 feet
class B CTB	.18
aggregate	.16
bituminous treated	.18
lime treated	.18
soil cement	.18

step 6: Use table 16.15 to determine the required pavement thickness. Table 16.15 lists the total required thickness (without a safety factor) above a layer whose R-value is known. Find the column with the proper TI value. Find the row with the correct (nearest) thickness of the base. Move to the left to read the actual thickness of pavement layer. Do not interpolate closer than .05.

step 7: Select a class of aggregate subbase. This would normally be specified in the materials report. If a class 2 aggregate subbase is used, its R-value is 50.

Table 16.15 (1:7-651.2A)

Gravel Equivalents of Structural Layers in Feet

Actual Thickness of Layer (Feet)	Asphalt Concrete — Traffic Index (TI) / Gravel Equivalent Factor (G_f)											Class B CTB, BTB, LTB, CS	Class A CTB	Aggregate base	Aggregate subbase
	5 and below	5.5 6.0	6.5 7.0	7.5 8.0	8.5 9.0	9.5 10.0	10.5 11.0	11.5 12.0	12.5 13.0	13.5 14.0	14.5 15.0 & up				
	G_f 2.50	2.32	2.14	2.01	1.89	1.79	1.71	1.64	1.57	1.52	1.50	G_f 1.20	G_f 1.70	G_f 1.10	G_f 1.00
0.10	0.25	0.23	0.21	0.20	0.19	0.18	0.17	0.16	0.16	0.15	0.15				
0.15	0.38	0.35	0.32	0.30	0.28	0.27	0.26	0.25	0.24	0.23	0.22				
0.20	0.50	0.46	0.43	0.40	0.38	0.36	0.34	0.33	0.31	0.30	0.30				
0.25	0.63	0.58	0.54	0.50	0.47	0.45	0.43	0.41	0.39	0.38	0.37				
0.30	0.75	0.70	0.64	0.60	0.57	0.54	0.51	0.49	0.47	0.46	0.45				
0.35	0.88	0.81	0.75	0.70	0.66	0.63	0.60	0.57	0.55	0.53	0.52	0.42	0.60	0.39	0.35
0.40	1.00	0.93	0.86	0.80	0.76	0.72	0.68	0.66	0.63	0.61	0.60	0.48	0.68	0.44	0.40
0.45		1.04	0.96	0.90	0.85	0.81	0.77	0.74	0.71	0.68	0.67	0.54	0.77	0.50	0.45
0.50		1.16	1.07	1.01	0.95	0.90	0.86	0.82	0.79	0.76	0.75	0.60	0.85	0.55	0.50
0.55			1.18	1.11	1.04	0.98	0.94	0.90	0.86	0.84	0.82	0.66	0.94	0.61	0.55
0.60				1.21	1.13	1.07	1.03	0.98	0.94	0.91	0.90	0.72	1.02	0.66	0.60
0.65				1.31	1.23	1.16	1.11	1.07	1.02	0.99	0.97	0.78	1.11	0.72	0.65
0.70					1.32	1.25	1.20	1.15	1.10	1.06	1.05	0.84	1.19	0.77	0.70
0.75						1.34	1.28	1.23	1.18	1.14	1.12	0.90	1.28	0.83	0.75
0.80						1.43	1.37	1.31	1.26	1.22	1.20	0.96	1.36	0.88	0.80
0.85						1.52	1.45	1.39	1.33	1.29	1.27	1.02	1.45	0.94	0.85
0.90							1.54	1.48	1.41	1.37	1.35	1.08	1.53	0.99	0.90
0.95								1.56	1.49	1.44	1.42	1.14	1.62	1.05	0.95
1.00								1.64	1.57	1.52	1.50	1.20	1.70	1.10	1.00
1.05									1.65	1.60	1.57	1.26	1.79	1.16	1.05

NOTES: CTB is cement treated base. BTB is bituminous treated base. LTB is lime treated base. CS is soil cement.
For the design of road-mixed asphalt surfacing, use 0.8 of the gravel equivalent factors (G_f) shown above for asphalt concrete.

step 8: Repeat steps 4 and 5 using the R-value for the subbase.

step 9: Subtract the base GE value from step 5 from the subbase GE value.

step 10: Use table 16.15 to determine the required aggregate base thickness. Find the column for the base material and go down until the value from step 9 is found. Read all the way over to the left and read the actual thickness of the aggregate layer.

step 11: Use equation 16.21 with the R-value of the basement soil to get the gravel equivalent of the subbase.

step 12: Subtract the sum of the gravel equivalents for the actual thicknesses of pavement and base from the value obtained in step 12. Round to the nearest .05 foot. This is the thickness of the subbase.

Example 16.3

Design a flexible pavement lane over a basement soil with R-value of 10 to carry the following ADT:

2 axle trucks	ADT =	935
3 axle trucks		550
4 axle trucks		225
5 axle trucks		1025

step 1: EAL = (1380)(935) + (3680)(550) + (5880)(225) + (13,780)(1025)

= 18.8 EE6   (20-year repetitions)

step 2: TI = $(9.0)(18.8)^{.119}$ = 12.76 (say 12.5 rounding down)

step 3: Choose aggregate base with R-value of 78.

step 4: GE = .0032(12.5)(100 - 78) = .88 feet

step 5: For an aggregate base, the safety factor is .16 feet.
So, the gravel equivalent is .88 + .16 = 1.04 feet.

step 6: From table 16.15, go down the TI = (12.5 to 13.0) column until 1.02 is reached (close enough to 1.04). Move to the left to read .65 feet required asphalt concrete thickness.

step 7: Select class 2 aggregate subbase with R-value of 50.

step 8: GE = .0032(12.5)(100 - 50) = 2.0 feet

Add .16 feet safety factor = 2.16 feet.

step 9: 2.16 - 1.02 = 1.14 feet (required gravel equivalent for base)

step 10: In the next-to-the-last column (for an aggregate base) go down until 1.14 is found. The table goes to 1.16, so use that value. Move to the left and read the aggregate base layer to be 1.05 feet.

step 11: GE = .0032(12.5)(100 - 10) = 3.6

step 12: 3.6 - 1.02 - 1.16 = 1.42 feet
So, the subbase aggregate thickness is 1.42 feet

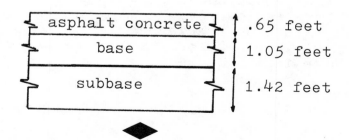

10. Roadway Detailing

The following geometric details are recommended.

* pavement width: 12 feet, all roadways and lanes
* crown slope: portland concrete cement: 2%
            bituminous mix pavement: 2%
            penetration treated earth and gravel: $2\frac{1}{2}\%$ - 3%
            unsurfaced, graded: $2\frac{1}{2}\%$ - 3%
* shoulders: to the right of traffic: 8 to 10 feet
          to the left of traffic:
               4 and 6 lanes: 5 feet
               8 lanes: 8 feet
* shoulder slope: 5% away from median
* maximum grade: 3% freeways
            6% other roads
* side slopes on adjacent cuts: freeways: 2:1 max (h:v)
                           other roads: $1\frac{1}{2}$:1 max (h:v)
* cut-to-right-of-way clearance: 10 feet minimum
                           50 feet maximum
                           20 feet for cuts 30 to 50 ft high
                           25 feet for cuts 50 to 75 ft high
                           (1/3) cut height above 75 feet
* divided median width: urban area freeways: 30 feet
                    rural area freeways: 46 feet
* median valley slopes: 10:1 to 20:1 (h:v)
* horizontal clearance to piers and walls: 30 feet desireable
                                      10 feet minimum
* vertical clearance: major structures: $16\frac{1}{2}$ feet
                  sign structures: 18 feet
                  pedestrial overcrossing: $18\frac{1}{2}$ feet

11. Queueing Models

Special Nomenclature

L expected system length (includes service)
L_q expected queue length
$p(n)$ probability of n customers in the system
s number of parallel servers
u mean service rate per server
W expected time in the system (includes service)
W_q expected time in the queue

λ mean arrival rate
ρ traffic intensity = (λ/u) and must be less than s

'Queue' is a technical word for a waiting line. Queueing theory can be used to predict the length of waiting time, the average time a customer can expect to spend in the queue, and the probability that some number of customers will be in the queue.

Many queueing models have been developed. Most of these models are fairly specialized and complex. However, two models are important enough to bear listing in this chapter. The relationships given below are for steady state operation, which means that the service facility has been open and in operation for some time.

A. General Relationships

The following simple relationships are valid for all queueing models.

$$L = \lambda W \tag{16.22}$$

$$L_q = \lambda W_q \tag{16.23}$$

$$W = W_q + (1/u) \tag{16.24}$$

$$\lambda < us \tag{16.25}$$

$$\text{average service time} = (1/u) \tag{16.26}$$

$$\text{average time between arrivals} = (1/\lambda) \tag{16.27}$$

B. The M/M/1 System

It is assumed that the following are true for the M/M/1 system.

* There is only one server ($s = 1$).

* The calling population is infinite.

* The service times are exponentially distributed with mean u. That is, the probability of a customer's remaining service time exceeding h (after already spending time with the server) is given by equation 16.28.

$$p(t>h) = e^{-uh} \tag{16.28}$$

Notice that equation 16.28 is independent of the time already spent with the server. This result holds true regardless of the elapsed service time. The specific service time distribution is

$$f(t) = ue^{-ut} \tag{16.29}$$

* The arrival rate is distributed as poisson with mean λ. The probability of x customers arriving in the next period is

$$p(x) = \frac{e^{-\lambda}\lambda^x}{x!} \tag{16.30}$$

The following relationships describe the M/M/1 system.

$$p(0) = 1 - \rho \tag{16.31}$$

$$p(n) = p(0)(\rho)^n \tag{16.32}$$

$$W = \frac{1}{(u-\lambda)} = W_q + \left(\frac{1}{u}\right) = \frac{L}{\lambda} \tag{16.33a}$$

$$W_q = \frac{\rho}{u-\lambda} = L_q/\lambda \qquad \text{16.33 b}$$

$$L = \frac{\lambda}{u-\lambda} = L_q + \rho \qquad \text{16.34}$$

$$L_q = \frac{\rho\lambda}{u-\lambda} \qquad \text{16.35}$$

Example 16.4

Given an M/M/1 system with u = 20 customers per hour and λ = 12 per hour, find the steady state value of W, W_q, L, and L_q. What is the probability that there will be 5 customers in the system?

$$\rho = \frac{12}{20} = .6$$

$$W = \frac{1}{20-12} = .125 \text{ hours}$$

$$W_q = \frac{.6}{20-12} = .075 \text{ hours}$$

$$L = \frac{12}{20-12} = 1.5 \text{ customers}$$

$$L_q = \frac{(.6)(12)}{20-12} = .9 \text{ customers}$$

$$p(0) = 1 - .6 = .4$$

$$p(5) = .4(.6)^5 = .031$$

◆

C. The M/M/s System

The same assumptions are used for the M/M/s system as were used for the M/M/1 system except that there are s servers instead of only 1. Each server has a mean service rate u. Each server draws from a single line so that the first person in line goes to the first (any) server that is available. Each server does not have its own line.

However, if customers are allowed to change the lines they are in so that they go to any available server, this model may also be used to predict the performance of a multiple server system where each server has its own line.

$$W = W_q + \left(\frac{1}{u}\right) \qquad \text{16.36}$$

$$W_q = L_q/\lambda \qquad \text{16.37}$$

$$L_q = \frac{p(0)(\lambda/u)^s \rho}{s!(1-\rho)^2} \qquad \text{16.38}$$

$$L = L_q + \rho \qquad \text{16.39}$$

$$p(0) = \frac{1}{\frac{(\rho)^s}{s!(1-(\rho/s))} + \sum_{j=0}^{s-1} \frac{(\rho)^j}{j!}}$$ 16.40

$$p(n) = \frac{p(0)(\rho)^n}{n!} \quad (n < s)$$ 16.41

$$p(n) = \frac{p(0)(\rho)^n}{s!s^{n-s}} \quad (n > s)$$ 16.42

Bibliography

1. Caltrans (State of California Department of Transportation), Highway Design Manual, Sacramento, CA 1975

2. County Engineers Association of California, League of California Cities, and Caltrans, "Flexible Pavement", 1979

3. Hay, William W., An Introduction to Transportation Engineering, John Wiley and Sons, Inc., New York, NY 1961

4. Institute of Traffic Engineers (John E., Baerwald, editor), Transportation and Traffic Engineering Handbook, Prentice-Hall, Inc., Englewood Cliffs, NJ 1976

5. National Academy of Sciences, National Research Council, Highway Research Board, Highway Capacity Manual, (Special Report 87, Publication 1328), Washington, D.C. 1965

Practice Problems: TRAFFIC ANALYSIS AND HIGHWAY DESIGN

Required

1. A flexible pavement with a 3" thick asphalt concrete surface layer is to be designed for a state primary road. The subgrade is well drained and is not considered susceptible to frost action. Give the recommended thicknesses of the base and subbase. Use the following information:

> maximum allowable load on a single axle: 18,000 pounds
> maximum aggregate size: 2"
> base: rolled stone with CBR = 90
> subbase: soil-aggregate with CBR = 40
> subgrade: CBR = 5

2. The outside lane of an interstate highway is being designed for a state where soil freezes to a depth of 5 inches. A flexible pavement is being considered. The projected life of the roadway is 20 years. The traffic which will use the roadway is as follows:

> mean ADT: 20,000 passenger cars and 2200 trucks (over 4 lanes)
> distribution: 75% in outside lane
> average gross truck load: 22,000 pounds
> average axle load: 8800 pounds
> axle load distribution:

| | % of ave. truck ADT | |
axle load (lbs)	single axle	tandem axle
under 8000	36.3	–
8000 - 16000	28.4	4.5
16000 - 20000	12.9	10.5
20000 - 24000		4.0
24000 - 30000		3.1
30000 - 34000		.3

The roadway is to be constructed of the following materials:

> pavement: high stability plant mix
> base: crushed stone
> subbase: coarse graded crushed stone treated with asphalt
> subgrade: CBR = 5

3. Two cars are moving at 60 mph going the same direction in the same lane. The cars are separated by 20 feet for each 10 mph of their speed. The coefficient of friction (skidding) between the tires and the roadway is .6. The reaction time is assumed to be .5 seconds. (a) If the lead car hits a parked truck, what is the speed of the second car when it hits the first (stationary) car? (b) At what speed does the rule of thumb of one car length per 10 mph become safe? (c) What should the rule actually be?

4. You have been hired by the owner of a demolished house to investigate the car crash that caused the damage. From the police report, you learn that the car was traveling down a 3% grade at an unknown speed. The skid marks are 185 feet long, and the pavement was dry at the time of the accident. The police report estimates from the visable damage to car and house that the initial speed of the car was 25 mph. The house owner doesn't believe the estimate of initial speed. You have the following test data from tests performed on level roadways:

| initial speed (mph) | : | 30 | 40 | 50 | 60 |
| coef. of friction | : | .59 | .51 | .45 | .35 |

(a) Find the minimum initial speed of the car. (b) If the police report was mistaken in assuming the road surface was dry, what was the minimum initial speed of the car?

5. One lane of a 2-lane road was observed for an hour during the day. The following data was gathered:

    average distance between front bumpers of successive cars: 80 ft
    average speed during the study: 30 mph
    space mean speed: 31 mph

(a) What is the average headway? (b) What is the density in vehicles per mile? (c) What is the traffic volume in vehicles per hour? (d) What is the maximum capacity of the lane? (e) Sketch a graph of the relationship between speed and density. Label the axes and indicate the region of unstable flow. (f) Sketch a graph of the relationship of speed and volume. Label the axes and indicate the jam density. (g) Sketch a graph of the relationship between volume and density. (h) Which is a more accurate parameter of traffic capacity: volume or density? Why? (i) What is the generally accepted capacity of one lane of a multi-lane freeway?

6. Two streets in the central business district are being investigated. The population of the metropolitan area is 250,000. The peak hour factor is .85. The signal cycle is 60 second, 2-phase.

parameter	First	Main (N)	Main (S)
green cycle	40 sec	30 sec	30 sec
yellow time	3 sec	3 sec	3 sec
parking	both sides	none	none
trucks	7%	5%	5%
buses	0%	0%	0%
left turns	10%	10%	10%
right turns	10%	0%	10%

(a) Find the maximum service capacity on First St. for service level E.
(b) Find the maximum service capacity on Main St. (N) for service level B.
(c) Find the maximum service capacity on Main St. (S) for service level E.

7. Two streets intersect at a stop sign: a 36 foot side street and a 44 foot arterial. There have been many complaints from users of the side street that traffic signals are needed at the intersection. Give a logical, step-by-step description of how you would investigate the problem. Discuss how you would arrive at your recommendations. If a signal is required, specify fixed timing or on demand type decision criteria.

NOTES

SURVEYING

Nomenclature

C	chord length, or a constant	ft
D	separation distance, or degree of a curve	ft, or degrees
E	error, or external distance	-, or feet
g	grade	%
h	height	ft
H	horizontal distance	ft
HI	height of the instrument above the ground	ft
I	intersection angle	degrees
k	number of observations	-
K	stadia interval factor	-
L	leg length	ft
LC	length of curve	ft, or stations
M	middle distance	
r	rate of grade change per station	%/station
R	rod reading, or radius	ft
s	sample standard deviation	-
S	sight distance	ft
T	tangent distance	ft
u	mean of a distribution	-
V	vertical distance	ft
w	weight	-
x	distance or location	stations
X	distance	stations

Symbols

θ	angle	degrees
α	angle	degrees
β	angle	degrees

Subscripts

a	actual
c	curvature
p	probable
r	refraction

1. Introduction

Surveying has traditionally been a part of civil engineering. However, most of the actual field work and office analysis is no longer actually done by engineers. Furthermore, most state licensing boards also license land surveyors. It is not surprising, therefore, that the P.E. examination contains questions that transcend simple trigonometry.

This chapter will enable you to prepare for the most common types of problems encountered in the NCEE examination. These types involve measurement errors, traverses, and curves.

2. Conversions (All chains and links are Gunter.)

Multiply	By	To Obtain
Acres	10	square chains
Acres	160	square rods
Acres	43,560	square feet
Chains	66	feet
Chains	100	links
Chains	.0125	miles
Feet	.01515	chains
Feet	.3048	meters
Feet	.01	stations
Feet	.3636	vara (California)
Inches	.1263	links
Links	7.92	inches
Links	.01	chains
Meters	3.2808	feet
Miles	80	chains
Miles	320	rods
Rods	.003125	miles
Square chains	.1009	acres
Square feet	2.296 EE-5	acres
Square rods	.00625	acres
Stations	100	feet
Vara (California)	33.0	inches

3. Basic Error Analysis

A. Measurements of Equal Weight

There are many opportunities for errors in surveying, although calculators and modern equipment have reduced the magnitudes of most errors. The purpose of error analysis is not to eliminate errors, but rather to estimate their magnitudes and to assign them to the appropriate measurements.

The expected value (also known as the most likely value or the probable value) of a measurement is the value which has the highest probability of being correct. If a series of measurements is taken of a single quantity, the most probable value is the average (mean) of those measurements. That is, if x_1, x_2, $\cdots$, x_n are values of some measurement, then the most probable value is

$$x_p = \frac{x_1 + x_2 + \cdots + x_n}{n}$$

17.1

For related measurements whose sum should equal some known quantity, the most probable values are the observed values corrected by an equal part of the total error. This is illustrated in example 17.1.

Example 17.1

The interior angles of a traverse were measured as 63°, 77°, and 41°. Each measurement was made once, and all angles were measured with the same precision. What are the most probable interior angles?

The sum total of angles should equal 180. The error in the measurements is (63 + 77 + 41 - 180) = +1°. Therefore, the correction required is -1° which is proportioned equally among the three angles. The most probable values are 62.67°, 76.67°, and 40.67°.

◆

Measurements of a given quantity are assumed to be normally distributed. If a quantity has a mean u and a standard deviation s, the probability is 50% that a measurement of that quantity will fall within the range of u ± (.6745)s. The quantity (.6745)s is known as the probable error. The probable ratio of precision is u/(.6745)s. The interval between the extremes is known as the confidence interval.

The standard deviation, s, is the small sample standard deviation used in chapter 1.

The probable error of the mean of k observations of the same quantity is given by equation 17.2.

$$E_{mean} = \frac{.6745s}{\sqrt{k}} = \frac{E_{total,\ k\ measurements}}{\sqrt{k}} \qquad 17.2$$

Example 17.2

12 tapings were made of a critical distance. The mean value was 423.7 feet with a standard deviation (s) of .31 feet. What are the 50% confidence limits for the distance?

From equation 17.2, the standard error of the mean value is

$$E_{mean} = \frac{(.6745)(.31)}{\sqrt{12}} = .06$$

Therefore, the probability is 50% that the true distance is within the limits of 423.7 ± .06 feet.

◆

Example 17.3

The true length of a tape is 100 feet. The most probable error of a measurement with this tape is .01 feet. What is the expected error if the tape is used to measure out a distance of one mile?

The number of tapings will be (5280/100) = 52.8, or 53 tapings. The most probable error will be (.01)($\sqrt{53}$) = .073 feet.

◆

B. Measurements of Unequal Weight

Some measurements may be more reliable than others. It is not unreasonable to weight each measurement with its relative reliability. Such weights may be determined subjectively, but more frequently, they are determined from relative frequencies of occurrence or from the relative inverse squares of the probable errors.

Example 17.4

An angle was measured five times by five equally competent crews on similar days. Two of the crews obtained a value of 39.77°, and the remaining three crews obtained a value of 39.74°. What is the probable value of the angle?

$$\theta = \frac{(2)(39.77) + (3)(39.74)}{5} = 39.75°$$

Example 17.5

A distance has been measured by three different crews. The lengths and their probable error intervals are given below. What is the most probable value?

$$\text{crew 1:} \quad 1,206.40 \pm .03 \text{ feet}$$
$$\text{crew 2:} \quad 1,206.42 \pm .05 \text{ feet}$$
$$\text{crew 3:} \quad 1,206.37 \pm .07 \text{ feet}$$

The sum of the squared probable errors is

$$(.03)^2 + (.05)^2 + (.07)^2 = .0083$$

The weights to be applied to the three measurements are

$$.0083/(.03)^2 = 9.22$$
$$.0083/(.05)^2 = 3.32$$
$$.0083/(.07)^2 = 1.69$$

The most probable length is

$$\frac{(1206.40)(9.22) + (1206.42)(3.32) + (1206.37)(1.69)}{9.22 + 3.32 + 1.69}$$

$$= 1206.40$$

The probable error and 50% confidence interval for weighted observations can be found from equation 17.3. x_i represents the ith observation and w_i represents its weight. The number of observations is n.

$$E_{p,weighted} = .6745 \sqrt{\frac{\Sigma[w(\bar{x}-x_i)^2]}{(\Sigma w)(n-1)}} \qquad 17.3$$

Example 17.6

What is the 50% confidence interval for the measured distance in example 17.5?

It is easier to work with the decimal part only.

$$\bar{x} = (\tfrac{1}{3})(.40 + .42 + .37) = .40 \text{ approximately}$$

i	x_i	$\bar{x}-x_i$	$(\bar{x}-x_i)^2$	w	$w(\bar{x}-x_i)^2$
1	.40	0	0	9.22	0
2	.42	-.02	.0004	3.32	.0013
3	.37	.03	.0009	1.69	.0015
				14.32	.0028

From equation 17.3,

$$E_{p,weighted} = .6745\sqrt{\frac{.0028}{(14.23)(3-1)}} = .0067$$

The 50% confidence interval is $1206.40 \pm .0067$

For related weighted measurements whose sum should equal some known quantity, the most probable weighted values are corrected inversely to the relative frequency of observation.

Example 17.7

The interior angles of a triangular traverse were repeatedly measured, with the results shown below. What is the most probable value for angle #1?

angle	value	number of measurements
1	63°	2
2	77°	6
3	41°	5

The total of the angles is $(63 + 77 + 41) = 181$. So $(-1°)$ must be divided among the three angles. These corrections are inversely proportional to the number of measurements. The sum of the measurement inverses is

$$(\tfrac{1}{2}) + (\tfrac{1}{6}) + (\tfrac{1}{5}) = .867$$

The most probable value of angle #1 is

$$63° + (\frac{\tfrac{1}{2}}{.867})(-1) = 62.42°$$

Weights may also be calculated when the probable errors are known. These weights are the relative squares of the probable errors. This is illustrated in example 17.8.

Example 17.8

The interior angles of a triangular traverse were measured, with the results shown below. What is the most probable value of angle #1?

angle	value
1	$63° \pm .01°$
2	$77° \pm .03°$
3	$41° \pm .02°$

The total of the angles is $(63 + 77 + 41) = 181$. So $(-1°)$ must be divided among the three angles. The corrections are proportional to the square of the probable errors.

$$(.01)^2 + (.03)^2 + (.02)^2 = .0014$$

The most probable value of angle #1 is

$$63 + \frac{(.01)^2}{.0014}(-1) = 62.93°$$

◆

C. Errors in Computed Quantities

When quantities with known errors are added or subtracted, the error of the result is given by equation 17.4, The squared errors under the radical are added regardless of whether the calculation is addition or subtraction.

$$E_{total} = \sqrt{E_1^2 + E_2^2 + E_3^2 + \cdots} \qquad 17.4$$

The error in the product of two quantities (x_1 and x_2) which have known errors (E_1 and E_2) is given by equation 17.5.

$$E_{product} = \sqrt{x_1^2 E_2^2 + x_2^2 E_1^2} \qquad 17.5$$

Example 17.9

The sides of a perfectly square rectangular section were determined to be $(1204.77 \pm .09)$ feet and $(765.31 \pm .04)$ feet respectively. What is the probable error in the area?

From equation 17.5,

$$E_{area} = \sqrt{(1204.77)^2(.04)^2 + (765.31)^2(.09)^2}$$

$$= 84.06 \text{ ft}^2$$

◆

4. Distance Measurement

Distance is frequently measured by means of a steel tape. Aside from the procedures required for error-free teamwork, there is nothing complex about using a measuring tape.

Lengths may be divided into 100 foot long sections called 'stations'. Interval stakes along an established line are ordinarily laid down at 100 foot intervals called 'full stations'. If a marker stake is placed anywhere else along the line, it is called a 'plus station'. Thus, a stake placed 1500 feet from a reference point is labeled "15+00" and a stake placed 1325 feet from a reference point is labeled "13+25".

Distance may also be measured tacheometrically (tachymetrically). This method involves sighting through a small angle at a distant scale. The angle may be fixed and the length measured (stadia method) or the length may be fixed and the angle measured (European method).

Stadia measurement consists of observing the apparent locations of the horizontal cross hairs on a distance stadia rod. The interval between the two rod readings is called the stadia interval or the stadia reading. This distance is directly related to the distance between the telescope and the rod. For rod readings R_1 and R_2 (both in feet), the separation distance is

$$D = K(R_2 - R_1) + C \qquad\qquad 17.6$$

C is determined by the manufacturer. It may be 1 foot (typical) or zero (as in the case of internal focusing telescopes). K is the stadia interval factor which usually has a value of 100.

Figure 17.1
Horizontal Stadia Measurement

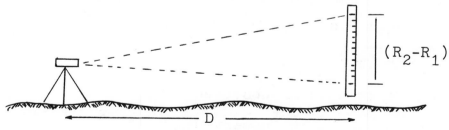

If the sighting is inclined, as it is in figure 17.2, it will be necessary to find both the horizontal and vertical distances. These may be found from equations 17.7 and 17.8. Notice that V is measured from the telescope to the sighting rod center. The actual elevation difference will require knowledge of the instrument height.

$$H = K(R_2-R_1)\cos^2\theta + C\cos\theta \qquad\qquad 17.7$$

$$V = \tfrac{1}{2}(K)(R_2-R_1)(\sin 2\theta) + C\sin\theta \qquad\qquad 17.8$$

Figure 17.2

Inclined Stadia Measurement

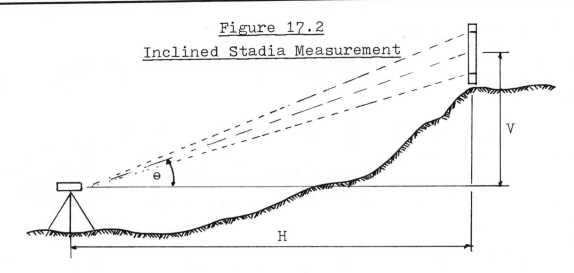

5. Elevation Measurement: Leveling

Leveling is the act of using an engineers level and rod to measure the vertical distance (the 'elevation') from an arbitrary level surface. Usually, the elevation is measured with respect to sea level.

A. Curvature and Refraction

If a level sighting is taken on an object with actual height h, the curvature of the earth will cause the object to appear taller by an amount h_c. In equation 17.9, D is measured in feet along the curved surface of the earth.

$$h_c = 2.4 \text{ EE-8}(D^2) \tag{17.9}$$

Atmospheric refraction will make the object appear shorter by an amount h_r.

$$h_r = 3.0 \text{ EE-9 } (D^2) \tag{17.10}$$

The corrected rod reading is

$$R_{corrected} = R_{observed} + h_r - h_c \tag{17.11}$$

$$= R_{observed} - (2.1 \text{ EE-8})(D^2) \tag{17.12}$$

Figure 17.3

Curvature and Refraction Effects

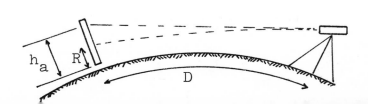

B. Direct Leveling

The most common method of determining the difference in elevations of two points is known as direct leveling. In using this method, a level is set up at a point midway between the two points whose difference in elevation is wanted. The vertical distances are observed by reading directly from the rod. Refer to figure 17.4 which uses the following nomenclature:

H_{A-B} the difference in elevations between points A and B

H_{A-L} the difference in elevations between points A and L

H_{L-B} the difference in elevations between points L and B

R_A the rod reading at A

R_B the rod reading at B

$h_{rc,A-L}$ the effects of curvature and refraction between points A and L

$h_{rc,B-L}$ the effects of curvature and refraction between points B and L

Figure 17.4

Direct Leveling

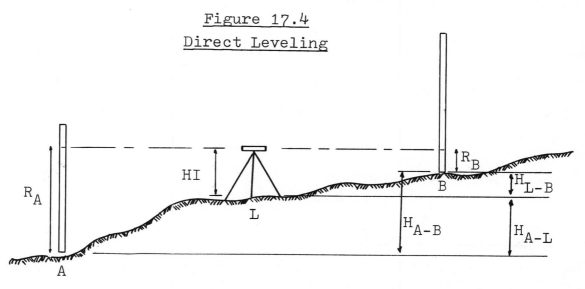

Then, it should be clear that

$$H_{A-L} = R_A - h_{rc,A-L} - HI \qquad 17.13$$

$$H_{L-B} = R_B - h_{rc,L-B} - HI \qquad 17.14$$

The difference in elevations between points A and B is

$$H_{A-B} = H_{A-L} - H_{L-B} = R_A - R_B + h_{rc,L-B} - h_{rc,A-L} \qquad 17.15$$

If the backsight and foresight distances are equal, then the effects of refraction and curvature cancel, resulting in equation 17.16.

$$H_{A-B} = R_A - R_B \qquad 17.16$$

C. Indirect Leveling

Indirect (trigonometric) leveling does not require a backsight
(although one can be taken to eliminate the effects of curvature and
refraction.) A case of indirect leveling is illustrated in figure
17.5 where the difference in elevations between points A and B is
needed. It is assumed that the distance AC has been determined.
Within the limits or ordinary practice, angle ACB is 90°.

Figure 17.5
Indirect Leveling

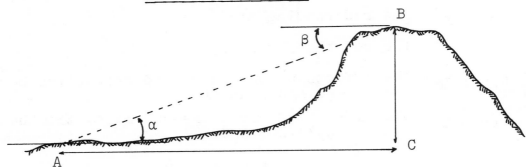

Including the effects of curvature and refraction,

$$H_{A-B} = AC(\tan\alpha) + 2.1 \text{ EE-}8(AC)^2 \qquad 17.17$$

If a backsight is taken from B to A and angle β is measured, then

$$H_{A-B} = AC(\tan\beta) - 2.1 \text{ EE-}8(AC)^2 \qquad 17.18$$

Adding equations 17.17 and 17.18 and dividing by 2,

$$H_{A-B} = \tfrac{1}{2}(AC)(\tan\alpha + \tan\beta) \qquad 17.19$$

D. Differential Leveling

Differential leveling is the consecutive application of direct
leveling to the measurement of large differences in elevation. There
is usually no attempt to exactly balance the foresights and backsights.
Thus, there is no record made of the exact locations of the level
positions. Furthermore, the path taken between points need not be
along a straight line connecting them, as only the height differences
are relevant.

If greater accuracy is desired without having to accurately balance
the foresight and backsight distances, it is possible to eliminate
most of the curvature and refraction error by balancing the sum of
the foresights against the sum of the backsights.

The following abbreviations are used with differential leveling.

BM: bench mark or monument
TP: turning point
FS: foresight
BS: backsight
HI: height of the instrument
 L: level position

Example 17.10

The following readings were taken during a differential leveling between bench marks 1 and 2. What is the difference in elevations between these two bench marks? Check your work.

Station	BS	HI	FS	Elevation
BM1	7.11			721.05
TP1	8.83		1.24	
TP2	11.72		1.11	
BM2			10.21	

The measurements are shown below.

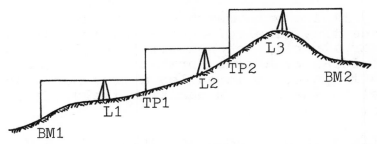

The first measurement is shown below in larger scale. The height of the instrument is: HI = 721.05 + 7.11 = 728.16

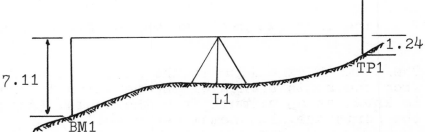

The second measurement is shown below. The height of the instrument is: HI = 728.16 + 8.83 - 1.24 = 735.75

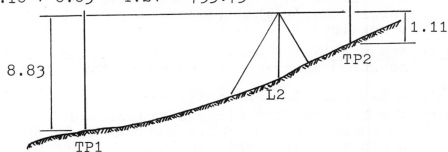

Similarly, the third measurement is shown below. The height of the instrument is: HI = 735.75 + 11.72 - 1.11 = 746.36. The elevation of BM2 is 746.36 - 10.21 = 736.15.

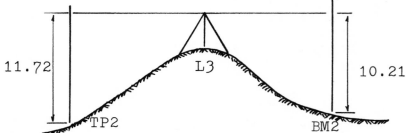

The difference in elevations is 736.15 - 721.05 = 15.1.

The backsight sum is: 711 + 8.83 + 11.72 = 27.66.

The foresight sum is: 1.24 + 1.11 + 10.21 = 12.56.

The difference is (27.66 - 12.56) = 15.1   (check)

◆

6. Angle Measurement

The direction of any line may be measured by means of an angle between it and some reference line. The reference line is known as a meridian. If the meridian is arbitrarily chosen, it is called an assumed meridian. If the meridian is a true north-to-south line passing through the true north pole, it is called a true meridian. If the meridian is parallel to the earth's magnetic field, it is known as a magnetic meridian.

A true meridian differs from a magnetic meridian by a declination ('magnetic declination' or 'variation'). If the north end of a compass points to the west of the true meridian, the declination is said to be a west declination. Otherwise, it is an east declination.

The variation of a line from its meridian may be given in several ways:

<u>Azimuths</u>: The azimuth of a line is the clockwise angle measured from the south branch of the meridian to the line. This is known as an azimuth from the south. (Azimuths from the north are also sometimes used.)

<u>Deflection Angles</u>: The angle between a line and its prolongation is a deflection angle. Such measurements must be labeled as 'right' (clockwise) or 'left' (counter-clockwise).

<u>Angles to the right</u>: The angle to the right is a clockwise angle measured from the preceding to the following line.

<u>Azimuths from the back line</u>: Same as angles to the right.

<u>Bearings</u>: The bearing of a line is referenced to the quadrant in which the line falls and the angle which the line makes with the meridian in that quadrant.

All of these methods of angle measurement are illustrated in figure 17.6.

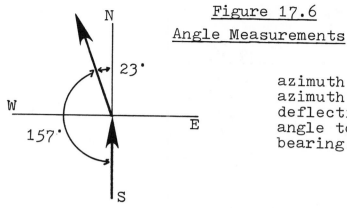

Figure 17.6
Angle Measurements

azimuth from the south: 157°
azimuth from the north: 337°
deflection angle: 23°L
angle to the right: 157°
bearing: N 23° W

7. Closed Traverses

A traverse is a series of straight lines whose lengths and deflection angles (or other angle measurements) are known. A traverse that comes back to its starting point is known as a closed traverse. The polygon that results from the closing of a traverse is governed by the following two requirements:

(a) The sum of the deflection angles is 360°.

(b) The sum of the interior angles of a polygon with n sides is (n-2)180°.

A. Adjusting Closed Traverse Angles

Due to errors, variations in magnetic declination, and local magnetic attractions, it is likely that the sum of angles making up the interior angles will not exactly equal (n-2)180. The following procedure may be used to proportion the angle error of closure among the angles.

step 1: Calculate the interior angle of each station from the observed bearings.

step 2: Subtract (n-2)180 from the sum of the interior angles.

step 3: Unless additional information (in the form of numbers of observations or probable errors) is available, assume the angle error of closure can be divided among all angles. Divide the error by the number of angles.

step 4: Find a line whose bearing is assumed correct. That is, find a line whose bearing appears unaffected by errors, variations in magnetic declination, and local attractions. Such a line may be chosen as one for which the forward and back bearings are the same. If there is no such line, take the line whose difference in foreward and back bearings is the smallest.

step 5: Start with the assumed correct line and add (or subtract) the error (divided by the number of angles) to each interior angle.

step 6: All bearings except the one for the assumed correct line are also corrected.

Example 17.11

Adjust the angles on the 4-sided closed traverse whose magnetic
bearings are shown below.

Line	Bearing
AB	N 25° E
BA	S 25° W
BC	S 84° E
CB	N 84.1° W
CD	S 13.1° E
DC	N 12.9° W
DA	S 83.7° W
AD	N 84° E

step 1: The interior angles are calculated from the bearings.

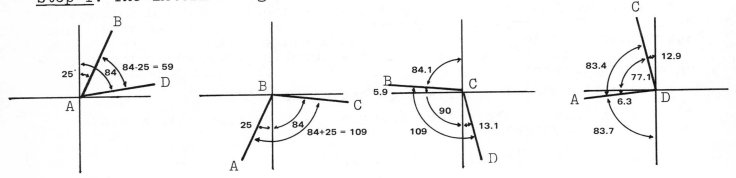

step 2: The sum of the angles is: 59 + 109 + 109 + 83.4 = 360.4. For
a 4-sided traverse, the sum of interior angles should be 360°, so
a correction of (-.4°) must be divided up among the four angles. The
before and after traverses are shown below.

Since the backsight and foresight bearings of line AB are the same,
it is assumed that the 25° bearings are the most accurate. The corrected
bearings are given below.

Line	Bearing
AB	N 25° E
BA	S 25° W
BC	S 83.9° E
CB	N 83.9° W
CD	S 12.8° E
DC	N 12.8° W
DA	S 83.9° W
AD	N 83.9° E

B. Latitudes and Departures

The latitude of a line is the distance which the line extends in a north or south direction. A line which runs towards the north has a positive latitude; a line which runs towards the south has a negative latitude.

The departure of a line is the distance which the line extends in an east or west direction. A line which runs towards the east has a positive departure; a line which runs towards the west has a negative departure.

Figure 17.7
Departures and Latitudes

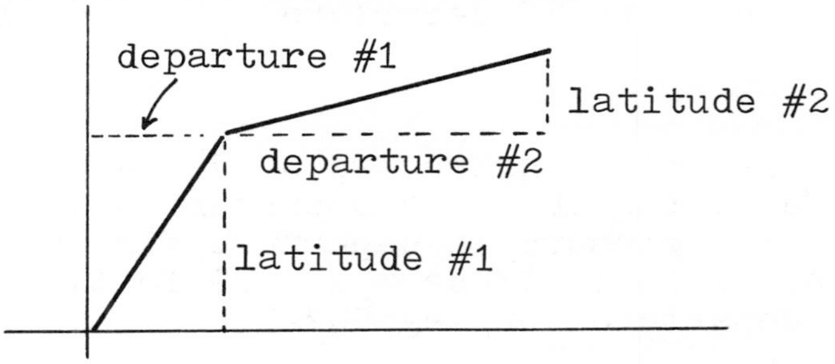

In a closed traverse, the algebraic sum of latitudes should be zero and the algebraic sum of departures should also be zero. These sums, which are distances in feet with actual values near zero, are called 'closure in latitude' and 'closure in departure' respectively. Traverses which have had their angles corrected will not necessarily have zero closures.

The traverse closure is the line which will exactly close the traverse. This is illustrated in figure 17.8. Since latitudes and departures are orthogonal, the closure in latitude and closure in departure may be considered as the rectangular coordinates to calculate the traverse closure length. The coordinates will have the opposite signs as do the closure in departure and closure in latitude. That is, if the closure in departure is positive, point A will lie to the left of point A', as shown in figure 17.8.

The length of a traverse closure is

$$L = \sqrt{\left(\substack{\text{closure in}\\\text{departure}}\right)^2 + \left(\substack{\text{closure in}\\\text{latitude}}\right)^2} \qquad 17.20$$

Figure 17.8
Traverse Closure

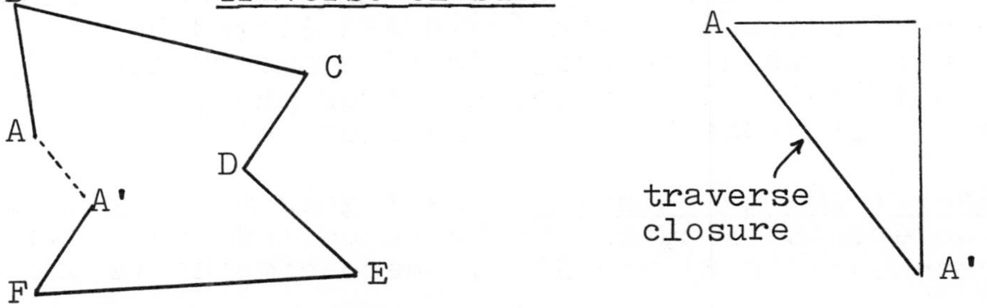

C. Adjusting Closed Traverse Latitudes and Departures

If a closed traverse has a traverse closure, then that closure must be divided among the various legs of the traverse. This correction requires that the latitudes and departures be known for each leg of the traverse.

The most common method used to balance the traverse legs is known as the 'Compass Rule'. This rule states that the correction to a leg of the traverse is to the total traverse correction as the leg length is to the total traverse length, with the signs reversed. For the leg correction in departure, for example,

$$\frac{\text{leg departure correction}}{\text{closure in departure}} = -\frac{\text{leg length}}{\text{total traverse length}} \quad 17.21$$

Example 17.12

A closed traverse was constructed of 7 legs, the total of whose lengths was 2705.13 feet. Leg CD has a departure of 443.56 and a latitude of 219.87. The total closure in departure for the traverse was +.41 feet; the total closure in latitude was -.29 feet. What are the corrected latitude and departure for leg CD?

The length of leg CD is

$$L_{CD} = \sqrt{(443.56)^2 + (219.87)^2} = 495.06 \text{ ft}$$

According to the compass rule,

$$\frac{\text{latitude correction}}{-.29} = -\left(\frac{495.06}{2705.13}\right)$$

Or, the latitude correction is .05 feet.

$$\frac{\text{departure correction}}{+.41} = -\left(\frac{495.06}{2705.13}\right)$$

Or, the departure correction is -.08 feet.

The corrected latitude is 219.87 + .05 = 219.92.
The corrected departure is 443.56 - .08 = 443.48 ft

◆

D. Reconstructing Missing Sides and Angles

If one or more sides or angles of a traverse is missing or cannot be determined by measurement, it will have to be reconstructed. The procedures for three common cases are listed below. These procedures draw primarily upon the subjects of geometry and trigonometry. They are not difficult. Of course, no error analysis or balancing of angles is possible if sides have been reconstructed.

case 1: One leg missing: One leg is missing in figure 17.9. However, the line EA can be reconstructed easily from its components E-E' and E'-A. These components are equal to the

sum of the departures and sum of the latitudes respectively, with
the signs changed. The angle can be determined from the ratio of
the sides, and the length E-A can be found from equation 17.20.

Figure 17.9

One Missing Leg

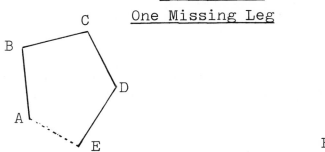

case 2: <u>Adjacent legs missing</u>: Figure 17.10 shows a traverse
that has two adjacent legs missing. The traverse may be closed
as long as some length/angle information is available. The
technique is to close the traverse using the method presented
in case 1 above. This will give the line D-A. Then, the triangle
E-A-D can be completed using whatever information is available.

Figure 17.10

Two Adjacent Legs Missing

case 3: <u>Two non-adjacent legs missing</u>: Figure 17.11 shows a
traverse with two non-adjacent legs missing. Since the latitudes
and departures of two parallel lines are equal, this can can be
solved by 'closing up' the traverse and shifting one missing leg
until it is adjacent to the other missing leg. This reduces the
problem to case 2.

Figure 17.11

Two Non-adjacent Legs Missing

E. Area of a Traverse

An area of a traverse may always be found by dividing the traverse into a number of geometric shapes (rectangles, triangles, etc.) and summing the area of each subdivision. If the coordinates of the traverse leg end points are known, the method of coordinates may be used.

The coordinates may be (x,y) coordinates referenced to some arbitrary set of axes, or they may be (departure, latitude) coordinates. The area calculation is simplified if the coordinates are written in the following form:

$$\frac{x_1}{y_1} \diagdown \frac{x_2}{y_2} \diagdown \frac{x_3}{y_3} \diagdown \frac{x_4}{y_4} \diagdown \frac{x_1}{y_1} \qquad (etc)$$

The, the area is

$$A = \tfrac{1}{2}(\Sigma \text{full line products} - \Sigma \text{dotted line products}) \quad 17.22$$

Example 17.13

Calculate the area of a triangle with coordinates of its corners given: (3,1), (5,1), and (5,7).

$$\frac{3}{1} \diagdown \frac{5}{1} \diagdown \frac{5}{7} \diagdown \frac{3}{1}$$

$$A = \tfrac{1}{2}((3)(1)+(5)(7)+(5)(1)-(1)(5)-(1)(5)-(7)(3))$$
$$= \tfrac{1}{2}(43 - 31) = 6$$

◆

8. Curves

Roads, rail lines, and water courses are usually designed in a straight line. Where a direction change is needed, a curve is used. The straight lines connected by a curve are known as 'tangents' or 'tangent lines'. A curve on level ground changing the direction of two tangents is known as a horizontal curve. Horizontal curves are usually arcs of circles.

Curves must also be used to connect roads and rail lines that change grade (slope). Such curves are called vertical curves. A curve that connects an upgrade tangent to a downgrade tangent is known as a crest curve, whereas a curve that connects a downgrade tangent to an upgrade tangent is known as a sag curve. Vertical curves are usually parabolic in shape.

Figure 17.12
Sag and Crest Vertical Curves

A. Elements of Circular Curves

The elements of circular curves and their standard abbreviations are given below and in figure 17.13.

R   radius of the curve
V   vertex of the tangent intersection point
PI   point of vertical intersection
I   interior angle
PC   point of curvature, the place where the first tangent ends and the curve begins
PT   point of tangency, the place where the curve ends and the second tangent begins
POC   any point on the curve
LC   length of the arc - the length of the curve from PC to PT
T   tangent distance from V to PC or from V to PT
C   the long chord - the straight distance from PC to PT
E   the external distance is the distance from V to the midpoint of the curve
M   the middle ordinate is the distance from the curve midpoint to the midpoint of the long chord
D   the degree of the curve (see explanation below)

The following alternative designations are also used.
PVI   point of vertical intersection (same as PI)
TC   a change from a tangent to a curve (same as PC)
CT   a change from curve to a tangent (same as PT)
BC   the beginning of a curve (same as PC)
EC   the end of a curve (same as PT)
TS   a change from tangent to spiral
SC   a change from spiral to curve
Δ the intersection angle (same as I)

Figure 17.13
Elements of a Circular Curve

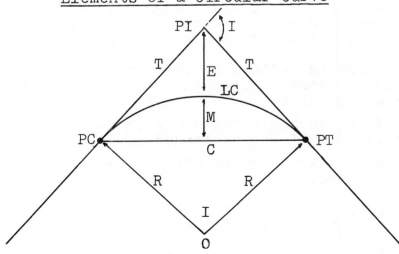

The relationships listed below may be used to solve problems involving circular curves. Some experience is necessary, however, to know the order of application of these relationships.

$$T = R \tan(\tfrac{1}{2}I) \qquad\qquad 17.23$$

$$E = R(\tan\tfrac{1}{2}I)(\tan\tfrac{1}{4}I) = R(\sec(\tfrac{1}{2}I) - 1) \qquad 17.24$$

$$M = R(1 - \cos\tfrac{1}{2}I) = \tfrac{1}{2}C(\tan\tfrac{1}{4}I) \qquad 17.25$$

$$C = 2R(\sin\tfrac{1}{2}I) = 2T(\cos\tfrac{1}{2}I) \qquad 17.26$$

$$LC = R(I \text{ in radians}) = R(I \text{ in degrees})(\tfrac{2\pi}{360}) = 100(\tfrac{I}{D}) \quad 17.27$$

The curvature of city streets, property boundaries, and some high-ways is specified by the radius, R. In this case, the length of curve (LC) is equal to the actual arc length. The curvature may also be specified (in degrees) by the 'degree of curve', D.

The degree of curve has two different meanings. (1) In most highway work, the length of the curve is understood to be the actual arc, and the degree of the curve is the angle subtended by an <u>arc</u> of 100 feet. (2) In railroad layout, the degree of curve is defined as the angle subtended by a <u>chord</u> of 100 feet.

When the radius is large, there is very little difference in the degrees of curve based on arcs and chords.

B. Horizontal Circular Curves: The Deflection Angle Method

Curves are usually designed in the office by an engineer. The surveyor must stake out the curve so that the road building crew knows where to put the road. Stakes should be put at the PC, PT, and at all full stations. If the curve is sharp, stakes may also be required at +25, +50, and +75 stations. The deflection angle method is the most common method used for staking out the curve. In this method, the curve distance is usually assumed to start from 00+00 at the PC.

The deflection angle is defined as the angle between the tangent and a chord. This is illustrated in figure 17.14. The deflection angles are calculated using the following theorems:

(1) The deflection angle between a tangent and a chord is measured by half the subtended arc.

(2) The angle between two chords is measured by half the subtended arc.

In figure 17.14, angle V-PC-A is a deflection angle between a tangent and a chord. From theorem (1), the angle is

$$V\text{-}PC\text{-}A = \tfrac{1}{2}\alpha \qquad\qquad 17.28$$

Angle α may be found from the following simple relationships:

$$\frac{\alpha}{360} = \frac{\text{arc length (PC-A)}}{2\pi R} \qquad\qquad 17.29$$

or $\qquad \dfrac{\alpha}{I} = \dfrac{\text{length (PC-A)}}{LC} \qquad\qquad 17.30$

The chord length (PC-A) is given by equation 17.31.

$$\text{PC-A} = 2R\sin(\tfrac{1}{2}\alpha) \qquad\qquad 17.31$$

Thus, the entire curve can be laid out from the PC by sighting the angle (V-PC-A) and taping the distance PC-A. The PC and PT may be found by solving for T and starting at V.

Figure 17.14
Circular Curve Deflection Angle

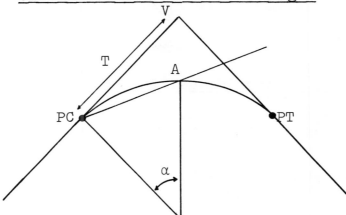

Example 17.14

A circular curve is to be constructed with a 225 foot radius and an interior angle of 55°. Determine where the stakes should be placed if the separation between stakes along the arc is 50 feet. Specify deflection angles and chord lengths.

The length of the curve is given by equation 17.27.

$$LC = (225)(55)(\tfrac{2\pi}{360}) = 215.98 \text{ ft}$$

The last stake will be (215.98 - 200) = 15.98 from the next to the last stake. The central angle for an arc of 50 feet is given by equation 17.29.

$$\alpha_{degrees} = (\tfrac{360}{2\pi})(\tfrac{50}{225}) = 12.732°$$

12.732° goes into 55° 4 times with a remainder of 4.072°. From equation 17.31, the required chord lengths are

$$(2)(225)\sin(\tfrac{12.732}{2}) = 49.90 \text{ ft}$$

$$(2)(225)\sin(\tfrac{4.072}{2}) = 15.98 \text{ ft}$$

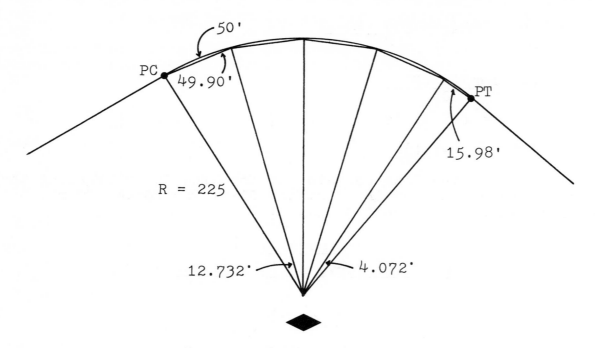

C. Horizontal Circular Curves: The Arc Basis

If the degree of curve, D, is specified as the angle subtended by an arc of length 100 feet, the radius of the curve can be found from equation 17.32.

$$R = (\frac{360°}{D°})(\frac{100}{2\pi}) = \frac{5729.6}{D°} \qquad\qquad 17.32$$

Example 17.15

An interior angle of 8.4° is specified for a horizontal curve. The PI station is 64+27.46'. Use the arc basis with a 2° curve to locate the PC and PT stations.

From equation 17.32,

$$R = (\frac{360}{2})(\frac{100}{2\pi}) = 2864.79 \text{ ft}$$

From equations 17.23 and 17.27,

$$T = 2864.79(\tan(\frac{8.4}{2})) = 210.38$$

$$LC = (2864.79)(8.4)(\frac{2\pi}{360}) = 420.00$$

Then, the PC and PT points are located.

$$PC = (64+27.46) - (2+10.38) = 62+17.08$$

$$PT = (62+17.08) + (4+20.00) = 66+37.08$$

D. Horizontal Circular Curves: The Chord Basis

If the degree of curve, D, is specified as the angle subtended by a chord of 100 feet, the radius of the curve can be found from equation 17.33.

$$\sin(\tfrac{1}{2}D) = (\frac{50}{R}) \qquad\qquad 17.33$$

This method is often used in railroad and highway surveys. Another customary practice on railway surveys is to number the stations around the curve as a continuation of the tangent line. Therefore, PC is not chosen as 00+00, nor does the PC normally correspond to a full station.

E. Length of Circular Horizontal Curve for Stopping Distance

A horizontal curve on level ground is shown in figure 17.15. A typical design problem is to design a curve (specify a radius) which will simultaneously provide the required sight stopping distance while maintaining a clearance from a roadside obstruction. Sight stopping distances for any speed can be found from chapter 16.

Figure 17.15

Stopping Distance

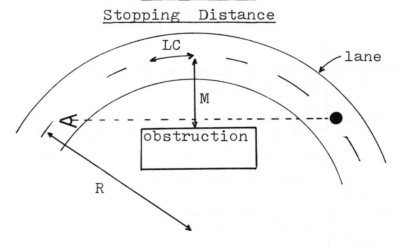

The governing equations are given below. In this analysis, the stopping sight distance and length of curve are the same. Note that the angles are given as degrees.

$$LC = \frac{R}{28.65}(\arccos(\frac{R-M}{R})) \qquad\qquad 17.34$$

$$M = R(1 - \cos(\frac{28.65LC}{R})) \qquad\qquad 17.35$$

F. Elements of Vertical Curves

Vertical curves are used to change the grade of a highway or railway. Equal-tangent parabolic curves are usually used for this purpose. A vertical sag curve connecting two grades is shown in figure 17.16. Since the grades are very small, the actual arc length of the curve is approximately equal to the chord length BVC-EVC.

Figure 17.16
A Vertical Curve

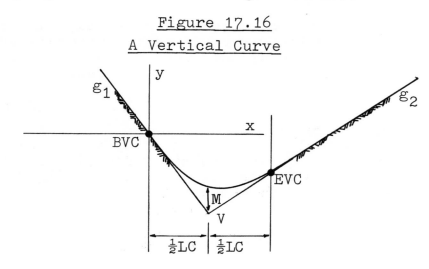

The following standard and optional abbreviations are used. Note the change in the definition of LC.

LC   the horizontal length of the curve in stations
g_1 the grade from which the stationing starts, in percent
g_2 the grade towards which the stationing heads, in percent
V   the vertex - the intersection of the two tangents
PVI   same as V
BVC   beginning of the vertical curve
EVC   end of the vertical curve
PVC   same as BVC
PTT   same as EVC
M   the middle ordinate

A vertical parabolic curve is completely specified by the two grades and the curve length. Alternately, the rate of grade change per station may be used in place of the curve length.

$$r = \frac{(g_2 - g_1)}{LC}$$

17.36

The equation of an equal tangent parabolic curve is given below. x is measured in stations beyond BVC. y is measured in feet, with the same reference point used to measure all elevations.

$$y = (\frac{r}{2})x^2 + g_1x + \text{(elevation of BVC)}$$

17.37

The maximum or minimum elevation will occur at the turning point. The turning point is not located directly above V, but is found at

$$x = -g_1/r \quad \text{(in stations)}$$

17.38

The middle ordinate is calculated as

$$M = \frac{(g_1 - g_2)(LC)}{8} \qquad 17.39$$

Example 17.16

A vertical crest curve with a length of 400 feet is to connect grades
of +1.0% and -1.75%. The vertex is located at station 35+00 and it
has an elevation of 549.20 feet. What are the elevations of the BVC,
EVC, and at all full stations on the curve?

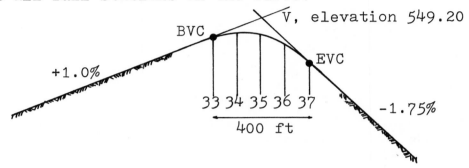

The elevation at BVC is 549.20 - 1(2) = 547.20

The elevation at EVC is 549.20 - 1.75(2) = 545.70

$$r = \frac{-1.75 - 1}{4} = -.6875 \text{ percent}$$

$$\tfrac{1}{2}r = -.3438$$

The equation of the curve is (using equation 17.37)

$$y = -.3438x^2 + x + 547.20$$

At station 34, x = (34-33) = 1,

$$y_{34} = -.3438(1)^2 + 1 + 547.20 = 547.86$$

Similarly, $$y_{35} = -.3438(2)^2 + 2 + 547.20 = 547.82$$

$$y_{36} = -.3438(3)^2 + 3 + 547.20 = 547.11$$

◆

G. Length of Vertical Curves for Sight Distances

The curve length should be longer than the safe passing or stopping
sight distances which are given in chapter 16. (Passing sight distance
is not relevant on multi-lane highways.) Table 17.1 may be used to
calculate lengths of curves.

Table 17.1 is used by calculating the curve length for both assumptions that S<LC and S>LC.

Table 17.1 (3:7-201.2 to 7-201.5)

Required Lengths of Curves on Grades

Assuming	Stopping Sight Distance (Crest Curves)	Passing Sight Distance (Crest Curves)	Stopping Sight Distance (Sag Curves)
S<LC	$\dfrac{(g_1-g_2)S^2}{1398}$	$\dfrac{(g_1-g_2)S^2}{3100}$	$\dfrac{(g_2-g_1)S^2}{400 + 3.5S}$
S>LC	$2S - \dfrac{1398}{g_1-g_2}$	$2S - \dfrac{3100}{g_1-g_2}$	$2S - \dfrac{400 + 3.5S}{g_2-g_1}$

Example 17.17

A car is traveling at 40 mph up a hill with a +1.25% grade. The descending grade is -2.75%. What is the required length of curve for proper stopping sight distance?

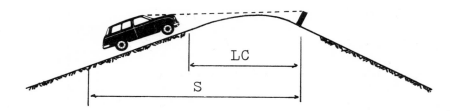

From chapter 16, the sight distance is 275 feet at 40 mph.

Using table 17.1 and assuming that 275>LC:

$$LC = 2(275) - \frac{1398}{1.25-(-2.75)} = 200.5$$

Using table 17.1 and assuming 275<LC:

$$LC = \frac{(1.25-(-2.75))(275)^2}{1398} = 216.4$$

Since 216.4 is less than 275, the second assumption is not valid. The required curve length is 200.5 feet.

H. Vertical Curves with Obstructions

If a curve is to have some minimum clearance from an obstruction, as in figure 17.17, the length of the curve, BVC, and EVC will generally not be known in advance. The problem of finding the curve length may be solved by using the following procedure. Note that d is in stations.

The following method may also be used when the vertical curve is placed over a culvert and a minimum cover depth is required.

Figure 17.17
A Curve With An Obstruction

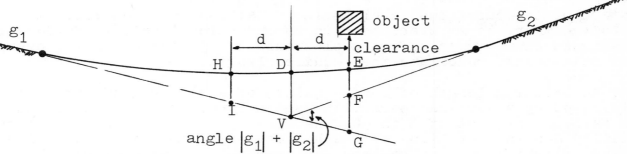

step 1: Calculate the elevation of point E:

$$(elevation)_E = (elevation)_{object} - clearance \qquad 17.40$$

step 2: Calculate distance EG. Remember, g_1 is negative as shown in figure 17.17.

$$EG = (elevation)_E - (elevation)_V - (d)(g_1) \qquad 17.41$$

step 3: Calculate the distance EF.

$$EF = (elevation)_E - (elevation)_V - (d)(g_2) \qquad 17.42$$

step 4: A vertical distance from a tangent line to a point on a curve is proportional to the square of the horizontal distance along the tangent. (See equation 17.37.) This, combined with the curve symmetry of the equal-tangent parabola (EF = HI in figure 17.17), lets us solve simultaneously for LC. Both d and LC are in units of stations in equation 17.43.

$$\frac{EG}{(\frac{LC}{2} + d)^2} = \frac{EF}{(\frac{LC}{2} - d)^2} \qquad 17.43$$

Bibliography

1. Bouchard, Harry, and Moffit, Francis, _Surveying_, 5th ed., International Textbook Company, Scranton, PA 1969

2. Breed, Charles B., _Surveying_, John Wiley & Sons, New York, NY 1942

3. Caltrans (State of California, Department of Transportation), _Highway Design Manual_, Sacramento, CA 1975

4. Caltrans (State of California, Department of Transportation), _Surveys Manual_, Sacramento, CA 1975

5. Davis, Raymond E., and Kelly Joe, W., _Elementary Plane Surveying_ 4th ed., McGraw-Hill Book Company, New York, NY 1969

Practice Problems: SURVEYING

Required

1. A downgrade of 4% meets a rising grade of 5% in a sag curve. At the start of the curve the level is 123.06 at chainage 4034+20. At chainage 4040+20 there is an overpass with an underside level of 134.06. If the designed curve is to afford a clearance of 15 feet under the overpass at this point, calculate the required length.

2. An existing length of road consists of a rising gradient of 1 in 20 followed by a vertical parabolic summit curve 300 feet long, and then a falling gradient of 1 in 40. The curve joins both gradients tangentially and the elevation of the highest point on the curve is 173.07 feet. Visability is to be improved over this stretch of road by replacing this curve with another parabolic curve 600 feet long. (a) Find the depth of excavation required at the mid-point of the curve. (b) Tabulate the elevations of points at 100 foot intervals on the new curve.

3. Two straights intersecting at a point B have the following azimuths: BA 270°, BC 110°. They are to be joined by a circular curve which must pass through a point D which is 350 feet from B. The azimuth of BD is 260°. Find the required radius, tangent lengths, length of curve, and setting out angle for a 50 foot chord.

Optional

4. Three points (A, B, and C) were selected on the centerline of an existing road curve as a first step in determining the curve radius. The telescope was set horizontally at point B. Readings were taken on a vertical staff at points A and C. The readings are summarized below. The instrument has constants of 100 and 0. (a) Calculate the radius of the circular curve. (b) If the trunnion axis was 4.7 feet above the road at B, find the gradients AB and BC.

staff at	horizontal bearing	stadia readings		
A	0.00°	4.851	3.627	2.403
C	195.57°	7.236	5.778	4.320

5. Four level circuits were run over four different routes to determine the elevation of a bench mark. The observed elevations and probable errors for each circuit are shown below. What is the most probable value for the elevation of the bench mark?

Route 1: 745.08 ± 0.03

Route 2: 745.22 ± 0.01

Route 3: 745.45 ± 0.09

Route 4: 745.17 ± 0.05

6. A transit with an interval factor of 100 and an instrument factor of 1.0 foot was used to take stadia sights. The instrument height was 4.8 feet. The location where the instrument was set up had an elevation of 297.8 feet. Using the data below, find the distance AB and the elevation of point B.

object	azimuth	rod interval (R_2-R_1)	middle hair reading	vertical angle
A	42.17°	3.22	5.7	-6.3°
B	222.17°	2.60	10.9	+4.17°

7. The balanced latitudes and departures of the legs of a closed traverse are given below. What is the traverse area?

leg	latitude	departure
AB	N 350	E 0
BC	N 550	E 600
CD	S 250	E 1200
DE	S 750	E 200
EF	S 550	W 1100
FA	N 650	W 900

8. Balance and adjust (to the nearest .1 ft) the traverse given below.

leg	bearing	length (ft)
AB	N	500.0
BC	N45.00°E	848.6
CD	S69.45°E	854.4
DE	S11.32°E	1019.8
EF	S79.70°W	1118.0
FA	N54.10°W	656.8

9. Determine the elevation of BM 11 and BM 12.

station	backsight	foresight	elevation
BM 10	4.64		179.65
TP 1	5.80	5.06	
TP 2	2.25	5.02	
BM 11	6.02	5.85	
TP 3	8.96	4.34	
TP 4	8.06	3.22	
TP 5	9.45	3.71	
TP 6	12.32	2.02	
BM 12		1.98	

10. A five-leg closed traverse is taped and scoped in the field, but obstructions make it impossible to collect all readings. It is known that the general direction of EA is easterly. Complete the table of information below.

leg	north azimuth	distance
AB	106.22°	1081.3
BC	195.23°	1589.5
CD	247.12°	1293.7
DE	332.37°	------
EA	------	1737.9

NOTES

DON'T GAMBLE!
These books will extract Every Last Point
from the examination for you

Engineering Law, Design Liability, and Professional Ethics

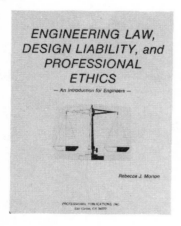

The most difficult problems are essay questions about management, ethics, professional responsibility, and law. Since these questions can ask for definitions of terms you're not likely to know, it is virtually impossible to fake it by rambling on. And yet, these problems are simple if you have the right resources. If you don't feel comfortable with such terms as comparative negligence, discovery proceedings, and strict liability in tort, you should bring **Engineering Law, Ethics, and Liability** with you to the examination.

None of this material is in your review manual. And, nothing from your review manual has been duplicated here.

8½" × 11", soft cover, 88 pages, $15.45 (includes postage).

(ISBN 0-932276-37-7)

Expanded Interest Table

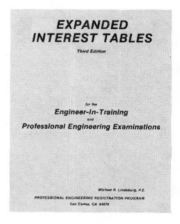

There's nothing worse than knowing how to solve problem but not having the necessary data. Engineering Economics problems are like that. You might know how to do a problem, but where do you get interest factors for non-integer interest rates? **Expanded Interest Tables** will prove indispensible for such problems. has pages for interest rates starting at ¼% and going 25% in ¼% increments. Factors are given for up to 1 years. There's no other book like it. So, if you want be prepared for an Engineering Economy problem wi 11.75% interest, you need **Expanded Interest Tables.**

8½" × 11", soft cover, 106 pages, $15.45, includir postage.

(ISBN 0-932276-35-0)

SUBJECT INDEX

Quick - I need additional study materials

Please rush me the review materials I have checked. I understand any item may be returned for a full refund within 30 days. I have provided my bank card number as method of payment, and I authorize you to charge your current prices against my account.

For the E-I-T Exam:

Solutions Manuals

( ) ENGINEER-IN-TRAINING REVIEW MANUAL
( ) QUICK REFERENCE CARDS
( ) MINI-EXAMS with solutions   ( )

For the P.E. Exams:
( ) CIVIL ENGINEERING REVIEW MANUAL   ( )
( ) SEISMIC DESIGN
( ) TIMBER DESIGN
( ) STRUCTURAL ENGINEERING PRACTICE PROBLEM MANUAL
( ) MECHANICAL ENGINEERING REVIEW MANUAL   ( )
( ) ELECTRICAL ENGINEERING REVIEW MANUAL   ( )
( ) CHEMICAL ENGINEERING REVIEW MANUAL
( ) CHEMICAL ENGINEERING PRACTICE PROBLEM SET
( ) LAND SURVEYOR REFERENCE MANUAL   ( )

Recommended for all Exams:
( ) EXPANDED INTEREST TABLES
( ) ENGINEERING LAW, ETHICS, AND LIABILITY

Ship to:

name _____

company _____

street _____ apt. no. _____

city _____ state _____ zip _____

daytime phone number _____

Charge to: (required for immediate processing)

VISA, MasterCard, or American Express number   expiration date
_____

name on the card
_____

signature

Send more information

Please send me descriptions and prices of all available E-I-T and P.E. review books. I understand there will be no obligation.

_____

_____

_____

_____

A friend of mine is taking the exam too. Send additional literature to:

_____

_____

_____

_____

We disagree

I think there is an error on page _____. Here is the way I think it should be.

Title of this book: _____

[ ] Please tell me if I am correct.

Contributed by (optional):

_____

_____

_____